SUPRAMOLECULAR CHEMISTRY OF ANIONS

SUPRAMOLECULAR CHEMISTRY OF ANIONS

Edited by
Antonio Bianchi
Kristin Bowman-James
and
Enrique García-España

New York • Chichester • Weinheim • Brisbane • Singapore • Toronto

About the Front Cover Design

Crystal structure of the chloride inclusion complex of the circular double helicate formed with iron(II) and a *tris*-bipyridine ligand.

From B. Hasenknopf, J.-M. Lehn, B. O. Kneisel, G. Baum, and D. Fenske, *Angew. Chew., Int. Ed. Engl.* **35**, 1838 (1996).

This book is printed on acid-free paper. ∞

Copyright © 1997 by John Wiley & Sons, Inc. All rights reserved.

Published simultaneously in Canada.

No part of this publication may be reproduced, stored in a retrieval system or transmitted in any form or by any means, electronic, mechanical, photocopying, recording, scanning or otherwise, except as permitted under Sections 107 or 108 of the 1976 United States Copyright Act, without either the prior written permission of the Publisher, or authorization through payment of the appropriate per-copy fee to the Copyright Clearance Center, 222 Rosewood Drive, Danvers, MA 01923, (508) 750-8400, fax (508) 750-4744. Requests to the Publisher for permission should be addressed to the Permissions Department, John Wiley & Sons, Inc., 605 Third Avenue, New York, NY 10158-0012, (212) 850-6011, fax (212) 850-6008, E-Mail: PERMREQ @ WILEY.COM.

Library of Congress Cataloging in Publication Data:

The supramolecular chemistry of anions / [edited by] Antonio Bianchi,
   Kristin Bowman-James, Enrique García-España.
      p.   cm.
   Includes bibliographical references and index.
   ISBN 0-471-18622-8 (alk. paper)
   1. Anions 2. Macromolecules. 3. Coordination compounds.
I. Bianchi, Antonio, 1956– . II. Bowman-James, Kristin, 1946–
III. García-España, Enrique.
QD474.S9525   1997
541.2'26—dc21

Printed in the United States of America

10 9 8 7 6 5 4 3 2 1

CONTENTS

Foreword—*Jean-Marie Lehn* ix

Preface—*Enrique García-España* xi

Contributors xiii

1. Physical Factors in Anion Separations 1

Bruce A. Moyer and Peter V. Bonnesen,
Oak Ridge National Laboratory, Oak Ridge, Tennessee

1.1	Introduction	1
1.2	Anion Radii, Hydration, and Solvation	4
1.3	Anion Transfer	9
1.4	Solvent Extraction of Anions	22
1.5	Resin Anion Exchange	34
1.6	Conclusions	40
1.7	Acknowledgments	40
1.8	References	41

2. Historical View on the Development of Anion Coordination Chemistry 45

Bernard Dietrich and Mir Wais Hosseini,
Université Louis Pasteur, Strasbourg, France

2.1	Introduction	45
2.2	Halide Complexation	47
2.3	Linear Anions	52
2.4	Trigonal Planar Anions	52
2.5	Tetrahedral Anions	54
2.6	Square Planar Anions	56
2.7	Octahedral Anions	56
2.8	Molecular Assemblies	58
2.9	Conclusions	59
2.10	References	59

3. Natural Anion Receptors: Anion Recognition by Proteins 63

Stefano Mangani, University of Siena, Siena, Italy
Marta Ferraroni, University of Florence, Florence, Italy

3.1	Introduction	63
3.2	Carboxypeptidase A	64
3.3	Superoxide Dismutase	70
3.4	Phosphate and Sulfate Binding Proteins	72
3.5	Citrate Synthase	74
3.6	Concluding Remarks	76
3.7	References	77

4. Artificial Anion Hosts. Concepts for Structure and Guest Binding 79

Franz P. Schmidtchen, Technische Universität München, Munich, Germany

4.1	Introduction	79
4.2	The Guests: Anions	80
4.3	The Concept of Artificial Anion Hosts	82
4.4	Positively Charged Anion Hosts	85
4.5	Electroneutral Hosts for Anions	118
4.6	References	135

5. Structural and Topological Aspects of Anion Coordination 147

Jerry L. Atwood, University of Missouri at Columbia, Columbia, Missouri
Jonathan W. Steed, King's College London, London, United Kingdom

5.1	Introduction	148
5.2	Protonated Polyamine-Based Receptors	150
5.3	The Guanidinium Moiety	171
5.4	Cyclophane Receptors	177
5.5	Neutral Receptors	177
5.6	Lewis Acid Receptors	183
5.7	Organometallic Receptors	195
5.8	Transition Metal Complexes	198
5.9	Polymetallic Receptors	200
5.10	Alkalide and Electride Salts	203
5.11	Anion Complexation in Biochemistry	207
5.12	Concluding Remarks	208
5.13	References	209

6. Thermodynamics of Anion Complexation 217

Antonio Bianchi, University of Florence, Florence, Italy
Enrique García-España, University of Valencia, Valencia, Spain

6.1	Introduction	217
6.2	A Simple Electrostatic Model for Ion Pairing	218
6.3	Anion Solvation	221
6.4	Energetics of Noncovalent Interactions	227
6.5	Covalently Bonded Anion Complexes	251
6.6	A Cautionary Word on Selectivity	253
6.7	Appendix: Methods for the Determination of Thermodynamic Parameters	259
6.8	References	266

7. Electrochemical Aspects of Anion Chemistry 277

Antonio Doménech Carbó, University of Valencia, Valencia, Spain

7.1	Introduction	278
7.2	The Nature of Electrode Processes	279
7.3	Review of Selected Electrochemical Techniques	282
7.4	Electrochemical Pattern of Anion Coordination	286
7.5	Electrochemical Analysis of Anion Coordination Equilibria	295
7.6	Electrochemical Analysis of Coordination Equilibria for Nonelectroactive Anions: Competitive Methods	304
7.7	The Significance of Electrochemical Data	308
7.8	Other Issues	313
7.9	References	314

8. Photochemistry and Photophysics of Supramolecular Systems Containing Anions 321

L. Moggi and M. F. Manfrin, University of Bologna, Italy

8.1	Introduction	321
8.2	Supramolecular Systems Involving Anionic Components	324
8.3	Ion Pairs	324
8.4	Adducts of Polyammonium Macrocycles	327
8.5	Other Studies	331
8.6	References	332

9. Anion Binding Receptors: Theoretical Studies 335

Joanna Wiórkiewicz-Kuczera and Kristin Bowman-James,
University of Kansas, Lawrence, Kansas

9.1	Introduction	335
9.2	Polyammonium Receptors	336

9.3	Calixarenes	346
9.4	Organoboron Macrocyclic Receptors	348
9.5	Organotin Macrocyclic Receptors	350
9.6	Electrides	350
9.7	Conclusion	351
9.8	References	353

10. Application Aspects Involving the Supramolecular Chemistry of Anions 355

*J. L. Sessler, P. I. Sansom, A. Andrievsky, and V. Kral,
University of Texas at Austin, Austin, Texas*

10.1	Introduction	355
10.2	General Classes of Synthetic Anion Receptors	356
10.3	Binding and Transport of Physiologically Important Anions	356
10.4	Anion Receptors in Catalysis	378
10.5	Anion Receptors in Analytical Chemistry	380
10.6	Waste Management	397
10.7	Anion Templated Reactions	401
10.8	References	403

11. Supramolecular Catalysis of Phosphoryl Anion Transfer Processes 421

Mir Wais Hosseini, Université Louis Pasteur, Strasbourg, France

11.1	Introduction	421
11.2	Molecular Recognition of Nucleotides	422
11.3	From Recognition to Transformation	428
11.4	Conclusion	443
11.5	References	444

Index   449

FOREWORD

Anion complexation was unrecognized until rather recently, while cation binding has been extensively studied for about a century. Although I was convinced from the start that there was a whole field to explore, the proposal made in the late 1970s (see *Acc. Chem. Res.* **11**, 49, 1978; *Pure Appl. Chem.* **50**, 871, 1978) that anion coordination chemistry was an identifiable and bona fide area of investigation was not accepted readily. But over the years, things have changed drastically. Thanks to the work of an increasing number of investigators, a body of results on anion binding, recognition, and transport, as well as on supramolecular catalysis involving anionic species, has accumulated. A variety of anion receptor molecules incorporating different types of anion binding sites has been devised, and their properties have been investigated. The recognition of anionic substrates of different shapes and charges has been achieved, as well as the catalytic transformation of substrates within supramolecular entities.

In harmony with the important role played by anionic species both in chemistry and in biology, anion coordination chemistry has become an established new domain of a generalized coordination chemistry. The field has been reviewed recently on several occasions. The present volume indicates that it has reached the status of a mature area of investigation. It provides a thorough overview of what has been realized and should inspire further developments. Indeed, this is just the start—anion coordination chemistry is a young field, wide open to future progress.

I wish to congratulate the editors and the authors for producing this high-level, very timely work, which should be of great value to chemists and biologists alike. It will help to further the cause and broaden the scope of anion coordination chemistry.

JEAN-MARIE LEHN

PREFACE

Supramolecular chemistry has developed in the past decades as a borderline scientific field destined to fill part of the conceptual gap between chemistry, physics, and biology. This chemistry, which bases its evolution in the Paul Ehlrich receptor concept and on the ideas of Alfred Werner on coordination, can be understood as an extended coordination chemistry in which the objects of coordination are not just metal ions but also organic cations, neutral organic or inorganic species, and anionic species of various types. Within this chemistry, the *supramolecular chemistry of anions* or *anion coordination chemistry* has been until now one of the aspects least explored. At first glance this seems a little bit surprising in view of the fundamental role anionic species play in many mineral or biological processes.

Therefore, at this stage it seemed necessary to organize a book with the aim of presenting the actual state of the art in this field, as well as the possible future developments and perspectives. To do this we have asked for the participation of different scientists in order to provide a wide panorama of the different aspects of anion coordination chemistry. The first chapter deals with presupramolecular anion chemistry, or how some of these ideas now attributed to the supramolecular era were extensively used in the past in phase-exchange or phase-transfer processes. The second chapter presents a historical point of view in which, starting from the finding of halide inclusion in 1967 by bicyclic diammonium receptors (katapinands), the most outstanding events that have marked this chemistry are presented. In the following two chapters are presented the most familiar types of natural and artificial molecules that can act as anion receptors. The concepts of preorganization and chemical design are introduced and discussed. The next chapters are devoted to structural, thermodynamic, electrochemical, and photochemical aspects of anion coordination. Owing to the importance of computer methods for the design of receptors and for the understanding of many host–guest relations, a chapter discussing this subject has been included. Finally, the last two chapters are devoted to catalytic aspects and to the perspectives and applications this chemistry is finding in many different fields.

Although some of the chapters on special themes have review characteristics, a comprehensive style has always been pursued. Therefore, we would be satisfied if this book could be of help both to researchers working in this or related fields and also to advanced science and particularly chemistry students.

We would like to thank very warmly all the chapter authors who have readily understood the importance of producing a book having the desired characteristics, and who have contributed exceptionally well-written works in all senses. We are also very grateful to Professor Jean-Marie Lehn for his eloquently written Foreword.

And last but not least, we would also like to acknowledge the patience of our families and co-workers, who, without directly participating in the book, have surely suffered from the time and travels devoted to achieving this goal.

Valencia, November 28, 1996

CONTRIBUTORS

Jerry L. Atwood, Department of Chemistry, University of Missouri—Columbia, Columbia, Missouri 65211

A. Andrievsky, Department of Chemistry and Biochemistry, University of Texas at Austin, Austin, Texas 78712

Antonio Bianchi, Department of Chemistry, University of Florence, Via Maragliano 75/77, 50144 Florence, Italy

Peter V. Bonnesen, Chemical and Analytical Sciences Division, Oak Ridge National Laboratory, Oak Ridge, Tennessee 37831-6119

K. Bowman-James, Department of Chemistry, University of Kansas, Lawrence, Kansas 66045

Antonio Doménech Carbó, Department of Inorganic Chemistry, University of Valencia, c/Dr. Moliner 50, 46100 Burjassot, Valencia, Spain

Bernard Dietrich, Institut Le Bel, Université Louis Pasteur, Rue Blaise Pascal 4, 67000 Strasbourg, France

Marta Ferraroni, Department of Chemistry, University of Florence, Via G. Capponi 6, 50121 Florence, Italy

Enrique García-España, Department of Inorganic Chemistry, University of Valencia, c/Dr. Moliner 50, 46100 Burjassot, Valencia, Spain

Mir Wais Hosseini, Institut Le Bel, Université Louis Pasteur, Rue Blaise Pascal 4, 67000 Strasbourg, France

V. Kral, Department of Chemistry and Biochemistry, University of Texas at Austin, Austin, Texas 78712

Stefano Mangani, Department of Chemistry, University of Siena, Pian de'Mantellini 44, 53100 Siena, Italy

Maria Francesca Manfrin, Department of Chemistry "G. Ciamician", University of Bologna, Via Selmi 2, 40126 Bologna, Italy

Luca Moggi, Department of Chemistry "G. Ciamician," University of Bologna, Via Selmi 2, 40126 Bologna, Italy

Bruce A. Moyer, Chemical and Analytical Sciences Division, Oak Ridge National Laboratory, Oak Ridge, Tennessee 37831-6119.

P. I. Sansom, Department of Chemistry and Biochemistry, University of Texas at Austin, Austin, Texas 78712

Franz P. Schmidtchen, Lehrstuhl für Organische Chemie und Biochemie, Technische Universität München, Lichtenbergstrasse 4, 85747 Garching, Germany

J. L. Sessler, Department of Chemistry and Biochemistry, University of Texas at Austin, Austin, Texas 78712

Jonathan W. Steed, Department of Chemistry, King's College, London, Strand, London WC2R 2LS, United Kingdom

J. Wiórkiewicz-Kuczera, Department of Chemistry, University of Kansas, Lawrence, Kansas 66045

SUPRAMOLECULAR CHEMISTRY
OF ANIONS

CHAPTER 1

Physical Factors in Anion Separations

BRUCE A. MOYER and PETER V. BONNESEN

1.1 Introduction
1.2 Anion Radii, Hydration, and Solvation
1.3 Anion Transfer
    1.3.1 Electrostatic Model of Anion Transfer
    1.3.2 Correlations of Ion Transfer with Physical and Empirical Properties
1.4 Solvent Extraction of Anions
    1.4.1 Hydration of Organic-Phase Species
    1.4.2 Liquid–Liquid Anion Exchange
    1.4.3 Ion-Pair Extraction
1.5 Resin Anion Exchange
    1.5.1 Selectivity between Anions of Like Charge
    1.5.2 Selectivity between Anions of Different Charge
1.6 Conclusions
1.7 Acknowledgments
1.8 References

1.1 INTRODUCTION

As dealt with at length in this volume, the topic of anion recognition encompasses an increasing body of knowledge within the area of chemical recognition. Whereas the focus of attention has recently been directed toward the synthesis and properties of receptors having preorganized Lewis acid functionalities for selective anion binding, it seemed appropriate to us to provide a brief introduction to those physical factors that have traditionally provided the basis for anion separations. It is certainly true that attaining good separations of

---

Supramolecular Chemistry of Anions, Edited by Antonio Bianchi, Kristin Bowman-James, and Enrique García-España.
ISBN 0-471-18622-8. © 1997 Wiley-VCH, Inc.

anions does not necessarily require a host–guest approach. In our zeal to explore novel anion receptors, it has been perhaps easy to overlook the progress that has been made with simple monofunctional solvent-extraction reagents familiar to analytical chemistry,[1–4] hydrometallurgy,[5–9] and the nuclear industry[10] since the 1940s. The physical and chemical phenomena associated with anion separations in such systems have been well described in a number of reviews and books.[11–16] Similarly, the field of resin anion exchange has grown in maturity.[13,17–19] While it may therefore be appreciated that highly effective anion separation processes have been devised based on traditional approaches, one may also appreciate that the means by which the underlying chemical driving forces may be manipulated to achieve a particular selectivity on demand are still rather limited. Basically, one has principles of solvation and electrostatics as tools, and these lead to anion separations based on anion properties such as size, hydrophobicity, and charge. Not surprisingly, the ordering within series of anions recurs persistently in many traditional separation systems, whether they are solvent or resin based and whether the anion-extraction equilibria involve true anion exchange or ion-pair extraction. Although limited changes in the ordering can be effected, gross reversals or peak selectivity have rarely been seen. Thus we may regard the traditional separation systems as representing a collective "baseline" by which it may be possible to judge the effectiveness and uniqueness of anion recognition by novel host molecules.

This chapter primarily describes anion-exchange separations by liquid–liquid extraction and resin systems. Parallels will be drawn with ion-pair extraction systems, which obey much the same selectivity relationships with regard to the anion transfer. Both anion-exchange and ion-pair extraction systems effect a spatial separation of anions from one phase to another, as opposed to homogeneous complexation processes studied commonly in coordination chemistry. Mass transfer of solute species across phase boundaries and a concomitant change in solute solvation characterize the majority of separation processes applicable to the liquid phase, and the solvation change strongly influences overall driving force and selectivity. The traditional anion-exchange reagents are typically ion pairs in which the cation is either restricted to a solvent phase because of the cation's hydrophobic character (solvent extraction) or immobilized on a polymer backbone (resin anion exchange). Such reagents may be viewed as "simple" in their composition, but their actual behavior has usually not turned out to be simple upon close inspection. In solvent extraction, ion pairs aggregate strongly because of their highly dipolar nature; the aggregates range from dimers to inverted micelles. Anions may include not only simple mineral acid anions but also complex anions, typically an assembly of aqueous anionic ligands coordinated to one or more metal cations. Although recent terminology was not available in the 1950s as such systems were beginning to be understood, we now can recognize such behavior as none other than "supramolecular." Certainly as we seek to develop novel supramolecular systems for separations and other uses, knowledge of the underlying behavior of the "simple" reagents can furnish a valuable foundation for advancement.

Writing this review naturally brought about the following question: Do the traditional anion-exchange systems "recognize" the anions that are selected? If selectivity stems predominantly from electrostatic interactions, as widely thought, then it depends fundamentally on the variables of ionic radius and dielectric constant. In principle, for spherical anions of a given charge type, size is the determining factor, and as we describe at length later, extractive preference effectively increases with increasing anion size. Thus no single anion is preferred, since it is generally possible to find a more highly preferred anion. In this case, the authors would argue against the use of the term *recognition*. Rather, the system exhibits "bias" in one direction or another. *Recognition* would, by contrast, be obtained in those systems that could be designed to depart from such baseline selectivity. As discussed in later chapters in this volume, such departures must entail some three-dimensional quality of a host or ligating species, entailing matching contours or bonding sites. Some of the distinguishing features of *bias* and *recognition* in chemical systems are compared in Fig. 1.1. Traditional anion-exchange systems lie near the left side of the figure. They involve primarily physical interactions and solvation without benefit of stable coordinate bonding. However, one could maintain that some of the traditional anion-exchange systems do indeed involve departures from the simple *bias* type of selectivity, and in such cases, one can point to specific effects such as steric interactions, multiple hydrogen bonding, or unique organization of the extraction complex. Such systems arguably entail primitive aspects of *recognition*. For most such systems, however, the authors will consciously avoid the term *recognition* when only *bias* is indicated. The terms *selectivity*, *preference*, or *discrimination* will be treated as all-inclusive.

As a matter of organization, this chapter will first outline some relevant electrostatic principles. Their use and limitations in devising and understanding separation schemes will then be discussed. We will present examples of anion exchange in liquid–liquid systems, defining "baseline" bias and pointing out a few specific departures from such bias. Anion exchange in resin systems

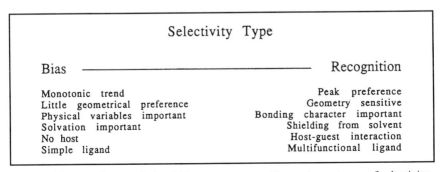

Fig. 1.1 Factors characteristic of *bias* versus *recognition*, extreme types of selectivity.

presents unique features, since the bound functional sites lack the mobility of the liquid–liquid systems and since other physical factors come into play. Discussion of the origin of bias in resin systems together with representative examples will conclude the chapter. In view of the huge, already well-reviewed volume of literature to draw from,[1-19] no attempt will be made to treat the topic of anion separations comprehensively. The reader is referred to this literature for specific information that has been cataloged. Rather, from the present vantage point of chemical progress, selected systems involving simple inorganic anions will be discussed here as a means of illustrating general principles. Unfortunately, the extraction of anionic metal complexes, a topic of considerable importance for industrial applications, is beyond the present scope and will only be treated superficially as basically an extrapolation of the behavior of smaller anions.

1.2 ANION RADII, HYDRATION, AND SOLVATION

Anion charge and size come to mind as primary properties by which solvent or resin systems may distinguish different anions. Table 1.1 presents ionic radii (r) and standard molar Gibbs hydration energies (ΔG_h^o) of common inorganic anions of varying charge and size.[20-29] Heat capacities,[30] entropies,[31] and enthalpies[32,33] of hydration are also available, as are other properties such as softness.[32] Since the concept of anion size partially loses definition in comparing multiatomic anions, thermochemical radii[21-24,26] derived mostly from lattice energies are also included in the table together with the familiar effective ionic crystallographic radii[20] where applicable. Generally, the crystallographic and thermochemical radii are mutually consistent,[34] notwithstanding differences that may reach as high as 0.010 nm (5%). Although most electrostatic calculations deal strictly with spherical ions, the thermochemical radii thus permit useful estimates to be made for the usual situation involving nonspherical ions. We will give preference here to the crystallographic radii in calculations, using the thermochemical radii only for multiatomic anions and implicitly assuming that anionic radii remain constant on transfer between phases. Despite the solid-state origin of the crystallographic and thermochemical radii, diffraction measurements and the results of molecular-dynamics simulations and Monte Carlo calculations have shown that the radii of ions in solution are not significantly different,[34] as suggested previously.[26] It should be borne in mind throughout this chapter that this assumption together with the treatment of multiatomic anions as spheres characterized by a thermochemical radius are only approximations. Comparisons of behavior always work best within families of chemically and geometrically similar anions.

The driving force for solvent-extraction and ion-exchange processes depends upon the state of the ion in the initial phase (here assumed aqueous) and the final phase. Since traditional reagents do not involve host molecules that completely desolvate the ion, considerable direct solvation of the ion in both phases can be presumed. For the most elementary discussion, it is easiest to begin by considering the case where the ion is completely solvated in one phase but not at all in the other. Since the source phase is water, we will begin with the process of anion hydration.

The hydration process refers to the transfer of the bare gas-phase ion to aqueous solution:

$$X^-(g) \rightarrow X^-(aq) \tag{1}$$

Table 1.1 gives the values of the standard molar Gibbs energy of hydration ΔG_h^o at 25°C based on the choice $\Delta G_h^o[H^+] = -1056$ kJ/mol. The choice of this reference value for the hydrogen ion follows from the reasonable assumption (TATB assumption) that the solvation of *t*etraphenyl*a*rsonium cation and *t*etraphenyl*b*orate anion are identical;[35] the TATB assumption will be used for all single-ion thermodynamic values employed in this review. Conceptually, we may view the reverse of the hydration process, namely, the removal of the ion from water to the gas phase, to be the most elementary extraction process as well as the first step in useful thermochemical cycles representing more complicated extraction processes. Although the "extraction" of most common anions into vacuum is obviously a highly unfavorable process, one may note that this process of removing all solvation exhibits extreme bias toward large anions in closely related families such as the halides. Comparing all the anions in Table 1.1, increasing anion size generally favors transfer of the anion to the gas phase. Exceptions to the trend are obvious, however, especially with ions having acidic protons, like $H_2PO_4^-$ and BO_2^- ($H_2BO_3^-$). Accordingly, the thermochemical radius as a single indicator somewhat imperfectly forecasts trends in anion hydration for a wide variety of inorganic anions. Notwithstanding the experimental limitations in measuring thermochemical radii and standard Gibbs hydration energies, the data thus suggest that primitive recognition factors may play a nonzero role even in a process as fundamental as hydration.

Ion solvation represents the analogous and more general process of transferring an ion from the gas phase to any pure solvent, including water:

$$X^-(g) \rightarrow X^-(S) \tag{2}$$

Much theoretical effort has been devoted to developing mathematical models describing the standard Gibbs energy change ΔG_s^o of this process for cations and anions of varying charge types and ionic radii.[36–40] From the simple Born model the electrostatic part of the standard Gibbs energy in kJ/mol may be obtained readily, treating ions as hard spheres in a continuous dielectric:

$$\Delta G_{Born}^o (kJ/mol) = \frac{Bz^2}{r'}\left(1 - \frac{1}{\varepsilon}\right) \tag{3}$$

where ε is the dielectric constant, z is the ionic charge, B is a temperature-independent constant equal to -69.47 kJ nm mol^{-1}, and r' is the ionic radius

TABLE 1.1. Ion Size and Hydration Data for Representative Anions

Anion[a] X^-	Ionic radius[b] (nm)	Thermoch. radius[c] (nm)	d_{M-O}[f] (nm)	$\Delta G_h(X^-)$ obs[g] (kJ mol^{-1})	$\Delta G_h(X^-)$ calc eq. 4[i] (kJ mol^{-1})	$\Delta G_h(X^-)$ calc lit[j] (kJ mol^{-1})
F^-	0.130	0.126		−465	−449	−345
OH^-	0.137	0.133		−430	−449	−345
HCO_3^-		0.156		−335	−388	−310
$CH_3CO_2^-$		0.162		−365	−375	−300
HCO_2^-		0.169		−395	−361	−290
Cl^-	0.181	0.172		−340	−338	−270
IO_3^-		0.181	0.181	−400	−338	−270
Br^-	0.196	0.188		−315	−314	−250
BrO_3^-		0.191		−330	−322	−260
CN^-		0.191		−295	−322	−260
NO_2^-		0.192	0.125	−330	−320	−255
N_3^-		0.195		−295	−316	−250
NO_3^-		0.196[d]	0.125	−300	−314	−275
$H_2PO_4^-$		0.200		−465	−308	−245
ClO_3^-		0.200		−280[h]	−308	−245
CNO^-		0.203		−365	−304	−240
HSe^-		0.205		−360	−301	−240
HS^-		0.207		−295	−298	−235
I^-	0.220	0.210		−275	−282	−220
SCN^-		0.213		−280	−290	−230
BF_4^-		0.232		−190	−268	−205
BO_2^-		0.240		−460	−259	−195
MnO_4^-		0.240	0.159	−235	−259	−195
ClO_4^-		0.240	0.144	−205	−259	−195
IO_4^-		0.249	0.179		−250	−170
ClO_2^-		0.250	0.158	−430	−249	−180
TcO_4^-		(0.252 est)[e]	0.171	(−251 est)[h]	−245	
ReO_4^-		0.260	0.174	−330[h]	−240	
Pic^-		0.329			−190	
Ph_4B^-		0.421		50	−149	15
CO_3^{2-}		0.178		−1315	−1399	−1300
S^{2-}	0.184	0.191		−1315	−1357	−1280
SO_3^{2-}		0.200	0.151	−1295	−1256	−1230
CrO_4^{2-}		0.240	0.163	−950	−1060	−1120
SeO_4^{2-}		0.243	0.152	−900	−1047	−1110
SO_4^{2-}		0.230	0.148	−1080	−1103	−1145
SiF_6^{2-}		0.259		−930	−986	−1075
$PtCl_6^{2-}$		0.313		−685	−823	−955
$PdCl_6^{2-}$		0.319		−695	−809	−945
PO_4^{3-}		0.238	0.152	−2765	−2413	−2835

[a] Ions are arranged in order of increasing thermochemical radii. Pic^- is picrate anion (2,4,6-trinitrophenolate), and Ph_4B^- is tetraphenylborate anion.
[b] Effective ionic radii from crystallographic data.[20]

(in nm) written for convenience here as $r' = r + \Delta_r$ to allow empirical modification of the Born model by application of a user-defined correction Δ_r. Under the hypothesis of a continuous dielectric and hard spheres ($\Delta_r = 0$), Eq. 3 exactly represents the electrostatic charging energy. Neglecting other effects discussed further below, the desired solvation energy ΔG_S^o under the stated assumptions then follows directly upon applying the practically negligible 7.9 kJ/mol correction term to account in principle for compression of the gaseous ions to 1 L at 25°C[21,36]:

$$\Delta G_S^o (\text{kJ/mol}) = \frac{Bz^2}{r'}\left(1 - \frac{1}{\varepsilon}\right) + 7.9 \quad (4)$$

Although we concern ourselves here only with the single temperature 25°C, Eq. 4 exhibits temperature dependence through ε and the compression factor. In applications of Eq. 4, r is usually taken to be the crystallographic radius if available or else the thermochemical radius. The best agreement with experiment results when Δ_r is an empirical constant equal to 0.080 nm for cations and 0.017 nm for anions in water[36] and other solvents.[41,42] The need for two values of Δ_r depending only on the sign of the ionic charge implies fundamentally different solvation of cations and anions. This has long been understood[43] to reflect the different orientation of solvent molecules around cations and ions.

Among the halides, the only truly spherical univalent anions listed in Table 1.1, the agreement of Eq. 4 with experiment may be seen to be excellent. Although in individual cases throughout the table Eq. 4 predicts the standard molar Gibbs hydration energy with an accuracy considerably worse than 10 kJ/mol, Table 1.1 shows that the overall agreement is still remarkably

---

[c] Thermochemical radii, unless otherwise noted, were taken from Ref. 21; most of the values are derived from lattice energies[21–24]; for monatomic ions, values were taken from Ref. 23.
[d] Although the nitrate ion radius was given as 0.179 nm in Refs. 21 and 23, the value of 0.196 from Ref. 26 was selected for inclusion in the table because of its better consistency with separation data presented in subsequent tables (see text).
[e] The value for TcO$_4^-$ anion was estimated by adding 0.083 nm (Δ_d) to d_{M-O}; Δ_d was assumed to be linearly related to d_{M-O} for MnO$_4^-$ and ReO$_4^-$, giving $\Delta_d = 0.083$ nm.
[f] Mean bond distances between central atom M and the oxygen atoms in selected oxyanions.[27,28]
[g] Standard molar Gibbs energies of hydration at 25°C taken from the tabulations in Refs. 21 and 25; values are based on the choice $\Delta G_h^o[\text{H}^+] = -1056$ kJ/mol under the assumption that the tetraphenylarsonium cation and tetraphenylborate anion have equal Gibbs hydration energies (TATB assumption); the uncertainty of the listed values is ±6 kJ/mol.
[h] A linear free-energy relationship based on anion-exchange equilibrium constants has been used to predict $\Delta G_h(X^-)$[29]; the values obtained for TcO$_4^-$, ClO$_3^-$, and ReO$_4^-$ were, respectively, −251 (shown in table), −280 (shown in table), and −234 kJ/mol.
[i] Standard molar Gibbs energies of hydration calculated from Eq. 4 using the ionic radius (if given) or the thermochemical radius and using $\Delta_r = 0.017$.
[j] Standard molar Gibbs energies of hydration reported for the single-shell model using the ionic radius (if given) or the thermochemical radius.[21]

good, considering the large magnitudes of the numbers involved. Further, Eq. 4 successfully predicts the gross dependence of ΔG_h^o on ion charge and ionic radius. On the other hand, a disagreement of even 10 kJ/mol between theory and experiment represents an ability to predict equilibrium constants with an accuracy of only two orders of magnitude. The discrepancies become even more worrisome in correlating solvation thermodynamic parameters with dielectric constant, which for different solvent types appears to be only a rough guide for predicting behavior, as discussed a bit later.

Improvements in the theory have long attacked the questionable assumptions of Eq. 4, including reliance on a hypothetical continuous dielectric, the use of an empirical constant (Δ_r), and the neglect of other effects.[36–40] Thorough reviews of the various approaches to correcting these inadequacies have been published.[37,38] For present purposes, a brief qualitative description of the single-shell model[21,39,40] will suffice to point out several important effects related to anion extraction; we will return to this model again, because it has in fact been applied to the prediction of ion partitioning.[44] Since solvent molecules are immobilized, or electrostricted, by the intense electric field about an ion, the solvation in the vicinity of an ion may conceptually be divided into shells, requiring a stepwise or continuous change in ε as the distance to the ion decreases. Assumption of a single shell gives rise to a manageable model that makes useful predictions, and according to this treatment, ΔG_S^o may be divided into three contributions:

$$\Delta G_S^o = \Delta G_{el}^o + \Delta G_{unsym}^o + \Delta G_{neut}^o \tag{5}$$

The 7.9 kJ/mol correction term used in Eq. 4 has been omitted because in practice it is incorporated automatically in the neutral term ΔG_{neut}^o. For the single-shell model, the electrostatic term ΔG_{el}^o is taken to be a summation, $\Delta G_{el1}^o + \Delta G_{el2}^o$. The first of these terms, ΔG_{el1}^o, is the electrostatic standard Gibbs energy change associated with the first shell, assumed to have a low dielectric constant (ε_1) and a thickness on the order of one solvent radius; different prescriptions for defining the exact thickness of the shell and the dielectric constant within it have been given.[39,40] The second electrostatic term ΔG_{el2}^o is the usual Born term given by Eq. 3, employing the bulk solvent dielectric constant ε and the ionic radius $r' = r + \Delta_r$, where Δ_r is now taken to be the shell thickness. The summation $\Delta G_{el1}^o + \Delta G_{el2}^o$ reduces to the expression

$$\Delta G_{el}^o (\text{kJ/mol}) = Bz^2 \left[\left(1 - \frac{1}{\varepsilon_1}\right)\left(\frac{1}{r} - \frac{1}{r'}\right) + \left(1 - \frac{1}{\varepsilon}\right)\left(\frac{1}{r'}\right) \right] \tag{6}$$

Since solvent molecules orient around cations differently than around anions, an unsymmetric solvation effect arises as explicitly taken care of by the term ΔG_{unsym}^o in Eq. 5; a theoretically sound basis for predicting its value, however, seems elusive at present.[21,39,43] Finally, a neutral term ΔG_{neut}^o accounts for solvation interactions involving the hypothetical ion having zero charge.

This neutralization term has been estimated empirically from Henry's Law constants of inert gases and hydrocarbons.[21,39,40,43] In the case of water, the neutral term gives a positive contribution to the standard molar Gibbs solvation energy and is thought to be dominated by cavity formation.[39] That is, the strong cohesive energy of water as an isolated effect tends to push the solute out in much the same way that aliphatic hydrocarbons are rejected. The neutral term in this case is small (probably $< 20\,\text{kJ/mol}$) for anions having radii less than 0.2 nm, but for large anions, the effect may rapidly become highly significant; for example, the calculated value of ΔG_h^o from Eq. 4 for the tetraphenylborate anion is $-149\,\text{kJ/mol}$, whereas the observed value is $50\,\text{kJ/mol}$. For nonaqueous solvents, however, the cohesive energies are typically much lower, and thus the pushing-out effect is overshadowed by attractive forces. As a consequence, the neutral term for nonaqueous solvents decreases from positive values and becomes negative for larger anions.[39,40] For the discussion presented here, the neutral term plays a limited role, but for researchers interested in large anionic complexes or large organic anions, this effect must be considered.

As shown in Table 1.1, the single-layer model actually does no better than the modified Born expression (Eq. 4) in predicting standard molar Gibbs hydration energies overall. In fact, for small ions, Eq. 4 predicts more reliably. Since the neutral term of the single-layer model allows ΔG_h^o to go to positive values, it must be used for the large ions. It will be seen that the single-layer model also must be used in calculating Gibbs transfer energies for solvents having appreciable dielectric constants. Thus, whereas the single-layer model offers little practical advantage in predicting values of ΔG_h^o here, it serves well in certain situations and further allows solvation to be understood in terms of its component effects.

1.3 ANION TRANSFER

1.3.1 Electrostatic Model of Anion Transfer

Since principles of electrostatics have been useful in understanding ion solvation, one may readily extend such ideas to ion-transfer processes between two liquid phases. The discussion above showed that anion radius and charge are fundamental ion properties upon which selectivity can be based, and we shall show now their expected influence in anion-extraction processes. The standard molar Gibbs energy of transfer $\Delta G_{\text{tr}[W \rightarrow O]}^o$ of an anion Y^- refers to the process of removing it from water into the gas phase as the bare ion and re-solvating it in a *dry* solvent S to give the net process:

$$Y^-(\text{aq}) \rightarrow Y^-(\text{S}) \tag{7}$$

To shorten notation while retaining the same meaning, we will henceforth often use ΔG_{tr}^o without the subscript $[\text{W} \rightarrow \text{O}]$ in reference to this process. The standard Gibbs energy change is given by:

$$\Delta G_{\text{tr}}^{\text{o}} = \Delta G_{\text{S}}^{\text{o}} - \Delta G_{\text{h}}^{\text{o}} \qquad (8)$$

A growing number of experimental values of $\Delta G_{\text{tr}}^{\text{o}}$ are available for cations and anions transferring to many different solvents.[42,45–47] Selected values have been listed in tables given later in this chapter for purposes of discussion. It may be noted by the reader that the literature on Gibbs energies of ion transfer deals almost exclusively with solvents having sufficiently high dielectric constants to dissociate ion pairs completely, at least in the limit of low, yet experimentally accessible, concentrations. Most such solvents have dielectric constants greater than ca. 10, and most are water miscible. However, a few water-immiscible solvents having $\varepsilon > 10$, such as 1,2-dichloroethane, 1-octanol, nitrobenzene, and o-nitrophenyloctylether, are important in anion extraction, and thus, to understand processes in them requires a general understanding of ion transfer in all high-dielectric-constant media. Table 1.2 summarizes some properties of selected polar solvents relevant to this discussion. More extensive listings of solvent properties are given in the literature cited in the table and in large compilations.[48] Of course, most solvents associated with anion extraction, especially as applied to process chemistry, employ low-polarity water-immiscible diluents, and we will later consider such systems, where ion-pairing phenomena play a central role.

Drawing from the discussion of ion solvation above, $\Delta G_{\text{tr}}^{\text{o}}$ may be predicted by applying the modified Born expression in Eq. 4 to estimate $\Delta G_{\text{S}}^{\text{o}}$ and $\Delta G_{\text{h}}^{\text{o}}$ in Eq. 8. Rearrangement then gives an expression applicable to both cations and anions:

$$\Delta G_{\text{tr}}^{\text{o}}(\text{kJ/mol}) = \frac{Bz^2}{r'} \left(\frac{1}{\varepsilon_{\text{W}}} - \frac{1}{\varepsilon_{\text{S}}} \right) \qquad (9)$$

Again, $B = -69.47\,\text{kJ}\,\text{nm}\,\text{mol}^{-1}$, and $r' = r + \Delta_r$, where $\Delta_r = 0.080$ nm for cations and 0.017 nm for anions. When $\varepsilon_{\text{S}} < \varepsilon_{\text{W}}$, as is true for most organic solvents of interest, Eq. 9 predicts unfavorable transfer of any ion to the organic phase. As in "extraction by vacuum," the process becomes less unfavorable with increasing ion size, in effect exhibiting strong size bias. Since the magnitude of the solvation of the ion in the solvent phase is still large, however, this size bias is considerably attenuated compared to extraction by vacuum. The attenuation rapidly increases with increasing dielectric constant, whence $1/\varepsilon_{\text{S}}$ approaches $1/\varepsilon_{\text{W}}$.

Although the modified Born expression given by Eq. 4 provides a useful means of calculating Gibbs hydration energies ($\Delta G_{\text{h}}^{\text{o}}$) of ions, the derived expression given by Eq. 9 poorly calculates Gibbs transfer energies ($\Delta G_{\text{tr}}^{\text{o}}$) as the dielectric constant increases above 10. Table 1.3 (line 1) gives experimental values of $\Delta G_{\text{tr}}^{\text{o}}$ for chloride ion transfer from water to a variety of solvents at 25°C. The solvents are arranged from left to right according to dielectric

TABLE 1.2. Properties of Selected Polar Solvents at 25°C

Solvent[a]	Water	1,1DCE	1,2DCE	py	BuOH	EtOH	TFE	HMPA	MeOH	PhNO$_2$	MeCN	MeNO$_2$	DMF	DMSO	PC
ε (diel. const.)[b]	78.4	10.0	10.3	12.2	17.5	24.4	26.3	29.4	32.2	35.1	36.4	37.0	37.0	46.5	64.5
V_S (cm^3/mol)[c]	18	85	79	81	92	59	72	178	41	103	53	54	77	71	85
r_S (nm)[d]	0.16	0.26	0.25	0.26	0.27	0.23	0.25	0.33	0.20	0.28	0.22	0.22	0.25	0.25	0.26
α(HBD)[e]	1.17	0.10	0.00	0.00	0.79	0.83	1.51	0.00	0.93	0.00	0.19	0.22	0.00	0.00	0.00
β(HBA)[e]	0.47	0.10	0.00	0.64	0.88	0.75	0.00	1.06	0.66	0.39	0.40	0.06	0.69	0.76	0.40
π^* (dipol. polariz.)[e]	1.09	0.48	0.81	0.87	0.47	0.54	0.73	0.87	0.60	1.01	0.75	0.85	0.88	1.00	0.83
E_T(30) (polarity)[f]	264.0	164.8	172.8	169.5	210.0	217.1	250.2	171.1	231.8	172.4	190.8	193.7	183.3	188.7	195.0
DN[g]	177.0	12.6	0.0	138.5	121.3	127.2	0.0	162.3	141.4	18.4	59.0	11.3	111.1	124.7	63.2
AN[h]	54.8	10.3	16.7	14.2	36.8	37.9	53.3	10.6	41.5	14.8	18.9	20.5	16.0	19.3	18.3

[a] Solvent abbreviations: 1,1DCE, 1,1-dichloroethane; 1,2DCE, 1,2-dichloroethane; py, pyridine; BuOH, 1-butanol; EtOH, ethanol; TFE, 2,2,2-trifluoroethanol; HMPA, hexamethylphosphorictriamide; MeOH, methanol; PhNO$_2$, nitrobenzene; MeCN, acetonitrile; MeNO$_2$, nitromethane; DMF, N,N-dimethylformamide; DMSO, dimethyl sulfoxide; PC, propylene carbonate.
[b] Dielectric constant of the dry solvent at 25°C taken from available sources.[25,47,89]
[c] Molar volume of the dry solvent at 25°C taken from available sources.[25,47,59,75]
[d] Solvent radius in nm defined as $r_S = [(V_S \times 10^{21})/(8N_{Av})]^{1/3}$, where N_{Av} is Avogadro's number and V_S (line 2) is expressed in cm^3/mol (Ref. 44).
[e] Kamlet–Taft hydrogen-bond donor (HBD), hydrogen-bond acceptor (HBA), and dipolarity–polarizability solvatochromic parameters taken from Ref. 47.
[f] Dimroth–Reichardt E_T solvatochromic parameter in kJ/mol taken from Ref. 53.
[g] Gutmann donor number (DN) in kJ/mol taken from Refs. 42 and 56; the values for TFE and 1,1-DCE were estimated from the correlation DN(kJ/mol) = $-3.8 + 163.9\beta$.[56]
[h] Gutmann acceptor number (AN) taken from Ref. 53; the value shown for 1,1-DCE was estimated from the correlation AN = $0.308E_T - 40.52$.[53]

constant in correspondence with Table 1.2. It may be seen that the experimental values of ΔG_{tr}^o have no obvious relationship to dielectric constant from $\varepsilon = 10$ to 65 (cf. Table 1.2). By contrast, Eq. 9 predicts that ΔG_{tr}^o should fall off to values approaching zero over this range, as shown in Table 1.3 (line 2). Similar same results can be obtained for any of the anions for which data are available.[47] The reader should note that these results do not necessarily reflect the usefulness of Eq. 9 at lower dielectric constants, as we will discuss later.

As mentioned in the previous section, a physically more justifiable approach than Eq. 9 to calculating ΔG_{tr}^o has been described employing a single-layer solvation-shell model. In this case, Eq. 5 provides the estimates of ΔG_S^o and ΔG_h^o for substitution in Eq. 8.[44] For calculating ΔG_{el}^o via Eq. 9, Abraham suggested setting $\varepsilon_1 = 2$ and the shell thickness Δ_r equal to the solvent radius r_S, calculated from the solvent molar volume V_S (cm^3/mol) according to the formula $r_S = [(V_S \times 10^{21})/(8N_{Av})]^{1/3}$, where N_{Av} is Avogadro's number.[40,44] According to this prescription, ΔG_{el}^o for any solvent can be readily calculated. It may be seen that the calculated values of ΔG_{el}^o in Table 1.3 (line 3) still indicate strong electrostatic-based solvation of chloride in all the solvents. As indicated above, the small positive neutral term ΔG_{neut}^o may be seen (line 4) to play almost no role for chloride. Following Abraham, the sum $\Delta G_{el}^o + \Delta G_{neut}^o$ was taken according to Eq. 5 without including the unsymmetric term, giving ΔG_S^o for both water and the various solvents (line 5). The net Gibbs transfer energies ΔG_{tr}^o based on the single-layer (SL) model are given in Table 1.3, line 6. Unlike the trend predicted by the modified Born treatment, the trend in ΔG_{tr}^o predicted by the single-layer model does not fall off to zero with increasing dielectric constant. Rather, the values decrease until the $1/\varepsilon$ term in Eq. 6 becomes negligible, whence the electrostatic contribution becomes only a function of the radius of the solvent molecules. Mainly because the predicted values do not disappear with increasing dielectric constant, the single-layer model gives better agreement with experiment. At low dielectric constants the modified Born and single-layer models nearly coincide.

Although the single-layer model improves the prediction of Gibbs transfer energies, it obviously has not accounted for all operative effects. It may be noted that, except for the alcohol solvents, the values in line 6 of Table 1.3 underpredict ΔG_{tr}^o for chloride ion by a substantial margin in most cases. Marcus suggested the use of the unsymmetric term ΔG_{unsym}^o in Eq. 5 for ion hydration to account for the different orientation of water molecules around cations as compared with anions.[21,39] An empirical formula ΔG_{unsym}^o(kJ/mol) $= 120(r/$nm$)z^3$ was proposed, which for chloride gives the value -22 kJ/mol. This correction moves the estimated value of the Gibbs hydration energy ΔG_h^o in Table 1.3 from -276 to -298 kJ/mol, closer to the observed value of -340 kJ/mol given in Table 1.1. The correction also moves the estimated Gibbs transfer energies ΔG_{tr}^o in Table 1.3 into better agreement with experiment, except for the alcohols. Although there is no guidance on estimating ΔG_{unsym}^o for any solvent other than water, one may expect that lack of strong hydrogen bonding ability would diminish the effect. Indeed, if the same value of

TABLE 1.3. Comparison of Observed and Calculated Standard Molar Gibbs Transfer Energies ΔG°_{tr} (kJ/mol) for Chloride Ion from Water into Various Pure Solvents at 25°C[a]

No.	Solvent[b]	Water	1,1DCE	1,2DCE	py	BuOH	EtOH	TFE	HMPT	MeOH	PhNO$_2$	MeCN	MeNO$_2$	DMF	DMSO	PC
1	ΔG°_{tr}(obs)[c]		58	52	34	29	20	−10	58	13	35	42	38	48	40	40
2	ΔG°_{tr} (calc, eq. 9)[d]		31	30	24	16	10	9	7	6	6	5	5	5	3	1
3	ΔG°_{el} (calc, Eq. 6)[e]	−293	−255	−256	−258	−261	−269	−267	−255	−276	−263	−273	−273	−268	−270	−268
4	ΔG°_{neut}(calc)[f]	17	11	11	11	10	13	12	1	15	9	14	14	11	12	11
5	ΔG°_{S} (calc. Eq. 5)[g]	−276	−244	−245	−247	−251	−256	−255	−254	−261	−255	−259	−259	−256	−258	−258
6	ΔG°_{tr} (calc SL)[h]		32	30	31	29	25	21	22	15	21	17	20	18	18	18
7	ΔG°_{tr} (calc SL)[i]		51	53	50	30	24	10	44	16	43	34	34	41	40	40

[a] Line numbers are provided for convenient reference from text; columns correspond identically to those shown in Table 1.2. Values shown in lines 3–5 were used in obtaining ΔG°_{tr} for lines 6 and 7.
[b] Solvent abbreviations: 1,1DCE, 1,2-dichloroethane; 1,2DCE, 1,2-dichloroethane; py, pyridine; BuOH, 1-butanol; EtOH, ethanol; TFE, 2,2,2-trifluoroethanol; HMPT, hexamethyl phosphoramide; MeOH, methanol; PhNO$_2$, nitrobenzene; MeCN, acetonitrile; MeNO$_2$, nitromethane; DMF, N,N-dimethylformamide; DMSO, dimethyl sulfoxide; PC, propylene carbonate.
[c] Observed standard molar Gibbs transfer energies ΔG°_{tr} at 25°C for chloride ion, in kJ/mol, based on the TATB assumption.[47]
[d] Calculated from the modified Born model, Eq. 9, using the tabulated value of ε (Table 1.2), $r = 0.181$ nm (Table 1.1), and $\Delta_r = 0.017$ nm.
[e] Calculated from the single-shell electrostatic model, Eq. 6, using $\varepsilon_1 = 2$ and $\Delta_r = r_S$ (Table 1.2).[44]
[f] Calculated essentially according to the recommendations of Abraham after conversion to the molar scale; using $r = 0.181$ nm for chloride ion, the following empirical relationships were used: ΔG°_{neut}(water) $= 29 - 86r - 530r^3$, $\Delta G^{\circ}_{S} = \Delta G^{\circ}_{S} - \Delta G^{\circ}_{h}$, respectively prescribed by Eq. 5, neglecting the unsymmetric term. ΔG°_{neut}(solvent) $= 33.6 - 0.572 V_S - 78.5r$; the former relationship was obtained by fitting ΔG°_{S} given in Ref. 40 versus r_S for inert gases and aliphatic hydrocarbons; the latter relationship follows from linear relationships $\Delta G^{\circ}_{S} = mr_S + c$ and the fact that m varies linearly with V_S.
[g] The value of ΔG°_{S} was obtained as the sum $\Delta G^{\circ}_{el} + \Delta G^{\circ}_{neut}$ (see lines 3 and 4, respectively prescribed by Eq. 5, neglecting the unsymmetric term.
[h] Calculated using the single-layer (SL) model via Eq. 8, $\Delta G^{\circ}_{tr} = \Delta G^{\circ}_{S} - \Delta G^{\circ}_{h}$, using values from line 5 for solvent and water.
[i] As discussed in the text, an adjusted single-layer model was used, where the unsymmetric term ΔG°_{unsym} was estimated and included in Eq. 5 for calculation of ΔG°_{S}; for water, the value -22 kJ/mol was used as prescribed in Ref. 21, and for the solvents this value was scaled by multiplying it by the Kamlet–Taft α parameter (Table 1.2)[47]; using the alternate values of ΔG°_{S}, ΔG°_{tr} was again calculated as the difference $\Delta G^{\circ}_{S} - \Delta G^{\circ}_{h}$.

−22 kJ/mol is applied only to the alcohols in full, much improved agreement is obtained. Even better is to scale the aysmmetric term of the solvent by the Kamlet–Taft parameter α, which measures solvent's hydrogen-bond donating ability.[49] The results after such scaling are given in the last line of Table 1.3 (SL' model). Although the good agreement with observed Gibbs transfer energies for chloride may be just coincidental, the improvement obtained suggests that hydrogen bonding ability may have a role to play in anion solvation. It also suggests that there may be alternate approaches to understanding solvation of ions. Further exploration of these points will be made in the next section.

Table 1.4 presents some experimental and calculated values for standard molar Gibbs energies of transfer $\Delta G^o_{tr[W \to O]}$ of nine anions from water to 1,2-dichloroethane 25°C. Included in the table are values of standard molar Gibbs energies of anion partitioning $\Delta G^o_{tr[W(O) \to O(W)]}$ taken from extraction experiments, where the two phases are mutually saturated. The anions are arranged in order of increasing ease of transfer. It may be seen that this essentially corresponds to the order of increasing anion size, as expected. If the recently tabulated thermochemical radius of nitrate anion is used (0.179 nm[21]), nitrate falls significantly out of line; in fact, nitrate ion consistently behaves more like bromide or chlorate ions in anion extraction, and we thus prefer to use 0.196 nm.[26] One may also note that the effect of mutual phase saturation is

TABLE 1.4. Standard Molar Gibbs Energies of Transfer and Partitioning of Anions from Water to 1,2-Dichloroethane at 25°C

Anion	r_X [a] (nm)	$\Delta G^o_{tr[W(O) \to O(W)]}$ [b] (kJ/mol)	$\Delta G^o_{tr[W \to O]}$ [c] (kJ/mol)	Calc Eq. 9 [d] (kJ/mol)	Calc SL [e] (kJ/mol)
F$^-$	0.130	58		39	43
Cl$^-$	0.181	46	54	30	31
Br$^-$	0.196	38	39	28	27
NO$_3^-$	0.196	34		28	27
ClO$_3^-$	0.200	33		27	26
SCN$^-$	0.213	26		25	23
I$^-$	0.220	26	25	25	22
ClO$_4^-$	0.240	17	17	23	17
Ph$_4$B$^-$	0.421	−35	−33	13	−34

[a] Taken from Table 1.1, column 3 (thermochemical radius[21]) unless data from column 2 (ionic data from crystallographic data) were available.
[b] Based on the standard Gibbs energies of partitioning of quaternary ammonium or tetraphenylarsonium salts from water to 1,2-dichloroethane at 25°C, extrapolated to infinite dilution whence complete ion-pair dissociation may be assumed; phases are mutually saturated; the TATB assumption was used to obtain the single-ion values.[50]
[c] From solubility measurements of salts into dry 1,2-dichloroethane at 25°C.[90,91]
[d] Calculated from the modified Born expression for single-ion transfer (Eq. 9) with $\varepsilon = 10.3$ (Table 1.2) and $\Delta_r = 0.017$ nm.
[e] Calculated based on the single-layer (SL) electrostatic model in the same manner described for line 6 in Table 1.3.[44]

relatively minor, affecting only chloride; this is consistent with the greater tendency of chloride to be hydrated,[50] effectively lowering the energy cost of solvation in the organic phase. Obviously, the process of ion transfer to either dry or moist 1,2-dichloroethane exhibits strong bias, making possible effective separations between certain anion pairs. For example, a separation of perchlorate from nitrate could in principle be effected with a driving force of 17 kJ/mol, equivalent to a separation factor of 10^3 in a single stage.

Although the electrostatic treatments have difficulty in dealing with the general dependence of Gibbs transfer energies for ions on the solvent dielectric constant, better predictability results in modeling the dependence on ionic radius. Since the theoretical calculations apply only to ion transfer between pure phases, the calculated values in the last two columns in Table 1.4 should be strictly compared with values of $\Delta G^o_{tr[W \to O]}$ when available in the fourth column. The fifth column gives the calculated value of $\Delta G^o_{tr[W \to O]}$ from the modified Born expression, Eq. 9, in the same manner as Table 1.3, line 2. The sixth column in Table 1.4 shows the prediction of the single-layer (SL) model with neglect of the unsymmetric contribution; the calculations correspond exactly to the method used in Table 1.3, line 6. As required by the mathematics, both columns of calculated values reproduce the correct order of anions. In the case of the modified Born model (Eq. 9), the predicted bias is lower than observed, since $\Delta G^o_{tr[W \to O]}$ is underpredicted for small anions and overpredicted for large ones. The modified Born model makes no provision for negative Gibbs transfer energies, and thus the model fails completely for tetraphenylborate. Nevertheless, for most of the anions, the estimates using Eq. 9 are not completely unreasonable, and this approach could be used cautiously for estimations.

As shown in Table 1.4, the single-layer model does better overall and through the neutral term in Eq. 5 takes care of the highly favorable transfer of tetraphenylborate. Since the cohesive energy of water greatly exceeds that of organic solvents, the neutral term becomes more negative with increasing anion size. This simply reinforces the size bias of the electrostatic effect. It may be seen that in the case of the smallest anions, however, the single-layer model falls short in its prediction of $\Delta G^o_{tr[W \to O]}$. It was suggested above that orientational effects involving hydrogen bonding for anions may have some role to play, and this would indeed seem to apply most strongly to the smallest anions. Since 1,2-dichloroethane has a relatively poor ability to hydrogen bond (cf. α parameters in Table 1.2), the reluctance of small anions to transfer into it would be expected to be accentuated. For such anions, an adjustment to the single-layer model such as used in Table 1.3, line 7 (SL′ model), might be applied for practical estimations.

1.3.2 Correlations of Ion Transfer with Physical and Empirical Properties

Before turning to a discussion of extractive processes, it is worthwhile to point out some other approaches to understanding ion solvation. We saw above that a purely electrostatic approach offers utility in rationalizing and even predicting the Gibbs energy of anion transfer. Even so, other considerations had to be invoked, albeit with some empiricism, for complete understanding. In view of such theoretical difficulties in predicting ion solvation quantitatively, especially as one compares different solvents, a number of authors have been led to ask what factors other than electrostatic effects might be involved. Donor–acceptor phenomena have often been identified. For example, it was found that standard molar Gibbs energies of transfer of chloride from acetonitrile to various solvents decrease linearly with increasing solvent acceptor number AN.[51] Intuitively, such a result makes sense. For an anion that can function as an electron donor, as chloride does in coordinating to metal ions (e.g., Hg^{2+}, Pd^{2+}, and Fe^{3+}), the more hospitable environment should be in the solvent possessing the greater acceptor ability. The electron-acceptor ability of a solvent, especially as related to its ability to donate hydrogen bonds, has in fact repeatedly been connected with anion solvation and chemical processes dependent upon anion solvation.[49,51–53] Unfortunately, many scales of donor–acceptor ability have been proposed, even with different nomenclature as to what is being donated and accepted, complicating the task of scale selection for practical purposes. Moreover, many of the scales correlate with one another and with various physical properties.

To facilitate the use of solvent scales for predictive purposes, statistical methods have been employed to identify the most significant independent factors governing ion solvation.[42,47,49] Some useful correlations have been obtained based on expressions of the form

$$\Delta G_{tr}^o(I) = \sum_j A_j(I)[P_j(S) - P_j(W)] \qquad (10)$$

where $P_j(S)$ is a solvent property, $P_j(W)$ is the corresponding property for water, and $A_j(I)$ are coefficients of fitting for each ion I. Typical solvent properties tried in such correlations are shown in Table 1.2. As mentioned above, a growing database of values of $\Delta G_{tr}^o(I)$, as sampled in Table 1.3, has been collected by Marcus for both cation and anion transfer to various polar organic solvents.[45,47]

In one study employing Eq. 10, four solvent properties were found to be most influential in a multiple regression analysis of the standard molar Gibbs energy of transfer $\Delta G_{tr}^o(I)$ of 16 ions from water to 13 solvents.[42] The four properties included the Dimroth–Reichardt $E_T(30)$ parameter,[53,54] the Gutmann donor number (DN),[55,56] the reciprocal of the dielectric constant $(1/\varepsilon)$,[57] and the square of the Hildebrand solubility parameter (δ^2).[58,59] The solvatochromic polarity parameter $E_T(30)$ (often denoted simply E_T, as will be hereafter adopted) is based on the transition energy of a charge-transfer band of a pyridinium N-phenolate betaine dye. Since the dye and its relatives possess a phenolate functionality and no electron acceptor sites, E_T reflects especially the acceptor properties of a solvent[53,54] and is closely related to the acceptor number[60] and Kamlet–Taft α parameter.[54,61] The solubility parameter squared δ^2 reflects the cohesive energy density of a solvent, essentially a measure of the work required to form a cavity in the solvent. In recognition of the strong electrostatic solvation energies as seen above, the regression was forced to accept $1/\varepsilon$ as a solvent property. The derived correlation was given by the expression

$$\Delta G_{tr}^o(I) = A_{E_T}(I)[E_T(S) - 264.0] + A_{DN}(I)[DN(S) - 177.0]$$
$$+ A_{1/\varepsilon}(I)[1/\varepsilon(S) - 1/78.54] + A_{\delta^2}(I)[\delta^2(S) - 2291] \quad (11)$$

$E_T(S)$(kJ/mol), DN(S)(kJ/mol), $1/\varepsilon(S)$, or $\delta^2(S)$(J/cm^3) are the solvent properties taken from tabulations. Table 1.5 gives the obtained coefficients of fitting $A_j(I)$ for five anions I; correlations allowing estimations of coefficients for other ions have been given.[52]

It may be seen from Table 1.5 that the most influential of the four parameters for the anions is E_T, indicating the importance of acceptor-type interactions in the solvation of anions. Neither DN nor, curiously, $1/\varepsilon$ contributes significantly to the calculated transfer energy for the anions; the values of A_{DN} and $A_{1/\varepsilon}$ are comparable to, or less than, their precision estimates. Also, A_{δ^2} is not very significant, though it gives a modest negative contribution to $\Delta G_{tr}^o(I)$. As an example, for Cl$^-$ transfer to nitromethane, the four terms in Eq. 10 corresponding to the solvent properties E_T, DN, $1/\varepsilon$, and δ^2 are, respectively, 58.6, -2.3, 1.1, and -17.3 kJ/mol, and these are summed to give $\Delta G_{tr}^o(Cl^-)_{calc} = 40.1$ kJ/mol, as compared with $\Delta G_{tr}^o(Cl^-)_{obs} = 38.7$ kJ/mol. For I$^-$ transfer to nitromethane, the analogous terms are 23.1, -0.7, 2.8, -7.9 kJ/mol, and $\Delta G_{tr}^o(Cl^-)_{calc} = 17.3$ kJ/mol compared with $\Delta G_{tr}^o(Cl^-)_{obs} = 19.0$ kJ/mol. Thus, the effect of solvent polarity as measured by E_T dominates but decreases with increasing anion size; A_{E_T} was in fact found to vary inversely

TABLE 1.5. Multiple Linear Regression Coefficients for Standard Molar Gibbs Energies of Transfer for Selected Anions Based on Solvent Properties E_T, DN, $1/\varepsilon$, and δ^2 ^{a}

Ion (I)	Variance S^2	A_{E_T}	A_{DN}	$10^{-2} A_{1/\varepsilon}$	$10^2 A_{\delta^2}$
Cl$^-$	12.4	-0.831 ± 0.081	0.014 ± 0.024	0.75 ± 1.24	1.07 ± 0.42
Br$^-$	13.2	-0.585 ± 0.084	0.017 ± 0.025	0.67 ± 1.28	0.59 ± 0.43
I$^-$	12.1	-0.327 ± 0.080	0.004 ± 0.024	1.91 ± 1.23	0.49 ± 0.41
SCN$^-$	15.4	-0.30 ± 0.11	-0.001 ± 0.003	2.69 ± 2.26	0.61 ± 0.55
N$_3^-$	18.5	-0.74 ± 0.13	-0.005 ± 0.030	1.23 ± 1.52	0.129 ± 0.068

^{a} The standard molar Gibbs energy (kJ/mol) of the ion-transfer process from water to a dry solvent is given by $\Delta G_{tr}^o(I) = \sum_j A_j(I)[P_j(S) - P_j(W)]$ (Eq. 10), where $P_j(S)$ is the solvent property E_T (Dimroth–Reichardt parameter, kJ/mol),[53,54] DN (Gutmann donor number, kJ/mol),[55,56] $1/\varepsilon$ (reciprocal dielectric constant),[57] or δ^2 (Hildebrand solubility parameter squared, J/cm^3)[58,59]; $P_j(W)$ is the corresponding property for water ($E_T = 264.0$ kJ/mol, DN = 177.0 kJ/mol, $1/\varepsilon = 1/78.4$, or $\delta^2 = 2291$ J/cm^3)[42]; and A_j are coefficients of fitting for each ion I. Data were taken from Ref. 42.

with r', where Δ_r for both cations and anions was assigned an optimal collective value equal to 0.059 nm. On the other hand, the fourth term, corresponding to the property δ^2, becomes less negative as anion size increases, making an interpretation based on cavity formation difficult. By use of the fitting coefficients listed in Table 1.5, some measure of predictability for a wide range of solvents can be obtained, provided the solvent properties are known. By taking the difference of standard molar Gibbs energies of transfer for two ions as in Eq. 10, one may then proceed to make estimates of anion-exchange selectivities in different diluents, taking into account ion pairing if necessary (see below).

The importance of the hydrogen-bond-donor (HBD) ability of the solvent in anion transfer was demonstrated by use of the Kamlet–Taft solvatochromic parameters.[49] Standard molar Gibbs transfer energies of tetraethylammonium and tetramethylammonium halide salts from water to methanol were well correlated ($r^2_{corr} = 0.962 - 0.978$) with transfer energies to 13 or 14 solvents by linear relationships of the form

$$\Delta G^o_{tr}(R^+ X^-) = A_{\delta^2}[\delta^2(S) - 2293] + A_{\pi^*}[\pi^*(S) - 1.09] + A_{\alpha}[\alpha(S) - 1.17] \quad (12)$$

where π^* is the Kamlet–Taft dipolarity–polarizability parameter. In this case, the total Gibbs transfer energies included the contributions of both cation and anion.

In a more ambitious study examining single-ion transfer from water to organic solvents (TATB assumption), a large database of standard molar Gibbs energies of transfer and standard molar enthalpies of transfer was subjected to multiple linear regression analysis in terms of ion and solvent properties according to Eq. 10.[47] Thirteen anions and 26 solvents were considered. Multiple linear regression analysis showed that the transfer functions (ΔG^o_{tr} and ΔH^o_{tr}) of the anions were well predicted for eleven small anions by three solvent properties: the Kamlet–Taft polarizability–dipolarity (π^*) and hydrogen-bond-donor (α) parameters together with the solvent molar volume scaled by a factor ($V/100$). Two large anions, picrate and tetraphenylborate, required only two solvent properties: π^* and the squared Hildebrand solubility parameter scaled by a factor ($\delta^2/1000$). The solvatochromic parameters π^* and α measure the solvents' respective abilities to stabilize a dipole or charge and to donate hydrogen bonds. The latter ability becomes especially significant for alcohol solvents. Not included in the regression trials was the Dimroth–Reichardt E_T parameter, but it may be noted that E_T measures both acceptor (Lewis acidity) and dipolarity–polarizability properties and correlates well with π^* and α.[53,54] Table 1.6 gives the obtained linear regression coefficients $A_j(I)$ for ΔG^o_{tr}; similar coefficients were given for ΔH^o_{tr}.[47] As before, the regression coefficients provide a convenient means of estimating values of ΔG^o_{tr} and ΔH^o_{tr}; by taking differences as in Eq. 12, predictions for hypothetical anion-exchange reactions can be made (see below). For anions not tabulated explicitly, one may estimate the needed regression coefficients by use of equations relating the various $A_j(I)$ coefficients to ion radius, softness (σ), and molar refractivity.[47] In particular, inspection of Table 1.6 shows that A_α correlates inversely with ionic radius; the relation $A_\alpha = 74.7(0.1z/r) + 30.2\sigma$ was given. Many solvatochromic parameters have been tabulated, with rules for estimating more,[62] and thus Eq. 10 provides a powerful tool for estimating and understanding solvent effects in ion-transfer proceses.

Table 1.7 shows that ΔG_{tr}^o values estimated from the regression coefficients agree reasonably well with observed values corresponding to the transfer of 10 anions from water to four solvents. The reader may note that the observed and calculated Gibbs energies of transfer roughly decrease with increasing size for the three water-immiscible solvents (PhNO$_2$, 1,1-DCE, and 1-BuOH); SCN$^-$ is out of line. Indeed, notwithstanding some exceptions, it is remarkable how the ordering of anions based largely on size bias persists for 26 solvents of widely varying properties.[47] The strong hydrogen-bond donor, 2,2,2-trifluoroethanol (TFE) represents the major exception. Although data for TFE are sparse, the observed values for the halides suggest a reversal; this is strongly supported by the calculations, which indicate a probable overwhelming preference for fluoride. Although TFE mixes completely with water, this result implies that more hydrophobic fluorinated alcohols could well lead to effective fluoride separations by solvent-extraction techniques.

The listed contributions to the total calculated Gibbs energies for the four solvents in Table 1.7 prove revealing. When the solvent has poor HBD strength, the contribution may be seen to be highly positive and dominant for the inorganic anions. Moderate ability in 1-BuOH allows the other factors to become comparable in importance, and the strong HBD strength of TFE exceeding that of water leads to negative contributions to the overall Gibbs transfer energies. Mildly unfavorable, the contributions due to the solvent polarity and polarizability properties only become decisive with the large inorganic anions I$^-$ and ClO$_4^-$. The fact that solvent volume was selected by the regression analysis indicates that packing of solvent molecules around the solute anions plays some role. However, except for fluoride in a large negative sense and nitrate in a positive sense, the solvent volume contribution appears to be relatively small for the solvents shown. Finally, the cohesive energy density of the solvent measured by δ^2 dominates the Gibbs energies of transfer of the two large organic anions. It may be seen in Table 1.6 that, for the large, hydrophobic anions Pic$^-$ and Ph$_4$B$^-$, the parameter $A_{\delta^2/1000}$ increases with increasing anion size; in fact, such a correlation was parametrized.[47] Since the difference $\Delta(\delta^2/1000)$ is negative, the increasing anion size gives rise to large negative contributions, as shown in Table 1.7.

Recalling the outcome of the electrostatic calculations discussed above, the empirical correlations with solvent and ion properties come to almost an equivalent conclusion. One may compare the effect of the neutral term ΔG_{neut}^o in Eq. 5 to the anion-volume contribution found in Table 1.7. Further, the unsymmetric term ΔG_{unsym}^o in Eq. 5 appeared to have a possible connection with the HBD ability of the solvent. The statistical analysis did not, however, reveal an important role for the dielectric constant. Even when $1/\varepsilon$ was forced upon

TABLE 1.6. Multiple Linear Regression Coefficients for Standard Molar Gibbs Energies of Transfer for Selected Anions based on the Solvent Properties π^*, α, $V/100$, and $\delta^2/1000$[a]

Univalent ion	A_{π^*}	A_α	$A_{V/100}$	$A_{\delta^2/1000}$
F^-	−24.0	−88.7	−74.9	
$CH_3CO_2^-$	−17.0	−44.8	12.9	
Cl^-	−15.5	−30.3	11.9	
CN^-	−6.0	−41.1	−15.0	
N_3^-	−8.7	−24.1	11.7	
Br^-	−14.4	−20.4	11.6	
NO_3^-	−7.4	−2.9	21.4	
SCN^-	−8.5	−9.9	3.8	
I^-	−17.8	−8.9	9.1	
ClO_4^-	−29.1	−2.6	−4.9	
I_3^-	4.2	24.4	7.2	
Pic^-	2.3			3.9
Ph_4B^-	−25.5			24.6

[a] The standard molar Gibbs energy (kJ/mol) of the ion-transfer process from water to a dry solvent is given by $\Delta G_{tr}^\circ(I) = \sum_j A_j(I)[P_j(S) - P_j(W)]$, where $P_j(S)$ is the solvent property π^* (Kamlet–Taft polarizability–dipolarity),[47,49,62] α (Kamlet–Taft hydrogen-bond-donor acidity),[47,49,62] $V/100$ (molar volume scaled by a factor of 100), or $\delta^2/1000$ (squared Hildebrand solubility parameter in J/cm^3 scaled by a factor of 1000)[58,59]; $P_j(W)$ is the corresponding property for water ($\pi^* = 1.09$, $\alpha = 1.17$, $V/100 = 0.18$ cm^3/mol, or $\delta^2/1000 = 2.29$ J/cm^3)[47]; and A_j are coefficients of fitting for each ion I. Data were taken from Ref. 47.

the regression, it was essentially insignificant (Table 1.5), as might have been predicted from the poor correlation with dielectric constant evident in Cl$^-$ ion transfer in Table 1.3. Indeed, the poor predictive power of $1/\varepsilon$ in ion-transfer processes is widely recognized.[51,52] Although this may seem perplexing, the single-layer electrostatic model in fact effectively de-emphasizes $1/\varepsilon$ in determining the overall solvation energy, since the effective ion size felt by the bulk dielectric includes the electrostricted solvent shell around it. The much larger effective ion size entails a more modest electrostatic solvation energy through the Born equation (Eq. 3), as pointed out earlier,[47] and the overall contribution of the electrostatic energy in the relation $\Delta G_{tr}^\circ = \Delta G_S^\circ - \Delta G_h^\circ$ (Eq. 8) is consequently diminished. A final point to be made concerns the fact that the dielectric constant relates to the "amphoteric" properties of a solvent, and $\log \varepsilon$ in fact correlates linearly with DN and AN according to $\log \varepsilon = 0.0054\,DN + 0.0711\,AN + 0.2581$.[60] Thus the donor–acceptor properties incorporate the dielectric constant together with other chemical interactions less easy to quantify in simple terms. These include to an extent hard-soft considerations, though it has been noted that solvation of most common anions has been mostly governed by hard–hard interactions.[52] As amplified in reviews,[51,52] it should not seem surprising overall that chemically based parameters contain more statistically meaningful information for predicting ion-transfer properties than a single "pure" physical property such as the solvent dielectric constant.

TABLE 1.7. Estimation of Standard Molar Gibbs Energies of Transfer for Selected Anions from Water into Four Solvents[a]

		Contributions to ΔG_{tr}^o calc				ΔG_{tr}^o	ΔG_{tr}^o
Solvent	Ion	$A_{\pi^*}\Delta\pi^*$	$A_\alpha\Delta\alpha$	$A_{V/100}$ $\Delta(V/100)$	$A_{\delta^2/1000}$ $\Delta(\delta^2/1000)$	calc	obs
PhNO$_2$	F$^-$	1.9	103.8	−63.7		42.0	44.0
	Cl$^-$	1.2	35.5	10.1		46.8	35.0
	N$_3^-$	0.7	28.2	9.9		38.8	
	Br$^-$	1.2	23.9	9.9		34.9	29.0
	NO$_3^-$	0.6	3.4	18.2		22.2	24.0
	SCN$^-$	0.7	11.6	3.2		15.5	6.0
	I$^-$	1.4	10.4	7.7		19.6	18.0
	ClO$_4^-$	2.3	3.0	−4.2		1.2	10.0
	Pic$^-$	0.2			−7.0	−6.8	−3.0
	Ph$_4$B$^-$	2.0			−44.3	−42.2	−36.0
1,1-DCE	F$^-$	14.6	94.9	−50.2		59.4	
	Cl$^-$	9.5	32.4	8.0		49.8	58.0
	N$_3^-$	5.3	25.8	7.8		38.9	
	Br$^-$	8.8	21.8	7.8		38.4	43.0
	NO$_3^-$	4.5	3.1	14.3		22.0	
	SCN$^-$	5.2	10.6	2.5		18.3	
	I$^-$	10.9	9.5	6.1		26.5	31.0
	ClO$_4^-$	17.8	2.8	−3.3		17.3	22.0
	Pic$^-$	1.4			−7.6	−6.2	
	Ph$_4$B$^-$	15.6			−48.0	−32.4	−27.0
1-BuOH	F$^-$	14.9	33.7	−55.4		−6.8	
	Cl$^-$	9.6	11.5	8.8		29.9	29.0
	N$_3^-$	5.4	9.2	8.7		23.2	
	Br$^-$	8.9	7.8	8.6		25.3	24.0
	NO$_3^-$	4.6	1.1	15.8		21.5	
	SCN$^-$	5.3	3.8	2.8		11.8	
	I$^-$	11.0	3.4	6.7		21.2	22.0
	ClO$_4^-$	18.0	1.0	3.6		15.4	22.0
	Pic$^-$	1.4			−6.8	−5.4	−2.0
	Ph$_4$B$^-$	15.8			−42.8	−27.0	−20.0
TFE	F$^-$	8.6	−30.2	−40.4		−62.0	
	Cl$^-$	5.6	−10.3	6.4		1.7	−10.0
	N$_3^-$	3.1	−8.2	6.3		1.3	
	Br$^-$	5.2	−6.9	6.3		4.5	−8.0
	NO$_3^-$	2.7	−1.0	11.6		13.2	
	SCN$^-$	3.1	−3.4	2.1		1.7	
	I$^-$	6.4	−3.0	4.9		8.3	−8.0
	ClO$_4^-$	10.5	−0.9	−2.6		6.9	
	Pic$^-$	0.8			−6.7	−5.9	
	Ph$_4$B$^-$	9.2			−42.3	−33.1	

[a] The standard molar Gibbs energies (kJ/mol) of the ion-transfer process from water to four dry solvents were estimated by the relationship $\Delta G_{tr}^o(I) = \sum_j A_j(I)[P_j(S) - P_j(W)]$, employing regression coefficients from Table 1.6 and solvent parameters from Ref. 47. PhNO$_2$ is nitrobenzene, 1,1-DCE is 1,1-dichloroethane, 1-BuOH is 1-butanol, and TFE is 2,2,2-trifluoroethanol; the properties of these solvents are given in Table 1.2. Pic$^-$ is picrate and Ph$_4$B$^-$ is tetraphenylborate. Observed values were taken from Ref. 47.

To conclude this section, we find that statistical approaches reveal ion size among like-charged anions to be the most important ion property in determining the size bias of anion transfer. For solvents, the hydrogen-bond-donating ability proves decisive for small anions, and the cohesive energy density proves decisive for large, hydrophobic anions. Qualitatively at least, the statistical approach finds agreement with the single-layer electrostatic model in these respects. Both statistical and electrostatic models offer useful predictive possibilities for making many quantitative estimates of the driving force of anion-transfer processes, which we will now discuss in terms of anion solvent extraction and ion exchange.

1.4 SOLVENT EXTRACTION OF ANIONS

1.4.1 Hydration of Organic-Phase Species

Although the discussion above has emphasized anion transfer from water to pure organic solvents, actual extraction processes involve mutually saturated phases. When referring to the Gibbs energy change associated with single-ion transfer (Eq. 7) under conditions of mutual saturation, we refer preferably to the Gibbs energy of ion partitioning. Following Kolthoff,[63,64] we will unambiguously denote this quantity by $\Delta G^{\circ}_{tr[W(O)\rightarrow O(W)]}$. In general $\Delta G^{\circ}_{tr[W(O)\rightarrow O(W)]}$ does not equal $\Delta G^{\circ}_{tr[W\rightarrow O]}$, because of the changes in phase properties and solute solvation associated with the finite concentration of water in the solvent and solvent in the water. Naturally, the effect becomes more pronounced with increasing miscibility of the two phases. Solvents such as alcohols apparently entail partitioning of hydrated ions and give agreement with theory only when such hydration is taken into account.[44,65] However, in relatively "dry" solvents, classed by Marcus as those solvents that contain less than 0.13 mole fraction of water at saturation at 25°C,[66] it was shown that calculated values of $\Delta G^{\circ}_{tr[W\rightarrow O]}$ agreed reasonably well with observed values of $\Delta G^{\circ}_{tr[W(O)\rightarrow O(W)]}$.[44] Such solvents include 1,2-dichloroethane, dichloromethane, chloroform, *o*-dichlorobenzene, chlorobenzene, benzene, and nitrobenzene. The data given in Table 1.4 demonstrate that, indeed, the presence of water in the solvent has little effect on the partitioning of large anions. Clearly for Cl^- ion, however, water influences partitioning, and it has been found that Cl^- in water-saturated 1,2-dichloroethane (23°C) associates with an average of 2.3 water molecules.[67] In nitrobenzene, Cl^-, Br^-, and I^- ions were found to associate, respectively, with an average of 3.3, 1.8, and 1.0 water molecules.[67]

For the remainder of the chapter, we shall restrict our attention primarily to "dry" solvents and largely overlook the effect of water, dropping all distinction between ion transfer and ion partitioning. This will be done cautiously, recognizing that in a strict treatment anion interactions with water in the organic phase do take place and must be taken into account. Unfortunately, hydration phenomena have not been sufficiently understood generally to enable one to extrapolate reliably from transfer thermodynamics to partitioning phenomena, and much remains to be learned. Despite this uncertainty, the reader will see that agreement between prediction based on transfer principles and experimental results on anion extraction is often reasonable without an explicit accounting of hydration effects.

1.4.2 Liquid–Liquid Anion Exchange

Since liquid–liquid separation processes entail zero net charge transfer, the anion transfer in Eq. 7 requires either countertransfer of an equivalent negative charge or cotransfer of an equivalent positive charge. If a cation M^+ cotransfers, the process is often called ion-pair extraction. It will be shown later that anion selectivity in ion-pair extraction essentially obeys the same relationships that apply to anion exchange. Focusing then on anion exchange, a univalent countertransfer process corresponding to the anion transfer in Eq. 7 may be written

$$X^-(S) \to X^-(\text{aq}) \qquad (13)$$

The net anion-exchange process of Eqs. 7 and 13 becomes

$$Y^-(\text{aq}) + X^-(S) \rightleftharpoons Y^-(S) + X^-(\text{aq}) \qquad (14)$$

and the standard Gibbs energy change is given by

$$\Delta\Delta G^o_{\text{tr}} = \Delta G^o_{\text{tr}}(Y^-) - \Delta G^o_{\text{tr}}(X^-) \qquad (15)$$

If each of the two terms on the right-hand side of Eq. 15 are estimated by use of Eq. 9, one obtains

$$\Delta\Delta G^o_{\text{tr}}(\text{kJ/mol}) = B\left(\frac{1}{r'_Y} - \frac{1}{r'_X}\right)\left(\frac{1}{\varepsilon_W} - \frac{1}{\varepsilon_S}\right) \qquad (16)$$

Above it was shown that Eq. 9 cannot be employed reliably when the dielectric constant exceeds ca. 10, and neither should Eq. 16 be employed under these conditions. In this case, either the single-layer electrostatic treatment or a statistical approach could be readily used to supply the needed estimates of $\Delta G^o_{\text{tr}}(Y^-)$ and $\Delta G^o_{\text{tr}}(X^-)$ to be used in Eq. 15. Many experimental values of $\Delta G^o_{\text{tr}[W \to O]}$ could be simply taken from the literature,[47] and as seen in Table 1.4, values of $\Delta G^o_{\text{tr}[W(O) \to O(W)]}$ also exist that apply directly to extractive systems.[50] Thus anion-exchange selectivity in such systems could be readily predicted based on all the considerations discussed above. Overall, one expects size bias towards large anions to predominate as the major selectivity criterion (strong HBD solvents excepted), subject to some usually gentle perturbations based on the nature of the solvents and ions specifically involved in a given system.

In practice, most anion-exchange systems involve solvents having dielectric constants lower than 10, thereby introducing ion-pairing effects in the solvent phase.[25] Often, a large quaternary ammonium cation R^+ is employed, where the cation's bulk prevents its significant distribution to the aqueous phase. Such a system is depicted in Fig. 1.2. One may represent this process as the ion-pair analog of Eq. 14 as follows:

$$Y^-(\text{aq}) + R^+X^-(\text{S}) \rightleftharpoons R^+Y^-(\text{S}) + X^-(\text{aq}) \quad (17)$$

To describe the standard Gibbs energy change of this process, Eq. 16 must be augmented to include terms accounting for the dissociation of the ion pair R^+X^- and the formation of the ion pair R^+Y^-:

$$\Delta\Delta G_{\text{tr}}^\circ = \Delta G_{\text{tr}}^\circ(Y^-) - \Delta G_{\text{tr}}^\circ(X^-) + \Delta G_{\text{ip}}^\circ(R^+Y^-) - \Delta G_{\text{ip}}^\circ(R^+X^-) \quad (18)$$

Treatments of Fuoss[68] or of Bjerrum[69] may be applied to estimate these ion-pairing terms.[70] Alternatively, an approximation based on Coulomb's Law has been applied, which together with Eq. 16 for the ion-transfer terms permits a simple form to be obtained[71,72]:

$$\Delta\Delta G_{\text{tr}}^\circ(\text{kJ/mol}) = B \left\{ \frac{1}{\varepsilon_S} \left[\left(\frac{1}{r'_X} - \frac{1}{r'_Y} \right) + 2 \left(\frac{1}{r_Y + r_R} - \frac{1}{r_X + r_R} \right) \right] \right.$$
$$\left. + \frac{1}{\varepsilon_W} \left(\frac{1}{r'_Y} - \frac{1}{r'_X} \right) \right\} \quad (19)$$

In the original expression, the radius correction $r' = r + \Delta_r$ was not used; however, we have retained it, since its use produces slightly better agreement with experiment (see below). It may be seen that with an infinitely large cation radius Eq. 19 reduces to Eq. 16. Despite the approximations involved and the

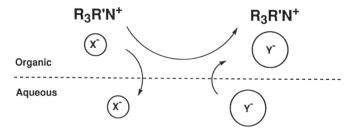

Fig. 1.2. Anion-exchange process involving a large aqueous anion Y^- being exchanged for a smaller organic-phase anion X^-. The organic-phase cation is depicted as a quaternary alkylammonium cation and should have sufficient bulk to prevent its loss to the aqueous phase. Common examples include trioctylmethylammonium and tetraheptylammonium chloride salts. Of course, alkylphosphonium, arsonium, or sulfonium salts may be employed, and large cationic metal complexes have been used.[95,106]

focus only on electrostatic effects, Eqs. 16 and 19 offer simplicity by employing as variables only the solvent dielectric constant, the radii of the two anions X^- and Y^-, and for Eq. 19 only, the radius of the cation R^+.

The plots shown in Fig. 1.3 employ Eq. 19 to illustrate the dependence of the anion-exchange equilibrium constant (expressed as $\log K_{\text{exch}} = -\Delta\Delta G^\circ_{\text{tr}}/2.303RT$) on the radius of the initial aqueous anion Y^- for a number of different cation radii. As an example, let the initial organic-phase anion X^- be chloride, and let the diluent be benzene ($\varepsilon = 2.467$ for water-saturated benzene at $24°C^{71}$). For aqueous anions Y^- smaller than chloride ion, the exchange may be seen to be unfavorable, while the value of $\log K_{\text{exch}}$ increases rapidly with the radius of Y^-. For aqueous anions Y^- larger than chloride ion, the exchange exhibits little bias if the cations are small. However, large cations such as the tetraheptylammonium cation ($r_R = 0.580$ nm^{71}) provide strong bias in favor of anions of increasing size. Preference for an anion of radius as low as 0.25 nm increases by six orders of magnitude from the smallest to the largest cation shown, reaching a value highly sufficient for practical anion-exchange separations. *Bias thus increases with increasing cation size.* For a hypothetical cation of infinite radius, the process becomes identical

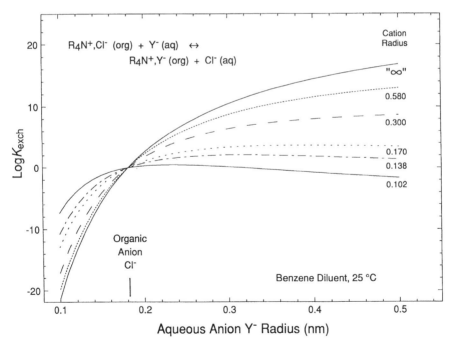

Fig. 1.3. Dependence of the anion-exchange equilibrium constant (expressed as $\log K_{\text{exch}} = -\Delta\Delta G^\circ_{\text{tr}}/2.303RT$) on the radius of the initial aqueous anion Y^- for a number of different cation radii. In this example, the initial organic-phase anion X^- is chloride, and the diluent is benzene ($\varepsilon = 2.467$ for water-saturated benzene at $24°C$).[71]

to the free-ion case in Eq. 16 and provides the maximum bias available. *It may be concluded that the effect of ion pairing is to decrease size bias.*

Figure 1.4 depicts the hypothetical effect of the diluent dielectric constant on $\log K_{\mathrm{exch}}$ according to Eq. 19, where the aqueous and organic anions are chosen to be ReO_4^- and Cl^- and the cations are the same as in Fig. 1.3. It may be seen that bias in favor of the large perrhenate anion increases to substantial values for large cations as the dielectric constant decreases below 5. To obtain the maximum bias, one obviously should choose alkane ($\varepsilon = 2$) or similarly low-permittivity diluents. By contrast, diluents having dielectric constants above 10 provide little bias in favor of perrhenate. *Thus decreasing dielectric constant increases bias.* It must be remembered that, as the dielectric constant increases over the entire range, actual anion-exchange processes may be expected to undergo a transition from complete ion pairing (Eq. 17) to the free-ion case (Eq. 14). Thus the plotted lines correspond to hypothetical cases up to that for infinite cation radius (free ions) with low dielectric constant and for finite cation radius with high dielectric constant. Overall, to increase bias, one must employ low-permittivity solvents, and to counteract the bias-weakening effect of ion pairing, one must choose the largest cation possible.

Table 1.8 shows that anion-exchange selectivity qualitatively follows roughly

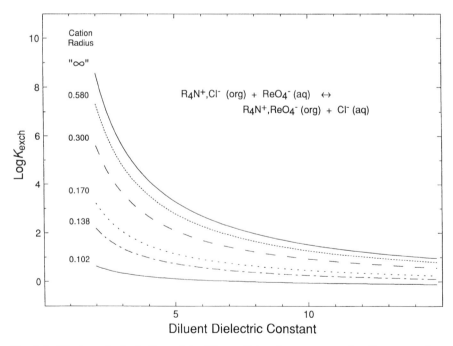

Fig. 1.4. The hypothetical effect of the diluent dielectric constant on $\log K_{\mathrm{exch}}$ according to Eq. 19, where the aqueous and organic anions are chosen to be ReO_4^- and Cl^- and the cations are the same as in Fig. 1.3.

TABLE 1.8. Observed Order of Preference for Anions in Anion-Exchange (AX), Acid Extraction (AE), or Ion-Pair (IP) Extraction Processes[a]

Cation	Diluent	AX or IP	Observed order	Ref.
Oc_4N^+	toluene	AX	$OH^- < F^- < AcO^- < HCO_3^- < HSO_4^- < Cl^- < Br^- < BzO^- < NO_3^- < I^- < ClO_4^-$	92
Hx_4N^+	xylene-hexone	AX	$PO_4^{3-} < CO_3^{2-} < SO_4^{2-} \ll Cl^- < NO_3^- < ClO_3^- \ll ClO_4^- < [Co(NH_3)_2(NO_2)_4]^-$	93
Oc_3MeN^+	chloroform	AX	$SO_4^{2-} \ll Cl^- < Br^- < I^- < ClO_4^- < [Co(NO_2)_2(C_4H_7N_2O_2)_2]^-$	94
Oc_3MeN^+	chlorobenzene	AX	$SO_4^{2-} \ll Cl^- < Br^- < NO_3^- < I^- < ClO_4^- < Fe(CN)_6^{3-}$	73
Hx_4N^+	hexone	AX	$Fe(CN)_6^{3-} \ll M_{D4h}(CN)_4^{2-} < M_{Td}(CN)_4^{2-} < ClO_4^- < M(CN)_2^- < M_{Td}(CN)_4^-$	95
Oc_3NH^+	cyclohexane	AE	$PO_4^{3-} < SO_4^{2-} \ll Cl^- < Br^- < NO_3^- < I^- < ClO_4^- < TcO_4^-$	96
Ph_4M^+	chloroform	IP	$AsO_4^{3-} < PO_4^{3-} < Cl^- < BrO_3^- < Br^- < NO_3^- < SCN^- < I^-$	13, 97–100
$[(DCC6)M]^+$	dichloromethane	IP	$F^- < Cl^- \le OH^- < Br^- < NO_3^- < I^- < ClO_4^-$	101
$[(DCC6A)M]^+$	chloroform	IP	$Br^- < NO_3^- < SCN^- < I^- < ClO_4^-$	80
$[(DBC6)K]^+$	m-cresol	IP	$SO_4^{2-} < Cl^- < Br^- < I^- < NO_3^- < AcO^- < F^-$	102

[a] The anion extraction selectivity is given for R^+X^- in various organic diluents. Oc_4N^+ is tetraoctylammonium; Hx_4N^+ is tetrahexylammonium; Oc_3MeN^+ is trioctylmethylammonium; Oc_3NH^+ is trioctylammonium; Ph_4M^+, where M = P, As; similar behavior was seen with Ph_3S^+ and Ph_3Sn^+; $[(DCC6)M]^+$ is the complex of alkali metal cations Li^+–Cs^+ with the crown ether dicyclohexano-18-crown-6; $[(DCC6A)M]^+$ is the complex of alkali metal cations K^+–Cs^+ with the crown ether cis, syn, cis-dicyclohexano-18-crown-6; $[(DBC6)K]^+$ is the complex of potassium with the crown ether dibenzo-18-crown-6; $[Co(NH_3)_2(NO_2)_4]^-$ is known as Erdmann's salt; $[(C_4H_7N_2O_2)_2]^{2-}$ is dimethylglyoximato; BzO^- is benzoate; hexone is methylisobutyl ketone.

the size bias predicted by the electrostatics model (Eqs. 16 and 19) for quaternary ammonium, phosphonium, and arsonium extractants. Large anions of a given charge type are clearly preferred. Anions of higher charge without compensating increase in size are rejected. (The discrepancy in the positioning of the hexacyanoferrate anion has been noted but remains unresolved.[73]) Moreover, the ordering persists in different solvents and for variously structured cations, including trialkylammonium cations formed upon acid extraction and even crown ether complexes of alkali metal cations formed upon the extraction of alkali metal salts. Interestingly, anion exchange favors tetrahedral versus square planar tetracyano $M(II)$ dianions, suggesting that a flat geometry in this case gives an effectively smaller ionic radius.

Equation 19 has been used successfully to predict standard molar Gibbs energy changes for anion-exchange processes for several inorganic anions when the cation is the tetraheptylammonium cation and benzene is the diluent.[71] The assumption was made that the effect of the water dissolved in the organic phase is negligible other than to increase slightly the bulk dielectric constant. Table 1.9 summarizes the results. The reader may note that the order of anion exchange, namely $Cl^- < Br^- < NO_3^- < ClO_4^- < ReO_4^-$, follows the general ordering seen in Table 1.8. Moreover, the observed exchange constants $\log K_{exch}$ indicate strong size bias; a separation of perrhenate in a nitrate-containing matrix could in principle be effected with a separation factor greater than 10^3. In making use of such electrostatic calculations, one typically faces the question of the appropriate radii to employ, and the choices made have an appreciable effect on the calculated energies, since the calculated $\log K_{exch}$ values depend steeply on ionic radius (Fig. 1.3). As shown in Table 1.9, use of a simple geometric rule for estimating the radii of oxyanions gave reasonably good results, but different values of the anion radii (including a larger one for perrhenate) and the tetraheptylammonium cation have since been published. Employing the thermochemical radii from the third column of Table 1.1 considerably worsens the agreement between observed and calculated values, but employing the correction $\Delta_r = 0.017$ nm improves the agreement. Further improvement is obtained by employing the more reliable Stokes radius 0.506 nm for the cation.[74] Considering that the radii used in the final column of Table 1.9 are the best values available, the calculated results agree well with the observed $\log K_{exch}$ values.

Several other examples of anion exchange have been summarized in Table 1.10, showing that Eqs. 16 and 19 have some utility in predicting anion-exchange bias in low-permittivity solvents. In each of the systems examined, a common anion Y^- was used as a reference against which the anion-exchange process was measured for a series of anions. To preserve uniformity for purposes of comparison, all systems reported in Table 1.10 refer to the process in which the reference anion is the initially aqueous anion. In some cases, the actual reference anion was the initially aqueous anion, and the reported $\log K_{exch}$ values were listed directly. In other cases, the reference anion was the initially organic-phase anion, and for these cases, the negative of the

TABLE 1.9. Observed and Calculated Constants for Exchange of Perrhenate Anion (Y^-) for Organic-Phase Anions (X^-) as Tetraheptylammonium Salts in Benzene[a]

X^-	Y^-	$\log K_{exch}$ obs	$^b \log K_{exch}$ calc[b] $r = $ cryst or geom $\Delta_r = 0.000$ $r_R = 0.580$	$^c \log K_{exch}$ calc[c] $r = $ thermochem $\Delta_r = 0.000$ $r_R = 0.580$	$^d \log K_{exch}$ calc[d] $r = $ cryst or therm $\Delta_r = 0.017$ $r_R = 0.580$	$^d \log K_{exch}$ calc[d] $r = $ cryst or therm $\Delta_r = 0.017$ $r_R = 0.506$
Cl^-	ReO_4^-	5.0	0.181 5.9	0.172 8.0	0.181 5.7	0.181 5.4
Br^-	ReO_4^-	3.7	0.195 4.2	0.188 5.9	0.195 4.2	0.195 4.0
NO_3^-	ReO_4^-	3.1	0.201 3.6	0.196 5.0	0.196 4.2	0.196 4.0
ClO_4^-	ReO_4^-	0.1	0.222 1.7	0.250 0.6	0.250 0.5	0.250 0.5
ReO_4^-	ReO_4^-		0.245 0.0	0.260 0.0	0.260 0.0	0.260 0.0

[a] The anion-exchange process is shown in Fig. 1.2 and Eq. 17. Data were taken from Scibona and co-workers[71]; observed values of $\log K_{exch}$ were collected at 24°C. Calculation of $\log K_{exch}$ employed Eq. 19, where the dielectric constant of water-saturated benzene was estimated[71] to be 2.467 and $r' = r + \Delta_r$. Four separate calculations were done according to different selections of ionic radii r, correction Δ_r, and tetraheptyl ammonium cation radius r_R; all units in nm. The cation radius was either 0.580 nm as estimated in Ref. 71 or the Stokes radius 0.506 nm.[74]
[b] The value of r was taken to be the crystallographic radius for Cl^- and Br^- or estimated geometrical radius for the oxyanions; the latter was taken to be the distance from the center atom to a line drawn between two oxygen atoms plus the van der Waals radius of the oxygen atom (0.14 nm).[71]
[c] The value of r was taken to be the thermochemical radius taken from Table 1.1.
[d] The value of r was taken to be either the crystallographic radius from Table 1.1 for Cl^- and Br^- or else the thermochemical radius taken from Table 1.1.

TABLE 1.10. Comparison of Observed Anion Exchange Constants with Values Calculated from the Electrostatic Approach[a]

Diluent[b]	CCl_4	benzene	benzene	benzene	benzene	benzene	xylene	$CHCl_3$	$CHCl_3$	$CHCl_3$	PhCl	MIBK
Dielectric Constant[c]	2.3	2.5	2.5	2.5	2.5	2.5	2.5	5.4	5.4	5.4	5.8	13.1
Aqueous anion Y^-[d]	ReO_4^-	ReO_4^-	Cl^-	Cl^-	Cl^-	Cl^-	E^-	E^-	E^-	Cl^-	Cl^-	E^-
r (nm) Y^-[e]	0.26	0.26	0.18	0.18	0.18	0.18	0.25	0.25	0.25	0.18	0.18	0.25
Cation R^+[f]	Do_3NH^+	Hp_4N^+	$R'NH_3^+$	$R''DoNH_2^+$	$R''DoNH_2^+$	Do_3NH^+	Hx_4N^+	Hx_4N^+	Oc_3MeN^+	$CeMe_3N^+$	Oc_3Me^+	Hx_4N^+
r (nm) R^+[g]/Footnote	0.15 h	0.51 h	0.15 i	0.15 i	0.15 i	0.15 i	0.47 j	0.47 k	0.48 k	0.24 l	0.48 m	0.47 k

r (nm)	Anion X^-							log K (kJ/mol)					
0.181	Cl^-	obs	1.73	4.96	0.0	0.0	0.0	0.0		3.95	3.24	0.0	0.0	
		calc	1.2	5.4	−0.42	−0.45	−0.40		4.8	2.1	1.3		−1.34	
0.196	Br^-	obs	0.70	3.66	−0.4	−0.4	−0.4						−0.4	
		calc	0.7	4.0										
0.196	NO_3^-	obs	0.81	3.08					3.4	1.5	0.9		−1.81	0.5
		calc	0.7	4.0	−0.4	−0.4	−0.4	−0.4	3.13	2.46	2.15	−1.10	−0.4	3.69
									3.4	1.5	0.9			0.5
0.200	ClO_3^-	obs	0.6	3.7					2.53	2.11	1.78		3.44	
		calc			−0.5	−0.5	−0.5	−0.5	3.7	1.3	0.8	−0.5	−0.4	0.5
0.220	I^-	obs	0.3	2.2	−0.80	−1.17	−1.47	−1.47				−2.53	−3.32	
		calc	0.28	0.11	−0.8	−0.8	−0.8	−0.8	1.7	0.7	0.4	−0.9	−0.9	0.3
0.250	ClO_4^-	obs	0.0	0.5	−1.0	−1.0	−1.0	−1.0	−0.12	0.31	0.12	−2.83	−4.47	0.80
		calc							0.0	0.0	0.0	−1.3	−1.2	0.0

| Literature Reference | [103] | [71] | [72] | [72] | [72] | [72] | [93] | [93] | [93] | [104] | [73] | [93] |

[a] The anion-exchange process under consideration is shown in Fig. 2. A reference anion Y^- initially in the aqueous phase is exchanged for an organic-phase anion X^-. In each column, the reference anion is held constant, and the exchange constants are evaluated for different anions. Columns are arranged in order of increasing dielectric constant. Observed values apply to the conditions of each experiment, which are all close to 25°C at various ionic strengths; calculated values, however, apply to infinite dilution.
[b] Xylene isomer was unspecified; PhCl is chlorobenzene; MIBK is methyl isobutyl ketone or hexone.
[c] Values were taken from Ref. 25 and corrected following earlier workers[71] for the slight effect due to the soluble water in the diluent; no correction was made for MIBK, which would require a significant[25] (and thus questionable) correction because of its appreciable saturation water content (0.11 mole fraction).
[d] E^- is the Erdmannate anion $[Co(NH_3)_2(NO_2)_4]^-$.
[e] Taken from Table 1.1 or estimated. See Ref. 105 for Erdmannate anion.
[f] The following abbreviations are used for alkyl substituents: Do = dodecyl; Hp = heptyl; Hx = hexyl; Oc = octyl; Ce = cetyl; Me = methyl; My = myristyl; Bz = benzyl. $R' = -CR_1R_2R_3$, where $R_1 + R_2 + R_3 = 11-14$ carbon atoms (Amberlite® LA-2); $R'' = -CR_1R_2R_3$, where $R_1 + R_2 + R_3 = 17-23$ carbon atoms (Primene® JMT).
[g] Radii estimated as described in the text.
[h] Log K values corrected to infinite dilution (ionic strength = 0 M; $[R^+]_{org}$ = 0 M); T = 22–27°C.
[i] Ionic strength = 0.019 M (NaX, HX); $[R^+]_{org}$ = 0.015 M; T = 23°C.
[j] For $X^- = NO_3^-$ and ClO_4^- ionic strength = 0.9 M (Na_2SO_4); for $X^- = ClO_3^-$ ionic strength = 0.8 M (Na_2SO_4); $[R^+]_{org}$ = 0.00066 M; T = ambient.
[k] For $X^- = NO_3^-$ and ClO_4^- ionic strength = 0.9 M (Na_2SO_4); for $X^- = ClO_3^-$ ionic strength = 0.8 M (Na_2SO_4); $[R^+]_{org}$ = 0.001 M; T = ambient.
[l] Ionic strength = $1 \times 10^{-5} - 1 \times 10^{-4}$ M; $[R^+]_{org}$ = 1×10^{-5} M; T = ambient.
[m] Ionic strength = $1 \times 10^{-5} - 1 \times 10^{-4}$ M; $[R^+]_{org}$ = 1×10^{-5} M; T = 20°C.

reported $\log K_{\text{exch}}$ value was listed in the table. For the 1°, 2°, and 3°, ammonium salts, the $\log K_{\text{exch}}$ values were obtained by difference from acid-extraction measurements (see next section).

As before, the different solvents in Table 1.10 are arranged from left to right according to their dielectric constants, from 2.3 (CCl_4) to 13.1 (methylisobutyl ketone). Anions are arranged in rows according to their size. Radii of tetraalkylammonium cations were taken from Ref. 74, and for unsymmetrical alkylammonium cations, the radius of the tetraalkylammonium cation formed from the smallest alkyl group was used. Thus, for trioctylmethylammonium cation, the radius used was that of tetramethylammonium. For trioctylammonium, the radius of the ammonium ion (NH_4^+) was used.[21]

Table 1.10 shows that for solvents having dielectric constants less than 3, Eq. 19 predicts $\log K_{\text{exch}}$ within approximately half a log unit (ca. ±3 kJ/mol). Tetraheptylammonium salts in benzene or tetrahexylammonium salts in xylene both exhibit strong size bias, as expected from Table 1.9. The large perrhenate (ReO_4^-) or Erdmannate ($[Co(NH_3)_2(NO_2)_4]^-$) anions exchange readily with the smaller anions Cl^-, Br^-, and NO_3^-. As may be seen from Table 1.10, Eq. 19 predicts less size bias than is observed in chloroform and diluents of even higher dielectric constant. As we saw in the case of predicting anion transfer, the modified Born model upon which Eq. 19 is based becomes increasingly inaccurate as the dielectric constant increases.

Replacing a long-chain alkyl group with hydrogen greatly diminishes size bias, suggesting that the anion approaches the cation closely. This may be seen by comparing the results for tetraalkylammonium cations and trialkylammonium cations in CCl_4, benzene, and xylene (Table 1.10). In fact, direct hydrogen bonding between the ammonium cation and anion ($R_3NH^+ \cdots X^-$) has been clearly demonstrated by a number of techniques, especially by IR spectrophotometry.[75] The fact that the anion-exchange constants are closely predicted by Eq. 19 suggests that the hydrogen bonding does not significantly alter size bias, but small effects may be seen. For example, increasing alkyl substitution of the ammonium ion causes little change in the unfavorable exchange of Br^- ion by Cl^- ion, but exchange of I^- by Cl^- ion becomes progressively less favorable. This follows the expectations of progressively decreasing acidity of the N–H bond, which leads to progressively less stable hydrogen bonds with Cl^- ion.

1.4.3 Ion-Pair Extraction

It was suggested above that anion-exchange and ion-pair extraction are governed by essentially the same considerations with regard to the anion. In view of the extensive development above, we will devote only enough attention to the topic of ion-pair extraction to establish this relationship. Since liquid–liquid separation processes entail zero net charge transfer, cotransfer of a cation M^+ suffices to meet this requirement. The resulting process is often called ion-pair extraction. The metal-ion cotransfer process for the univalent case may be written

$$M^+(\text{aq}) \rightleftharpoons M^+(\text{S}) \tag{20}$$

The net ion-pair extraction process of Eqs. 7 and 20 becomes

$$M^+(\text{aq}) + Y^-(\text{aq}) \rightleftharpoons M^+(\text{S}) + Y^-(\text{S}) \tag{21}$$

and the standard Gibbs energy change is given by

$$\Delta\Delta G^o_{\text{tr},MY} = \Delta G^o_{\text{tr}}(M^+) + \Delta G^o_{\text{tr}}(Y^-) \tag{22}$$

Note that M^+ may also be the hydrogen ion H^+, in which case the process may be referred to as acid extraction.

If the solvent has a dielectric constant of ca. 10 or less, ion pairing becomes appreciable,[25] and then the physically observed ion-pair extraction process may be written

$$M^+(\text{aq}) + Y^-(\text{aq}) \rightleftharpoons M^+Y^-(\text{S}) \tag{23}$$

The standard Gibbs energy change for the process now involves an ion-pairing term

$$\Delta\Delta G^o_{\text{tr},MY} = \Delta G^o_{\text{tr}}(M^+) + \Delta G^o_{\text{tr}}(Y^-) + \Delta G^o_{\text{ip}}(M^+Y^-) \tag{24}$$

Many ion-pair extraction processes are known to follow Eqs. 21 and 22 for the case of univalent metal ions.[13,14,16,107] It is but a simple matter to extend such equilibrium expressions to multivalent cases. Evaluation of Eq. 24 can be made in exactly the same manner already discussed, employing for the ion-transfer terms the modified Born approach or single-layer model, for example.

Although Eq. 24 possesses terms dependent upon the metal ion, the anion selectivity for ion-pair extractions can still be examined by comparing the extraction of a series of salts having a common cation. In one study, the extraction of the quaternary ammonium and tetraphenylarsonium salts of nine anions in 1,2-dichloroethane ($\varepsilon = 10.4^{25}$) was measured. Results were extrapolated to infinite dilution, such that the obtained Gibbs energies of salt partitioning correspond to the free-ion case of Eq. 21, thus eliminating the effect of ion pairing. A relative scale of Gibbs energies of transfer $\Delta G^o_{\text{tr}[W(O)\to O(W)]}$ of anions could then be set up, and this was in fact shown to be independent of the cations employed. Employing the TATB assumption, the scale was converted to an absolute scale, as shown in Table 1.4.

It may be appreciated that the process of ion-pair extraction is doubly unfavorable compared to the case of single-ion transfer for small, equal-sized ions. Not surprisingly, observable extraction of ion pairs is confined to the polar solvents and large ions, and for metal salts, high salt concentrations are usually needed to drive the extraction. If the cations are metal or hydrogen ions, then the solvents usually must also have good electron-pair donor properties. As was seen above, the solvent properties most favorable for anion extraction are the electron-pair acceptor strength and hydrogen-bond (HBD) donor strength. Solvent properties that favor anion *selectivity* are another matter, however. We saw above that greatest size bias occurs when the solvent lacks HBD strength. Thus, an inherent limitation in ion-pair extraction for anion separation involves a compromise between the need for extraction power and the need for selectivity. Factors controlling efficiency and selectivity in ion-pair extraction processes in which M^+ is an alkali metal cation have been thoroughly reviewed.[107]

A more useful form of ion-pair extraction entails augmenting the extraction processes given in Eqs. 21 and 23 with an extractant B whose function is to complex the cation:

$$M^+(\text{aq}) + Y^-(\text{aq}) + bB(\text{S}) \rightleftharpoons MB_b^+(\text{S}) + Y^-(\text{S}) \qquad (25)$$

$$M^+(\text{aq}) + Y^-(\text{aq}) \rightleftharpoons MB_b^+ Y^-(\text{S}) \qquad (26)$$

In this manner, the problem of transferring the cation to the solvent phase is taken care of by the coordination offered by B, effectively making the cation transfer, and hence the overall process, more favorable. One may note that the role of the anion remains the same as before, and thus the typical bias expected in anion-transfer processes is expected. Having a more favorable extraction, however, potentially allows one to choose solvents that promote anion selectivity rather than efficiency. The choices for the extractant B include simple extractants such as tributylphosphate and alkylamines or multidentate ionophores such as crown compounds. Since this topic exceeds the scope of this review, the reader is referred to a large body of solvent-extraction literature.[1,4,13,14,16,108]

Despite its inherent difficulty in obtaining both efficiency and selectivity in anion separations, ion-pair extraction offers the advantage of ease of reversibility. In process chemistry, a useful separation has not generally been achieved unless the extraction step can be followed by a practical back-extraction or stripping step. Anion exchange presents a potential difficulty in that stripping is generally accomplished by another anion-exchange step involving a more extractable anion or a high concentration of a replacement anion. Notwithstanding the applied usefulness of anion-exchange technology, regeneration of the extractant often results in another separation problem or in the generation of waste. By contrast, ion-pair extraction can be reversed by contact with water.

1.5 RESIN ANION EXCHANGE

As was mentioned above, anion exchange in resin systems presents unique features, since the bound functional sites lack the mobility of the liquid–liquid systems. Nevertheless, some of the factors that govern anion selectivity in solvent extraction, such as anion radius, charge, and hydration, are also operative in resin systems. The scope of resins discussed here will be limited to organic polymers (mainly styrenic backbone), where the anion-exchange sites are either of the weak- or strong-base variety. Weak-base sites are amines (usually tertiary) that can be protonated under acidic conditions; thus the anion-exchange site is active only at low pH. Strong-base sites are generally either amines or phosphines that have been quaternized and thus act as anion exchangers at any pH. Certain strong-base resins can be further subdivided into Type I and Type II resins, where Type I resins contain alkyltrimethylammonium functionality and Type II resins contain alkyldimethyl ethanolammonium functionality[76] (see Fig. 1.5 for some common examples of weak- and strong-base resins).

The discussion of anion selectivity in resins will be divided into two areas: selectivity between anions of the same charge (e.g., perchlorate vs. chloride) and selectivity between anions of different charge (e.g., sulfate vs. nitrate). The discussion will be focused mostly on strong-base anion-exchange resins, and how structural and functional group modifications of the resin give rise to the achievement of these selectivities. In addition, the discussion will focus mostly on inorganic anions, though the relative exchangeability of certain organic anions will be noted for comparative purposes.

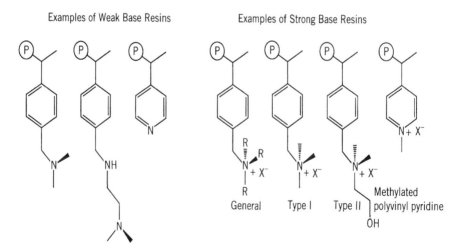

Fig. 1.5. Examples of weak-base and strong-base anion-exchange resins. The circled "P" represents the polymer backbone.

1.5.1 Selectivity between Anions of Like Charge

Factors that contribute to the achievement of *bias* of one anion over another in resin anion exchange include both the size and the hydration energy of the anion. Just as was observed in solvent extraction, large anions with low hydration energies (e.g., iodide, perrhenate) will associate with large, hydrophobic cations preferentially over smaller anions with higher hydration energies (e.g., chloride, nitrate). Or in other words, as was shown in Fig. 1.3, large cations will be *biased* toward anions of increasing size. The anion selectivity observed in solvent extraction using quaternary ammonium compounds (such as trioctylmethyl ammonium or tetraheptyl ammonium) forms a starting point for predicting anion-exchange selectivity by strong-base anion-exchange resins; many examples were shown above in Table 1.8. In another example, the affinity sequence observed for anion exchange via trioctylmethyl ammonium cation in xylene or chlorobenzene solvent[29] followed the order $Cl^- < Br^- < NO_3^- < ClO_3^- < I^- < ClO_4^-$, which generally follows the trend predicted on the basis of hydration energy ($\Delta G_h(X^-)$) values shown in Table 1.1. A similar trend was observed for the extraction of various anions using tetraoctyl ammonium cation in toluene[77]: $OH^- < F^- < CH_3CO_2^- < HCO_3^- < Cl^- < Br^- <$ benzoate $< NO_3^- < I^- < ClO_4^-$. The *general* trend is increasing affinity with decreasing hydration energy, though there are slight discrepancies in this analysis when comparing nonspherical anions (e.g., oxoanions such as nitrate) with halogens, and there are reversals from the expected trend when comparing certain classes of transition metal and main group oxoanions (vide infra). Nevertheless, this general trend is also observed for Type I and Type II (Table 1.11) strong-base resins.[78]

The resin selectivity sequence follows the order predicted by increasing size and decreasing hydration energy extremely well for the spherical halide ions F^-, Cl^-, Br^-, and I^-, for both types of resins. A few "reversals" from the expected bias are observed for the affinities of very basic anions like hydroxide and phenoxide with the Type II resins, which contain the hydroxyethyl group. The stronger affinity of these anions for the Type II resin is undoubtedly due to a strong hydrogen-bonding interaction between the basic anion and the hydroxyl group.

In a very interesting experiment,[79] an attempt was made to correlate the affinity sequence of various anions with the resin Dowex 1-X8 (a Type I resin, 8% cross-linked), with the distribution of the benzyltrimethylammonium salts of these same anions between an aqueous phase and various organic phases. It was generally found that the distribution of the benzyltrimethylammonium salt to the organic phase (e.g., *n*-pentanol) paralleled the resin ion-exchange affinity, though there was one noteworthy anomaly. As shown in Table 1.12, while perchlorate is retained better on the resin than thiocyanate, thiocyanate was *solvent-extracted* slightly better than perchlorate. (The reader will recall that data in Table 1.7 showed a lower observed ΔG_{tr}^o for thiocyanate than perchlorate into nitrobenzene, even though perchlorate has the lower calculated

TABLE 1.11. Relative Affinities of Various Anions for Polystyrene-Based Strong-Base Anion-Exchange Resins

Anion	Relative affinity[a]		$\Delta G_h(X^-)$ obs (kJ/mol)
	Type I	Type II	
Hydroxide (reference)	1.0	1.0	−430
Fluoride	1.6	0.3	−465
Acetate	3.2	0.5	−365
Formate	4.6	0.5	−395
Iodate	5.5	0.5	−400
Bicarbonate	6.0	1.2	−335
Chloride	22	2.3	−340
Nitrite	24	3	−330
Bromate	27	3	−330
Bromide	50	6	−315
Nitrate	65	8	−300
Chlorate	74	12	−280
Bisulfate	85	15	
Phenoxide	110	27	
Iodide	175	17	−275
Citrate	220	23	
Salicylate	450	65	
Benzenesulfonate	500	75	

[a] Relative affinity data taken from Ref. 78. Hydration data (where available) were taken from Table 1.1.

TABLE 1.12. Distribution of Benzyltrimethyl Ammonium Salts of Selected Anions (RA) to *n*-Pentanol: Comparison of the Anion Selectivity to Hydration Energy and to the Selectivity in Dowex 1-X8 Resin[a]

Anion	Distribution to *n*-pentanol (M)	$\Delta G_h(X^-)$ obs (kJ/mol)	K_c of Dowex 1-X8
Iodate	0.136	−400	0.198
Bromate	0.442	−330	0.918
Chloride	0.595	−340	1
Bromide	1.17	−315	3.14
Chlorate	1.32	−280	3.35
Iodide	4.11	−275	5.04
Thiocyanate	9.26	−280	6.89
Perchlorate	6.02	−205	56.3

[a] Extraction and resin data taken from Ref. 79. The distribution was calculated according to the equation $1/K_a = [RA]_S/[R^+]_W[A^-]_W$, where $[RA]_S$ is the concentration of RA in the solvent layer, and $[R^+]_W$ and $[A^-]_W$ are the concentrations of cation R^+ and anion A^- in the aqueous layer, respectively (this is the inverse of what was shown in Ref. 79, where K_a was tabulated). $[R^+]_W$ was always equal to $[A^-]_W$ in the experiment. Hydration data are taken from Table 1.1 of this work.

ΔG_{tr}^o value.) A possible explanation for the observed difference between the ion-exchange and solvent-extraction results in this system may involve differences in the polarizability of the anions and how each anion interacts with the organic cation in the solvent employed (n-pentanol). However, it should be noted that in many other solvent-extraction systems (such as the extraction of potassium salts by crown ethers in chloroform solution[80]), perchlorate is extracted preferentially over thiocyanate.

The basicity of the anion is an additional important qualitative parameter in explaining observed affinity orders, particularly with regard to cases where the observed affinity sequence deviates from what would be predicted on the basis of size or (calculated) hydration energy alone. For oxyacids of type HMO_3, HMO_4, H_2MO_3, and H_2MO_4, where M represents a family of elements going down a column in the periodic table and becoming more electropositive with increasing atomic weight, the acidities decrease as expected, and thus the strengths of the conjugate bases (anions) increase.[18,81] Thus, for the four sequences of oxyanions $\{ClO_3^-, BrO_3^-, IO_3^-\}$, $\{MnO_4^-, TcO_4^-, ReO_4^-\}$, $\{SO_3^{2-}, SeO_3^{2-}, TeO_3^{2-}\}$, and $\{CrO_4^{2-}, MoO_4^{2-}, WO_4^{2-}\}$, the basicity of the anion (and thus how strongly the anion interacts with water) increases in each sequence. These series do show the expected decrease of affinity for Type I resins: $IO_3^- < BrO_3^- < ClO_3^-$ [18,81,82]; $ReO_4^- < TcO_4^- < MnO_4^-$ [18,81,83]; $TeO_3^{2-} < SeO_3^{2-} < SO_3^{2-}$ [18,84] and $WO_4^{2-} < MoO_4^{2-} < CrO_4^{2-}$.[18,81,85] The affinity sequence does not follow what would be expected on the basis of size, however. The affinity sequence does follow what would be expected on the basis of the observed hydration energies, which increase as one progresses from Mn to Re, rather than decrease, as do the calculated hydration energies (using Eq. 4). It is quite possible that the discrepancy between the observed and calculated hydration energies is a consequence of the basicity of the anion, which is not taken into account in Eq. 4.

In examining the competing affinity of the anion for the external aqueous phase versus the resin phase, it is often useful to consider the differences in ion–water (ion hydration) and water–water (water-structure) interactions between the resin and aqueous phases.[17,18,81] The more basic the anion is, the more it prefers the dilute external phase, and ion–water interactions dominate. For weaker bases (anions of strong acids), which are often large anions, ion hydration takes on less importance compared to the choice of disrupting the more ordered hydrogen-bonded water structure of the external phase versus disrupting the generally less-ordered aqueous environment within the resin. In other words, a large anion is squeezed out of the external solution into "the less structured resin phase (pore size permitting), as such behavior maximizes the ion–water and water–water interactions in the total system."[81]

At this point, one might ask whether any structural changes can be made to the resin that could alter or modulate the biases or affinity sequences observed in the "baseline" Type I resin systems. It turns out that for anion exchange by a strong-base resin of type polymer–$CH_2N(R)_3^+$, the exchange potential one anion has over another can be modulated by adjusting the size of the alkyl group R on the amine group of the exchange site and by adjusting the degree of cross-linking of the polymer chains. In general, increasing the size of the alkyl group will make the exchange site more hydrophobic, which will tend to diminish the affinity for the smaller, more hydrated anions (e.g., Cl^-) relative to larger, less hydrated anions (e.g., I^-). Thus the natural affinity for iodide over chloride (or perchlorate over nitrate) can be enhanced by increasing the hydrophobicity of the exchange site. Where the water–water interactions are more important (i.e., for less basic, generally larger anions), the increased hydrophobicity of the exchange site decreases the water structuring in the resin phase so that the favorability of the transfer of the anion from the structured external phase to the resin phase is enhanced. This can be seen in Table 1.13, which illustrates the changes in relative retentions of various anions as the alkyl group is increased from methyl to octyl.[85] Note that as the chain length of the alkyl group increases, the affinity for the larger anions with lower hydration energies (especially iodide) increases dramatically, relative to the smaller, more hydrated anions, but that the relative ordering of affinity or *bias* for this group of anions remains the same (no reversals). Thus to obtain a higher separation factor or degree of selectivity for iodide over chloride, the trihexyl or trioctyl ammonium resin would be appropriate.

The extent to which the polymer lattice is networked together by cross-linking (usually by the addition of divinyl benzene, DVB) can also affect the affinity for one anion over another. It has been observed that increasing the degree of cross-linking (from 0.5% to 20% DVB) will enhance the affinity for

TABLE 1.13. Relative Retentions of Selected Anions on Trialkylammonium Resins of Varying Size of the Alkyl Group[a]

Anion	$\Delta G_h(X^-)$ obs (kJ/mol)	Alkyl group					
		Methyl	Ethyl	Propyl	Butyl	Hexyl	Octyl
Fluoride	−465	0.66	0.70	0.69	0.71	0.68	0.69
Nitrite	−330	0.82	0.82	0.86	0.90	0.89	0.98
Chloride	−340	1.0	1.0	1.0	1.0	1.0	1.0
Bromide	−315	1.20	1.19	1.25	1.34	1.32	1.41
Nitrate	−300	1.30	1.32	1.38	1.54	1.63	1.72
Chlorate	−280	1.53	1.55	1.56	1.73	1.92	2.15
Iodide	−275	2.51	2.48	3.05	3.82	5.00	> 5.0
$t_{R(Cl^-)}$ min		8.3	8.8	9.5	8.9	10.6	9.4

[a] Data taken from Ref. 85. The resins were prepared by amination of polychloromethylstyrene with the appropriate trialkyl amine. A solution containing all anions of interest was passed through a column containing each resin, eluting with benzoate ion, and the relative retention times for each anion determined and normalized to the retention time of chloride ion ($t_{R(Cl^-)}$). All resins were of low ion-exchange capacity (0.027 ± 0.001 mequiv/g).

less hydrated anions over more hydrated anions. An explanation for the enhanced selectivity is that cross-linking rigidifies the polymer chains so that the functional groups within the network are less able to congregate in hydrated domains. That is, the flexibility in a lightly cross-linked polymer may allow several sites to fold toward one another, along with several waters of hydration to form a hydrated zone amenable to hydrated anions; rigidifying the polymer chains by cross-linking prevents the formation of these hydrated zones, so that the affinity for more hydrated anions then decreases relative to less hydrated anions. Once again, the underlying factor is that large, less hydrated anions are "squeezed out" of the structured external phase more efficiently into a resin phase with a water structure made less structured by the effect of cross-linking.

Finally, making an exchange site more hydrophilic and increasing the number of hydroxyl groups can sometimes cause reversals in affinity. A trimethyl or tributyl ammonium resin prefers molybdate anion (MoO_4^{2-}) over tungstate anion (WO_4^{2-}), in accordance with the slightly higher basicity of tungstate (vide supra). However, the affinity is reversed for a methyldiethanol ammonium resin (tungstate preferred).[86] This stronger affinity of tungstate versus molybdate is likely due to the stronger hydrogen bonding interaction with the hydroxyl groups of the resin from the higher basicity of tungstate, in much the same way that phenoxide and hydroxide are strongly attracted to dimethyl–ethanol ammonium ("Type II") resins, as was discussed above.

1.5.2 Selectivity between Anions of Different Charge

Selectivity between anions of differing charge (such as between sulfate and nitrate) can vary substantially with not only resin structure, but also with the concentration of the anion and the pH of the aqueous solution. Sulfate/nitrate selectivity issues are of particular importance with regard to potable water treatment. In quaternary ammonium resins such as Type I resins, even at dilute concentrations, one would predict that nitrate would be favored over sulfate due to sulfate's higher hydration energy. In fact, a divalent ion such as sulfate is generally preferred over a monoanion such as nitrate by Type I exchangers in dilute solution; thus, removing nitrate from water containing low levels ($< 0.010 M$) of both nitrate and sulfate can be a challenge. For weak-base resins, the preference for sulfate remains high even at ionic strengths above $0.6 M$.[87] The bisulfate–sulfate equilibrium could be an important factor here: Bisulfate would associate with a weak base (amine) exchange site, and in fact that acid–base reaction could shift the equilibrium towards formation of more bisulfate, with ostensibly more sulfate being retained on the resin as bisulfate (e.g., polymer–$CH_2NR_2H^+SO_4H^-$). The order of preference of resins with amine functional groups for sulfate has been found to be $-NH_3^+ > -NRH_2^+ > -NR_2H^+ > -NR_3^+$.[87]

The uptake of a divalent anion such as sulfate requires the presence of two closely spaced positive charges, and if this requirement is met, then sulfate will be preferred due to what has been termed "electroselectivity."[86] This preference for divalent anions can be diminished, and in fact the selectivity can be reversed, if the positive sites are spread apart. Modifications to the resin that will suppress the affinity for divalent anions include increasing the length of the alkyl chain (R group) of the amine on the strong-base exchange site, increasing the spacing between the exchange sites (the fixed-charge separation distance), and increasing the degree of cross-linking.[87,88] Increasing the chain length of the alkyl group (e.g., from trimethyl to tripropyl) will separate the positive charges sufficiently to reverse the sulfate/nitrate selectivity, so that nitrate is preferred. Inserting a "spacer" (e.g., styrene monomer) between the exchange sites has a similar effect. Increasing the cross-link density rigidifies the resin matrix, preventing the positive charges from moving close enough for effective interaction with the dianion. For achievement of high monoanion/dianion selectivities, often all three of these modification are used in combination.

1.6 CONCLUSIONS

The concept of size bias has been introduced as a primitive selectivity type in anion separations. Since the solvation of anions tends to involve primarily hard–hard interactions, principles of electrostatics have successfully explained many anion-transfer processes, as explored herein. Useful predictions based on simple expressions having functional dependence only on ionic radii and solvent dielectric constant may be made. Alternatively, statistical methods have revealed other underlying factors that could be used to advantage to perturb or even reverse the direction of size bias. Such factors could be argued to introduce weak recognition, though without geometrical constraints. Although many useful applications, including large industrial processes, have been developed based on size bias in solvent extraction of anions or resin anion exchange, recognition approaches stand to create many new, even revolutionary, separation processes. Treating the principles outlined in this review as a "baseline," advances in anion recognition can henceforth be critically evaluated.

1.7 ACKNOWLEDGMENTS

This research was sponsored by the Division of Chemical Sciences, Office of Basic Energy Sciences, U.S. Department of Energy, under contract number DE-AC05-96OR2264 with Oak Ridge National Laboratory, managed by Lockheed Martin Energy Research Corp. The authors would like to thank C. F. Coleman for helpful comments in the preparation of this contribution.

1.8 REFERENCES

1. G. H. Morrison and H. Freiser, *Solvent Extraction in Analytical Chemistry*; John Wiley & Sons: New York, 1957.
2. C. F. Coleman, C. A. Blake, J., and K. B. Brown, *Talanta* **9**, 297–323 (1962).
3. F. L. Moore, "Liquid–Liquid Extraction with High-Molecular Weight Amines," National Academy of Sciences, 1960.
4. A. K. De, S. M. Khopkar, and R. A. Chalmers, *Solvent Extraction of Metals*; Van Nostrand–Reinhold: London, 1970.
5. C. F. Coleman, K. B. Brown, J. G. Moore, and D. J. Crouse, *Industrial Engineering Chemistry* **50**, 1756–62 (1958).
6. C. F. Coleman, *Nuclear Science and Engineering* **17**, 274–86 (1963).
7. G. M. Ritcey, A. W. Ashbrook, *Solvent Extraction, Principles and Applications to Process Metallurgy*; Elsevier: New York, 1984; Vol. Part I.
8. J. D. Miller and M. B. Mooiman, *Separation Science and Technology* **19**, 895–909 (1984–1985).
9. M. Cox, In *Science and Practice of Liquid–Liquid Extraction*; J. D. Thornton, Ed.; Clarendon Press: Oxford, 1992; Vol. 2; pp. 1–101.
10. C. F. Coleman, In *Process Chemistry*; Pergamon Press: Oxford, 1969; Vol. 4; pp. 233–85.
11. V. S. Shmidt and E. A. Mezhov, *Russian Chemical Reviews* **34**, 584–99 (1965).
12. V. S. Shmidt, *Amine Extraction*; Keter Press: Jerusalem, 1971 (trans. from Russian).
13. Y. Marcus and A. S. Kertes, *Ion Exchange and Solvent Extraction of Metal Complexes*; Wiley Interscience: New York, 1969.
14. T. Sekine, *Solvent Extraction Chemistry: Fundamentals and Applications*; Marcel Dekker: New York, 1977.
15. E. Högfeldt, In *Developments in Solvent Extraction*; S. Alegret, Ed.; Ellis Horwood: Chichester, 1988; pp. 36–86.
16. *Principles and Practices of Solvent Extraction*; J. Rydberg, C. Musikas, and G. R. Choppin, Eds.; Marcel Dekker: New York, 1992.
17. D. Reichenberg, In *Ion Exchange, A Series of Advances*; J. Marinsky, Ed.; Marcel Dekker: New York, 1966; Vol. 1; pp. 268–73.
18. R. M. Diamond and D. C. Whitney, In *Ion Exchange, A Series of Advances*; J. Marinsky, Ed.; Marcel Dekker: New York, 1966; Vol. 1; pp. 277–302.
19. C. E. Harland, *Ion Exchange: Theory and Practice*; 2nd ed.; The Royal Society of Chemistry: Cambridge, 1994.
20. R. D. Shannon, *Acta Crystallographica* **A32**, 751–67 (1976).
21. Y. Marcus, *Journal of the Chemical Society, Faraday Transactions* **87**, 2995–99 (1991).
22. J. E. Huheey, *Inorganic Chemistry: Principles of Structure and Reactivity*; 2nd ed.; Harper & Row: New York, 1972, p. 77.
23. H. D. B. Jenkins and K. P. Thakur, *Journal of Chemical Education* **56**, 576–77 (1979).
24. A. F. Kapustinskii, *Quarterly Reviews* **10**, 283–94 (1956).

25. Y. Marcus, In *Principles and Practices of Solvent Extraction*; J. Rydberg, C. Musikas, and G. R. Choppin, Eds.; Marcel Dekker: New York, 1992; pp. 21–70.
26. W. L. Masterton, D. Bolocofsky, and T. P. Lee, *Journal of Physical Chemistry* **76**, 2809–15 (1971).
27. B. J. Hathaway, In *Comprehensive Coordination Chemistry: The Synthesis, Reaction, Properties & Applications of Coordination Compounds*; G. Wilkenson, R. D. Gillard, and J. A. McCleverty, Eds.; Pergamon Press: Oxford, 1987; Vol. 2; pp. 413–34.
28. B. Krebs and K.-D. Hasse, *Acta Crystallographica* **B32**, 1334–37 (1976).
29. V. S. Shmidt, K. A. Rybakov, and V. N. Rubisov, *Russian Journal of Inorganic Chemistry* **27**, 855–57 (1982).
30. M. H. Abraham and Y. Marcus, *Journal of the Chemical Society, Faraday Transactions 1* **82**, 3255–74 (1986).
31. Y. Marcus, *Journal of the Chemical Society, Faraday Transactions 1* **82**, 233–42 (1986).
32. Y. Marcus, *Thermochimica Acta* **104**, 389–94 (1986).
33. Y. Marcus, *Journal of the Chemical Society, Faraday Transactions 1* **83**, 339–49 (1987).
34. Y. Marcus, *Chemical Reviews* **88**, 1475–98 (1988).
35. Y. Marcus, *Journal of the Chemical Society, Faraday Transactions 1* **83**, 2985–92 (1987).
36. D. A. Johnson, *Some Thermodynamic Aspects of Inorganic Chemistry*; 2nd ed.; Cambridge University Press: London, 1982.
37. V. S. K. Markin and A. G. Volkov, *Electrochimica Acta* **34**, 93–107 (1989).
38. J. Barthel, H.-J. Gores, G. Schmeer, and R. Wachter, In *Physical and Inorganic Chemistry*; Springer-Verlag: New York, 1983; pp. 32–144.
39. Y. Marcus, *Pure and Applied Chemistry* **59**, 1093–101 (1987).
40. M. H. Abraham and J. Liszi, *Journal of the Chemical Society, Faraday Transactions 1* **74**, 1604–14 (1978).
41. Y. Marcus, E. Pross, and J. Hormadaly, *Journal of Physical Chemistry* **84**, 2708–15 (1980).
42. S. Glikberg and Y. Marcus, *Journal of Solution Chemistry* **12**, 255–70 (1983).
43. D. D. Eley and M. G. Evans, *Transactions of the Faraday Society* **34**, 1093–112 (1938).
44. M. H. Abraham and J. Liszi, *Journal of Inorganic and Nuclear Chemistry* **43**, 143–51 (1981).
45. Y. Marcus, *Pure and Applied Chemistry* **55**, 978–84 (1983).
46. Y. Marcus, *Rev. Anal. Chem.* **5**, 53–137 (1980).
47. Y. Marcus, M. J. Kamlet, and R. W. Taft, *Journal of Physical Chemistry* **92**, 3613–22 (1988).
48. J. A. Riddick, W. B. Bunger, and T. K. Sakano, *Organic Solvents: Physical Properties and Methods of Purification*; 4th ed.; Wiley-Interscience: New York, 1986; Vol. II.
49. R. W. Taft, J.-L. M. Abboud, M. J. Kamlet, and M. H. Abraham, *Journal of Solution Chemistry* **14**, 153–86 (1985).

50. J. Czapkiewicz and B. Czapkiewicz-Tutaj, *Journal of the Chemical Society, Faraday Transactions I* **76**, 1663–68 (1980).
51. U. Mayer, *Pure and Applied Chemistry* **51**, 1697–712 (1979).
52. G. Gritzner, *Pure and Applied Chemistry* **60**, 1743–56 (1988).
53. C. Reichardt, *Solvents and Solvent Effects in Organic Chemistry*; 2nd ed.; VCH: Weinheim (Federal Republic of Germany), 1990.
54. C. Reichardt, *Chemical Reviews* **94**, 2319–58 (1994).
55. V. Gutmann, *The Donor–Acceptor Approach to Molecular Interactions*; Plenum: New York, 1978.
56. Y. Marcus, *Journal of Solution Chemistry* **13**, 599–624 (1984).
57. Y. Marcus, *Ion Solvation*; Wiley: Chichester, 1985.
58. J. H. Hildebrand and R. L. Scott, *The Solubility of Nonelectrolytes*; 3rd ed.; Reinhold Publishing Corp.: New York, 1950.
59. A. F. M. Barton, *Handbook of Solubility Parameters and Other Cohesion Parameters*; CRC Press: Boca Raton, 1983.
60. R. Schmid, *Journal of Solution Chemistry* **12**, 135–52 (1983).
61. Y. Marcus, *Journal of Solution Chemistry* **20**, 929–44 (1991).
62. M. J. Kamlet, J.-L. M. Abboud, M. H. Abraham, and R. W. Taft, *Journal of Organic Chemistry* **48**, 2877–87 (1983).
63. I. M. Kolthoff, *Analytical Chemistry* **51**, 1R–22R (1979).
64. I. M. Kolthoff, M. K. J. Chantooni, and W. Wang, *J. Chem. Eng. Data* **38**, 556–59 (1993).
65. Y. Marcus, *Pure and Applied Chemistry* **54**, 2327–34 (1982).
66. Y. Marcus, *Solvent Extraction and Ion Exchange* **10**, 527–38 (1992).
67. T. Kenjo and R. M. Diamond, *Journal of Inorganic and Nuclear Chemistry* **36**, 183–88 (1974).
68. R. M. Fuoss, *Journal of the American Chemical Society* **80**, 5059–61 (1958).
69. N. Bjerrum, *Kgl. Danske Videnskab. Selskab, Mat.-fys. Medd.* **7**, 108 (1926).
70. R. A. Robinson and R. H. Stokes, *Electrolyte Solutions*; Butterworths: London, 1959.
71. G. Scibona, J. F. Byrum, K. Kimura, and J. W. Irvine, Jr., In *Solvent Extraction Chemistry*; D. Dyrssen; L.-O. Liljenzin, and J. Rydberg, Eds.; North-Holland Publishing Company: Amsterdam, 1967; pp. 399–407.
72. G. Scibona, F. Orlandini, and P. R. Danesi, *Journal of Inorganic and Nuclear Chemistry* 1701–6 (1967).
73. J. Itoh, H. Kobayashi, and K. Ueno, *Analytica Chimica Acta* **105**, 383–90 (1979).
74. B. S. Krumgalz, *Journal of the Chemical Society, Faraday Transactions I* **78**, 437–49 (1982).
75. B. A. Moyer, *Solvent Extraction and Ion Exchange* **6**, 1–37 (1988).
76. R. L. Albright and P. A. Yarnell, In *Encyclopedia of Polymer Science and Engineering*; Vol. 8, Wiley-Interscience: New York, 1985; p. 347.
77. M. L. Navtanovich, L. A. Dzhanashvili, and V. L. Kheifets, *J. Gen. Chem.* **48**, 1755–58 (1978), and cited Ref. 1.

78. I. M. Abrams and L. Benezra, In *Encyclopedia of Polymer Science and Technology*; Vol. 7, Interscience, New York, 1967, p. 719, 720.
79. T. Kenjo and T. Ito, *Bull. Chem. Soc. Jpn.* **41**, 1757–60 (1968).
80. U. Olsher, M. G. Hankins, Y. D. Kim, and R. A. Bartsch, *J. Amer. Chem. Soc.* **115**, 3370–71 (1993).
81. B. Chu, D. C. Whitney, and R. M. Diamond, *J. Inorg. Nucl. Chem.* **24**, 1405–15 (1962).
82. D. A. Aveston, D. A. Everest, and R. A. Wells, *J. Chem. Soc.* 231–39 (1958).
83. R. W. Atterbury and G. E. Boyd, *J. Amer. Chem. Soc.* **72**, 4805–6 (1950).
84. A. Iguchi, *Bull. Chem. Soc. Jpn.* **31**, 748–52 (1958).
85. R. E. Barron and J. S. Fritz, *Journal of Chromatography* **284**, 13–25 (1984).
86. R. E. Barron and J. S. Fritz, *Journal of Chromatography* **316**, 201–10 (1984).
87. D. Clifford and W. J. Weber, Jr., *Reactive Polymers* **1**, 77–89 (1983).
88. S. Subramonian and D. Clifford, *Reactive Polymers* **9**, 195–209 (1988).
89. D. R. Lide, *Handbook of Organic Solvents*; CRC Press: Boca Raton, 1995.
90. M. H. Abraham, A. F. Danil de Namor, and R. A. Schultz, *Journal of Solution Chemistry* **5**, 529 (1976).
91. M. H. Abraham and A. F. Danil de Namor, *Journal of the Chemical Society, Faraday Transactions I* **72**, 955 (1976).
92. M. Ivanov, L. M. Gindin, G. N. Chichagova, M. Ivanov, Khim. Protsessov Ektstr., 3rd Mater. Conf. Khim. Ekstr. (1969), 1972, p. 207 (cited in Ref. 106).
93. W. E. Clifford and H. Irving, *Analytica Chimica Acta* **31**, 1–10 (1964).
94. H. M. N. H. Irving and A. H. Nabilsi, *Analytica Chimica Acta* **41**, 505–13 (1968).
95. H. M. N. H. Irving and A. D. Damodaran, *Analytica Chimica Acta* **53**, 267–75 (1971).
96. G. E. Boyd and Q. V. Larson, *Journal of the Chemical Society* **64**, 988–96 (1960).
97. R. Bock, H. T. Niederauer, and K. Behrends, *Z. Anal. Chem.* **190**, 33 (1962).
98. R. Bock and G. M. Beilstein, *Z. Anal. Chem.* **192**, 44 (1963).
99. R. Bock and G. Jainz, *Z. Anal. Chem.* **198**, 21 (1963).
100. R. Bock and C. Hummel, *Z. Anal. Chem.* **198**, 176 (1963).
101. V. V. Yakshin, V. M. Abashkin, and B. N. Laskorin, *Doklady Akademii Nauk SSSR* **252**, 373–76 (1980).
102. Y. Marcus and L. E. Asher, *Journal of Physical Chemistry* **82**, 1246–54 (1978).
103. G. Scibona, R. A. Nathan, A. S. Kertes, and J. W. Irvine, Jr., *Journal of Physical Chemistry* **70**, 375–79 (1966).
104. H. K. Biswas and B. M. Mandal, *Analytical Chemistry* **44**, 1636–40 (1972).
105. J. A. Caton and J. E. Prue, *Journal of the Chemical Society*, 671 (1956).
106. H. M. N. H. Irving and J. Hapgood, *Analytica Chimica Acta* **119**, 207–16 (1980).
107. B. A. Moyer and Y. Sun, In *Ion Exchange and Solvent Extraction*; Vol. 13, J. A. Marinsky and Y. Marcus, Eds.; Marcel Dekker, New York, 1997, pp. 295–391.
108. B. A. Moyer, In *Molecular Recognition: Receptors for Cationic Guests*; G. W. Gokel, Ed., Vol. 1, Comprehensive Supramolecular Chemistry, J. E. D. Davies, D. D. MacNicol, F. Vögtle, and J.-M. Lehn, Eds., Pergamon, Elsevier, Oxford, 1996; Ch. 10, pp. 377–416.

CHAPTER 2

Historical View on the Development of Anion Coordination Chemistry

BERNARD DIETRICH and MIR WAIS HOSSEINI

2.1 Introduction
2.2 Halide Complexation
2.3 Linear Anions
2.4 Trigonal Planar Anions
2.5 Tetrahedral Anions
2.6 Square Planar Anions
2.7 Octahedral Anions
2.8 Molecular Assemblies
2.9 Conclusions
2.10 References

The aim of this chapter is to provide the reader a short view on the major steps of the expansion of the field of anion coordination chemistry. As all the various aspects of the area will be largely developed in the other chapters, only the key steps will be mentioned here.

2.1 INTRODUCTION

Anions play a variety of roles in both organic and mineral worlds. For example, 70–75% of substrates and cofactors engaged in biological processes are negatively charged species. Despite their ubiquitous roles, perhaps due to intrinsic complications associated with them (i.e., solubility, hydration energy,

---

Supramolecular Chemistry of Anions, Edited by Antonio Bianchi, Kristin Bowman-James, and Enrique García-España.
ISBN 0-471-18622-8. © 1997 Wiley-VCH, Inc.

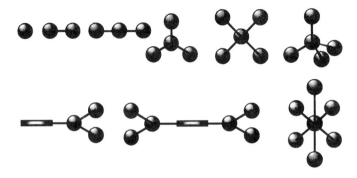

Fig. 2.1. Geometrical shapes of anions in order of increasing complexity.

large size, and numerous geometries), the recognition of anions by synthetic receptor molecules still remains a fabulous playground for supramolecular chemists. Nevertheless, the complexation of anions by synthetic receptor molecules has recently been recognized and developed as a new area of coordination chemistry.

In this chapter, we follow a scheme based on increasing complexity in terms of geometry (Fig. 2.1) that covers, to a certain extent, the chronological development of anion coordination chemistry. After covering spherical monocharged halide anions, we shift to molecular species such as the linear triatomic azide anion, trigonal carboxylates, dicarboxylate substrates, and tetrahedral species such as perchlorate, sulfate, and phosphate and its derivatives, and finally we will cover the molecular recognition of complexed anions such as metal tetrachloride and metal hexacyanide derivatives.

Almost 30 years ago, in 1967, two manuscripts were submitted for publication by C. J. Pedersen on April 13 and by C. H. Park and H. E. Simmons on November 13, all three authors affiliated with the du Pont de Nemours Company, to the *Journal of the American Chemical Society*. Pedersen's historical paper dealt with the first series of crown ethers, including the by-now-classical 18-crown-6 **2.1**, and their ability to bind cations.[1] The second manuscript reported the first synthetic organic ligands of the bicyclic diammonium type **2.2**, displaying halide complexation.[2]

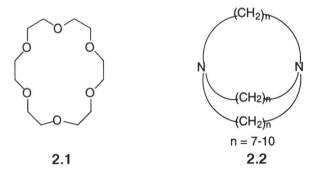

Thus, in terms of chemical archaeology, it is worth noting that synthetic molecules capable of binding cations and anions were discovered at the same time. Nevertheless, whereas the impact of Pedersen's publication on cation complexation was apparently instantaneous, since without long delay the literature was invaded by a tremendous number of crown ether–type complexing agents, the same was not the case for the Park and Simmons report. In other words, whereas the coordination chemistry of cations has been extensively studied, anion coordination still remains relatively unexplored.

2.2 HALIDE COMPLEXATION

The halide complex formed with the diammonium salt (**2.2**) was named *katapinate* (in Greek: swallow up, engulf) by Park and Simmons. The origin of the complexing properties was clearly postulated by the authors: "The stability of the katapinate ions must arise in part from the high positive potential of the hole with respect to anions and from hydrogen bonding within the cavity, [$^+$N–H···Cl$^-$H–N$^+$] or [$^+$N–H···Cl$^-$···H–N$^+$]; it is not unlikely that the latter structure with two hydrogen bonds is involved."[2] The encapsulation of the halide (Fig. 2.2) was firmly established a few years later by the X-ray analysis of the Cl$^-$ katapinate formed by the diprotonated diazabicyclo[9,9,9] (**2.2**-2H$^+$, Cl$^-$).[3] Although katapinates, as mentioned above, are indeed inclusion complexes, nevertheless their stability constants are rather low (K_S slightly higher than 10^2 M^{-1} in 50% deuterated trifluoroacetic acid) and their selectivity towards halides rather modest (Cl$^-$/Br$^-$ ~ 8).[4]

Some 8 years after this premier, in 1976 the binding ability towards halides of the tetraprotonated macrotricyclic ligand (**2.3**–4H$^+$) was described (Fig. 2.3).

Although the spherical ligand **2.3** was designed to bind large alkali (K$^+$, Rb$^+$, Cs$^+$) and ammonium (NH$_4^+$) cations,[5] it has been observed that the fully protonated tetraammonium compound **2.3**–4H$^+$ was able to bind halides with great affinity for Cl$^-$ (log K_S > 4 in H$_2$O). Furthermore, this compound showed a selectivity factor of 10^3 between Cl$^-$ and Br$^-$ anions.[6] The X-ray crystal analysis (Fig. 2.3) of the (**2.3**–4H$^+$,Cl$^-$) complex revealed that the

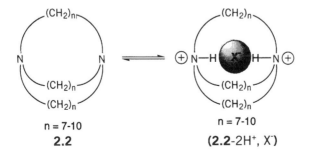

Fig. 2.2. Reaction of encapsulation of halide ion by the diammonium katapinands.

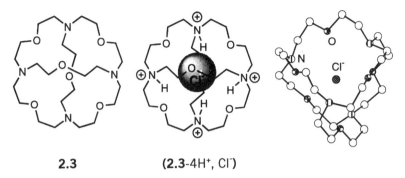

2.3   **(2.3-4H$^+$, Cl$^-$)**

Fig. 2.3. Binding of halides by the macrotricyclic ligand **2.3**.

chloride anion was located within the cavity of the protonated receptor interacting with the latter through a tetrahedral array of four $^+$N–H···Cl$^-$ hydrogen bonds.[7] Binding of Cl$^-$ by **2.3**–4H$^+$ was also established by ^{35}Cl–NMR studies.[8] In order to give a complete picture of the binding properties of the compound **2.3**, one may notice that diprotonated **2.3**–2H$^+$ has been shown to form an inclusion complex with a water molecule.[9]

In 1977 the quaternary ammonium analogues of compound **2.3**, the macrotricyclic receptor **2.4**, as well as other similar compounds bearing polyethylene fragments connecting the ammonium centers, were reported (Fig. 2.4). These ligands were also shown to bind halides but with much lower affinity. The highest stability constants (log K_S = 2.25 in H$_2$O) were measured for bromide and iodide.[10] The crystal analysis of the iodide complex with compound **2.5** confirmed the inclusion of the anion in the center of the ligand's cavity.[11]

The receptor molecules described above were mainly devoted to the Cl$^-$, Br$^-$, and I$^-$ anions. The first stable fluoride inclusion complex reported was obtained with the protonated macrobicyclic ligand **2.5**, also called "bis-tren" (Fig. 2.5).

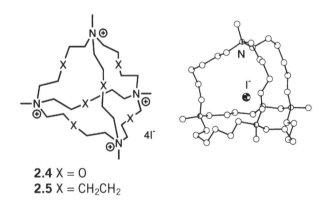

2.4 X = O
2.5 X = CH$_2$CH$_2$

Fig. 2.4. Binding of iodide ion by the quarternary ammonium macrotricycle **2.4**.

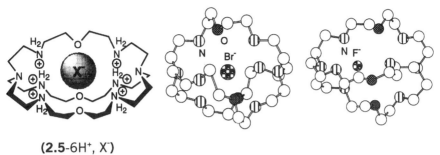

(2.5-6H⁺, X⁻)

Fig. 2.5. Depiction of the binding of halides by the macrobicyclic ligand **2.5** (bis-tren) where size considerations govern where the halide is bound in the cavity.

This rather versatile compound was synthesized in 1977 for several purposes, in particular for its ability to form dinuclear complexes with metal cations.[12] The hexaprotonated form of **2.5** (**2.5**–6H⁺) was shown to form stable complexes with F^-, Cl^-, Br^-, and I^-.[13] The stability constants in water, expressed as log K_S, were measured to be 4.1, 3.0, 2.6, and 2.15, respectively.[14] The inclusive nature of the halide complexes was established by X-ray analysis for F^-, Cl^-, and Br^- anions (Fig. 2.5). The structural analysis not only revealed that in all three cases the halide was located within the cavity of the hexaammonium receptor, but also allowed to have a close look at the mismatch between the size of the anion and the shape and the dimension of the receptor's cavity. In particular, in the case of F^- anion the dramatic mismatch is shown on the X-ray structure in Fig. 2.5.[14]

In 1989, based on the above-mentioned observed mismatch between the size of fluoride anion and the dimensions of the ligand **2.5** cavity, the bicyclic compound **2.6**, possessing a more suitable cavity for F^-, was designed and prepared (Fig. 2.6). The size complementarity between the anion and the ligand's cavity was again established by X-ray analysis of the (**2.6**-6H⁺,F⁻)

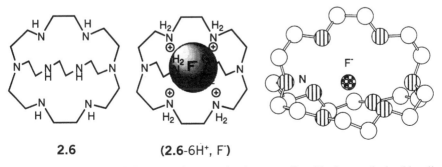

2.6   (**2.6**-6H⁺, F⁻)

Fig. 2.6. Illustration of size complementarity between fluoride ion and the bicyclic ligand **2.6**.

complex.[15] Thus, as a result of the almost perfect complementarity between the receptor and the substrate, in aqueous solution an extraordinarily high stability constant (log $K_S = 11.2$) and a remarkable F^-/Cl^- selectivity (ca. 10^8) was observed.[16]

Following the same strategy, compounds **2.7**[17] and **2.8**[18] were designed in order to have in hand receptor molecules displaying high affinity for Cl^- and I^- anions, respectively. Although in terms of affinity the goal was reached for both compounds, unfortunately the obtained selectivity between different halides was rather poor.

Many other halide receptors have been described, such as macrocyclic tetramines for F^-,[19] fluorinated crown ethers for F^-,[20] cascade halide binding to mononuclear or dinuclear cationic cryptates formed with ligand **2.5** and analogues,[21] and protonated "expanded porphyrins."[22] In particular, complexation of F^- by the ligand **2.9** was established in the solid state by X-ray studies (Fig. 2.7).[22]

During the past decade a new class of neutral receptors such as compounds **2.10–2.13** were developed; these ligands contain Lewis acids as binding sites. In acetonitrile, chloride complexation by the ligands **2.10** has been studied by ^{119}Sn-NMR, which revealed 1:1 and 1:2 ligand:chloride stoichiometries.[23] The

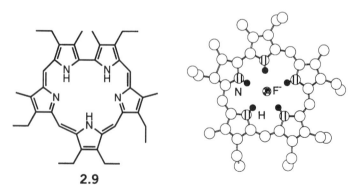

Fig. 2.7. Drawing of the fluoride complex of an expanded porphyrin, based on the crystal structure.

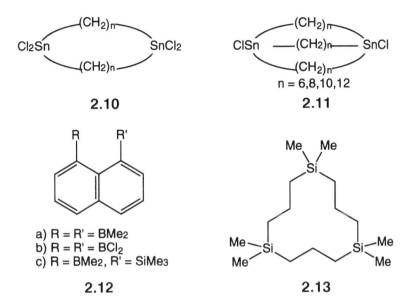

distannamacrobicycles **2.11** was shown to bind only one halide, suggesting an encapsulation of the anion.[24]

Although the inclusion of F$^-$ in and of Cl$^-$ by **2.11** (with $n = 6$ and 8, respectively) has been confirmed in the solid state by X-ray studies (Fig. 2.8),[25] the stability constants of these complexes in solution were found to be rather weak.[26] The class of boron-containing ligands **2.12** has been described and the complexation of H$^-$ by **2.12a**, of Cl$^-$ by **2.12b**, and of F$^-$ by **2.12c** has been established.[27]

Silicon has been also incorporated into macrocyclic receptors. Compounds such as the silacrown **2.13** have been shown to transport chloride and bromide anions across an organic layer.[28] Mercury has also been considered; a mercury carborand derivative was shown to bind Cl$^-$. The X-ray structure study of the

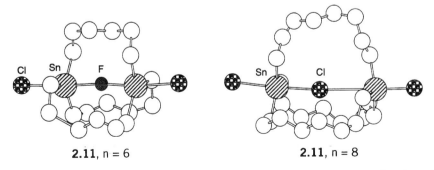

Fig. 2.8. Lewis acid distannamacrobicycles shown binding to halides as a function of cavity size.

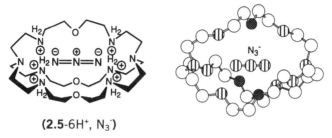

Fig. 2.9. Linear binding of azide ion by the hexaprotonated macrobicyclic receptor **2.5**.

complex revealed that the anion was bound to all four mercury centers.[29] Further details on the Lewis acids containing ligands may be found in other chapters and Ref. 30.

2.3 LINEAR ANIONS

So far, we have only considered the binding of spherical monatomic halide anions. Recognition of molecular anions was first reported in the case of azide binding by the hexaprotonated macrobicyclic receptor **2.5** (Fig. 2.9). The almost perfect shape and size complementarity between the cavity of the protonated ligand **2.5-6H$^+$** and the linear triatomic N_3^- anion was established by an X-ray study of the (**2.5-6H$^+$**,N_3^-) complex in the solid state and in aqueous solution by stability constant measurements (log K_S = 4.3).[13,14]

2.4 TRIGONAL PLANAR ANIONS

Although the binding of nitrate anion by the tetraprotonated hexacyclen **2.14-4H$^+$** has been established in aqueous solution (log K_S = 2.4), in the solid state, however, no convincing arguments in favor of a peculiar binding of nitrate by the ligand have been obtained.[31] In 1986, the more elaborate and suitable receptor **2.15** for nitrate anion has been reported. In aqueous solution, a rather

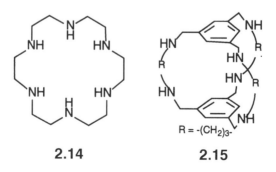

high affinity (log K_S = 4.35) of the hexaprotonated ligand **2.15**-6H$^+$ for NO$_3^-$ was measured. Again, although a clear 1:1 stoichiometry was demonstrated by NMR studies in solution supporting the inclusion of the anion by the receptor, in the solid state the X-ray analysis revealed that the anion was not located within the cavity of the receptor.[32]

The binding of the carbonate anion by triprotonated penta- or hexaazamacrocycles has been described as early as 1982. The stability constants reported in aqueous solution are moderate (log K_S in the range 2–4),[33] and no structural features are available for the above-mentioned complexes.

Carboxylate anions, owing to their fundamental biological roles, have been important targets, and many receptors capable of recognizing them have been described. We will describe here only representatives examples.

Dealing with monocarboxylate anions, only few receptors based on guanidinium group as binding site have been designed. The latter cationic site, owing to the high protonation constant of the guanidine group (pK of ca. 13.5), exists in solution over a wide pH range. Furthermore, because of its ability to form two zwitterionic hydrogen bonds with anions, the guanidinium unit may be considered as the best-adapted unit for carboxylate recognition. Guanidinium-containing receptor molecules such as **2.16** based on a rigid framework have been largely used for carboxylate complexation. By appropriate synthetic strategies the (**2.16**-SS) and (**2.16**-RR) enantiomers were obtained and shown to perform chiral recognition of optically active carboxylate derivatives.[34]

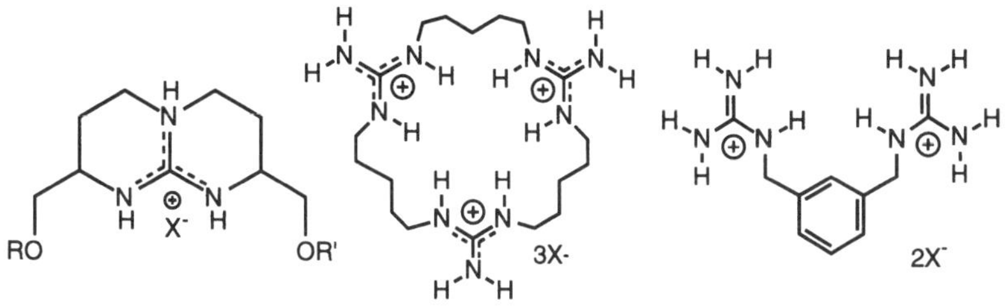

A large number of receptors for di- or polycarboxylates have been described. Many partially or totally protonated polyazamacrocycles or -macrobicycles bind polycarboxylates such as oxalate, malonate, succinate, tartrate, maleate, fumarate, citrate, and benzenetricarboxylate anions.[14,35–38] Linear molecular recognition of dicarboxylate anions (Fig. 2.10) was achieved using ditopic hexaazamacrocycles such as **2.19**.[39]

A more elaborate receptor, such as the bicyclic compound **2.20** for dicarboxylate dianion, has been recently reported (Fig. 2.11). The inclusion of the terephthalate dianion within the cavity of the protonated **2.20** in the solid state was established by an X-ray study.[39]

Lipophilic polyazamacrocycles have been used as anion carriers.[40] In recent years many other ditopic receptors have been described. In particular, receptor

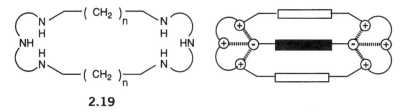

2.19

Fig. 2.10. Linear recognition of dicarboxylate anions by ditopic hexaazamacrocycles such as **2.19**.

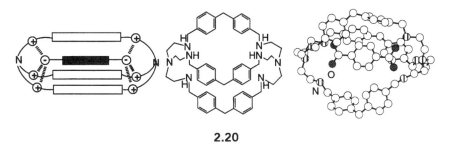

2.20

Fig. 2.11. Drawing based on the crystal structure of the terphthalate complexation by **2.20**.

molecules based on quaternary ammonium,[41] amide groups capable of hydrogen bonding,[42] diuranyl-containing metallomacrocycles,[43] and guanidinium[44] have been reported.

2.5 TETRAHEDRAL ANIONS

Phosphates and their derivatives such as nucleotide mono-, di-, and triphosphates are biologically important anions. In 1978, the binding of the phosphate anion by guanidinium-containing receptor molecules was achieved. In particular, compound **2.17** bearing three guanidinium units was shown to bind PO_4^{3-} anion in aqueous solution with a rather low affinity (log K_S = 2.4).[45] Recently, other tris-guanidinium macrocycles have been synthesized.[46] Several noncyclic di-, tri-, and tetraguanidinium ligands displaying stable complexes with many phosphate derivatives have been also described.[47] Many bis-guanidinium ligands based on **2.18** have been described.[48] These compounds are well adapted to the binding of several types of phosphate derivatives, including natural nucleotides.[48–50] A receptor molecule possessing a central guanidinium core and a side chain capable of hydrogen bonding was reported for cyclic AMP.[51]

Phosphate binding was also achieved by polyammonium macrocycles. A large number of protonated macrocyclic polyammonium ligands exhibit strong

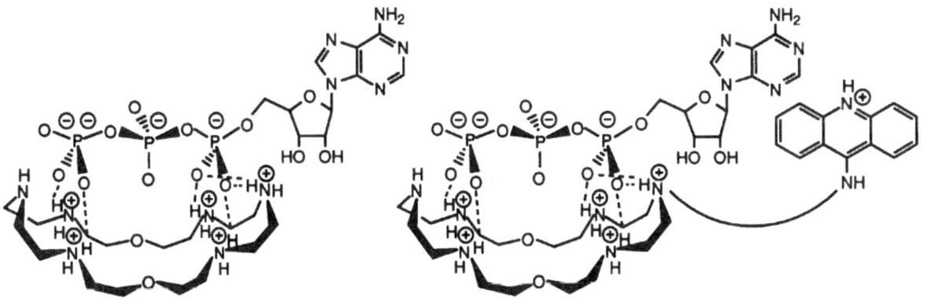

Fig. 2.12. Proposed mode of binding of ATP with **2.21** and binding enhancement by introduction of an acridine moiety to give **2.22**.

binding of AMP^{2-}, ADP^{3-}, ATP^{4-}.[52–54] Using polyammonium receptors, calculi dissolution based on $(PO_4)_3Ca_2$ solubilization has been studied.[55] Among various polyamines investigated, the ditopic macrocyclic hexaamine $[24]N_6O_2$ **2.21** (Fig. 2.12) was found to bind strongly and selectively nucleotide polyphosphates via electrostatic interactions and hydrogen bonding between the cationic binding sites (ammonium groups) of the receptor and the negatively charged polyphosphate groups of the substrate.[53]

In order to increase the binding ability **2.21** of nucleotide polyphosphates, compound **2.22** (Fig. 2.12) bearing an acridine derivative and capable of simultaneous interaction with both the polyphosphate chain of the nucleotide and their nucleic base moiety was reported.[56] Compound **2.22** was also shown to bind strongly to DNA plasmid pBR 322 at 10^{-6} M, probably via a double type of interaction involving both intercalation and electrostatic interactions with the phosphate groups.

Since compound **2.21** was found to be the most powerful complexing agent for ATP, it has been attached covalently to polystyrene calibrated beads **2.23**. This polymer-supported receptor was shown to uptake nucleotides, in particular ATP, at neutral or slightly acidic conditions and to release it back into solution under alkaline conditions.[57]

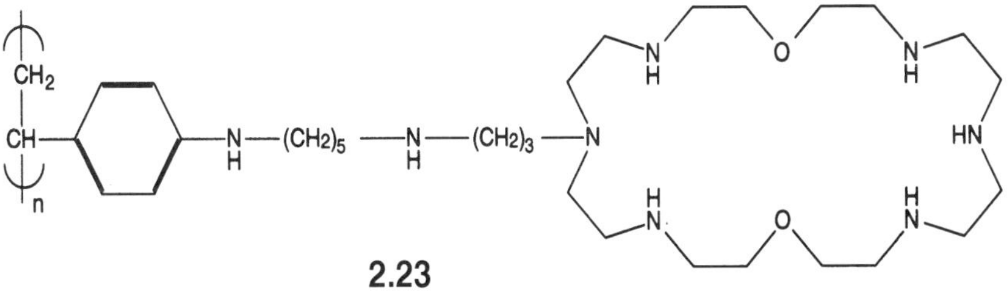

We have so far focused on molecular recognition of various substrates, in particular of nucleotide polyphosphates, by preorganized receptors. In addition to molecular recognition, supramolecular catalysis and reactivity as well as transport processes represent other basic functional features of supramolecular chemistry. Supramolecular catalysis, the chemical transformation of the bound substrate, involves first a binding step for which molecular recognition is a prerequisite, followed by the transformation of the complexed species, and finally the release of the product with regeneration of the catalytic unit. In the case of nucleotide polyphosphates, the supramolecular catalysis of phosphoryl transfer processes involved in both bond-breaking and bond-making reactions has been demonstrated using the above-mentioned versatile compounds **2.21** and **2.22**.[58] These aspects will be discussed in Chapter 11.

Many other phosphate receptors have been published in recent years.[59–62]

Complexation of the sulfate anion by protonated polyazamacrocycles[63,64] and -macrobicycles[14] has also been studied.

Perchlorate complexation has apparently not been considered thus far to be an important target, and it was rather surprising to discover that the hexaprotonated compound forms in the solid state an inclusion complex with perchlorate anion. The inclusive nature of the complex (**2.24**-6H$^+$,ClO$_4^-$) was established by an X-ray study, which indeed revealed that the anion was enclosed in the center of the ligand's cavity.[65]

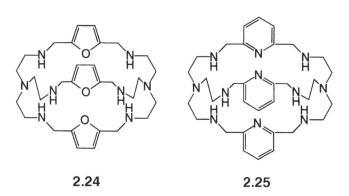

2.24   **2.25**

2.6 SQUARE PLANAR ANIONS

Although the complexation of transition metal anionic species was reported earlier (see below), the binding of square planar complexes such as PdCl$_4^{2-}$ by the protonated polyazamacrocycle **2.26** has recently been reported (Fig. 2.13). Again the inclusive nature of the complex was demonstrated by an X-ray analysis.[66]

2.7 OCTAHEDRAL ANIONS

Although unexpected, the complexation of SiF$_6^{2-}$ anion by the protonated compound **2.25** was observed in the solid state.[65] Investigations on the

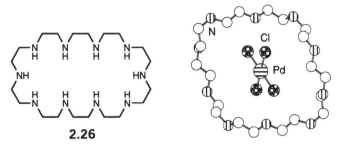

Fig. 2.13. Binding of the square planar $PdCl_4^{2-}$ within the macrocyclic cavity by a large polyammonium macrocycle, **2.26**.

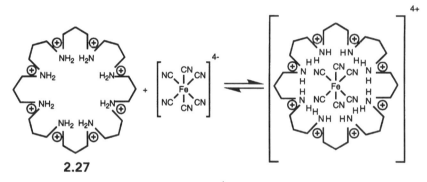

Fig. 2.14. Binding of the octahedral $Fe(CN)_6^{4-}$ by the polyammonium macrocycle **2.27**.

coordination chemistry of anions has led to the development of several types of organic receptor molecules capable of binding various anionic transition metal complexes.[67] Among these receptors protonated polyazamacrocycles such as **2.27** have been shown to form stable and selective complexes with metal hexacyanide anions such as $Fe(CN)_6^{4-}$, $Fe(CN)_6^{3-}$, $Ru(CN)_6^{4-}$ and $Co(CN)_6^{3-}$.[35] It is worth noting that the complexation of these complex anions leads to supercomplexes. Indeed these supramolecular species may be considered as complexes of complexes: The central cation forms a complex with cyanide anions, and the resulting anionic species is in turn complexed by the polyammonium macrocycles (Fig. 2.14). This second complexation process is, in fact, the organization of the second coordination sphere around the central transition metal cation.

Since the formation of these supercomplexes results in the modification of the second coordination sphere around the metal, the electrochemical[68] as well as photochemical[69] properties of the complexed anion were shown to be dependent upon and adaptable to the structure of the polyammonium receptor.

2.8 MOLECULAR ASSEMBLIES

More recently, the concept of molecular recognition of anions was used in the solid state for the construction of large-size molecular assemblies (10^{-6}–10^{-4} m scale). Using an iterative process based on the assembly of complementary exomodules in which the interaction sites are outwardly oriented, molecular rods, tapes,[70] and sheets[71] were obtained. In the case of rods and tapes, the individual components were the dicationic bisamidinium **2.28** and tere- and isophthalate dianions (Fig. 2.15).

The structures of the above-mentioned molecular assemblies in the solid state were established by X-ray analysis, which revealed that the bisamidinium unit **2.28** was indeed well suited to recognize carboxylate anions through strong H bonding and electrostatic interactions. As a consequence of this molecular recognition process between the components, all anionic and cationic modules were interconnected, thus leading to one-dimensional solids (Fig. 2.16).

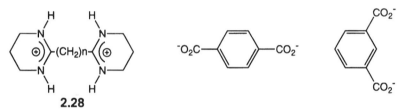

Fig. 2.15. Individual components including the bisamidinium ion and tere- and isophthalate ions for building large molecular assemblies.

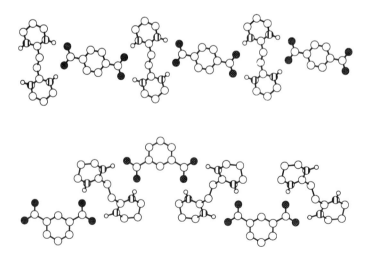

Fig. 2.16. Two structural pictures of molecular assemblies based on bisamidinium ion and tere- and isophthalate interactions.

2.9 CONCLUSION

Despite the important roles played by negatively charged species in natural and in unnatural worlds, the recognition of anions by synthetic receptor molecules still remains relatively unexplored. Although over the past 20 years only a rather small number of groups have investigated the coordination of anions with diverse approaches, it appears that over the past decade many other groups are taking up this challenge. A deep understanding of the different facets governing the interactions of anions with receptors may allow one to define a generalized coordination chemistry including both cations and anions.

2.10 REFERENCES

1. C. J. Pedersen, *J. Am. Chem. Soc.* **89**, 7017 (1967).
2. C. H. Park and H. E. Simmons, *J. Am. Chem. Soc.* **90**, 2431 (1968).
3. R. A. Bell, G. G. Christoph, F. R. Fronczck, and R. E. Marsh, *Science* **190**, 151 (1975).
4. H. E. Simmons, C. H. Park, R. T. Uyeda, and M. F. Habibi, *Trans. New York Acad. Sci.* **32**, 521 (1970).
5. E. Graf and J.-M. Lehn, *J. Am. Chem. Soc.* **97**, 5022 (1975); E. Graf, J.-P. Kintzinger, J.-M. Lehn, and J. LeMoigne, *J. Am. Chem. Soc.* **104**, 1672 (1982).
6. E. Graf and J.-M. Lehn, *J. Am. Chem. Soc.* **98**, 6403 (1976).
7. B. Metz, J. M. Rosalky, and R. Weiss, *J. Chem. Soc. Chem. Comm.*, 533 (1976).
8. J.-P. Kintzinger, J.-M. Lehn, E. Kauffmann, J. L. Dye, and A. I. Popov, *J. Am. Chem. Soc.* **105**, 7549 (1983).
9. J.-M. Lehn, *Acc. Chem. Res.* **11**, 49 (1978).
10. F. P. Schmidtchen, *Angew. Chem. Int. Ed. Engl.* **16**, 720 (1977); F. P. Schmidtchen, *Chem. Ber.* **114**, 597 (1981).
11. F. P. Schmidtchen and G. Müller, *J. Chem. Soc. Chem. Comm.* 1115 (1984).
12. J.-M. Lehn, S. H. Pine, E-I. Watanabe, and A. K. Willard, *J. Am. Chem. Soc.* **99**, 6766 (1977).
13. J.-M. Lehn, E. Sonveaux, and A. K. Willard, *J. Am. Chem. Soc.* **100**, 4914 (1978).
14. B. Dietrich, J. Guilhem, J.-M. Lehn, C. Pascard, and E. Sonveaux, *Helv. Chim. Acta* **67**, 91 (1984).
15. B. Dietrich, J.-M. Lehn, J. Guilhem, and C. Pascard, *Tetrahedron Lett.* **30**, 4125 (1989).
16. S. D. Reilly, G. R. K. Khalsa, D. K. Ford, J. R. Brainard, B. P. Hay, and P. H. Smith, *Inorg. Chem.* **34**, 569 (1995).
17. B. Dietrich, B. Dilworth, J.-M. Lehn, J.-P. Souchez, M. Cesario, J. Guilhem, and C. Pascard, *Helv. Chim. Acta*, **79**, 569 (1996).
18. M. W. Hosseini, J.-P. Kintzinger, J.-M. Lehn, and A. Zahidi, *Helv. Chim. Acta* **72**, 1078 (1989); S. Boudon, A. Decian, J. Fischer, M. W. Hosseini, J.-M. Lehn, and G. Wipff, *J. Coord. Chem.* **23**, 113 (1991).

19. E. Suet and H. Handel, *Tetrahedron Lett.* **25**, 645 (1984).
20. W. B. Farnham, D. C. Roe, D. A. Dixon, J. C. Calabrese, and R. L. Harlow, *J. Am. Chem. Soc.* **112**, 7707 (1990).
21. R. J. Motekaitis, A. E. Martell, and I. Murase, *Inorg. Chem.* **25**, 938 (1986); R. J. Motekaitis, A. E. Martell, I. Murase, J.-M. Lehn, and M. W. Hosseini, *Inorg. Chem.* **27**, 3630 (1988).
22. J. L. Sessler, M. J. Cyr, V. Lynch, E. McGhee, and J. A. Ibers, *J. Am. Chem. Soc.* **112**, 2810 (1990); J. L. Sessler, T. Morishima, and V. Lynch, *Angew. Chem. Int. Ed. Engl.* **30**, 977 (1991); M. Shionoya, H. Furuta, V. Lynch, A. Harriman, and J. L. Sessler, *J. Am. Chem. Soc.* **114**, 5714 (1992); J. L. Sessler, M. Cyr, H. Furuta, V. Kral, T. Mody, T. Morishima, M. Shionoya, and S. Weghorn, *Pure and Appl. Chem.* **65**, 393 (1993).
23. M. Newcomb, A. M. Madonik, M. T. Blanda, and J. K. Judice, *Organometallics* **6**, 145 (1987).
24. M. Newcomb, J. H. Horner, and M. T. Blanda, *J. Am. Chem. Soc.* **109**, 7878 (1987).
25. M. Newcomb, J. H. Horner, M. T. Blanda, and P. J. Squattrito, *J. Am. Chem. Soc.* **111**, 6294 (1989).
26. M. T. Blanda, J. H. Horner, and M. Newcomb, *J. Org. Chem.* **54**, 6294 (1989).
27. H. E. Katz, *J. Org. Chem.* **50**, 5027 (1985); H. E. Katz, *Organometallics* **6**, 1134 (1987); H. E. Katz, *J. Am. Chem. Soc.* **108**, 7640 (1986).
28. M. E. Jung and H. Xia, *Tetrahedron Lett.* **29**, 297 (1988).
29. X. Yang, C. B. Knobler, and M. F. Hawthorne, *Angew. Chem. Int. Ed. Engl.* **30**, 1507 (1991).
30. H. E. Katz, in *Inclusion Compounds*, J. L. Atwood, J. E. D. Davies, and D. D. MacNicol, Eds., 1991, Vol. 4.
31. J. Cullinane, R. I. Gelb, T. N. Margulis, and L. J. Zompa, *J. Am. Chem. Soc.* **104**, 3048 (1982).
32. D. Heyer and J.-M. Lehn, *Tetrahedron Lett.* **7**, 5969 (1986).
33. E. Kimura and A. Sakonaka, *J. Am. Chem. Soc.* **104**, 4984 (1982).
34. F. P. Schmidtchen, A. Gleich, and A. Schummer, *Pure and Appl. Chem.* **61**, 1535 (1989); C. Seel, A. Galan and J. de Mendoza, *Topics in Curr. Chem.* **175**, 101 (1995).
35. B. Dietrich, M. W. Hosseini, J.-M. Lehn, and R. B. Sessions, *J. Am. Chem. Soc.* **103**, 1282 (1981).
36. E. Kimura, A. Sakonaka, T. Yatsunami, and M. Kodama, *J. Am. Chem. Soc.* **103**, 3041 (1981).
37. M. W. Hosseini and J.-M. Lehn, *Helv. Chim. Acta* **71**, 749 (1988).
38. A. Bencini, A. Bianchi, M. I. Burguete, E. García-España, S. V. Luis, and J. A. Ramirez, *J. Am. Chem. Soc.* **114**, 1919 (1992).
39. M. W. Hosseini and J.-M. Lehn, *J. Am. Chem. Soc.* **104**, 3525 (1982); M. W. Hosseini and J.-M. Lehn, *Helv. Chim. Acta* **69**, 587 (1986); J.-M. Lehn, R. Meric, J.-P. Vigneron, I. Bkouche-Waksman, and C. Pascard, *J. Chem. Soc. Chem. Comm.*, 62 (1991).
40. H. Tsukube, *J. Chem. Soc. Perkin Trans 1*, 615 (1985); B. Dietrich, T. M. Fyles, M. W. Hosseini, J.-M. Lehn, and K. C. Kaye, *J. Chem. Soc., Chem. Commun.*, 691 (1988); G. Brand, M. W. Hosseini, and R. Ruppert, *Helv. Chim. Acta* **75**, 721 (1992).

REFERENCES

41. F. P. Schmidtchen, *J. Am. Chem. Soc.* **108**, 8249 (1986).
42. E. Fan, S. A. Van Arman, S. Kincaid, and A. D. Hamilton, *J. Am. Chem. Soc.* **115**, 369 (1993).
43. S. M. Lacy, D. M. Rudkevich, W. Verboom, and D. N. Reinhoudt, *J. Chem. Soc. Perkin Trans 2*, 135 (1995).
44. P. Schiessl and F. P. Schmidtchen, *Tetrahedron Lett.* **34**, 2449 (1993).
45. B. Dietrich, T. M. Fyles, J.-M. Lehn, L. G. Pease, and D. L. Fyles, *J. Chem. Soc. Chem. Comm.*, 934 (1978).
46. R. Gross, G. Dürner, and M. W. Göbel, *Liebigs Ann. Chem.*, 49 (1994).
47. B. Dietrich, D. L. Fyles, T. M. Fyles, and J.-M. Lehn, *Helv. Chim. Acta* **62**, 2763 (1979).
48. R. Gross, J. W. Bats, and M. W. Göbel, *Liebigs Ann. Chem.*, 205 (1994).
49. F. P. Schmidtchen, *Tetrahedron Lett.* **30**, 4493 (1989).
50. R. P. Dixon, S. J. Geib, and A. D. Hamilton, *J. Am. Chem. Soc.* **114**, 365 (1992).
51. G. Deslongchamps, A. Galan, J. de Mendoza, and J. Rebek, Jr., *Angew. Chem. Int. Ed. Engl.* **31**, 61 (1992).
52. B. Dietrich, M. W. Hosseini, J.-M. Lehn, and R. B. Sessions, *Helv. Chim. Acta* **66**, 1262 (1983); see also Ref. 35.
53. M. W. Hosseini and J.-M. Lehn, *Helv. Chim. Acta* **70**, 1312 (1987).
54. E. Kimura, Y. Kuramoto, T. Koike, H. Fujioka, and M. Kodama, *J. Org. Chem.* **55**, 42 (1990).
55. E. Kimura, A. Watanabe, and H. Nihira, *Chem. Pharm. Bull.* **31**, 3264 (1983); E. Kimura, *Topics in Curr. Chem.* **128**, 113 (1985).
56. M. W. Hosseini, A. J. Blacker, and J.-M. Lehn, *J. Chem. Soc., Chem. Commun.*, 596 (1988); *J. Am. Chem. Soc.* **112**, 3896 (1990).
57. D. Cordier and M. W. Hosseini, *New. J. Chem.* **14**, 611 (1990).
58. M. W. Hosseini, *Bioorganic Frontiers*, H. Dugas, Ed., Springer-Verlag, 1993, Vol. 3.
59. H. Furuta, M. J. Cyr, and J. L. Sessler, *J. Am. Chem. Soc.* **113**, 6677 (1991).
60. D. M. Rudkevich, W. P. R. V. Stauthamer, W. Verboom, J. F. J. Engbersen, S. Harkema, and D. N. Reinhoudt, *J. Am. Chem. Soc.* **114**, 9671 (1992).
61. M. Dhaenens, J.-M. Lehn, and J.-P. Vigneron, *J. Chem. Soc. Perkin Trans. 2*, 1379 (1993).
62. D. H. Vance and A. W. Czarnik, *J. Am. Chem. Soc.* **116**, 9397 (1994).
63. R. I. Gelb, L. M. Schwartz, and L. J. Zompa, *Inorg. Chem.* **25**, 1527 (1986).
64. G. Wu, R. M. Izatt, M. L. Bruening, W. Jiang, H. Azab, K. E. Krakowiak, and J. S. Bradshaw, *J. Incl. Phenom. Mol. Recog. Chem.* **13**, 121 (1992).
65. G. Morgan, V. MacKee, and J. Nelson, *J. Chem. Soc. Chem. Comm.*, 1649 (1995).
66. A. Bencini, A. Bianchi, P. Dapporto, A. García-España, M. Micheloni, P. Paoletti, and P. Paoli, *J. Chem. Soc., Chem. Commun.*, 753 (1990).
67. M. W. Hosseini, *Perspectives in Coordination Chemistry*, A. F. Williams, C. Floriani, and A. Merbach, Eds. VCH, 1992.
68. F. Peter, M. Gross, M. W. Hosseini, J.-M. Lehn, and R. B. Sessions, *J. Chem. Soc., Chem. Commun.*, 1067 (1981); F. Peter, M. Gross, M. W. Hosseini, and J.-M. Lehn, *J. Electroanal. Chem.* **144**, 279 (1983).

69. M. F. Manfrin, N. Sabbatini, L. Moggi, V. Balzani, M. W. Hosseini, and J.-M. Lehn, *J. Chem. Soc., Chem. Commun.*, 555 (1984); M. F. Manfrin, L. Moggi, V. Castelvetro, V. Balzani, M. W. Hosseini, and J.-M. Lehn, *J. Am. Chem. Soc.* **107**, 6888 (1985).
70. M. W. Hosseini, R. Ruppert, P. Schaeffer, A. De Cian, N. Kyritsaka, and J. Fischer, *J. Chem. Soc., Chem. Commun.*, 2135 (1994); G. Brand, M. W. Hosseini, R. Ruppert, A. De Cian, J. Fischer, and N. Kyritsaka, *New. J. Chem.* **19**, 9 (1995).
71. M. W. Hosseini, G. Brand, P. Schaeffer, R. Ruppert, A. De Cian, and J. Fischer, *Tetrahedron Lett.* **37**, 1405 (1996).

CHAPTER 3

Natural Anion Receptors: Anion Recognition by Proteins

STEFANO MANGANI and MARTA FERRARONI

3.1 Introduction
3.2 Carboxypeptidase A
3.3 Superoxide Dismutase
3.4 Phosphate and Sulfate Binding Proteins
3.5 Citrate Synthase
3.6 Concluding Remarks
3.7 References

3.1 INTRODUCTION

The interaction of anions with proteins is of paramount relevance in the chemistry of life since it concerns essential aspects like the activity of enzymes, the transport of hormones, protein synthesis, and DNA regulation. An understanding of the structural rules underlying the engineering of the anion binding sites in proteins also has considerable impact in the design of molecules that may be used as therapeutic agents. A subject of such a complexity obviously cannot be compressed into a single chapter. Here we will limit ourselves to some examples of very specific recognition of anionic substrates by enzymes and transport proteins as established by X-ray crystallography and to highlight the nature of these interactions with an aim to understand only the simplest key factors responsible for the phenomenon.

Supramolecular Chemistry of Anions, Edited by Antonio Bianchi, Kristin Bowman-James, and Enrique García-España.
ISBN 0-471-18622-8. © 1997 Wiley-VCH, Inc.

3.2 CARBOXYPEPTIDASE A

Carboxypeptidase A (CPA) is a zinc enzyme of molecular weight 35,215 tailored to perform the hydrolytic cleavage of the terminal peptide (or ester) bond at the carboxylate end of polypeptide (or ester) substrates bearing hydrophobic (aromatic) side chains on the last residue. High-resolution X-ray crystal structures of the native enzyme and of several of its complexes with substrate analogues, transition-state analogues, and inhibitors are available. These studies have greatly contributed to the understanding of CPA catalytic mechanism and chemistry. Several reviews are available on this topic,[1-3] and we will not discuss it further.

Our interest here will be focused on the highly specific recognition of anionic substrates that has been achieved by this enzyme since it provides a prototypical example of host–guest interaction.

The active site of CPA may be divided into several subsites. The dead end of the cavity (S1' subsite) is the part of the enzyme dedicated to substrate recognition. In front of it lies the catalytic site (S1) hosting the essential zinc ion. Moving toward the enzyme surface at least two other subsites (S2, S3) able to interact with extended polypeptide substrates have been identified.[4]

The requirements for the recognition of a molecule by the CPA active-site cavity are very demanding. Figure 3.1 shows a stereo view of the CPA S1' site with a scheme of a minimal substrate. In order to be recognized by this site a molecule must resemble an α-amino acid and bear several chemical determinants, namely a negatively charged carboxylate group and an adjacent aromatic

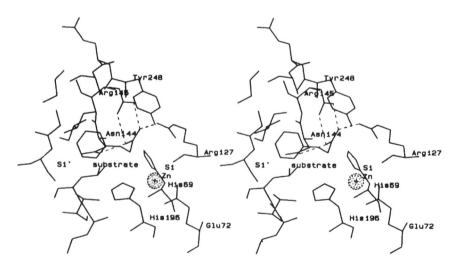

Fig. 3.1. Stereo view of the CPA active-site cavity showing the dead-end substrate site (S1') with a sketch of the minimal substrate bound, and the catalytic site (S1) hosting the metal ion.

(hydrophobic) side chain bound to a β-carbon. Extending the chain further from the α-carbon causes the substrate to interact with the metal site and with the outer sites S2 and S3.

The crystal structures of several CPA complexes with molecules like (−)2-benzyl-3-p-methoxybenzoylpropionic acid (BMBP),[5] 2-benzyl-3-formylpropanoic acid (BFP),[6] N-[[[(benzyloxycarbonyl) amino)]methyl]hydroxyphosphinyl]-1-phenylalanine (ZGP′),[7] D-phenylalanine,[8] and L-phenyllactate[9] have revealed the key features of the substrate binding pocket. As an example, Fig. 3.2 shows the binding of the L-phenyllactate inhibitor as found by crystallographic studies.[9]

The main recognition site suited to discriminate between an anionic and a cationic peptide end is the side chain of the Arg145 residue, located at the bottom of the pocket. This positively charged side chain is able to establish a "salt bridge" (two equivalent hydrogen bonds reinforced by electrostatic interaction) with the complementary substrate carboxylate group, thus providing specificity towards C-terminal peptides. Arg145 is the crucial residue used by the evolved enzyme to ensure substrate recognition. The carboxylate group of CPA substrates is held in place by two further hydrogen bonds received from Asn144 and by Tyr248 side chains. This latter is a beautiful example of "induced fit,"[10] since, upon substrate (inhibitor) binding, the Tyr248 side chain swings down to lock the substrate in the cavity from its position in the native enzyme about 14 Å away (see Fig. 3.3). Close to Arg145 lies a hydrophobic cavity whose shape is perfectly suited to accommodate the aromatic ring of the substrate side chain. This is the second feature of the CPA cavity determining its specificity. Figure 3.3 displays the amazingly perfect fit obtained by the hydrophobic wall of the cavity with the aromatic ring of the inhibitor L-phenyllactate. It must be underlined that such fit is obtained with a proper arrangement of different

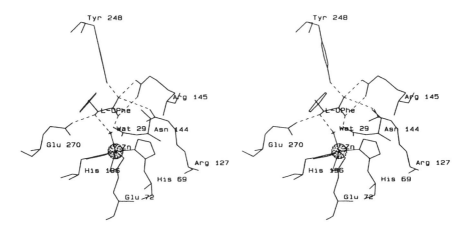

Fig. 3.2. Stereo plot showing the binding of the L-phenyllactate inhibitor to CPA as found in the crystal structure of the complex.[9]

NATURAL ANION RECEPTORS: ANION RECOGNITION BY PROTEINS

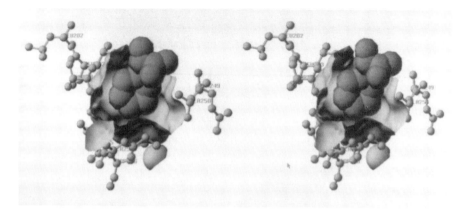

Fig. 3.3. Stereo view of the Connolly surface[23] of the CPA hydrophobic pocket showing the perfect fit between the substrate aromatic ring and the cavity. The phenyl ring is shown as a CPK model. The figure shows also the protein side chains constituting the hydrophobic wall. The Connolly surface has been traced using a probe of 1.4 Å radius.

hydrophobic side chains from different regions of the polypeptide. Finally the amino or hydroxyl groups of the inhibitor are engaged in hydrogen bonds with the S1 site residue Glu270 side chain. Table 3.1 reports the contacts occurring in the S1′ site between the enzyme and the L-phenyllactate inhibitor as found in the X-ray structure of the complex.[9]

The product of peptide hydrolysis, L-phenylalanine, has been observed bound to the CPA cavity in the crystal structures of a hydrolyzed phosphonamidate inhibitor,[11] of an enzyme–substrate–product ternary complex,[12] and

TABLE 3.1. Enzyme–Inhibitor Interactions (Å) in the CPA–L-OPhe Complex[9]

L-OPhe atom	CPA atom	(Å)
Hydrogen bonds		
Carboxylate O1	Tyr248 OH	2.65
Carboxylate O1	Arg145 NH2	2.62
Carboxylate O1	Arg127	3.38
Carboxylate O2	Arg145 NH1	2.97
Carboxylate O2	Asn144 ND2	2.86
Hydroxyl O	Glu270 OE2	2.38
Hydroxyl O	Zinc-bound water	2.72
Hydrophobic contacts		
Phenyl C2	Ala250 CB	3.66
Phenyl C3	Ala250 CB	3.48
Phenyl C3	Ile243 CD1	3.93
Phenyl C4	Ile243 CD1	3.47
Phenyl C4	Gly253 CA	3.89
Phenyl C5	Ile243 CD1	3.82
Phenyl C6	Thr268 CG2	3.93

on the CPA–L-phenylalanine–azide ternary complex.[13] The comparison between the binding modes of the L-phenylalanine and L-phenyllactate inhibitors allows some interesting considerations. Despite their chemical similarity, the two inhibitors display different affinities for the enzyme and different binding. This fact is reflected by the different catalytic efficiency of the enzyme toward peptide and ester substrates. CPA is able to hydrolyze both peptides and esters. However, contrary to the common solution chemistry teaching that an ester bond is more easily hydrolyzed than a peptide bond, the enzyme-catalyzed peptide hydrolysis is faster than that of the corresponding ester substrate.[14] It has been demonstrated[14] that peptide and ester hydrolysis follow the same reaction pathway:

$$E + S \rightleftharpoons ES_1 \rightleftharpoons ES_2 \rightleftharpoons E + P$$

and that both peptide and esters bind to the active site prior to the rate-limiting step.[15–18] These studies demonstrate that the hydrolysis of esters occurs before the rate-limiting step, so that the intermediate ES_2 may be more correctly described as a ternary complex EP_1P_2, where P_1 and P_2 are the product of esterolysis or substrates for the reverse reaction. In contrast, peptide hydrolysis is either simultaneous or follows the rate-limiting step. Hence, the rate-limiting step of ester hydrolysis is the product release, while the rate-limiting step of peptide hydrolysis is the actual cleavage of the bond, thus reconciling enzyme chemistry with laboratory chemistry.

The structures of the CPA complexes with L-phenyllactate (product of ester hydrolysis) and with L-phenylalanine (product of peptide hydrolysis) indicate a likely explanation of the above phenomenon. The L-phenylalanine binding to CPA as observed and in the hydrolyzed phosphonamidate inhibitor[11] may be considered a better representation of the actual interaction between CPA and L-phenylalanine since in the azide complex, the binding of the inorganic anion changes the conformation of some of the active-site residues resulting in a different binding of the L-phenylalanine molecule.[13] Once bound to the S1' site, the α-OH group of the L-phenyllactate anion is engaged in two strong hydrogen bonds, one with the Glu270 side chain and the other with the zinc-bound water molecule. On the contrary, the L-phenylalanine product from BZF is not involved in hydrogen bonds through its positively charged α-NH$_3^+$ group, which is pointing away from the zinc site toward the entrance of the cavity. These results explain the higher inhibitory constant of L-phenyllactate with respect to L-phenylalanine (vide infra) and the different rates of the CPA-catalyzed hydrolysis, which can be attributed to the stabilization of the ester product by the extra hydrogen bond.[9] In other words, the structural observation can be nicely correlated with the kinetic studies on the CPA ester/peptide hydrolysis, which have proposed that the rate-limiting step in the CPA ester hydrolysis is the product release by the enzyme and not the hydrolytic step.[15] Furthermore, the measured difference in the inhibition constants K_i of

L-phenylalanine and L-phenyllactate of 55×10^{-4} M and 1.3×10^{-4} M, respectively,[14,19,20] results in a difference of the ΔG of binding of about 9.3 kJ mol^{-1} at 298 K, close to the energy of a medium-strength hydrogen bond.[21] [K_i is the constant relative to the dissociation equilibrium: EI $\Leftrightarrow$ E + I. In this case K_i has been measured by the inhibition of the CPA hydrolysis of the substrate carbobenzyloxyglycylglycyl-L-phenylalanine (peptide) or of the substrate O-(*trans*-cinnamoyl)-L-phenyllactate (ester)[20] (E = enzyme; I = inhibitor).]

In analyzing the interactions of small molecules with a protein, the entropic contribution to the Gibbs free energy of binding should be considered, since in the majority of cases it provides the most relevant part of the driving energy for the process. In the binding of the substrate to the protein the enthalpy gained in the enzyme–ligand interaction is counterbalanced by that lost in the breaking of the ligand–solvent interactions. On the contrary, on the entropy side, the overall translational and rotational entropies of the ligand are lost upon the interaction of the substrate with the protein. However, this entropy loss is more than compensated by that gained in the release of both the ligand-bound and protein-bound water molecules.[21] In the case of the binding of the inhibitors L-phenylalanine and L-phenyllactate to CPA, the occurrence of this phenomenon can be appreciated very precisely, since each molecule of the ligand displaces in binding four water molecules bound to the cavity in the native structure.[9]

The crystal structure of the ternary complex of CPA with L-phenylalanine and the azide anion[13] offers the chance to observe another anion binding site in the enzyme, namely, the azide binding site. The spectroscopic experiments of binding of small inorganic anions to CPA[22] demonstrated the existence of two separated binding sites for small anions. For example, azide binds to the first site with a $K_i = 35$ mM and to the second site with a $K_i = 1.5$ M. Only the occupancy of the low-affinity site by azide perturbs the electronic absorption of the cobalt-substituted CPA (Co^{2+} in place of Zn^{2+}), suggesting its closeness to the metal site. The same study also shows the occurrence of a synergistic effect in binding with the products of peptide and ester hydrolysis; that is, it is shown that the K_d of azide (the spectral K_d is only slightly different from the above-reported K_i) goes from 1.4 M to 4 mM in presence of 5 mM L-phenylalanine and that vice versa, the K_d for L-phenylalanine goes from 1.67 mM to 0.07 mM in the presence of 100 mM azide. [K_d is the spectrophotometrically determined constant of the equilibrium: EL $\Leftrightarrow$ E + L(E = enzyme, L = ligand).] The crystallographic analysis of the ternary complex CPA–L-phenylalanine–azide confirmed the presence of two azide binding sites and showed their location: the low affinity one at the metal and the other at the Arg145 positively charged side chain. This is an example of how a protein can recognize a very simple and small anion, a widespread and crucial problem occurring in nature. At variance with the CPA substrates, azide is lacking those chemical determinants able to provide strict specificity. However, the enzyme has the right cavity to host the anion. The first selection criterion is in this case the size of the cavity. Figure 3.4 shows the CPA cavity with a model of the bound azide. It can be seen that the

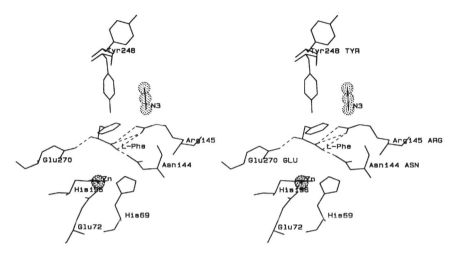

Fig. 3.4. Stereo plot of the ternary complex CPA–L-phenylalanine–azide showing the amino acid and only the azide molecule bound to the high-affinity, nonmetal site. The azide molecule is represented both as stick and spheres of arbitrary radius.

anion exactly fits its shape. Furthermore, there are crucial electrostatic interactions with the positive end of the Arg145 side chain and with the backbone carbonyl groups of Gly155 and Gln249 (see Table 3.2). The latter two are particularly interesting since they suggest that the enzyme stabilizes the polarized azide structure:

$$[N{\equiv}N^+{-}N^2]$$

The electric dipole present in this Lewis structure may favorably interact with the other dipole constituted by the positive charge of the Arg145 side chain and the negatively polarized ends of the Gly155 and Gln249 backbone carbonyls.

TABLE 3.2. Interaction Distances (Å) between CPA and the Azide Anion Bound to the Distal Site (RMSD = 0.3 Å) as Observed in the Crystal Structure of the CPA–L-Phenylalanine–Azide Complex[13]

CPA	azide atom	distance (Å)
$N\eta 1$Arg145	$N\alpha$	2.3
$N\eta 1$Arg145	$N\alpha$	3.6
$N\varepsilon$Arg145	$N\alpha$	2.4
OGly155	$N\beta$	2.6
OGln249	$N\gamma$	2.6

3.3 SUPEROXIDE DISMUTASE

The Cu,Zn superoxide dismutase enzyme (Cu,ZnSOD) is a very efficient catalyst of superoxide dismutation into hydrogen peroxide and dioxygen, being characterized by turnover numbers of the order of 10^6 s^{-1}.[24] The enzyme is a homodimer of about 32,000 MW composed of two identical subunits, each containing a Cu,Zn center. The copper ions are coordinated by four histidine residues and by a water molecule at the apex of a tetrahedrally distorted square pyramid. The zinc ions instead are in a tetrahedral environment provided by two histidine imidazole nitrogens, with an aspartate carboxylate oxygen and a further nitrogen from a histidinato moiety bridging copper and zinc.[25] During the catalytic cycle the essential copper ion undergoes alternate redox reactions between the +1 and +2 oxidation states maintaining in each state identical catalytic efficiency.[24]

The enzyme is specifically designed to attract in the active-site cavity anionic molecules through a positive electric field gradient,[26,27] which can be modified by site-directed mutagenesis to improve the steering of the substrate toward the active site.[28,29] Both the SOD substrate and inhibitors are monovalent anionic molecules, and understanding their interaction with the enzyme has obvious relevance toward the comprehension of the enzyme chemistry. Again the key residue in the Cu,ZnSOD active-site cavity is a positively charged arginine (Arg141) lying close to the catalytic copper. Mutations of this residue with neutral amino acids lead to a tenfold decrease of the enzyme affinity for anions.[30]

The interest in understanding SOD–anion interactions has led to a structural study of the Cu,ZnSOD–azide complex. Azide was the anion of choice because of its similarity in charge and electronic distribution with the substrate superoxide anion. The azide binding site has been structurally characterized in both Cu(II),Zn and Cu(I),Zn superoxide dismutases.[31,32] The complex with cyanide anion has been characterized only for the oxidized enzyme.[33] The structural data have contributed to the understanding of the mode of binding of anions to the active-site cavity in both copper oxidation states. The two crystallographic determinations of the oxidized and reduced Cu,ZnSOD-azide complexes have been performed by two laboratories on enzymes from different sources and on different crystalline forms. The results are very interesting since the azide anion displays dissimilar binding modes in the two cases.

In the oxidized enzyme the azide anion is found coordinated to the Cu(II) ion with a Cu(II)–N(azide) distance of 1.96 Å. The binding of the anion to the metal causes a displacement of one of the donor histidines (His-46) by about 0.5 Å. The binding of azide to the reduced enzyme is completely different since the maximum approach of the anion to the cuprous atom is at 2.9 Å, indicating a very weak interaction mostly electrostatic in nature. The anion is at a hydrogen bonding distance from Arg143 and from a water molecule in the cavity. Figure 3.5 shows the azide binding site in Cu(I),Zn superoxide dismutase as it appears

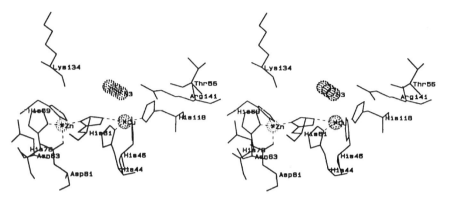

Fig. 3.5. The active site cavity of Cu(I),ZnSOD with the bound azide anion from the X-ray structure of the complex. The azide is represented as stick and sphere model.[32].

in the crystal structure. This site has been proposed by the authors as the anion binding site in the reduced enzyme.[32]

The different binding of azide in the two structures is perfectly consistent with the chemistry occurring at the copper atom. In the oxidized enzyme, the strong-base azide replaces the water molecule in the Cu(II) coordination sphere, making a covalent bond with the metal due to the more pronounced Lewis acid character of the metal in the 2+ oxidation state. In the reduced enzyme the copper–histidine coordination distances remain unchanged with respect to the unligated enzyme, and only the water molecule has disappeared from the copper coordination sphere. In this case the weaker tendency of azide to coordinate to the cuprous ion is overbalanced by the favorable noncovalent interactions with the cavity. The result is a different behavior of the anion as expected, considering the reduced charge of the metal and its tendency to prefer four instead of five coordination.

If the azide mimics the mode of binding of superoxide in the oxidized and reduced forms of the enzyme, then some aspects of the superoxide dismutation mechanism as it is commonly accepted have to be reconsidered. The mechanism was proposed in 1982 by Tainer et al. following the spectroscopic and structural evidence available at the time.[34] In this mechanism the copper ion is bound in both oxidation states to the incoming superoxide substrate. Furthermore, upon reduction, the histidinato bridge is broken from the copper side, leaving the Cu(I) tricoordinate. The recent crystallographic determinations of the reduced enzyme[35] and of its complexes[33] do not support this view. The above mechanism may be reviewed in light of the alternative mechanism proposed by Osman and Basch in 1984 based on quantum-mechanical calculations on model systems.[36] The computational approach shows that the Cu(II)–O_2^- complex is indeed very stable. The authors propose that the copper reduction may occur through an outer-sphere electron transfer between the Cu(II)–O_2^- complex and a second substrate molecule bound in the cavity. This mechanism does not imply the breaking of the histidinato bridge, and the protons needed by the complete reaction may be provided by a chain of water molecules linking the cavity to the bulk solution. The azide behavior found in the Cu(II) and Cu(I),ZnSOD–azide complexes may model the shuttling of the superoxide anion in and out the copper coordination sphere following the transfer of electrons from and to the metal site.

3.4 PHOSPHATE AND SULFATE BINDING PROTEINS

Other beautiful examples of small inorganic anion recognition by proteins have been recently provided by the crystal structures of several complexes, among them the transport proteins phosphate binding protein (PBP) and sulfate binding protein (SBP) with their respective bound anions. The specificity of the phosphate and sulfate recognition has been developed by the members of a family of bacterial periplasmic proteins devoted to the transport of small molecules like sugars and oligopeptides. Their function is to bind the anion tightly once it has passively diffused across the cell outer membrane.[37,38] The crystal structure of PBP[39,40] with bound orthophosphate has been determined at high resolution (1.7 and 1.8 Å, respectively) on crystals obtained at pH 4.5 and 6.2. The electron density corresponding to the phosphate anion in the 1.7 Å structure is extraordinarily well resolved and allows the unambiguous positioning of the anion and the direct observation of the hydrogen bonding network responsible for the anion binding. The phosphate binding site is located in a deep cleft between the two similarly folded globular domains constituting the protein (MW 34,400). The phosphate is found buried in the cavity 8 Å from the protein surface. Table 3.3 reports the hydrogen bonds between the phosphate and PBP.

There are 12 hydrogen bonds holding the anion. Seven of them involve NH groups from the main chain or from arginine side chains. The remaining five

TABLE 3.3. Hydrogen Bonds between Phosphate and PBP[39]

Phosphate atom	PBP residue	Group	Distance (Å)
O1	Thr10	N	2.81
O1	Thr10	O	2.69
O1	Arg135	N1	2.84
O2	Arg135	N2	2.90
O2	Ser139	O	2.77
O2	Thr141	N	2.83
O2	Thr141	O	2.68
O3	Ser38	N	2.67
O3	Ser38	O	2.63
O3	Gly140	N	2.76
O4	Phe11	N	2.92
O4	Asp56	O2	2.45

bonds originate from two serine and two threonine side-chain OH groups and an oxygen from the carboxylate group of the Asp56 side chain. This latter is the crucial interaction responsible for the selectivity of the protein. The presence of a negatively charged carboxylate allows the protein to discriminate between protonated and deprotonated anions. The carboxylate group of Asp56 can only act as a hydrogen bond acceptor. The only other group in the cavity able to accept hydrogen bonds is the OH group of Ser38 side chain, which, in turn, donates its proton to the O2 of Asp56. As a result of the presence of those two groups, the protein is able to bind selectively mono- or dihydrogenphosphate and not fully deprotonated tetrahedral anions like sulfate and tungstate, which will be repelled by the carboxylate negative charge. This situation is analogous to that observed in the phosphate-inhibited CPA.[41] At a pH around neutrality CPA binds phosphate with a $K_i = 0.4$ mM.[42] Only phosphate may bind to the CPA active site, whereas other coordinating anions like CNO^-, SCN^-, and Cl^- cannot replace the water/hydroxyl group in the Zn(II) coordination sphere (vide supra and refs. 42 and 43).

The crystal structure of the CPA–phosphate complex has shown that the phosphate is coordinated to zinc and that it is engaged in a strong hydrogen bond with the deprotonated carboxylate group of Glu270;[41] hence, the anion must be at least monoprotonated. Also in this case, the presence of the carboxylate regulates the access of anions to zinc coordination. The Asp56 and Ser38 side chains of PBP are spatially close to concentrate the negative charge of the bound anion on the opposite side, where the positively charged side chain of Arg135 appears in the site. In this way the protein is able to direct the protonated anion toward a perfect docking into the cavity.

In order to bind sulfate, a specialized protein, the sulfate binding protein (SBP), has been developed in bacterial transport systems.[44] Contrary to PBP, SBP is able to bind only fully deprotonated anions. The 2.0 Å resolution crystal structure of *Salmonella typhimurium*[45,46] shows the sulfate anion bound to a cavity similar to that of PBP. Figure 3.6 shows the sulfate anion bound to its site in the protein. The sulfate molecule is held by seven hydrogen bonds received from backbone peptide NH, serine OH, and Trp NH groups (Table 3.4). The only difference with the PBP cavity is the absence of carboxylate side chains and of other groups able to act as hydrogen bonding acceptors.

Besides the protonation state, more subtle differences may contribute in the discrimination of sulfate from phosphate anions. A survey of the Cambridge Crystallographic Database for structures of small molecules containing phosphonyl (R–PO_3^{2-}) and sulfonyl groups (R–SO_3^-)[47] has been conducted to analyze the stereochemistry of the hydrogen bonds involving these two groups as acceptors. The study has evidenced a difference in the behavior of the two anions. The P=O···H angle is 118.9° ± 10.9°, and the phosphonyl–hydrogen bond donor interactions are scattered around the group with a preference for the *gauche* stereochemistry. The above value is similar to that found for phosphinyl (R–P=O_2–R') interactions,[48] which display also a similar distribution of bonds with a preference for *anti* stereochemistry. The geometry

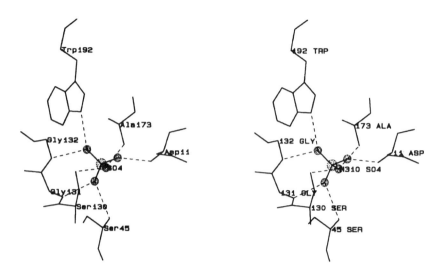

Fig. 3.6. Stereo view of the binding of the sulfate anion to SBP,[46] showing the hydrogen bonds between the anion and the protein.

TABLE 3.4. Hydrogen Bonds between Sulfate and SBP[46]

Sulfate	SBP residue	Group	Distance (Å)
O1	Ser130	O	2.65
	Ala173	N	2.77
O2	Gly131	N	2.83
	Ser45	N	2.87
O3	Gly132	N	2.70
	Trp192	N1	2.81
O4	Asp11	N	2.8

of sulfonyl hydrogen bond donor interactions differs significantly from that of phosphonyl and phosphinyl groups, the average S=O···H angle being 127.9° ± 12.9°. The distribution of the bonds is more densely clustered and displays the tendency towards an *eclipsed* geometry. The difference in the behavior of the two kinds of anions as hydrogen bond acceptors may play an important role in tailoring the specificity of binding sites that can be converted one into the other simply by a little change in the spatial distribution of the hydrogen bond donor groups.

3.5 CITRATE SYNTHASE

The relevance of the recognition of phosphate groups by a protein involves widespread processes in biological chemistry as, for example, RNA hydrolysis by RNAse enzymes, ATP synthesis and hydrolysis by ATP synthases, and DNA cleavage and repair by nuclease and polymerase enzymes. Phosphate groups are also present in ubiquitous enzyme cofactors like acetylcoenzyme A.

The crystal structure at 1.7 Å resolution of the enzyme citrate synthase[49] offers the possibility of viewing the binding sites of both the citrate anion and the coenzyme A. Additionally, one can see their spatial relationship, which can be correlated with the oxaloacetate and acetylCoA binding sites, respectively. These two species are the substrate and the coenzyme, respectively, in citrate synthesis, which is the actual function of the enzyme. Figure 3.7 is taken from the refined structure of the monoclinic model of the enzyme[49] and shows the citrate and the cofactor bound to the enzyme.

The tricarboxylic acid is tightly bound only to the "closed" form of citrate synthase. Also in this case the pocket is deeply buried within the enzyme molecule, and the citrate functional groups are all engaged in hydrogen bonds. Citrate is bound by three histidine and three arginine residues, one of them coming from a symmetry-related monomer. The citrate is completely surrounded by the protein so that only two of its oxygens are accessible to the solvent in the absence of coenzyme A. It is worth noting that the citrate binding site is located at the positive end of the electric dipoles originated by two α-helices, which certainly have a role in directing the approach and the binding of the substrate. The approach of the SH group of CoA to citrate occurs through the narrow channel connecting the O3,O4 oxygens to the solvent. In the open form of the enzyme only weak electron density has been found in the citrate binding site, indicating the loosening of the interaction.

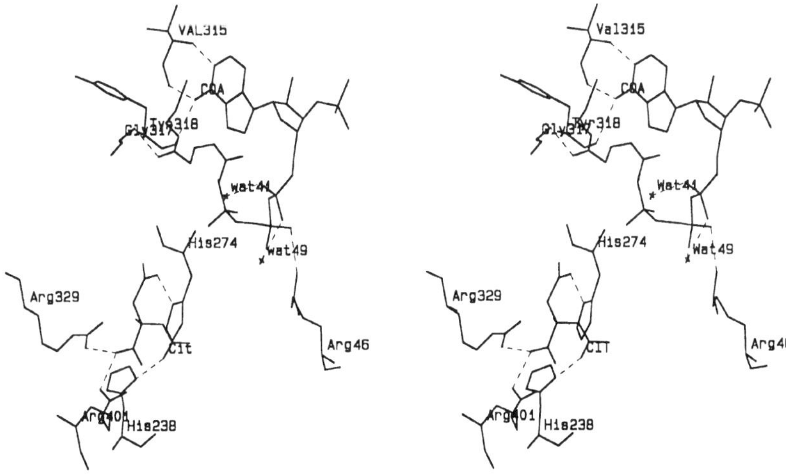

Fig. 3.7. Stereo view of the citrate and coenzyme A bound to the citrate synthase enzyme,[49] displaying the hydrogen bonding scheme and the contacts between the protein, the substrate, and the cofactor.

The coenzyme A binding to the enzyme is reported in Fig. 3.7. The cofactor electron density obtained from the crystallographic experiment is well defined and allows the unambiguous tracing of the molecule except for the disordered sulfur-bearing tail. The cofactor adopts a compact conformation as a result of internal hydrogen bonds and stacking between the adenine ring and one of its two peptide linkages of the pantothenic arm. Also in this case arginine residues are found to bind the three charged carboxylate moieties through salt bridges. It is remarkable the way the enzyme recognizes the adenine aromatic ring: a section of the main chain (Fig. 3.7) is wrapped around the edge of the ring by means of three specific hydrogen bonds to adenine N1 and N10 atoms. The contribution of van der Waals forces to the binding of the cofactor is evidenced by the several contacts with the protein. Once again when the anionic substrate is a large molecule bearing several chemical determinants as in this case, it becomes clear the relevant role played by hydrophobic interactions in the binding. These interactions are also evidenced by the contacts between a valine residue (Val314) present in the "adenine recognition loop" and the aromatic ring.

3.6 CONCLUDING REMARKS

The few examples of protein–anion interactions provided above have been sufficient to evidence some of the factors influencing the process of recognition and binding. It has been possible to appreciate that enzymes and proteins have evolved to match the chemical determinants of their substrate molecules. The enormous possibilities of different conformations available to a polypeptide backbone and to its side chains (despite the rigidity of the peptide bond) allow a protein to build a perfect site to host its substrate. We have seen examples of how both hydrophobic interactions and hydrogen bonds are used to recognize a molecule and the different strategies used depending on the type of substrate. A gradient of electrostatic potential can be used to attract an anion into the active-site cavity, then the size of the cavity and the type of residues in it may be used to select the anion. This is what is happening when the substrate molecule is small and devoid of particular features, as in the case of the superoxide substrate of Cu,ZnSOD. A protein with larger substrates has the possibility of engineering a more complex site using the different types of noncovalent interactions: hydrophobic, hydrogen bonds, and electrostatic attraction between opposite charges, as in the cases of CPA and citrate synthase. Finally, the high degree of selectivity offered by the examples of PBP and SBP is indicative of how finely the protein–substrate interaction can be tuned. The Protein Data Bank[50] contains many structures of proteins with phosphate and sulfate anions bound. This demonstrates the abundance of anion binding sites in proteins, but the large majority of them are nonspecific and nonfunctional anionic sites. Only the need for functionality has evolved the specific binding observed in PBP and SBP.

3.7 REFERENCES

1. D. W. Christianson and W. N. Lipscomb, *Acc. Chem. Res.* **22**, 62 (1989).
2. D. W. Christianson, *Adv. Protein Chem.* **42**, 281 (1991).
3. S. Mangani, P. Carloni, and P. Orioli, *Coord. Chem. Rev.* **120**, 309 (1992).
4. D. C. Rees and W. N. Lipscomb, *Proc. Natl. Acad. USA* **78**, 5455 (1981).
5. D. C. Rees, R. Honzatko, and W. N. Lipscomb, *Proc. Natl. Acad. USA* **77**, 3288 (1980).
6. D. W. Christianson and W. N. Lipscomb, *Proc. Natl. Acad. USA* **82**, 6840 (1985).
7. D. W. Christianson and W. N. Lipscomb, *J. Am. Chem. Soc.* **110**, 5560 (1988).
8. D. W. Christianson, S. Mangani, G. Shoham, and W. N. Lipscomb, *J. Biol. Chem.* **264**, 12849 (1989).
9. A. Teplyakov, S. Mangani, P. Orioli, and K. Wilson, *Acta Cryst. D* **49**, 534 (1993).
10. D. E. Koshland, *Proc. Natl. Acad. Sci. U.S.A.* **44**, 98 (1958).
11. D. W. Christianson and W. N. Lipscomb, *J. Am. Chem. Soc.* **108**, 545 (1986).
12. D. W. Christianson and W. N. Lipscomb, *J. Am. Chem. Soc.* **109**, 5536 (1987).
13. S. Mangani and P. Orioli, *Inorg. Chem.* **31**, 365 (1992).
14. A. Galdes, D. S. Auld, and B. L. Vallee, *Biochemistry* **22**, 1888 (1983).
15. D. S. Auld, A. Galdes, K. F. Geoghean, B. Holmquist, R. A. Martinelli, and B. L. Vallee, *Proc. Natl. Acad. Sci. U.S.A.* **81**, 5041 (1984).
16. K. F. Geoghegan, A. Galdes, R. A. Martinelli, B. Holmquist, D. S. Auld, and B. L. Vallee, *Biochemistry* **22**, 2255 (1983).
17. K. F. Geoghegan, A. Galdes, G. Hanson, B. Holmquist, D. S. Auld, and B. L. Vallee, *Biochemistry* **25**, 4669 (1986).
18. A. Galdes, D. S. Auld, and B. L. Vallee, *Biochemistry* **25**, 646 (1986).
19. B. L. Kaiser and E. T. Kaiser, *Proc. Natl. Acad. Sci. U.S.A.* **64**, 36 (1969).
20. L. D. Byers and R. Wolfenden, *Biochemistry* **12**, 2070 (1973).
21. A. Fersht, *Enzyme Structure and Mechanism*, W. H. Freeman and Co., Reading and San Francisco, 1977, pp. 227–243.
22. R. Bicknell, A. Schaeffer, I. Bertini, C. Luchinat, B. L. Vallee, and D. S. Auld, *Biochemistry* **27**, 1050 (1988).
23. M. L. Connolly, *Science* **221**, 709 (1983).
24. J. A. Tainer, E. D. Getzoff, K. M. Beem, J. S. Richardson, and D. C. Richardson, *J. Mol. Biol.* **160**, 181 (1982).
25. J. A. Fee and C. Bull, *J. Biol. Chem.* **261**, 13000 (1986).
26. E. D. Getzoff, J. A. Tainer, M. M. Stempien, G. I. Bell, and R. A. Hallewell, *Proteins* **5**, 322 (1989).
27. I. Klapper, R. Hagstrom, R. Fine, K. Sharp, and B. Honig, *Proteins* **1**, 47 (1986).
28. E. D. Getzoff, D. E. Cabelli, C. L. Fisher, H. E. Parger, M. S. Viezzoli, L. Banci, and R. A. Hallewell, *Nature* **358**, 347 (1992).
29. J. Shen and J. A. McCammon, *Chem. Phys.* **158**, 191 (1991).
30. I. Bertini, A. Lepori, C. Luchinat, P. Turano, *Inorg. Chem.* **30**, 3363 (1991).
31. K. Djinovic-Carugo, F. Polticelli, A. Desideri, G. Rotilio, K. S. Wilson, and M. Bolognesi, *J. Mol. Biol.* **240**, 179 (1994).

32. S. Mangani, W. Rypniewski, M. Ferraroni, B. Bruni, P. Orioli, and K. S. Wilson, manuscript in preparation.
33. C. Djinovic-Carugo, A. Battistoni, M. T. Carri', F. Polticelli, A. Desideri, G. Rotilio, A. Coda, and M. Bolognesi, *FEBS Lett.* **349**, 93 (1994).
34. J. A. Tainer, E. D. Getzoff, J. S. Richardson, and D. C. Richardson, *Nature* **306**, 284 (1983).
35. W. Rypniewski, S. Mangani, B. Bruni, P. Orioli, M. Casati, and K. S. Wilson, *J. Mol. Biol.* **251**, 282 (1995).
36. R. Osman and H. Basch, *J. Am. Chem. Soc.* **106**, 5710 (1984).
37. F. A. Quiocho, *Annu. Rev. Biochem.* **55**, 287 (1986).
38. F. A. Quiocho, *Phil. Trans. R. Soc.* **B341**, 341 (1990).
39. H. Luecke and F. A. Quiocho, *Nature* **347**, 402 (1990).
40. J. J. He and F. A. Quiocho, *Science* **251**, 1497 (1991).
41. S. Mangani, M. Ferraroni, and P. Orioli, *Inorg. Chem.* **33**, 3421 (1994).
42. A. C. Williams and D. S. Auld, *Biochemistry* **25**, 94 (1986).
43. I. Bertini, A. Donaire, L. Messori, and J. Moratal, *Inorg. Chem.* **29**, 202 (1990).
44. A. B. Pardee, *J. Biol. Chem.* **241**, 5886 (1966).
45. J. W. Pflugrath and F. A. Quiocho, *Nature* **314**, 257 (1985).
46. J. W. Pflugrath and F. A. Quiocho, *J. Mol. Biol.* **200**, 163 (1988).
47. Z. F. Kanyo and D. W. Christianson, *J. Biol. Chem.* **266**, 4264 (1991).
48. R. S. Alexander, Z. F. Kanyo, L. E. Chirlian, and D. W. Christianson, *J. Am. Chem. Soc.* **112**, 933 (1990).
49. S. Remington, G. Wiegand, and R. Huber, *J. Mol. Biol.* **158**, 111 (1982).
50. F. C. Bernstein, T. F. Koetzle, G. J. B. Williams, E. F. Meyer, M. D. Brice, J. R. Rodgers, O. Kennard, T. Shomanouchi, and M. Tasumi, *J. Mol. Biol.* **112**, 535 (1977).

CHAPTER 4

Artificial Anion Hosts. Concepts for Structure and Guest Binding

F. P. SCHMIDTCHEN

4.1 Introduction
4.2 The Guests: Anions
4.3 The Concept of Artificial Anion Hosts
4.4 Positively Charged Anion Hosts
    4.4.1 Azonia Compounds
    4.4.2 Oligopyrrole-Derived Receptors
    4.4.3 Guanidinium-Based Receptors
    4.4.4 Miscellaneous Cationic Hosts for the Complexation of Anions
4.5 Electroneutral Hosts for Anions
    4.5.1 Hosting of Anions by Poly Lewis Acids
    4.5.2 Lewis Acidic Hosts Connected by Covalent Bonds
    4.5.3 Lewis Acidic Hosts Based on Metal Cation Coordination
    4.5.4 Anion Hosts Operating by Ion–Dipole Binding
4.6 References

4.1 INTRODUCTION

Anions, like any other molecular species, experience attractive forces with their environment irrespective of its chemical nature. Thus the association of anions with some other structurally defined entity in a now called host–guest relationship appears as a natural consequence of this basic interaction and is nothing special at all. On a quantitative basis, however, one observes great differences in the strength of this interaction, which poses the question for the origin of these nonuniform effects. In principle the measurable quantity of interaction of two binding partners under consideration, that is, the association constant K_S, reflects and comprises all direct mutual interactions as well as the changes in the

Supramolecular Chemistry of Anions, Edited by Antonio Bianchi, Kristin Bowman-James, and Enrique García-España.
ISBN 0-471-18622-8. © 1997 Wiley-VCH, Inc.

environment, for example, in the solvent. Both contributions are heavily dependent on the covalent structures of the partners, but it seems fair to state that the attempts to modify or redesign the direct interaction component are far more abundant.

In general all compounds capable of binding another molecular species with a somewhat higher affinity than what must be expected from their fundamental molecular properties are termed molecular hosts. Of course this is only an operational definition and in addition is even subject to the experimental conditions. As a corollary, a compound of proven utility as a host in one solvent may completely fail to complex the same guest species in a different, more competitive solvent. Other insightful definitions for host compounds have appeared in the literature[1–11] that relate the host–guest association event to multiple dedicated interaction modes. This view of what to call a host becomes increasingly problematic the more subtle and sophisticated our methods to pinpoint dedicated interactions are. Moreover, as today's usage of the term *host* is synonymous with *complexing agent*, we will use this expression here in the broadest sense.

Much of the attention currently being paid to the investigation of host–guest interactions of anions derives from the dominant role anions play in the biological living world. The chemistry of life to a large degree happens in an aqueous environment, which also is the privileged solvent to stabilize negatively charged species. Thus it appears only natural that metabolites contain anionic substructures if they are to be solubilized in or confined to an aqueous compartment. Import and export across the boundaries and all structural manipulations within them then require dedicated host molecules,[1–3] which, by virtue of a defined set of noncovalent interactions, influence basic physicochemical properties (e.g., transport across membranes) or even direct the chemical reactivity with any possible result for reaction rate and selectivity. These shining examples, only a few of which have been studied in sufficient detail to realize how noncovalent interactions can be used to shape the molecular properties of anions, have inspired and challenged molecular architects to look out for and design artificial hosts for anions.[4–10] The main goal is to achieve analogous qualitative performance as in the natural archetype by using similar interaction principles, however, on more stable, more easily accessible and less enigmatic parent structures.

Under this shell the discussion will be limited to hosts acting in the solution phase, excluding host–guest relations observed solely in the gas phase (e.g., in mass spectrometers) or in crystal packings. Also only host systems having a unique covalent structure will be covered, so that micellar, vesicular, and macroionic systems are not considered either.

4.2 THE GUESTS: ANIONS

The feature distinguishing anions from other guest species is their negative electrostatic charge. Therefore, this is the prime property to address if specific complexing agents for these species are to be constructed. Aside from the halides and alkalides, which are monatomic and have a spherical charge distribution, all others, and especially the biologically important members, have distinct covalent structures and are frequently complexes by themselves with concomitant stereoelectronic properties.[12] These form the basis for the distinction between, for example, sulfate and hydrogen phosphate,[13,14] which is vital to all biological processes involving these oxoanions (see Chapter 3).

Due to the polyatomic structure of most anions, their sizes are much larger compared to common metal cations.[15] Table 4.1 gives some values for ionic radii of more common anions, although these numbers vary considerably with the method of determination, probably owing to the nonspherical shape.[16]

According to one of nature's most fundamental laws, the Coulomb law, anions associate with cations or multipoles, if the latter are oriented appropriately. This phenomenon comes in several variants and has been termed *ion pairing*.[17] It is so universal and pronounced that one almost never observes singular anionic species except in strongly solvating solvents under high-dilution conditions. In fact, at present a popular effort is to design noncoordinating anions,[18,19] with the understanding that the tendency for coordination to

TABLE 4.1. Ionic Size and Experimental Enthalpies and Gibbs Free Energies of Hydration for Selected Ions[21]

	r (pm)	H_{hy} (kJ mol^{-1})	G_{hy} (kJ mol^{-1})
F$^-$	133	−510	−465
Cl$^-$	181	−367	−340
Br$^-$	196	−336	−315
I$^-$	220	−291	−275
HCOO$^-$	156	−432	−335
NO$_3^-$	179	−312	−300
H$_2$PO$_4^-$	200	−522	−465
ClO$_4^-$	250	−246	−430
CO$_3^{2-}$	178	−1397	−1315
SO$_3^{2-}$	200	−1376	−1295
SO$_4^{2-}$	230	−1035	−1080
PdCl$_6^{2-}$	319	−730	−695
PO$_4^{3-}$	238	−2879	−2765
Li$^+$	69	−531	−475
Na$^+$	102	−416	−365
K$^+$	138	−334	−295
Cs$^+$	170	−283	−250
NH$_4^+$	148	−329	−285
(C$_2$H$_5$)$_4$N$^+$	337	−73	0
Ca^{2+}	100	−1602	−505
Zn^{2+}	75	−2070	−1955
Al^{3+}	53	−4715	−4525
Fe^{3+}	65	−4462	−4265
La^{3+}	105	−3312	−3145
Th^{4+}	100	−6057	−5815

cations should depend both on anion structure and on solvation. Solvation of anions itself is dominated by electrostatic interactions with solvent molecules,[20,21] and huge amounts of energy are involved in these interactions, particularly in hydroxylic solvents, which can form multiple and strong hydrogen bonds to the anion. The thermodynamic parameters of hydration given in Table 4.1 indicate that this binding mechanism stabilizes anions to a greater extent than it does cations of equal charge density. Of course, hydrogen bonding is only one component of solvation, and other contributions (dispersion, solvophobic interactions, etc.) gain in weight the smaller and more delocalized the charge density of the anion is and the more polarizable the solvent molecules become. However, a good demonstration of how powerful and effective hydrogen bonding solvation of anions may be is provided by the observation that two anions may overcome Coulombic repulsion to form stable dimeric aquo complexes in water.[22]

Essential to hydrogen bonding stabilization of anions is their Lewis-base character.[23] The presence of lone electron pairs serving as H-bond acceptor sites is a matter of course in anions apart from a few exotic exceptions (AlH_4^-, BPh_4^-, $closo$-$B_{12}H_{12}^{2-}$, etc.). Their Lewis basicity, however, varies within broad limits. Nevertheless, Lewis basicity is a common feature of anions and may be used as a basic interaction type in the construction of anion hosts.

4.3 THE CONCEPT OF ARTIFICIAL ANION HOSTS

We have emphasized already that anion binding is easy to achieve in principle. For a compound to qualify as a molecular host, however, it is mandatory to show discrimination between various guests offered for association. The selectivity observed is thought to emerge from a more or less structured molecular interaction of the binding partners, and should be amenable to characterization and optimization by prudent engineering of the host structure. With this ultimate goal in mind a few fundamental points of host–guest relationships must be considered:

- Which guests are to be selected by binding to the molecular host?
- In which solvent is the host–guest complexation to take place?
- Does the complexation serve a particular purpose?

In answering the first question, the mutual recognition pattern of host and guest must be defined with the aim of maximizing discrimination of similar guest species. This addresses the structural peculiarities and integral molecular properties including all types, numbers, and topologies of functional groups to be bound by the host. Clearly, this issue is at the heart of host–guest interactions and thus has attracted the major attention. It is also the point most easily investigated by computer modeling[24,25] in simple docking experiments, although one cannot overemphasize the necessity to proceed beyond this stage if any meaningful comparisons to experimental observables are planned.

A necessary extension, for instance, is the incorporation of solvent in the calculations, and this meets the second point in the list. The solvent shells around the binding partners will unavoidably be altered on complexation, and this can either be a costly process in terms of the net free energy or restructuring the solvent may rather favor complex formation. In any case, the solvent will affect complex stability even to the extent that the association constant may vary by orders of magnitude depending on the shear size of solvent molecules.[26] One can easily foresee that consideration of the change in the solvation characteristics will pay off greatly in rational host design, although solid calorimetric data on the contributions of individual substructures to the overall solvation energy are still scarce at present.

The analysis of the third question seems trivial at first sight. The answer, however, requires accounting whether host–guest binding is part of a functional process, that is, a vectorial guest transport across a membrane, some information transfer in a signaling chain, or even a catalytic reaction. In all these processes the overall selectivity is related to a rate, which is a kinetic phenomenon. The simple description of selectivity as a ratio of binding constants breaks down in these cases, as is true for the great majority of biological processes, which typically operate far from thermodynamic equilibrium. A big difference in binding free energy of two competing guests at the ground-state level is not necessarily mirrored by the observed selectivity in the overall process. This is most obvious in host systems mimicking enzymes, where strong substrate binding is detrimental to catalysis.[27] Rather the intrinsic interaction energy of host and guest must be utilized to lower every kinetic barrier along the functional trajectory including the binding event itself. One may translate this into the claim that utmost selectivity for one particular guest will be achieved by a host compound capable of organizing all binding groups in optimal complementarity to the transitory changing structural determinants of the guest. In view of the paucity of reliable experimental information on the dynamics of hosts and guests, it is not surprising that deliberate engineering of their time-dependent complementarity has not been undertaken. Hints from molecular modeling[24] and experimental observations in enzyme–substrate complexes[28] suggest, however, an important role for correlated motions, which remain to be verified in designed artificial hosts.

Irrespective of the type, kinetic or thermodynamic selectivity depends on the difference in the total interaction between a host and an ensemble of potential guests. This energy difference can only be large if the absolute values of the interaction energy are high, that is, if rather stable complexes are formed. One can intuitively feel that the total interaction energy will increase the greater the number of independently recognizable epitopes. Based on this argument, one would predict that binding constants will be higher the larger and more diversely structured the guest is and the better its entire size, geometry, and topological pattern of functional groups can be scanned by the host. Binding strength in addition is also influenced by the conceptual design of the host.

There is good reason to believe that a rigid host with all its anchor groups preorganized for guest binding should show the strongest binding.[11]

Especially for larger guest species this notion meets with exceeding difficulties in synthesis, because a set of anchor groups would have to be placed in predetermined topology and orientation in space. The convergence of "sticky" binding sites toward a binding center requires construction of macropolycyclic frameworks which are inherently difficult to build or even modify, if the first design proves not optimal (Fig. 4.1). Moreover, rigid molecular skeletons bear the risk of slow guest exchange kinetics, which is very undesirable if some application is on the list of wishes. An alternative strategy for host design places various anchor groups on a linear chain molecule and leaves it to a peculiar folding process to arrange the binding functions in the appropriate layout in space. Given the same type and number of anchor groups, this latter approach will display inferior selectivity towards competing guests, because part of the intrinsic interaction energy is utilized to organize the host into a defined three-dimensional structure and is no longer available for guest discrimination. However, in addition to the close analogy to biological host–guest systems (proteins, nucleic acids, polysaccharides are all linear copolymers acquiring a distinct tertiary structure by informed folding), foldable artificial hosts hold the virtues of comparably easy synthesizability and ready modifiability (shortening or extension of the chain, alteration of the sequence of anchor groups, etc.) and do not suffer from slow guest exchange rates. The analysis of host–guest interactions—in particular the correlation of binding data and complex structures—is not as obvious as with completely rigid host systems, where results from X-ray crystallography can be transferred to the solution phase with high confidence. Nevertheless, the practical advantages including the

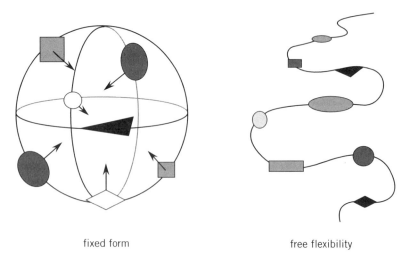

Fig. 4.1. Opposing concepts of host design: fixed form versus free flexibility.

straightforward adaptation to experimental limitations (solubility, aggregation behavior, etc.) make the concept of foldable hosts a valuable and useful option.

The quality of host–guest binding is generally measured by regression calculation of an association constant K_S applying the primary data to a suitably chosen binding model. Experience tells us that literature holds a strong bias towards the 1:1 stoichiometric model. Neglecting to test alternative models or to choose correct concentration relations in combination with the human fascination of hitting high numbers may easily yield K_S values that are in error by more than an order of magnitude, even though every care was taken to obtain correct experimental data.[29] This is particularly problematic when integral physicochemical parameters are measured (cf., conductivity,[30] reaction enthalpy[31] or pH value[32]) that are not observably related to the complex structure (as, for example, NMR; see **6.7.2**). Thus a fair measure of caution or even scepticism must accompany the interpretation of K_S values. The most significant and reliable binding data beyond doubt arise from trend analyses obtained on ensembles of guests with systematically varied structures.

4.4 POSITIVELY CHARGED ANION HOSTS

4.4.1 Azonia Compounds

Cationic hosts capable of forming ion pairs with anions in solution are most easily prepared by protonation of suitable basic compounds. Since many anions possess some basic properties as well, host–guest binding in these cases depends on relative proton affinities in interconnected multiple equilibria. In water as a solvent protonation equilibria are readily established, and the corresponding pK_a values of individual groups may be determined from titration curves. It is no surprise, therefore, that water was the solvent of choice to study anion binding to a great variety of protonated polyaza compounds. Due to its high dielectric permittivity of $\epsilon = 78.46$ and high hydrogen bond donor–acceptor abilities, electrostatic ion pairing is hampered to the extent that significant association at moderate concentrations ($\lesssim 0.1\,M$) can only be observed with multiply charged species. On the other hand, the mutual interaction rapidly increases with charge size so that great thermodynamic stability can be attained on complexation of highly charged ions. Of course, the state of protonation depends on pH—and as a general rule several different protonated species make up the ensemble of hosts at any given value. As a corollary the host–guest association observed in the actual experiment is an integral event that may be divided by regression calculation into singular contributions from the species involved. In this case the respective association constants may contain relatively large errors. Moreover, it is not straightforward to derive complex structures from the analysis of trends in the binding constants (see **6.7.2**). Relating the binding effect to host structure is essential for selectivity development, which constitutes the basis of molecular design.

In this sense it was like a quantum jump when Park and Simmons[33] observed a very peculiar ion pairing process: When the bicyclic ammonium salts **4.1** were mixed with the heavy halide salts, qualitative NMR experiments were compatible with the view that the anion penetrates into the central cavity of the host, forming a noncovalent encapsulation complex in which the topology of host and guest were well defined. This view was also supported by a correlation of cavity and guest sizes and by the finding that complexation required passage of a respectable activation barrier as would be plausibly expected for a molecular penetration process. Much later anion encapsulation into the interior of host **4.1** was proven by an X-ray crystal structure.[34]

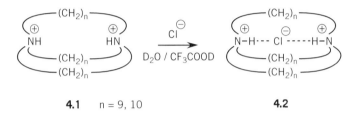

4.1 n = 9, 10   **4.2**

This serendipitous finding proved seminal for the development of host–guest chemistry with anions. Fostered by the interest in azacrown ethers and cryptands as cation complexones, the synthesis of this class of compounds, which, after protonation, could serve as anion hosts, was pushed to a high standard.[35,36] The general scheme followed the original work of Stetter,[37] who condensed an open-chain α,ω-diamine with an α,ω-dicarboxylic acid chloride under high dilution conditions to obtain the macrocyclic bisamide in moderate to reasonable yields (Scheme 4.1). Reduction with LiAlH$_4$ or better with BH$_3$*THF afforded the secondary amines, which were either used directly or elaborated further by alkylation or acylation.

Scheme 4.1

In contrast to open-chain polyamines like spermine or spermidine, which are known[38] to bind to phosphate anions or polyanions in water at neutral pH, but most likely adopt a flexible extended conformation, polyprotonated azacrown ethers possess a greater charge density and thus a greater predisposition to anion binding. A logical extension of this idea would incorporate as many ammonium sites as synthetically feasible in close proximity to each other in order to maximize the electrostatic attraction for the negatively charged guest. The pK_a values for the two most acidic protonation steps in the series **4.3a–c** show that very acid pH values are necessary to protonate di- and tricationic ammonium salts if the separation of the cationic centers is less than a propane spacer unit. For the complexation of biologically relevant anions under physiological pH conditions, the compression of host volume by using ethylene spacers is not a viable route. Instead, the accumulation of charge has been pursued by enlargement of the macrocycle[40,41] or the attachment of amino-functionalized side chains to smaller azacrown ethers.[42] Typical examples are the macrocycles **4.4**, **4.5**, and **4.6**, which were prepared by stepwise alkylation, reduction, and finally ring closure of suitably substituted tosylamides.[39]

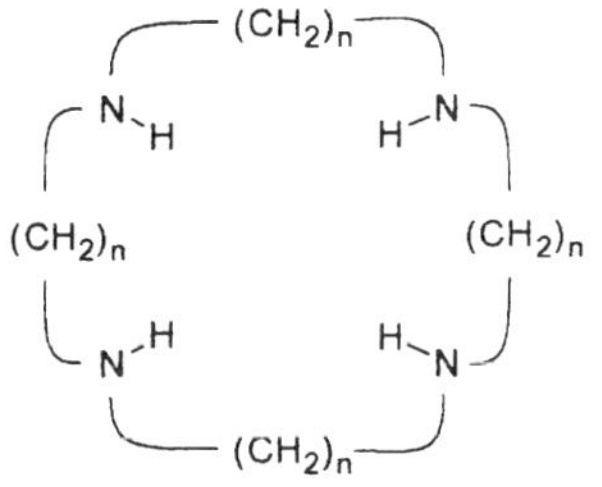

	n	pK_{a3}	pK_{a4}
4.3a	2	1.7	< 1
4.3b	3	6.9	5.4
4.3c	4	10.6	8.9

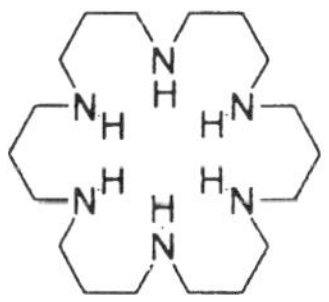

4.4 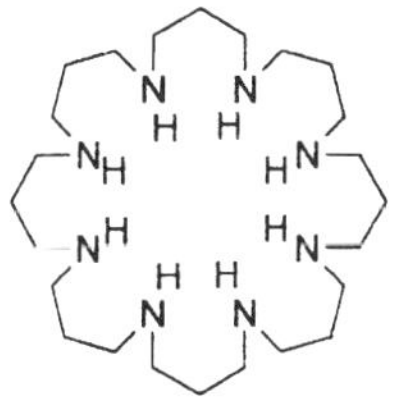   **4.5** 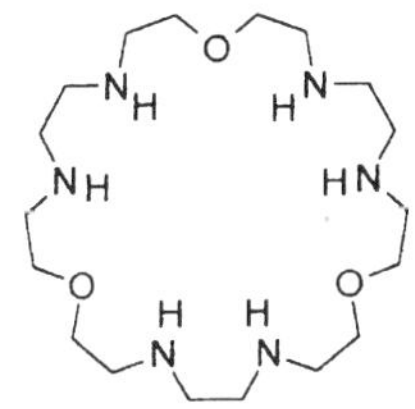   **4.6**

Table 4.2 contains binding data that characterize these compounds as strong complexones with a variety of anions in water. From the general trend that complex stability increases with guest charge, one can infer the dominance of coulombic interactions. This is also supported by only a moderate dependence of stabilities on structure. For instance, the greatest discrimination between any two guest anions of the same charge amounts to a factor of 60. The biggest preference of an anion for a particular host equals a factor of only 20 (AMP binding **4.4** or **4.6**). Although cases with substantially higher selectivities (up to a factor of 1000) have been reported,[43] the observation of mediocre guest discrimination is widespread in this class. This may originate from a great variety of host–guest binding modes, all having very similar energies and from conformational flexibility, which allows the host to adapt its structure easily to

TABLE 4.2. Selected Stability Constants Log K_S ($\pm$ 0.2) for Anion Binding by Polyammonium Macrocycles 4.4–4.6 in Water [0.1 M $(CH_3)_4N^+Cl^-$], as Determined by pH-Metric Titration and Regression Calculation[40]

	4.4·6H$^+$	4.5·8H$^+$	4.6·6H$^+$
oxalate^{2-}	3.8	3.7	4.7
sulfate^{2-}	4.0	4.0	4.5
fumarate^{2-}	2.2	2.9	2.6
squarate^{2-}	3.2	3.6	3.4
citrate^{3-}	4.7	7.6	5.8
1,3,5-benzene tricarboxylate^{3-}	3.5	6.1	3.8
Co(CN)$_6^{3-}$	3.9	6.0	3.3
adenosine monophosphate^{2-} (AMP)	3.4	4.1	4.7
ADP^{3-}	6.5	7.5	7.7
ATP^{4-}	8.9	8.5	9.1

the topological needs of the guest. That symmetry relations play a minor role is also suggested by the complexation of ATP to the ornithine-derived macrocycles **4.7** and **4.8**.[44] In spite of the C_3 symmetry of H bonding sites in **4.7**, which have a complementary topology to the terminal phosphate group of ATP (**4.7a**), this favorable setup is not reflected by enhanced binding. ^{31}P-NMR reveals a 1:1 stoichiometric interaction of the γ-phosphate group with either host, but **4.7** appears to form the more stable complex.

Thorough investigation of the binding features of hexacyclen **4.9** as a

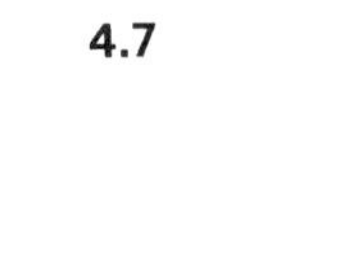

4.7

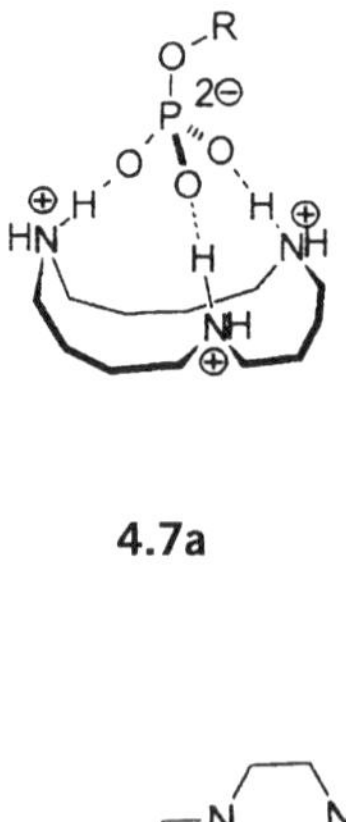

4.7a

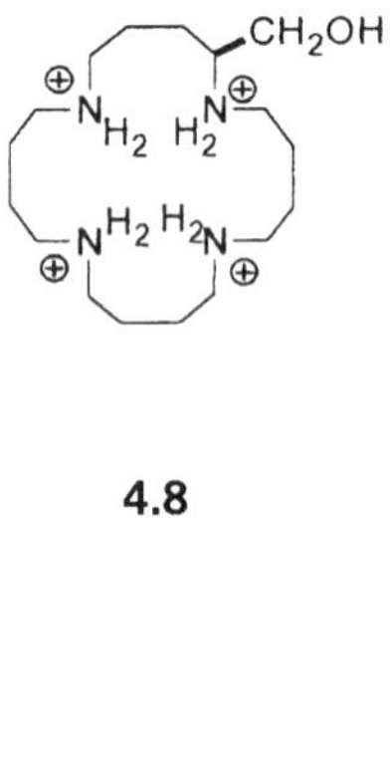

4.8

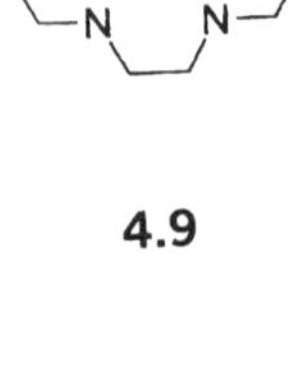

4.9

tetraprotonated cation offered a clue to the weak influence of structural factors: van't Hoff plots allowed for the Gibbs free energy of association with chloride, bromide, and a number of oxoanions to be split into its enthalpy and entropy contributions. Gelb, Zomba et al.[45,46] found the amazing result that association is essentially entropy driven in water. Undoubtedly, there is a strong enthalpic interaction of the charged host–guest binding partners. But this serves to release strongly bound solvent molecules from either binding partner, so that the enthalpic effects almost cancel. The gain in entropy resulting from the generation of freely movable solvent molecules outbalances the inherently negative entropy of host–guest association with the effect that complex stability increases with rising temperature. Enthalpy–entropy compensations are quite common in host–guest associations in strongly solvating solvents[47,48] and testify to solvent reorganization being of main importance rather than direct mutual interactions of host and guest. Nevertheless, although we currently do not have an intimate understanding of the interdependence of structure and complex stability, subtle features are apparent and significant and may be rationalized when a trend analysis of a receptor series is available. When complex stability in water in an ensemble of nitrogen macrocycles with some transition metal complex anions is probed, one observes a monotonous decrease up to a certain ring size.[49] Enlarging the ring further yields a sudden stability enhancement, which, in combination with supportive evidence from photochemical measurements and some X-ray crystal structures,[50] can be interpreted as a switch to a novel type of complex: In the interaction with $[Pt(CN)_4]^{2-}$ taken as an example (Fig. 4.2)[51] the anion may approach the faces of the highly charged macrocycles. If the macrocycle is of insufficient size, the guest will at best reach a perching position due to steric repulsion featuring nonoptimal contacts to the positive centers. Beyond a certain threshold size, which evidently

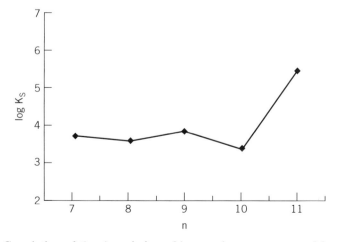

Fig. 4.2. Correlation of the size relation of host and guest as reported by the $\log K_S$ value in a series of polyammonium macrocycles complexing tetracyanoplatinate.

depends on guest dimensions, the anion can slip into the center of the macrocyclic ring and enjoy maximal stabilization mirrored by an increase in complex stability.

In addition to this encapsulation, the presence and topology of H-bonding donor sites are of crucial importance. Of course, this influence will most readily be seen with basic H-bond acceptor guests. But even $Co(CN)_6^{3-}$, a notoriously poor H-bond acceptor, shows no complexation at all in solution or in the crystal when all the nitrogen atoms in $([12]aneN_4)^{4+}$ (**4.10**, $n = 4$) are methylated to form the completely quaternized cation $[Me_8[12]aneN_4]^{4+}$.[52]

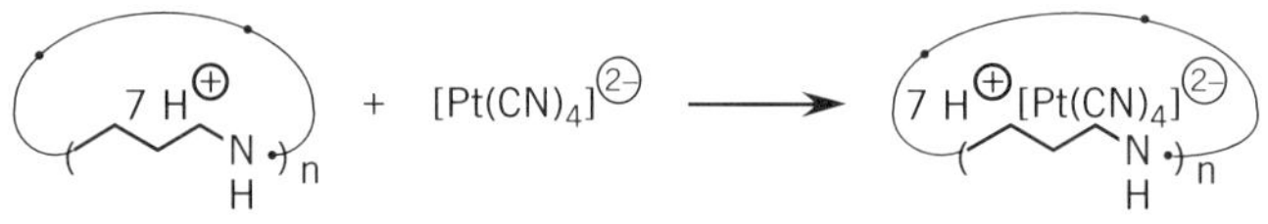

4.10 **4.11**

Another striking example for the obvious participation of a subtle array of H bonds is provided by the hydrolysis and transphosphorylation catalyzed by the oxoazacrown ether **4.12**.[53] This host, much more than closely related analogs, cleaves phosphoric anhydrides forming a covalent phosphoryl amidate intermediate in which the phosphoryl group is carried by the central nitrogen in one hemisphere of the macrocycle, **4.13**.[54,55] Although compared to the respective enzymes the rate acceleration over spontaneous hydrolysis is moderate at best ($\sim 10^2$; ATPase reaches a 10^{10} acceleration factor), and may be indicative of

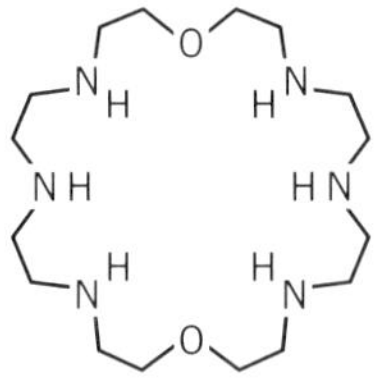

4.12

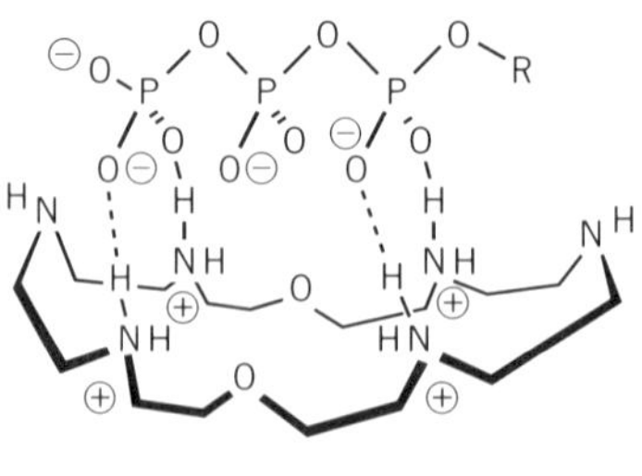

4.13 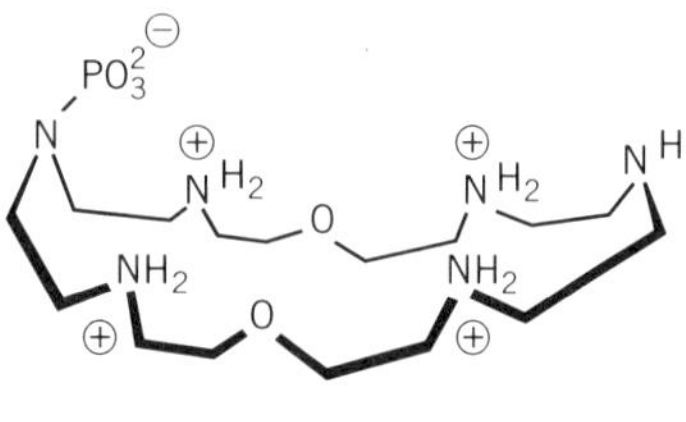 **4.14**

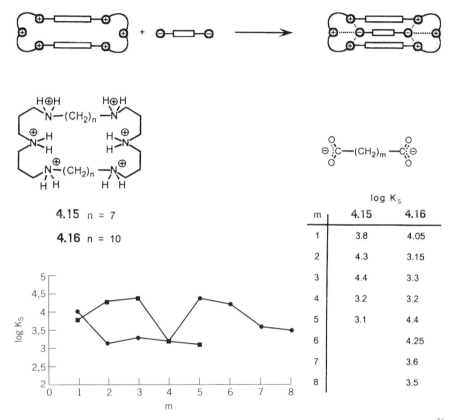

Fig. 4.3. Dimensional matching of polyammonium macrocycles and α-dicarboxylates.[56]

multiple nonproductive substrate binding modes, this simple model shows various aspects of true enzymes like turnover, saturation, inhibition, and a reasonably defined mechanism. It is at least plausible that all these properties are based on a common binding motif **4.14** mediated by a peculiar network of hydrogen bonds. (See Chapter 10 on applications for a more detailed treatment.)

Dimensional matching is held responsible for the size selectivity observed in the monocyclic hexamines **4.15** and **4.16** (Fig. 4.3).[56] When a series of dicarboxylates of incrementally increasing chain length m is complexed by these hosts in neutral aqueous solution in which they exist as hexaprotonated cations, the most stable complexes are formed with guests of intermediate length ($m = 3$ or 5). This selectivity is much less pronounced with open-chain polyamines, though even quaternary ammonium salts like **4.17** exhibit this type of dimensional selectivity to some extent.[57] Presumably, the simple macrocycles set up a size restriction, so that the carboxylic moieties can interact with both triammonium substructures only if their spacing is correct. Linear hosts cannot exert such a barrier due to their greater flexibility.

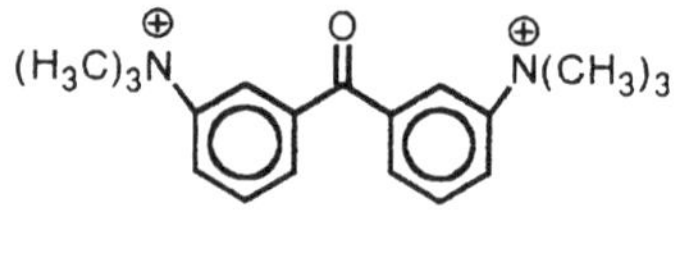

4.17

Some insight into complex structures may also be obtained by cautious interpretation of physical effects that are structure dependent. Electrochemical methods like cyclic voltammetry or polarography had been successfully applied to elucidate the stoichiometry and some association constants of polyammonium salts with cyano complexes of transition metals[58] or even oxoanions like carbonate[59,60] or phosphate[61] but could not disclose structural details. This, however, became possible when the photohydration of $Co(CN)_6^{3-}$ in the presence of macrocycles **4.18** and **4.19** was studied. The reduction in quantum yield reflected the barrier for escape of CN^- from the metal due to interaction with the macrocyclic ligand. A correlation with the number of cyano sites in direct contact with the ligand and the mode of association was thus derived.[62]

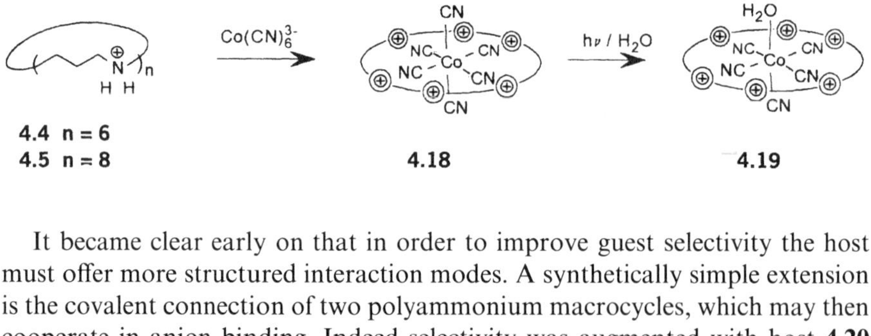

4.4 n = 6
4.5 n = 8

4.18

4.19

It became clear early on that in order to improve guest selectivity the host must offer more structured interaction modes. A synthetically simple extension is the covalent connection of two polyammonium macrocycles, which may then cooperate in anion binding. Indeed selectivity was augmented with host **4.20** binding phosphates or citrate, although the effect amounts to a factor 2 only when compared to the parent monocyclic host **4.21** of the same charge.[63]

A much greater improvement in stability and selectivity of anion complexation was achieved by rigidification of the binding sites. Elaborating on the conceptual plan of Park and Simmons,[33] the Lehn group arrived at the bicyclic cryptands **4.22–4.24**, which, in their penta- or hexaprotonated forms, eagerly complex a variety of well-solvated anions in aqueous solution (Table 4.3).[64,65]

The analysis of the K_S values, such as the comparsion of oxalate to malonate, indicated that strong complexation must be due to an encapsulation process, in which the guest anion invades the molecular cavity and is held there by an oriented set of H bonds. This was confirmed by several X-ray crystal structures[64] showing that the ellipsoidal cavity and topology of nitrogen sites

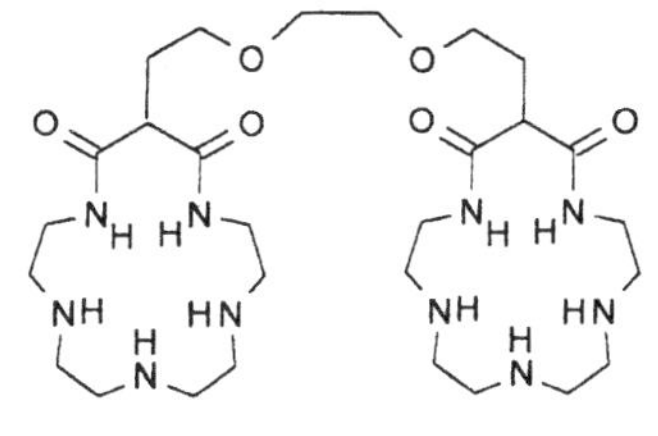

POSITIVELY CHARGED ANION HOSTS   **93**

4.20 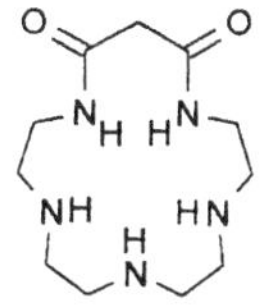   **4.21** 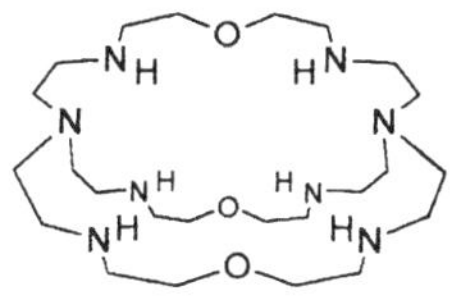   **4.22**

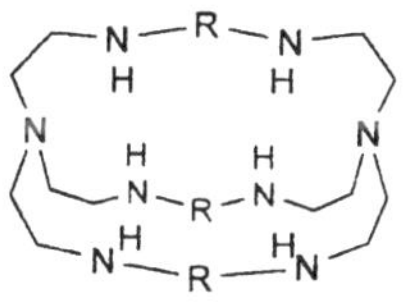

	R
4.23	-CH$_2$-CH$_2$-
4.24	-CH$_2$-CH$_2$-CH$_2$-

provide optimal complementarity to azide anion, resulting in extraordinarily high complex stability. The halides can interact less well with the extended cavity, and the monotonous decrease of K_S from fluoride to iodide testifies to the importance of H bonding as the main attractive binding force.

The putatively minute change in structure from **4.22** to **4.24**, which does not alter the host dimensionality but may influence the cavity shape and certainly the solvation, leads to the reverse order in halide complex stability. At the same time the absolute binding affinity is diminished with any anion except iodide.

Comparing monocyclic and bicyclic hosts, one can conclude that the sensitivity of host–guest binding to structural variation with respect to selectivity as well as to overall binding strength is more visible and pronounced the better the host structure is defined. Pushing this concept even further, a macrotricyclic aza crown ether **4.25** was constructed containing the aesthetically

TABLE 4.3. Host–Guest Association Constants (log K_S) of Anions with Bicyclic Polyammonium Salts as Determined by pH Titration in Water at 25°C (0.1 M NaOTos)[64,66]

Anion	22·6H$^+$	24·6H$^+$
F$^-$	4.10	
Cl$^-$	3.00	1.70
Br$^-$	2.60	2.20
I$^-$	2.15	2.40
N$_3^-$	4.30	
SO$_4^{2-}$	4.90	4.20
oxalate^{2-}	4.95	4.50
malonate^{2-}	3.10	2.85
AMP^{2-}	3.85	
ATP^{4-}	8.00	

4.25

appealing arrangement of 4 nitrogens in a tetrahedral topology imposed on an octahedron of 6 oxygen sites. Originally, this molecule was constructed as an alkali metal complexone,[67] but it was soon discovered that **4.25** when protonated forms very stable anion complexes as well.[68] By virtue of an array of 4 H bonds converging onto the center of the molecular cavity expanded by the electrostatic repulsion of the positive charges, a chloride ion can be encapsulated as shown by the X-ray structure.[69] Fluoride and bromide, but not iodide or any of the polyatomic anions, may complex in a similar way. Again as pointed out above, a favorable topology of binding interactions, which is enforced by a rigid and almost undistortable molecular framework, guarantees both high association constants and unprecedented selectivity. The discrimination of bromide versus chloride exceeds a factor of 10^3, and nitrate, having similar size and H-bonding abilities to Cl$^-$, but even a smaller hydration free energy, is totally excluded. The apparent lack of shape complementarity for this anion might explain this result. The disadvantage of well defined rigid hosts becomes apparent from molecular mechanics and dynamics calculations[70]: If chloride or bromide approach **4.25**·4H$^+$ from infinity, they first form an external complex at one face of the tetrahedral host. Penetration into the cavity then requires a substantial barrier (31.8 or 49.8 kJ mol^{-1} for Cl$^-$ and Br$^-$, respectively) to be overcome, reflecting mainly the Pauli repulsion of the electron clouds. In the gas phase both anions experience greater stabilization in the encapsulated configuration, but bromide only to a marginal extent, so that on these grounds inclusion complexation of bromide is less likely and has in fact not unambiguously been proven. Aqueous solution MD calculations for Cl$^-$ encapsulation by **4.25**·4H$^+$ suggest that desolvation of the anion and the deformation of the host are the main factors to slow down guest binding.[71]

A major limitation in the use of protonated polyaza hosts for anion binding is their confinement to a somewhat restricted acid pH region. Moreover, a change in solvent can cause partial deprotonation and thus impair anion complexation. Since a good share of the total interaction energy is contributed by H bonding in these systems, it was not at all self-evident that peralkylation of all nitrogen sites to form quaternary ammonium salts would still give useful anion hosts. The macrotricyclic quaternary ammonium salts **4.26** and **4.27**,[72,73]

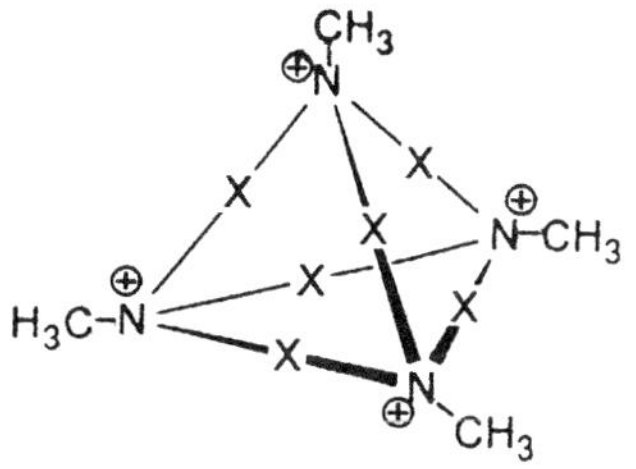

	X
4.26	-(CH$_2$)$_6$-
4.27	-(CH$_2$)$_8$-

prepared by methylation of the parent tertiary amines obtained by the strategy depicted in Scheme 4.2, however, were shown to form complexes of strict 1:1 stoichiometry in water with a variety of anions.[74] NMR data as well as a crystalline complex of **4.26** with iodide[75] revealed inclusion complexation of the guests. With CPK models the cavities were estimated as spheres having diameters of 4.6 Å (**4.26**) and 7.6 Å (**4.27**). Thus, iodide $[d(I^-) = 4.5 \text{ Å}]$ fits very snugly into the smaller tetrahedral host, which is unable to complex larger organic anions like *p*-nitrophenolate. These anions, in contrast, bind to the bigger host **4.27**, providing more evidence for an encapsulation process. Having a chemically quite inert host with pH-independent binding power at hand, the behavior of reacting systems under the influence of host–guest binding was studied.

The main possibilities are shown in Fig. 4.4 and comprise two alternatives: In monomolecular reactions the anionic substrate may form a complex with the host and thereby experience a change in the molecular environment that may translate into a change in transformation rate. As the ground state and transition state of a reaction will be affected by complexation to different degrees, one may expect catalyses as well as reaction inhibitions to be possible. Bimolecular reactions likewise can show inhibition, if the second reaction partner cannot reach the first substrate, which is included in the cavity and shielded against attack. However, if the cavity is large enough to accommodate both substrates, one may observe accelerations because of the simple entropic effect resulting from the collection of the partners in a smaller volume than they had in bulk solution. This concentration effect that is responsible for the rate enhancement seen by micelles as well[76] can be complemented by enthalpic stabilization of the transition state, if the host interacts more favorably with it than with the ground state. A number of reactions running through highly delocalized anionic transition states (i.e., nucleophilic aliphatic and aromatic substitutions[77]) are catalyzed by host **4.27** with rate acceleration factors reaching 1700 in some cases. The smaller host **4.26** invariably inhibits these reactions,

Scheme 4.2

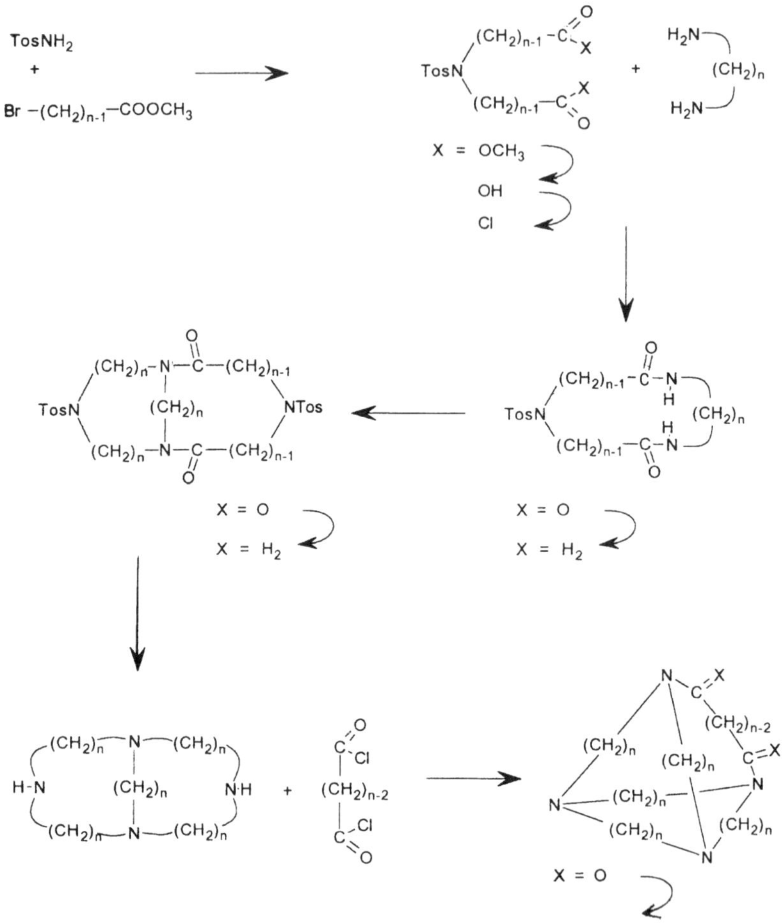

providing a solid argument that the catalyses happen inside the cavity of **4.27**.[78] The kinetic analysis suggested a rapid equilibrium random order process, the rate-determining step being the conversion of a ternary complex of the two substrates and the host into the corresponding product complex.[79] The monomolecular decarboxylation of 6-nitrobenzisoxalole-3-carboxylate is a clean reaction that is easy to follow by UV spectroscopy and is catalyzed by a variety of artificial and protein hosts including catalytic antibodies.[80] The quaternary ammonium host **4.27** not only accelerates the rate of decarboxylation by 100-fold, matching β-cyclodextrin in this respect, but also shows positive cooperativity kinetics.[81] Cooperativity was traced back to 1:2 host–guest complex that formed with a higher K_{ass} than the complex with 1:1 stoichiometry. The catalysis produced by host **4.27** mimics true enzymes in many respects. Substrate and chemoselectivity, but, of course, no stereoselectivity, have been observed along

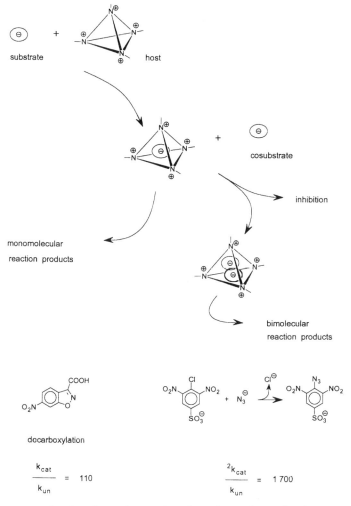

Fig. 4.4. Examples for alternative modes of catalyses shown by the quaternary cage receptors **4.26** and **4.27**.

with saturation kinetics, inhibition, cooperativity, and turnover. Since **4.27** contains no catalytically active functionality whatsoever, all rate effects must arise from the change in solvation on complexation (i.e., a microsolvent effect). Elaboration of the terahedral cations **4.26**, **4.27** into ditopic hosts for amino acid zwitterions[82] or biogenic ammonium salts (**4.28a** or **4.28b**, respectively)[83] furnished some insight on the advantage of assembling various anchor groups into an open-chain receptor molecule. More quantitative data were obtained using the ditopic host **4.29** complexing a series of dianionic dimensional probes (Fig. 4.5).[84] The extra interaction possible with **4.29** in comparsion with the monotopic analog **4.27** translates into a factor of 3 in the binding constant.

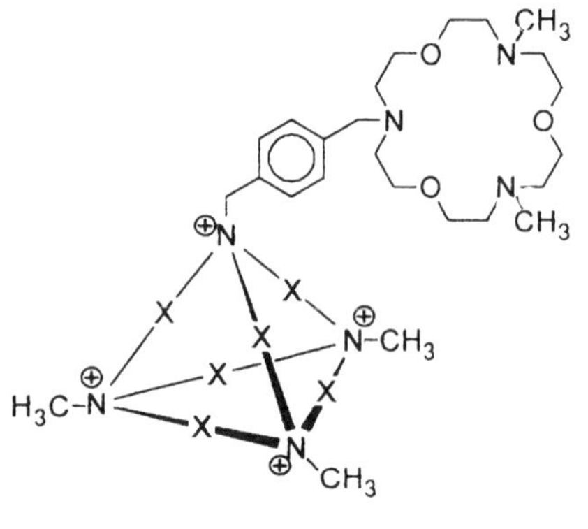

4.28   X = -(CH$_2$)$_6$-

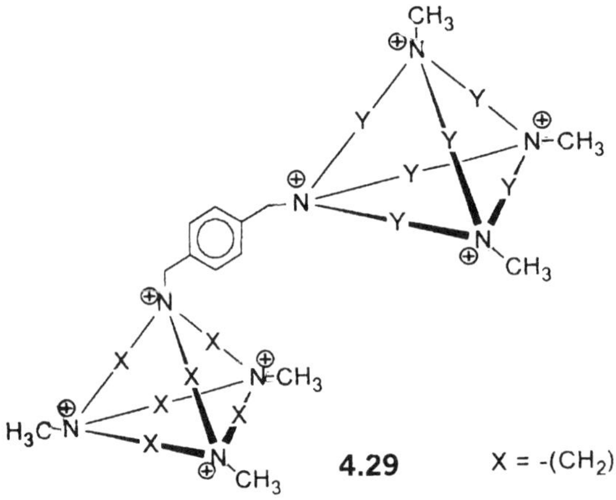

4.29   X = -(CH$_2$)$_6$-
Y = -(CH$_2$)$_8$-

The solvation of anions is dominated by electrostatics, but not to the extent that other factors like van der Waals interactions, steric influences, or solvophobic forces can be neglected. This is particularly of importance if large and subtly structured anions that might accommodate the negative charge on a localized moiety within the global structure are to be bound to a host in a well-structured solvent like water. The host structures of choice are cationic cyclophanes, since Koga[85] had discovered that the macrocycle **4.30** forms well-defined inclusion complexes with some aromatic guest compounds in aqueous solution and Tabushi[86] had demonstrated catalysis of ester hydrolysis by the quaternary cyclophane **4.31**. Many variants of these parent structures have appeared in the literature, and the field has been extensively reviewed.[87] The common binding principle in water appears to rest on a superposition of the hydrophobic effect and coulombic interactions.[88] Thus very stable host–guest complexes were obtained with substrates consisting of an aromatic ring system attached to some anionic moiety like in a nucleotide.[89] Guests having the negatively charged substructure connected to an aliphatic residue form much weaker complexes.[90] Since uncharged analogs of the guests (e.g., nucleosides) are also bound with considerable strength, the charge interaction is more like an adjuvant modulating the more fundamental hydrophobic binding. Selectivity

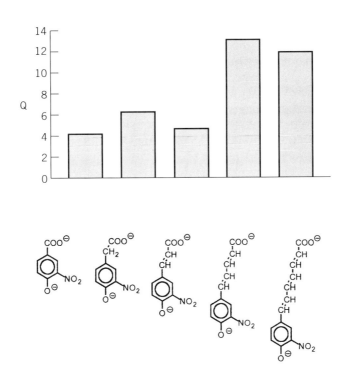

Fig. 4.5. Host–guest complexation of **4.27** and **4.29** to an ensemble of dimensional guests. Binding to the ditopic receptor **4.29** is augmented threefold if the guest can span the distance between the anchor groups.

and complex stability therefore are dictated preferentially by the shape complementarity or steric fit of the hydrophobic moieties. Frequently, an anionic function has been introduced into the guest solely to warrant sufficient solubility of the compound.

Because biologically relevant anions often contain a combination of hydrophobic and charged moieties, the design of hosts satisfying both requirements is an attractive goal. It is well known that planar electron-deficient aromatics like the nucleotide bases stack in the face-to-face mode to positively charged extended aromatic ring systems. When incorporated into a cyclophane framework, this principle can yield very potent hosts for binding nucleotides in water.

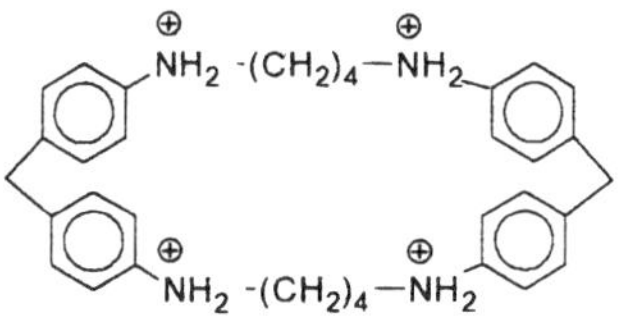

4.30

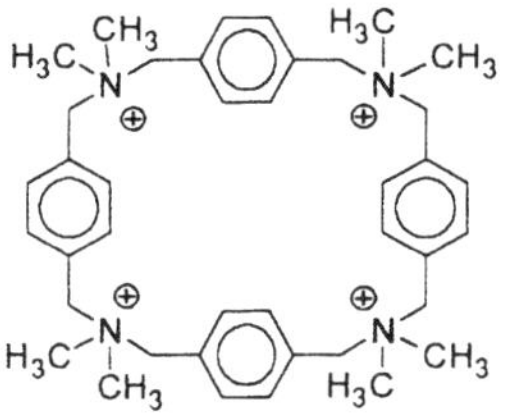

4.31

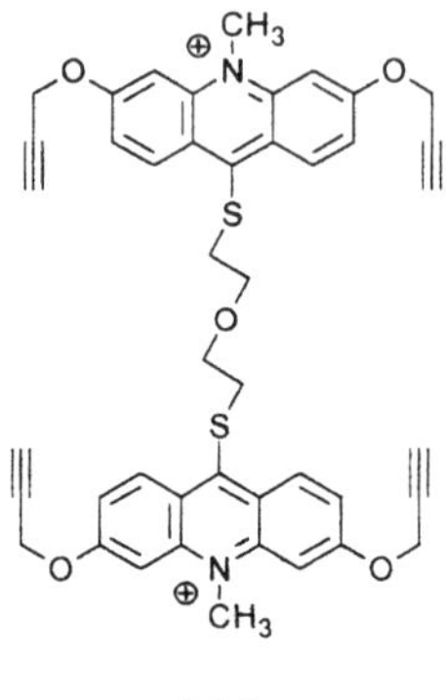

4.32

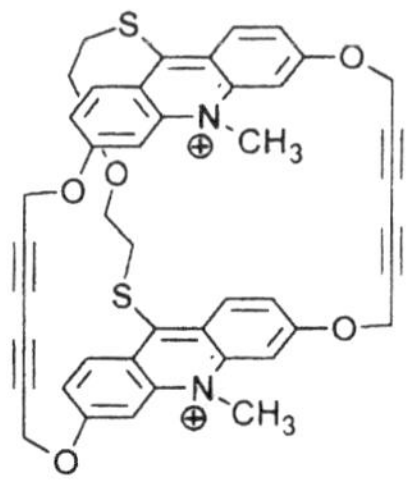

4.33

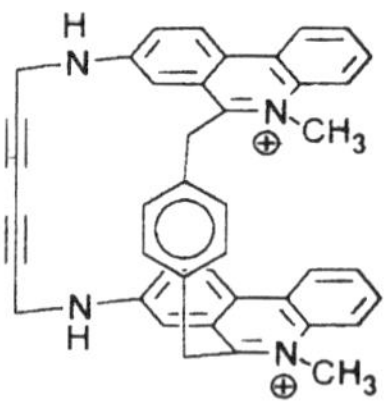

4.34

The acridinium cyclophane **4.33**, for instance, obtained by Glaser coupling from the open-chain precursor **4.32**[91] binds a great variety of planar aromatic carboxylates and nucleotides with log K_S values ranging from 4 to 7.[92] There is good evidence from UV, stoichiometry, and X-ray data that true inclusion complexes are formed in which the guest is held by stacking interactions in the cavity. The exceedingly rigid architecture of **4.33** becomes apparent in an amazing observation: Though the virtue of organizing two acridinium units to allow formation of a sandwich structure is reflected by a >100-fold increase in K_S compared to a monoacridinium salt, the open-chain compound **4.32** invariably is the best host in this series. This amazing result giving the less organized host **4.32** an edge over the cyclized and more rigid counterpart **4.33** points to a misfit of the cavity and the guest. The spacing of the flat walls in the host seems a little too far to make optimal contacts with the guest, and rigidity prohibits the necessary collapse that is possible with the more flexible host **4.32**. This series has been elaborated[94] to include phenanthridinium hosts **4.34**, which form the most stable nucleotide complexes known at present.[93] Binding constants, however, are almost unaffected by the size of the anionic charge, indicating the prime importance of hydrophobic/stacking interactions.

Open-chain analogs (*seco*-cyclophanes) have also been prepared in a rich variety[95] but in general form weaker complexes with aromatic polycarboxylic anions than their cyclic counterparts. The bridging of 1,3,5-substituted benzene by chains containing amino functionalities yields the polycyclic molecules **4.35**–**4.37**, which can be solubilized in water by protonation.[96] NMR-titration experiments and pH-metric determinations confirmed clean 1:1 complex formation of **4.35** with numerous small inorganic anions like NO_3^-; Cl^-, and SO_4^{2-} with log K_S values ranging from 2.5 (monovalent) to 6.0 (divalent anions). The large shifts of NMR signals obtained with nitrate as a guest, the slow host–guest exchange kinetics observed, and other NMR data leave little doubt that true inclusion complexation occurs in solution. Surprisingly, this could not be confirmed by X-ray crystal structures. In the solid state the salts all have the anions hydrogen bonded at the outside of the molecular cages, suggesting a molecular rearrangement happening on dissolution.

4.35   **4.36**   **4.37**

The elaboration of the monocyclic cyclophanes **4.30** and **4.31** into polycycles yields the aesthetically appealing cubic azaparacyclophanes **4.38** and **4.39**.[97,98] They dissolve in water at pH 4 to form ammonium tetracations, which can complex a variety of anionic fluorescent probes like ANS **4.40**. The study of host–guest binding with these hosts is severely hampered by very broad NMR signals originating from slow conformational interconversions. However, judged from the guest selectivities obtained in comparsion with analogous hosts having partially opened cages, one must conclude that the guest penetrates into the molecular cavities of **4.38** and **4.39**. The major driving force is the hydrophobic interaction, since omission of the anionic charge in the fluorescent

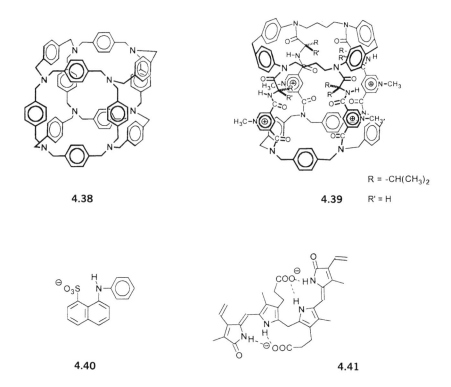

4.38   4.39   R = -CH(CH₃)₂   R' = H

4.40   4.41

probe **4.40** (log K_S = 5.2) diminishes binding by a factor of 2 only, whereas deletion of the anilino moiety in **4.40** leads to a drop in K_S by 3 orders of magnitude (log K_S = 2.0). The chirality of the amino acid–derived spacers in **4.39b** induces a helical twist in the entire molecule. As was shown by CD measurements, this helicity serves to bind an enantiomeric form of the anionic bilirubin guest **4.41** with preference over the other. Bilirubin is known to exist as a rapidly interconverting mixture of helical conformers in solution.

4.4.2 Oligopyrrole-Derived Receptors

Metal complexes with chelating ligands that leave the metal center coordinatively unsaturated may further associate anionic species in a kinetically labile manner. Very frequently one observes binding preferences that reflect an intrinsic selectivity. Particularly prominent examples are provided by polypyrrol complexes like metalloporphyrins and corrins that have been studied for anion-selective sensing and transport.[99] Due to the dedicated single-point mode of interaction between the binding partners, these systems hardly fit into our definition of a host–guest relationship. But in a few examples the interactions have been purposely extended beyond the unique first-sphere ligation of the cationic metal center: Kuroda et al. have supplemented a rhodium(III) porphyrin with two quaternary ammonium moieties in **4.42**.[100] In addition to

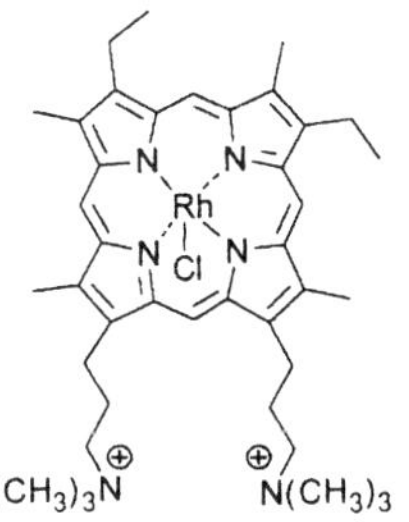

4.42

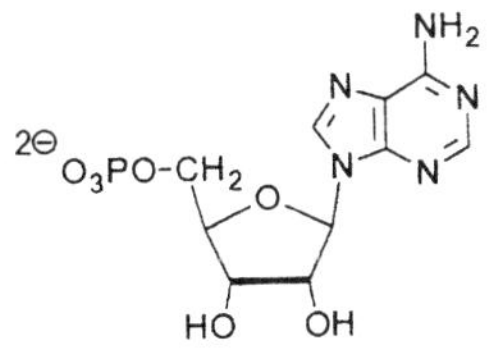

4.43

conveying water solubility, the high cationic charge prevents these flat molecules from dimerization by stacking adhesion in water. This sets the stage to complex adenine nucleotides. The nucleotide heterocyclic base is believed to coordinate to the metal, leaving the phosphate group free for coulombic interaction with the ammonium cations. The major contribution to the overall K_S value, however, is delivered by metal ligation (estimated to -13.4 kJ/mol), whereas the contribution of coulombic attraction in the case of AMP^{2-} **4.43** amounts to -3.3 kJ/mol only.

In contrast to the metal complexes, the decomplexed porphyrin ligands do not bind to anions significantly.[101,102] This is presumably due to the small size of the porphyrin cavity, which does not allow chelation even to the simplest anions. The rational remedy was the expansion of the porphyrin cavity by incorporation of more pyrrolic or other spacer moieties. The chemistry of ring-extended porphyrins[102–104] brought up polypyrrolic systems, among which the sapphyrins were shown to exhibit anion receptor properties. Sapphyrins **4.46** consist of a planar pentapyrrolic skeleton with aromatic character in which three N–H bonds converge to the center of a cavity of ca. 5.5 Å diameter. Owing to the basicity of the macrocycle ($pK_{a1} = 3.5$; $pK_{a2} = 9.5$), this array can be supplemented by two additional N–H bonds to give a stiff collar of H-bond donor sites that is perfectly preorganized for anion encapsulation. The sapphyrins were characterized for the first time nearly 30 years ago[105] and require considerable synthetic effort for preparation. The Sessler group elaborated a convergent synthetic scheme from **4.44** and **4.45**, and greatly improved the

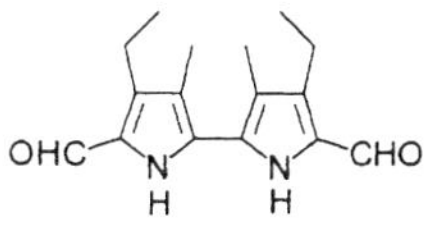

4.44

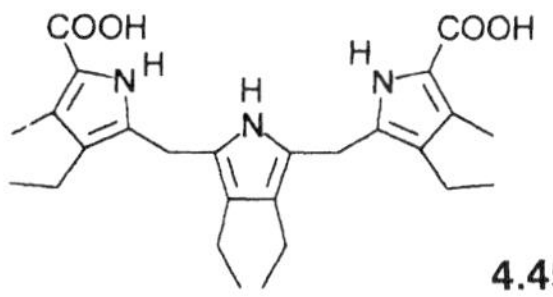

4.45

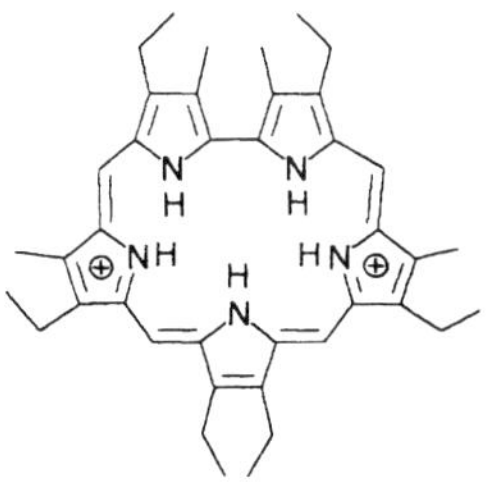

4.46

accessibility of the intermediates.[106] For them it was thus gratifying to discover by serendipity that diprotonated sapphyrin **4.46** formed a very stable ($K_S = 1 \times 10^5$ M^{-1}) complex with fluoride even in methanol solution.[107]

The X-ray crystal structures (see also Chapter 5) show the fluoride anion completely encircled by the aromatic dicationic macrocycle. Obviously, this is an energetically very favorable arrangement, since the heavier halides, chloride and bromide, which cannot form these inclusion-type complexes for steric reasons, are discriminated against by more than 1000-fold. Though all the oxoanions are too big for encapsulation as well, they may form chelation-type complexes. As evidenced by various X-ray structures,[102a] phosphate esters, for instance, bind with one oxygen atom of the anionic phosphoryl moiety in a perching position over the center of the macrocycle. In this way hydrogen bonding to all N–H donor sites is possible, enabling reasonably stable complexes with these guests in noncompetitive solvents. This can be exploited, for example, for transporting nucleotides from one aqueous solution into another separated by a CH$_2$Cl$_2$ liquid membrane employing sapphyrins as carrier molecules.[108] Other promising applications include nucleotide chromatographic separations on stationary phases containing sapphyrin in a covalent linkage to the solid support[109] or the binding to single-stranded or double-stranded DNA.[110] Further elaboration of the sapphyrin concept was undertaken either towards multitopic anion recognition as in the case of the simple covalently joined dimer **4.47** or by variation of the macrocyclic cavity itself. The dimer **4.47**

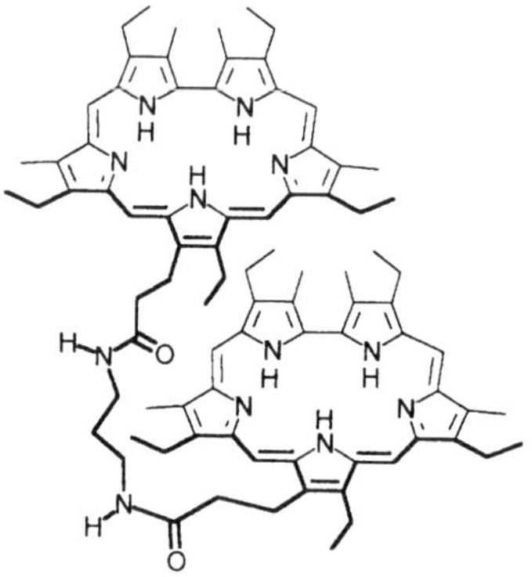

4.47

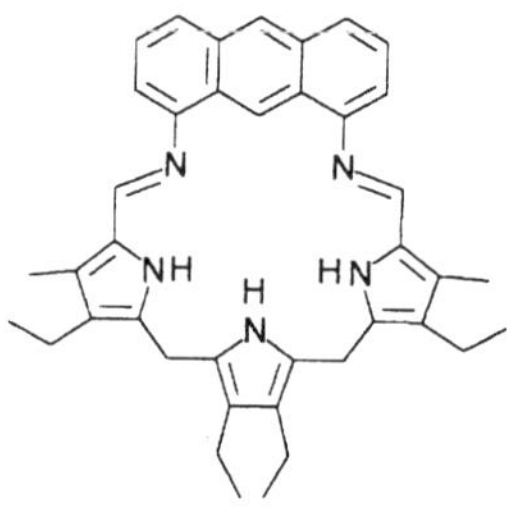

4.48

proved well suited for the recognition of dicarboxylic anions,[111] whereas the incorporation of nonpyrrolic aromatic spacers like in the anthraphyrin **4.48** leads to a widened cavity that when diprotonated now shows stronger binding of chloride over fluoride in CH$_2$Cl$_2$. Nevertheless, anthraphyrin **4.48** may serve as an excellent carrier for fluoride, outmatching even sapphyrin by a factor of six.[112]

A different approach to use porphyrins for anion binding was followed by Sanders et al.[113] The covalent connection of three porphyrin macrocycles

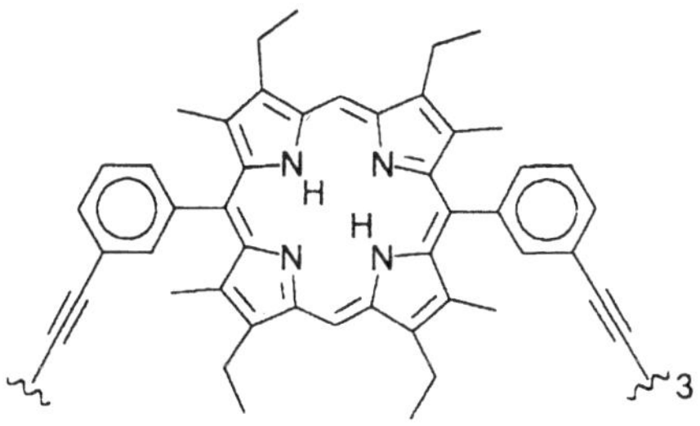

4.49

produced the giant cage molecule **4.49** that can be converted by acid into the hexaprotonated cation $[H_6 \cdot \mathbf{4.49}]^{6+}$. FAB mass spectra of 3-nitrobenzylalcohol solutions of this cation also containing the cluster anions $[PW_{12}O_{40}]^{3-}$, $[SiW_{12}O_{40}]^{4-}$, or $[Os_{10}C(CO)_{24}]^{2-}$ showed signals of 1:1 complexes in addition to the peaks derived from the free host. As no complexes with a variety of small inorganic anions that should be at least as volatile as those with the big clusters could be detected, one can conclude that the special complementarity in size, charge, and shape of $[H_6 \cdot \mathbf{4.49}]^{6+}$ and the cluster anions causes their noncovalent association in solution. Apparently, this system still holds the record in size for guest encapsulations.

4.4.3 Guanidinium-Based Receptors

The abundant involvement of arginine in the binding of anionic substrates to proteins[114,115] fostered the suspicion early on that interaction of a guanidinium ion with common oxoanions must hold special virtues. Much later it was concluded from site-directed mutagenesis experiments affecting the active sites of certain enzymes that in the protein environment the energetic stabilization of a carboxylate by the guanidinium side chain of arginine outmatches the analogous interaction with the primary ε-ammonium group of lysine by as much as -21 kJ/mol.[116] The reason for this tremendous difference apparently originates in the peculiar binding pattern (Fig. 4.6) featuring two strong parallel

Fig. 4.6. Schematic binding patterns of the guanidinium group with oxoanions as observed in many X-ray crystal structures.

hydrogen bonds superimposed on an attractive coulombic interaction. This is also the recurring motif seen in the X-ray crystal structures of enzyme complexes with oxoanionic substrates as well as in simple guanidinium salts.[115,117]

The guanidinium moiety is one of the most hydrophilic functional groups known.[118] Solvation by water is so efficient that despite the favorable binding pattern, ion pairing with carboxylates and phosphates in aqueous solution is negligible ($K_S < 5$ M^{-1}).[119] Bridging by water molecules may even allow the electrostatic repulsion to be overcome and lead to face-to-face dimerization of two guanidinium cations.[120] In spite of these solvation properties, which aggravate any attempt to make use of guanidines in artificial receptors, the attractive features of this group, combined with the secure knowledge of its successful participation in natural host–guest binding, lured quite a number of researchers to design abiotic guanidinium host compounds. Along with the unique binding pattern that ensures a predictable host–guest alignment, the main benefits offered by the guanidinium group are its pH-independent binding power and amenability to structural elaboration. The extreme basicity of guanidine (pK_a 13.5), which is conserved or even enhanced by prudent substitution,[121] guarantees a fixed protonation state and opens the entire range of accessible pH values for study. On the other hand, the well-developed chemistry of introducing, protecting, or modifying the guanidino group greatly eases its incorporation into designed receptors.

In an effort to compare anion binding abilities to analogous azacrown ethers, the macrocyclic guanidinium salts **4.50** and **4.51** were produced.[122] The association with PO$_4^{3-}$ as determined by pH-metric titration in water was disappointingly weak [log K_S = 2.2 (**4.50**) and 2.5 (**4.51**)], leading to the conclusion that electrostatic interactions dominate anion binding with these hosts. A more extensive and thorough study[123] comprising about a dozen different polyguanidinium salts further supported this view. Owing to the lower charge density, guanidinium hosts show inferior binding of carboxylate or phosphate

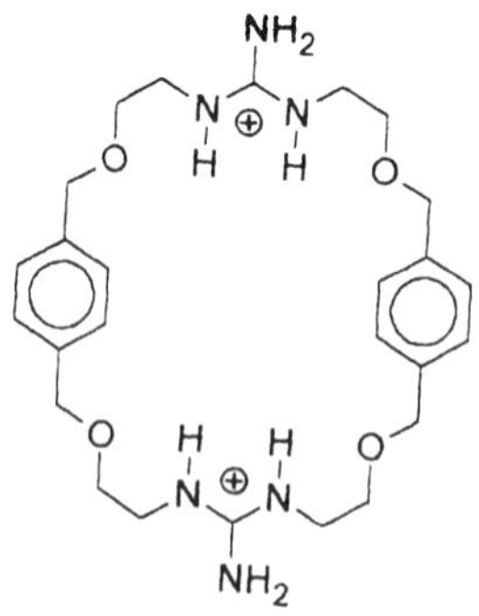

4.50

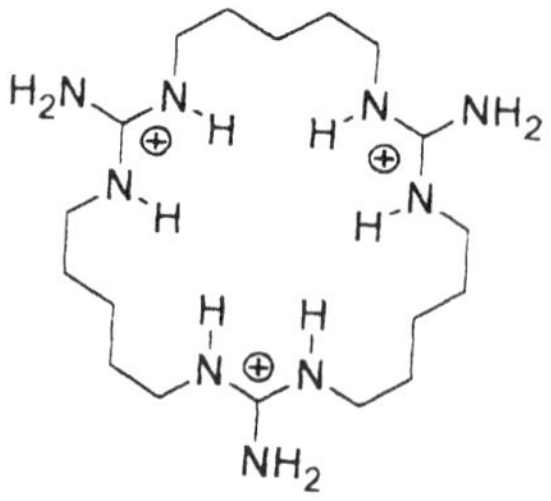

4.51

POSITIVELY CHARGED ANION HOSTS

anions than the corresponding ammonium analogs of the same basic skeleton and charge. However, subtle structural effects were noted.

Increasing interest in mimicking the enzymatic cleavage of phosphoric ester[124] initiated the design of artificial receptors that would bind to a phosphate monoanion. At the beginning some primitive *bis*-guanidinium salts like **4.52** were tested and found to bind and accelerate transesterifications of activated phosphodiesters up to a factor of 5000.[125] With increasing structural sophistication of the host (e.g., **4.53**,[126] **4.54**[127]), phosphate binding could stand up to more competitive aqueous solvation conditions, but showed complex binding equilibria of higher host–guest stoichiometries when probed in organic solvents.[126] With respect to catalytic efficiency in phosphate ester hydrolysis these simple *bis*-guanidinium salts cannot match the power of metalloenzyme mimics.[128]

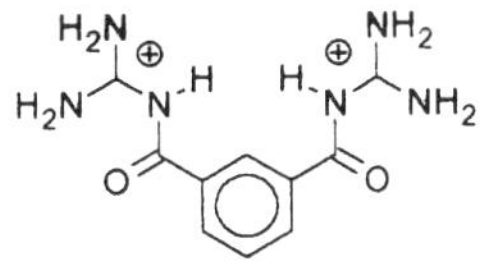

4.52

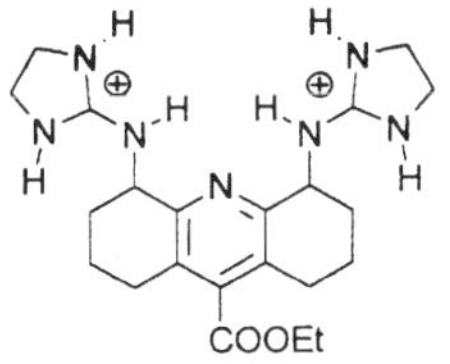

4.53

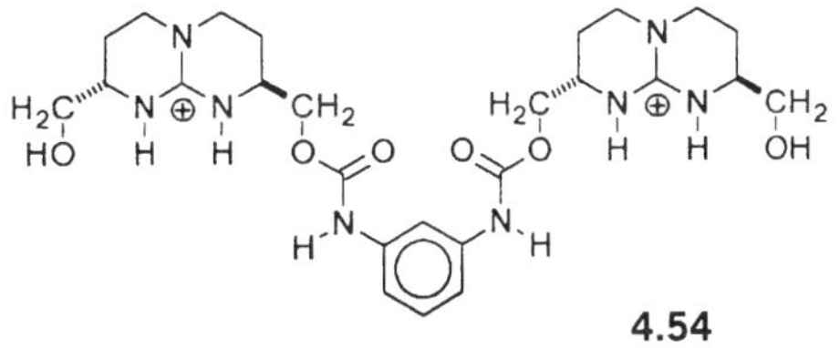

4.54

In order to circumvent the ambiguity of binding modes possible with **4.52** or **4.53**, the guanidinium function may be incorporated into a bicyclic framework. This design holds several virtues over the noncyclic analogs. As depicted in Fig. 4.7, the host to oxoanion binding in the characteristic and energetically favorable pattern can happen in only one mode, which, as a corollary, leads to a well-defined orientation and distance relationship in the complex. If there is any guest binding detectable, one knows precisely the position of the guest relative to the host structure. In addition, the insertion of the guanidine moiety into an almost strainfree bicycle renders the host chemically even more stable and basic than the parent guanidine. The accumulation of hydrophobic hydrocarbon residues is expected to hamper hydration of the charged moiety and lower the

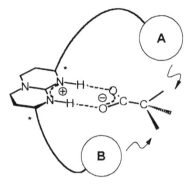

Fig. 4.7. Host–guest binding pattern of bicyclic chiral guanidines with oxoanions.

dielectric constant. Both influences would be beneficial to oxoanion binding. The favorable aspects of bicyclic guanidinium anchor groups were recognized nearly 2 decades ago when symmetrically tetrasubstituted derivatives became available.[129] Extensive use of these host structures, however, was feasible only with the first synthesis of a functionalized chiral analog (Fig. 4.8).[130] Subsequently, the preparation was supplemented by another more reliable route.[131]

The easy modification by attachment of more binding groups opened access to work towards enantioselective recognition of suitable substrates employing entirely attractive interactions with the guest species rather than relying on repulsive barriers for enantiodiscrimination, which inherently oppose the binding event.

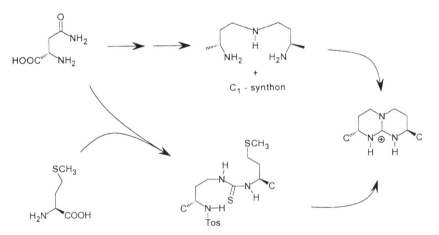

Fig. 4.8. Schematic routes for the synthesis of chiral bicyclic guanidinium anchor groups. The target compounds can be obtained by cyclization of an open-chain triamine in combination with a C_1-synthon[130,132] or by a 4-step–one-pot reaction of the unsymmetrical thiourea.[131]

Even without any additional binding complement, the bicyclic guanidinium moiety deserves classification as a host compound. The tetrasubstituted derivative **4.55a** forms an ion pair with *p*-nitrobenzoate of extraordinary stability in chloroform ($K_S = 1.4 \times 10^5$ M^{-1}). The acetate salt of **4.55b** in the X-ray crystal structure showed the anticipated host–guest binding pattern (cf. Fig. 4.7), which was part of an even greater hydrogen bonding network.[133] The chiral analog **4.56a** formed diastereomeric complexes with a number of chiral aromatic carboxylates in CHCl$_3$.[134] A two-point attractive interaction comprising the guanidinium–carboxylate salt bridge and π stacking of the aromatic moiety of the guest to the naphthalene residues of this host was considered responsible for enantio recognition. This rationale also explained the moderate preference (17% de) for extraction of N-acetyl or BOC-L-tryptophane from aqueous solutions containing the racemic derivative into CHCl$_3$ using **4.56a**. However, even the parent anchor group **4.56b** offers a dissymmetric binding site that allows detection of diastereomeric host–guest association with aliphatic carboxylic salts like D,L-N-acetylalanine or D,L-2-methylbutyrate in acetonitrile.[135] The bulky silylether substituents apparently suffice to shape a chiral environment that can be sensed even by lean guest species.

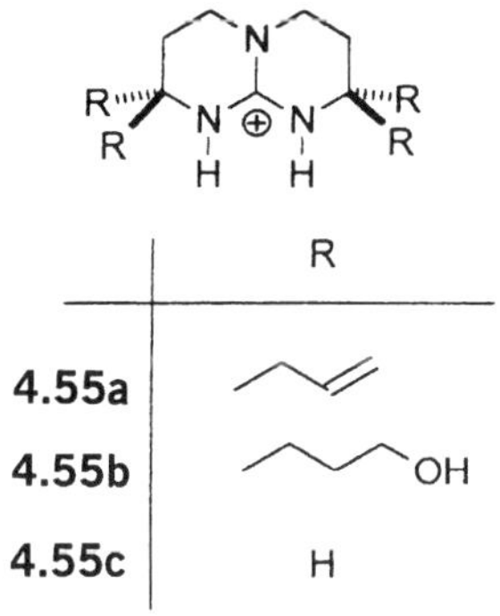

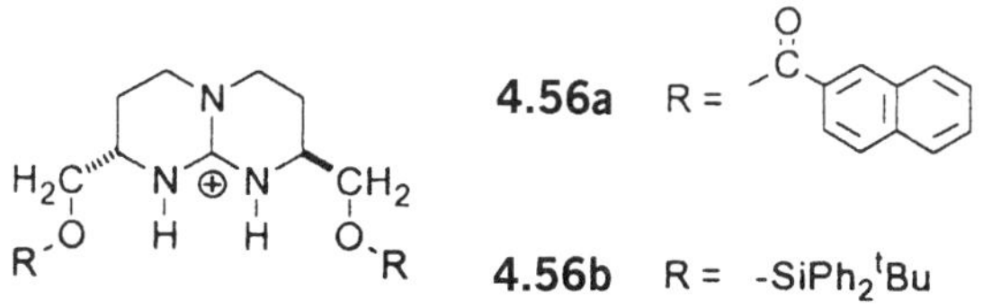

The idea that suitable anionic transition states might be complexed and thereby stabilized by **4.56b** and related bicyclic guanidines led to their testing in Michael reactions.[136] The addition of nitroethane to methylvinylketone is efficiently catalyzed by **4.55c**, and an intermediate structure of this reaction type, the nitronate bound by the bicyclic guanidinium host, has been crystallized and explored by X-ray structure determination.[136,137] In another vein, de Mendoza observed catalysis in the addition of pyrrolidine to furanone in

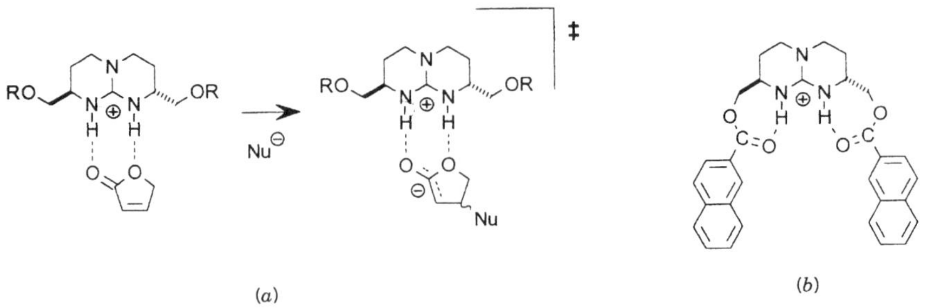

110 ARTIFICIAL ANION HOSTS. CONCEPTS FOR STRUCTURE AND GUEST BINDING

Fig. 4.9. (a) Concept of the catalysis of the Michael addition to an unsaturated lactone. The developing anionic charge is stabilized by the guanidinium host, resulting in a smaller activation barrier for addition. (b) Presumed molecular conformation of **4.56a** in CHCl$_3$.

chloroform (Fig. 4.9).[138] Only **4.56b** but not the ester **4.56a** displayed weak rate acceleration factors up to 8.4. The ester **4.56a** probably adopts a conformation in which a pair of intramolecular hydrogen bonds prohibits a host–guest interaction with the lactone substrate. The transfer of chirality from the optically active guanidinium catalyst to the product, while in principle possible, was not observed.

The proven utility of bicyclic guanidines as anchors for oxoanionic species triggered the search for more selective analogs. With this aim several alternative syntheses in particular for benzoannellated bicyclic guanidine derivatives were elaborated,[139] which among others furnished compound **4.57**. The conjugation of the nitrogen sites into the aromatic moieties renders this compound much less basic than ordinary guanidines. But in slightly acidic solution this rigid host when incorporated into a liquid membrane acts as an electrochemical sensor for hydrosulfite with impressive selectivity, sensitivity, and detection range.[140]

Another approach towards more specific anion receptors used the parent bicyclic guanidine **4.56b** as the fundamental building block. Elaboration into polytopic hosts that recognize more than one structural epitope of the guest by specific interactions requires progressive attachment of more anchor groups.

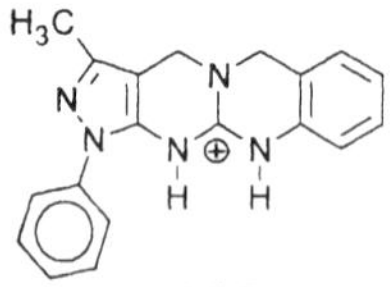

4.57

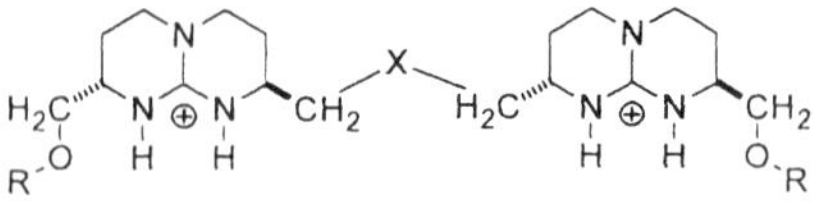

	X	
4.58	-SiPh$_2$^{t}Bu	
4.59	H	
4.60	H	

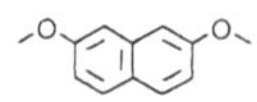

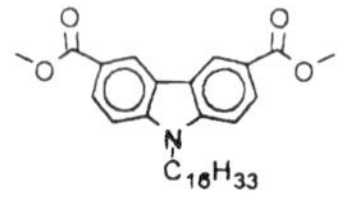

The initial step in this direction is the connection of two guanidinium units by a linear spacer module to give a flexible and foldable ditopic host. This basic design has been realized in the construction of compounds **4.58–4.60**,[141,142] which by virtue of their chirality and the planarity of the spacer may fold on binding an oxoanionic guest to arrange the main planes of the bicyclic framework perpendicular to each other. In combination with the N–H bond donor sites converging towards the binding center, this process produces an optimal array for interaction with tetrahedral oxoanions (Fig. 4.10). Thus it does not come as a surprise that these hosts bind phosphate and its biologically important ester/anhydride derivatives (ATP, NAD, other mono- and dinucleotides) with considerable strength in water ($\log K_S = 2$–3).

In an attempt to elucidate some principal binding properties of **4.58**, the K_S values with a series of dicarboxylic guests of increasing dimensionality and directionality were determined.[143] Though a highly flexible host, **4.58** exhibited a marked preference for binding malonate over its shorter- or longer-chain analogs. Geometric isomers, however, can hardly be distinguished by **4.58**. Probing the influence of host flexibility on binding with an ensemble **4.61–4.64** of mannitol-derived spacer units did not show the expected response. Incremental rigidification by freezing internal rotations gave varying association constants rather than yielding a monotonous increase in binding of a squarate guest. Obviously, spacer flexibility is not a limiting factor in guest binding.[144]

The transport of hydrophilic phosphates across diffusion barriers like membranes presents an ambitious target as an application of a selective host. Towards this end the linear *bis*-guanidinium compounds **4.65–4.67** were prepared, all containing the same primary anchor groups, spacer unit, and hydrophobic tail to render them soluble in an organic phase. They differ, however, in their terminal substituent (hydroxy versus silylether protecting group) and building block connection (ester versus amide). As illustrated in Fig. 4.11, these structural variations greatly influence their extraction properties.[145]

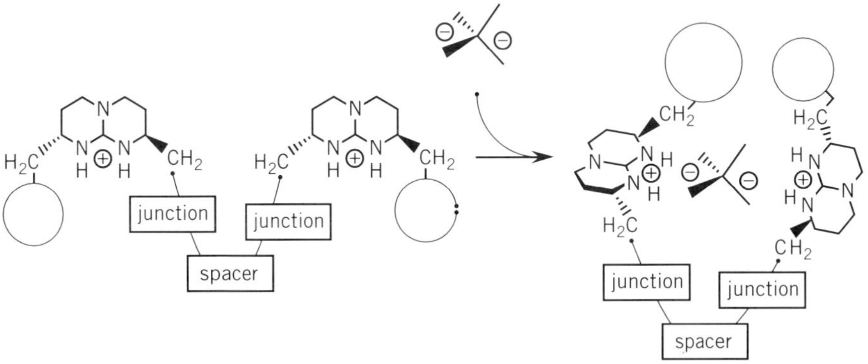

Fig. 4.10. Assumed host–guest binding pattern of linear *bis*-guanidinium hosts (e.g., **4.58–4.64**) and tetrahedral oxoanions.

112 ARTIFICIAL ANION HOSTS. CONCEPTS FOR STRUCTURE AND GUEST BINDING

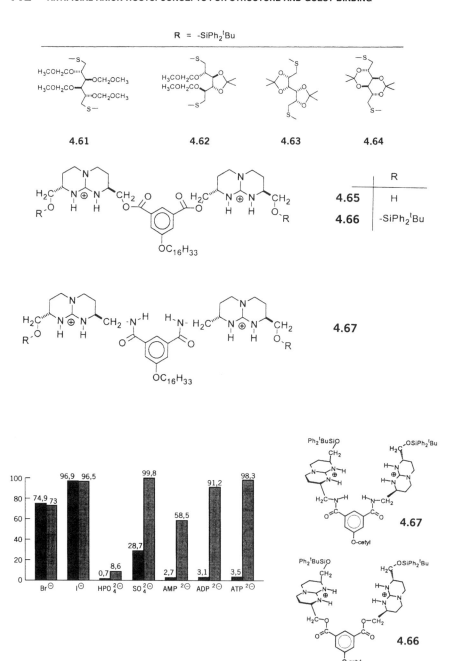

Fig. 4.11. Liquid–liquid extraction of anions in water [10^{-4} M] by host compounds **4.66–4.77** (10^{-3} M in CHCl$_3$, pH 7.4).

The amide **4.67** is the only host compound capable of extracting oxoanions into chloroform from very dilute aqueous solutions. In fact, sulfate is extracted best (> 99.8%), but well-hydrated nucleotides can also be readily transferred to the organic layer, so that **4.67** qualifies as the best vehicle for this purpose at present.

In order to improve on nucleotide extraction selectivity, de Mendoza endowed the guanidinium anchor group with a uracil moiety, which could interact by base pairing with the nucleotide guest.[146] Regrettably, this host proved too hydrophilic to be useful for nucleotide extraction, though the anticipated multiple interaction mode was evidenced by NMR. An even better interaction scheme was provided in host **4.69**, which has the uracil moiety of **4.68** replaced by a tweezerlike Kemp acid derivative.[147] This was supposed to bind preferentially the nucleobase adenine by a combination of Hoogsteen and Watson–Crick hydrogen bonding arrays.[148] As expected, **4.69** showed some preference for *cyclo*-adenosine monophosphates over the guanosine analogs in two-phase extractions. From the 1:1 extraction stoichiometry, NMR evidence obtained from ROESY spectra, which are sensitive to molecular

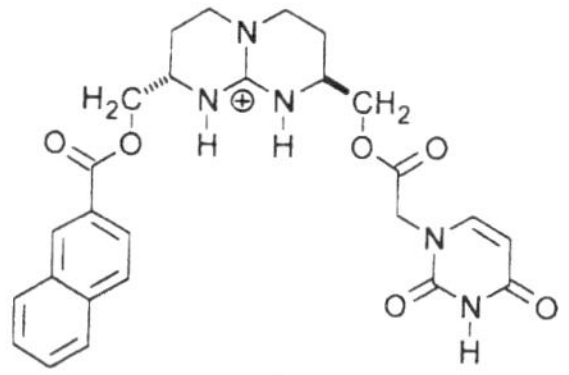

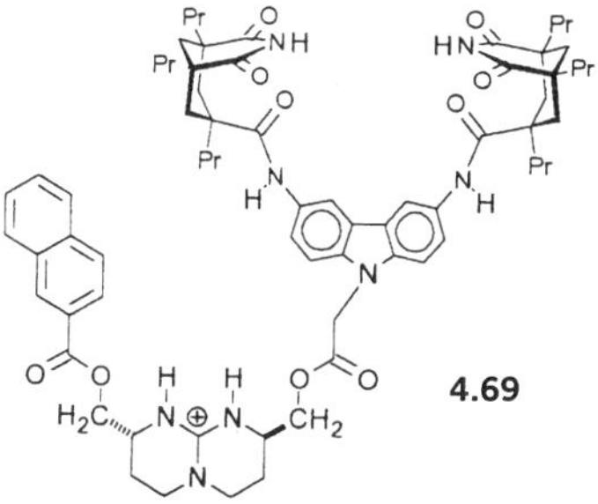

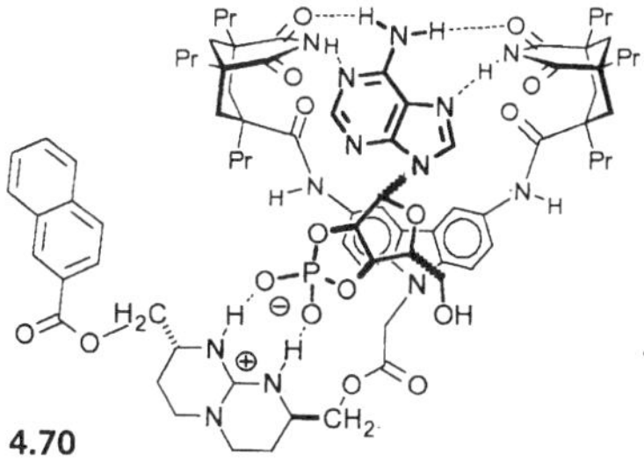

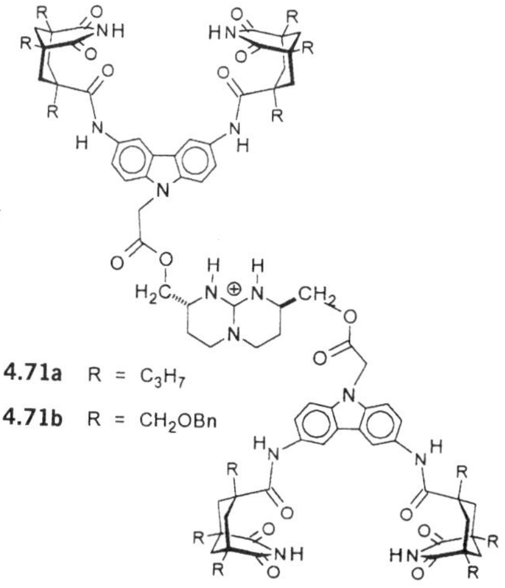

114 ARTIFICIAL ANION HOSTS. CONCEPTS FOR STRUCTURE AND GUEST BINDING

proximity, and comparsion with truncated analogs lacking one or the other anchor group, the host–guest interaction pattern depicted in **4.70** was deduced. There the mandatory guanidinium–phosphate diester salt bridging is supplemented by π stacking of the nucleobase to the carbazole and the network of hydrogen bonds responsible for adenine recognition. Replacement of the naphthoic ester in **4.69**, which apparently is not involved in guest binding by another adenosine recognition group, furnished the C_2-symmetric host **4.71a**.[149] The combination of two nucleoside anchor groups connected to the guanidinium moiety proved advantageous for complexation and extraction of nucleotides having exactly the complementary set of functions like the dinucleoside phophate dApA. Even longer nucleic acids up to a molecular weight of 25000 D were transferred to a dichloroethane phase by the similar host **4.71b** provided they contained stretches of adenosine nucleotides.[150] U-tube transport studies indicate, however, that this is an obligatory, but not sufficient, requirement for successful extractions. These results promise that by easily conceivable extension of this basic concept to give linearly connected multitopic hosts, the sequence-selective transport of oligonucleotides across biological membranes may become feasible.[151]

Bicyclic guanidines have also been used for the recognition of the carboxylate function of α-amino acids. Deviating from the usual strategy that employs binding of the amino acid side chain as the prime target for supramolecular interactions, de Mendoza introduced host **4.72** having anchor groups for the exceedingly well-hydrated α-ammonio-carboxylate moiety in addition to an aromatic naphthalene ester suitable for π stacking to an appropriate side-chain function of the guest.[152] In single liquid–liquid extraction experiments from neutral aqueous solutions the selectivity for aromatic amino acids like phenylalanine or tryptophane was established, supporting a three-point host–guest

4.72   **4.73**

interaction mode. Moreover, this host displayed amazingly high enantioselectivity extracting the L-enantiomer with ca. 80% ee.[153] Molecular modeling disclosed[153] that the peculiar ion pairing with the guanidinium receptor function contributes about half of the total enthalpic interaction, whereas the remaining energy is provided by the azacrown (one-third of total) and naphthalene (one-sixth of total) substructures. In a more systematic study Gloe and Schmidtchen[154] prepared the related host **4.73** having the putatively better triazacrown ether[155] for complexing the primary ammonium group attached with a chemically more stable thioether bridge to the guanidine. They could show by phase transfer of radioactive amino acids that a clean 1:1 host–guest stoichiometry governs the extraction process. Even quite hydrophilic species like glycine and serine, which have never been able to be extracted by other artificial hosts, were carried into the organic phase by **4.73**. The pH dependence showing an optimum around pH 9 suggested that it is the zwitterionic guest species undergoing phase transfer. Compared to **4.72**, host **4.73**, lacking the planar aromatic surface present in **4.72**, exhibits weaker enantioselection (40% ee with Phe).

4.4.4 Miscellaneous Cationic Hosts for the Complexation of Anions

Cyclic oligomeric α-glycosides of glucose (cyclodextrines) **4.74** (β-cyclodextrin) represent classic examples as host compounds for molecular recognition in water.[156] They provide toroidal cavities of variable size that offer a hydrophobic environment rimmed by arrays of highly hydrophilic hydroxy groups. As a general rule guest molecules not well hydrated in water, but of the correct complementary size to fit into the molecular cavity, will associate with this class of host compounds. In line with this expectation some inorganic anions like ClO_4^-, I^-, and SCN^-, but not better hydrated species (CH_3COO^-, Cl^-, SO_4^{2-}), form weak complexes (K_S 10–50 M^{-1}) with α- or β-cyclodextrins (6 or 7 glucose units respectively) (Fig. 4.12).[157]

With the advent of reliable methods to modify further these polyfunctional hosts in a regioselective fashion,[158] the attachment of more and well-positioned additional binding functions became feasible. Particularly prominent in this respect was the introduction of amino groups in the 6'-position, which after protonation could interact by salt bridging with anionic substructures of the guest. The placement of 3 ammonio functions in a C_3-symmetrical manner on

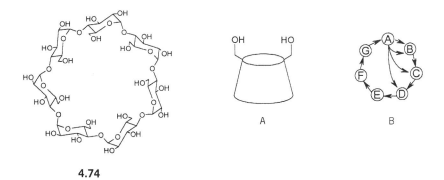

4.74

Fig. 4.12. Shorthand description of β-cyclodextrin (7 glucose units) viewed from the side (A) to show the toroidal cavity with the primary 6'-OH functions located on the smaller rim. Projection B along the main axis visualizes the three possible regioisomeric homodisubstituted derivatives.

the small rim of permethylated α-CD gave the host **4.75** the capability of binding benzylphosphate at pH 7.0 at least 1000-fold better than either of its constituents, benzylalcohol or hydrogen phosphate.[159] A somewhat weaker synergism of binding interactions was observed when β-CD was modified with two imidazole heterocycles forming a coordinatively unsaturated zinc complex of **4.76**. Binding of cyclohexane-1,4-dicarboxylate as tentatively sketched in **4.77** was found to outmatch complexation by the parent ligand **4.76** by a factor of 6.6.[160] Similar results were obtained for the combination of an azacrown ether with β-CD complexing alkali metal salts of *p*-nitrophenolate in DMF, too.[161]

4.75   4.76   4.77

bis-Imidazolyl cyclodextrins can also catalyze the cleavage of anionic catechol phosphates (Fig. 4.13) and thus mimic certain nucleases. More than the moderate rate enhancement (ca. eightfold) it was the impressive regioselectivity of phosphate cleavage that made this model resemble the real enzymes. Proton transfer in the general acid–base catalysis of this process could be correlated to the relative spatial disposition of the imidazole heterocycles.[162]

Owing to the accumulation of positive charge in the vicinity of the molecular cavity, amino cyclodextrins may serve as molecular hosts for nucleotides.[163] Quite dramatic association constants were calculated for complexation of fully protonated heptamethylamino-β-CD **4.78** and deprotonated nucleotides, ATP^{4-} hitting the highest mark ($K_S = 3 \times 10^6$ M^{-1}). Though the actual

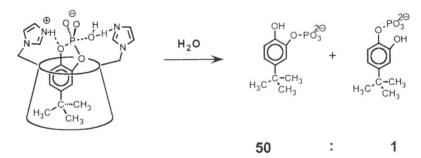

50   :   1

Fig. 4.13. Phosphodiesterase models: Regioselective cleavage of *m*-^{t}butylcatechol phosphate by A,B-*bis*-imidazolyl-β-CD. The buffer-catalyzed hydrolysis produces the monoesters in equal amounts.

POSITIVELY CHARGED ANION HOSTS 117

contribution of these equilibria to the apparent host–guest association at any given pH is only minor due to the minute concentrations of the individual protonation states, the extreme K_S values reflect the maximum intrinsic interaction energy. The comparison of [**4.78a**·$7H^+$] and its A,D-disubstituted methylammonio analog **4.78b** revealed subtle differences in the response to structural variations in the nucleotide (the base, the sugar moiety, points of connection of the phosphate ester), but underlined the general trend, also found in other host systems, that accumulation of charge is deleterious to guest selectivity. Apparently, different guests may find multiple configurations of very similar energy in the multipole electrostatic field of a highly charged host, so that structural subtleties have no bearing on the overall association.

Aiming at the application of artificial anion hosts using electrochemical sensing of negatively charged guests, a number of systems containing a reversible redox couple in proximity to some anion binding moiety have been constructed.[164] In most cases the extreme sensitivity of the redox potential of

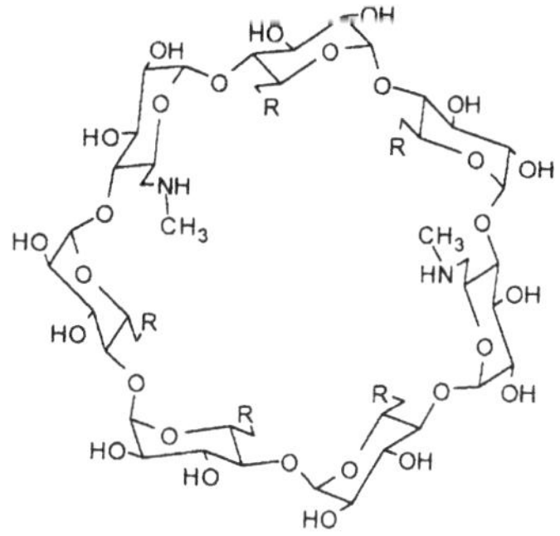

4.78a R = NHCH$_3$
4.78b R = OH

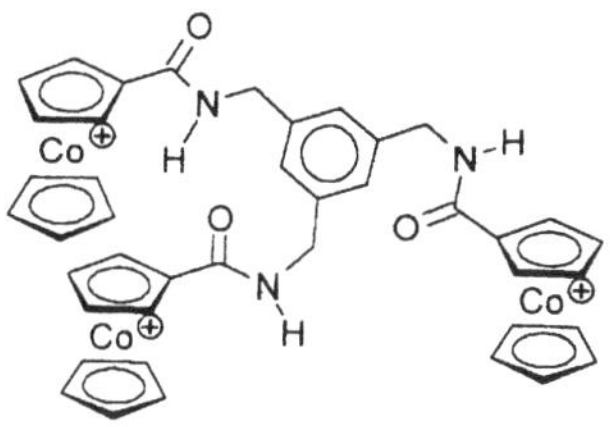

4.79

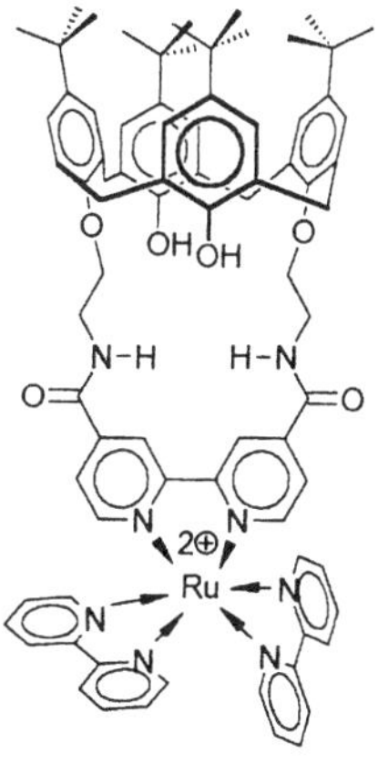

4.80

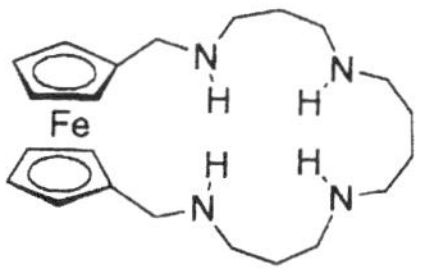

4.81

certain cationic transition metal complexes on the molecular environment can be exploited to detect anion association. The design of **4.79**[165] and **4.80** [166] as prototypical examples in this class was primarily governed by the idea of maximizing the electrochemical signaling of the ion-pairing event. Screening of quite a number of ligands in their response to the presence of inorganic anions (Cl^-, Br^-, NO_3^- or $H_2PO_4^-$) or even the organic dicarboxylate adipate revealed profound qualitative differences in noncompetitive solvents (CH_3CN, acetone). But also in water the detection of biologically important substrates like ATP was feasible. At pH 6.5 the ferrocenyl ligand **4.81** exists in its diprotonated form and may ion pair with ATP, preponderantly present in its trianionic state, to give rise to a cathodic shift of 60–80 mV of the ferrocene redox couple.[167] Clearly the elaboration of this basic design to incorporate more dedicated receptor functions holds a promising perspective.

4.5 ELECTRONEUTRAL HOSTS FOR ANIONS

4.5.1 Hosting of Anions by Poly Lewis Acids

The success of the concept of crown ethers in complexing even very weakly coordinating cations, that is, the placement of multiple Lewis-basic moieties in a preorganized molecular skeleton, can be adapted to the host–guest complexations of certain anions, too. Since most anions possess lone electron pairs, they qualify as Lewis-basic species that may interact with complementary Lewis acid structures. As was found in crown ether chemistry, the accumulation of binding sites in a topologically well-set framework is expected to maximize the binding strength. The prime virtues of this approach of anion host design, which requires a reciprocal layout of the anchor groups commonly used in crown ether chemistry and thus was termed *anticrown chemistry*,[168] reside in the electroneutrality of the corresponding hosts and in the intrinsically more selective binding mode. Electroneutral hosts do not face the problem of competitive counterion binding, which is unavoidable with cationic hosts and frequently spoils all efforts to arrive at a decent guest selectivity in the latter class. Moreover, some potential applications, such as potentiometric anion sensing, rather, call for uncharged selective receptors in order to ease the generation of a potential difference between two phases and arrest the host to the hydrophobic one. Compared to pure coulombic attractions, which just sense the size and distance of charge, the Lewis base–Lewis acid interaction depends on peculiar molecular properties of the directly bonded partners (stereoelectronics, symmetry of molecular orbitals, softness, back bonding ability, etc.). The construction of suitable poly-Lewis-acidic hosts thus may use an additional tool that is not available for the cationic counterparts. On the other hand, Lewis-acid hosts have to combat the natural competition of solvents with their dedicated guests. Most solvents (except for hydrocarbons) are quite Lewis basic as well and in general exceed the molar concentration of a guest anion by several

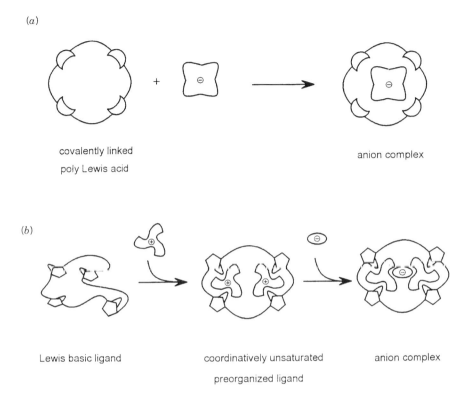

Fig. 4.14. Two principal alternatives for incorporation of Lewis acidic sites into anion hosts: (a) covalent connection into the host skeleton; (b) cation–ligand coordination.

orders of magnitude. Therefore, solvation design (i.e., the exclusion of solvent from the binding site while retaining sufficient solvent interaction energy to keep the host and host–guest complexes in solution) is of great importance. Of course, this is more difficult the smaller and more basic the solvent molecules are. But the examples of natural metalloproteins processing small inorganic anions (e.g., superoxide dismutase, carbonic anhydrase, etc.) clearly show this to be possible and even suggest that it might well be the preferable concept, if chemical manipulation of small anions is envisaged.

The incorporation of Lewis acid substructures into host molecules has been undertaken by two alternative routes (Fig. 4.14): The anchor groups can be embedded into the host by covalent bonds requiring an intimate knowledge of the chemical reactivity of the Lewis acid site and a stepwise synthetic plan. In the second alternative an organic ligand is constructed, and it is left mostly to the last step to implant the Lewis acid metal cation by a straightforward complexation reaction that, however, leaves the cation coordinatively unsaturated. So far the latter approach has been reserved for transition metal cations such as Lewis acids, while covalent incorporation used elements from main groups III and IV or mercury, all reputed to form stable covalent bonds.

4.5.2 Lewis Acidic Hosts Connected by Covalent Bonds

In analogy to the construction of proton sponges, in which amino groups that are basic and rigidly held in close proximity cooperate to bind a proton with outstanding affinity, Katz introduced the 1,8-disubstituted naphthalene **4.82** having two boron Lewis acids juxtaposed.[169] As anticipated, **4.82** is a reagent that abstracts hydride ion from potassium hydride to give a borohydride complex that is virtually inert against moderately strong acids or benzaldehyde. The X-ray crystal structure reveals that hydride is in a bridging position between both boron atoms. Though it is not a symmetrical bridging interaction, both Lewis acidic centers are pyramidalized, reflecting their participation in hydride anion bonding, thereby explaining the unusual thermodynamic and kinetic stability. In much the same way fluoride and hydroxide anions can be complexed, justifying the view that **4.82** acts as a simple yet effective bidentate receptor for small anions. One borane group may be replaced by a trimethylsilyl function without destruction of fluoride binding capability.[170] These observations are in accord with molecular orbital (AM 1) calculations[171] of the organoboron macrocyclic host **4.83** binding hydride, fluoride, chloride, or oxide anions. All boron atoms in the neutral hosts are sp^2 hybridized. Anion encapsulation occurs with a reduction in B–B distance and a partial rehybridization $sp^2 \rightarrow sp^3$ of one or more boron atoms. In an extension of this concept a boron Lewis acid center was supplemented by an alkali cation binding crown

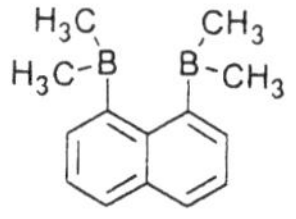

4.82

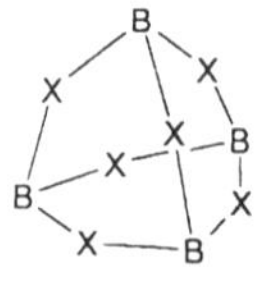

4.83 X = -(CH$_2$)$_{2-4}$

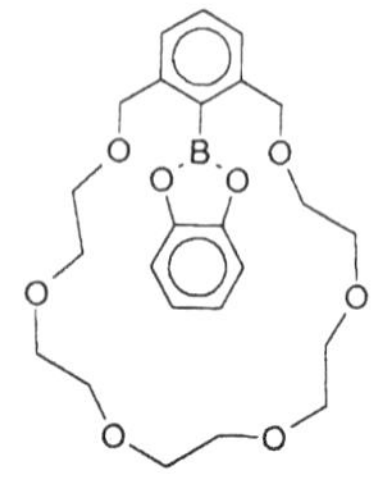

4.84

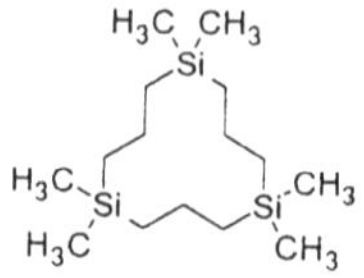

4.85

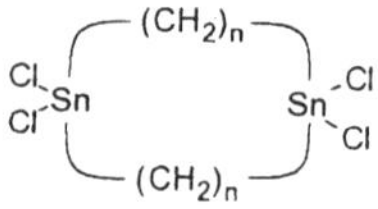

4.86 n = 8, 10, 12

moiety to give **4.84**.[172] With this compound potassium fluoride, but not chloride or bromid, respectively, are solubilized in dichloromethane. The crystal structure proves coordination of fluoride to the boron atom, whereas the K^+ ion interacts with the crown ether as well as in an intermolecular fashion with the boron fluoride of a neighboring molecule. Qualitative evidence that even weakly Lewis acidic sila macrocycle **4.85** exhibits selective interaction in transport studies[173] promoted investigations of the more acidic germanium(IV)[174] or tin(IV)[175] analogs. In particular, tin(IV) macrocycles like **4.86** had already been characterized to bind chloride anion in acetonitrile solution.[176] However, chloride affinity was only marginally increased by a factor of 2 over the corresponding open-chain pendant. Since binding was almost independent of ring size and 1:1 and 1:2 host–guest equilibria of comparable stability were established, the cooperativity of binding sites could only be minor.

A considerable improvement in selectivity was found on switching to the bicyclic tin compound **4.87a**, which, as evidenced by an X-ray structure, exists as the out–out isomer only.[177] The molecular framework is reminiscent of the first chloride binding artificial host, the bicyclic bisammonium salt of Park and Simmons,[33] which could encapsulate chloride into the molecular cavity by virtue of coulombic attraction assisted by two linear hydrogen bonds. The metalla macrocycle **4.87a** binds fluoride exclusively, exhibiting discrimination against chloride by a factor of 10^5, although the association constant in $CDCl_3$ was estimated to be K_S $1-2 \times 10^4$ M^{-1}.[178] In the solid state the fluoride is seen to occupy the cavity of **4.87a**, bridging both Lewis acidic tin centers.[178] In contrast the bigger host **4.87b** forms a Cl^- encapsulation complex, but the anion is

```
         ┌─(CH₂)ₙ─┐
  Cl–Sn            Sn–Cl
         ├─(CH₂)ₙ─┤
         └─(CH₂)ₙ─┘
```

4.87a n = 6

4.87b n = 8

bound to one tin atom only. This is also apparent from ^{119}Sn-NMR spectra in solution, which suggest a hopping process of the anion from one tin atom to the other with an activation barrier of 22.2 kJ/mol. As a corollary of monotopic binding, the K_S value of chloride binding by **4.87b** is of the same order of magnitude as with a simple tributyltin chloride. The effect of encapsulation binding, however, is visible on comparing cavity sizes. Smaller or larger host compounds **4.87** ($n = 7, 10, 12$) form weaker complexes with Cl^- than **4.87b**.[179] Absolute binding affinity can be enhanced by incorporation of more Lewis-acidic sites into the host structure, provided these sites can participate simultaneously in anion binding. This seems to apply for the macrotricyclic tetratin compound **4.88**. Not only does this host give only one ^{119}Sn-NMR signal on titration with chloride, indicating rapid host–guest exchange kinetics, but clean

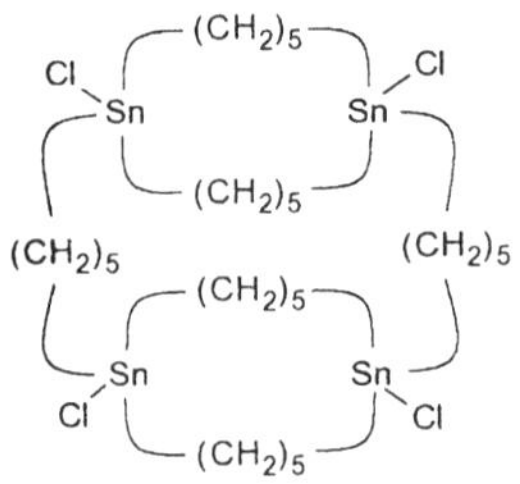

4.88

1:1 stoichiometry is also observed, even in considerable excess of the guest anion. The K_S was determined as 500 M^{-1} (CDCl$_3$), representing an increase in binding energy of ca. 8.4 kJ/mol over the bicyclic analog **4.87b**.[180] Based on the fundamental observation that simple organic tin(IV) compounds have a pronounced intrinsic affinity towards phosphate,[181] a number of ditopic derivatives have been developed and tested for sensing this anion.[182] Very good factors of discrimination against even highly hydrophobic guests like ClO$_4^-$ or SCN$^-$ (which always present serious competition problems in membrane processes) were found and, in combination with the favorable response time, open a bright perspective to arrive at a truly applicable chemical sensor, if the organic frame embedding the tin(IV) centers can be tailored appropriately.

Mercury, being a transition metal capable of forming quite stable covalent bonds to carbon, can also be utilized with advantage as an architectural element in the assembly of artificial hosts. It can contribute an array of 2 linear covalent bonds, leaving 2 empty *p* orbitals perpendicular to each other and the covalent junctions. A peculiar feature of a mercury Lewis acidic site is that upon binding a Lewis base this basic stereochemical arrangement is not affected, so that guest complexation does not distort the host skeleton on stereoelectronic grounds. A first example of chloride binding by a simple organic mercury compound was provided by *ortho*-phenylenedimercury dichloride **4.89**, which in the crystal

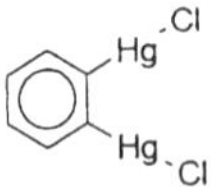

4.89

shows the chloride anion ligated by 4 mercury centers.[183] Elaboration of this principle yielded the ten-membered pentamercura macrocycle **4.90**.[184] This planar molecule very easily forms chloride and bromide 2:1 complexes, hosting the anions above and below the main plane equidistant to all mercury sites. The distance of the chlorides perching on both sides of the macrocycle is considerably shorter than the sum of their ionic radii, testifying to strong binding interactions with the mercury Lewis acids.

ELECTRONEUTRAL HOSTS FOR ANIONS **123**

4.90

On reacting dilithio orthocarborane with $HgCl_2$, a cyclic tetramer **4.91** was obtained as the chloride complex in the exceptionally high yield of 75%.[185] Solution NMR studies and the X-ray crystal structure confirmed an almost ideal coordination of 4 mercury atoms surrounding the central chloride in the same plane. Halide obviously functions as a template in the macrocyclization, because reacting the lithiated carborane in the presence of $Hg(OAc)_2$ yields a cyclic trimer. The halide complexes of the tetramer are rather stable, but lose the halide on reaction with silver ions. The uncomplexed ligand then folds into a

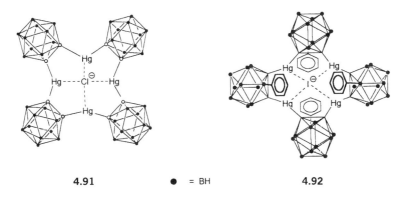

4.91   ● = BH   **4.92**

more compact conformation in which all mercury and carbon atoms are lined as on the seam of a tennis ball.[186] Titration of the free ligand with iodide in acetone following the ^{199}Hg-NMR resonance gave evidence for successive formation of 1:1 and 1:2 host–guest complexes, all of which are in slow exchange on the NMR time scale. With chloride, however, rapidly exchanging complexes were observed. For complexation in solvents having higher solvating power, it is of vital importance to exclude solvent molecules from the Lewis acidic sites. A step in this direction was the incorporation of phenyl rings that sterically limit access to the binding cavity. However, when substituents were introduced with the carborane building block, one must expect the formation of maximally four stereoisomers differing in their relative configuration (the situation is much like the one observed with calix[4]arene). Employing HgI_2 in the macrocyclization, only one isomer **4.92** was formed, which had the 1,3-alternate array of phenyl groups, so that the arene rings of opposing carborane clusters cover the same face of the mercura macrocycle.[187] With $HgCl_2$ instead, a mixture of isomers was obtained, from which three distinct complexes could be characterized. All contain the chloride ion almost in plane with the Hg atoms, whereas in the iodide complex the anion occupies a perching position above the main plane. Though these structures look quite complicated and fragile at first sight, their synthesis with respect to further structural variations appears reliable, fast, and gives good yields. Taken together with the chemical stability and other favorable properties like solubility in most organic solvents and ready detectability of the complexation event by ^{199}Hg-NMR, this should pave the way for applications both in basic organic chemistry and technology.

4.5.3 Lewis Acidic Hosts Based on Metal Cation Coordination

When organic chelating ligands coordinate to metal cations, a mismatch of coordination sites may eventually arise that, however, does not spoil the entire complexation process and may still give thermodynamically stable species. If the potential donor atoms in a ligand outnumber the coordination sites of a metal cation and in addition cannot be arranged to satisfy its coordination needs due, for example, to flexibility restrictions, one observes as a rule polynuclear complexes.[189] Very frequently these possess open coordination sites that may be filled by anions. This anion binding process is a widespread phenomenon and is basically a remedy to arrive at a stable structure demanded on coordinative and electrostatic grounds that would lead to crystallization and allow characterization by X-ray structure determination. From the standpoint of host–guest chemistry, anion complexation there is somewhat adventitious, but the great potential was recognized early on.[190]

The discussion here must be limited to a few representative examples, and the selection certainly contains an arbitrary component because the interface to the more classical fields of coordination chemistry of metals in their first and second coordination spheres is not at all clear cut but rather blurred. The goal is to tailor ligands in order to construct structurally well-defined complexes with appropriate metal cations which contain the Lewis acidic sites in a molecular environment that would impose high barriers for anion discrimination. Of essential importance is the prudent selection of metal cations, since they have to meet multiple requirements. Being the points of direct interaction with the negatively charged guest, they can exert their intrinsic binding preferences, which very often can be modulated by the chelating ligand in a predictable manner. In addition, the cation of choice should bring in a precise geometry of ligation that would induce a structure-forming effect on the organic ligand as well as place and orient the Lewis acid sites within this structure. On top it would be desirable to have guest binding and overall complex formation as independent processes; that is, the metal complex with the ligand should be kinetically stable while anion binding itself should occur in rapid exchange. Fortunately, this latter caveat is met by most of the chelating ligands used so far. Based on these principles, the lower transition metal cations, notably copper(I), copper(II), iron, manganese, cobalt, nickel, and ruthenium have been preferred, although even the uranyl cation or main group metals have been successfully used in this approach as well (see below), and there is no obvious limitation to employ elements from the entire range of the periodic table.

An instructive example for what was termed *cascade anion binding* is provided by the Cu(II)–bistren system **4.93**.[191] Insight into the multiple equilibria contributing to anion binding in this system rests on very thorough pH-titration data. In spite of the high precision obtainable in these measurements, the method only senses an integral response of the system on a deliberate addition of aqueous base. This response has to be split by a multiparameter computer fit into individual contributions of up to 10 or more equilibria of a mutually compensating nature. Of course, this limits the accuracy of the respective association constants. Moreover, these binding data a priori do not yield information about complex structures. The correlation of structure and binding strength rather must emerge from a trend analysis of purposefully introduced variation of the ligand. However, if then a consistent picture of all binding events can be advanced, the principles derived therefrom have higher reliability and greater predictive power than direct structural studies on a single example. In the Cu(II)–bistren system the conclusions drawn are backed by extensive investigations in the acyclic and monocyclic series.[192,193] The dinuclear complex **4.93** binds chloride in water with $K_S = 3550$ most likely by insertion as a bridging ligand between the copper centers as sketched in **4.94**. This configuration must be well preferred, because the mononuclear doubly protonated analogous complex (same total charge as **4.93**) binds Cl$^-$ more than 100-fold more weakly. The difference between the dinuclear and mononuclear copper complexes with hydroxide as the guest anion is even greater: The factor of 10^8 observed in this case reflects the particular difficulty of hydroxide to associate to the mononuclear copper center, which is hidden in the interior of the ligand. Hydroxide binding is aggravated by 10^3-fold relative to the uncomplexed aquo-Cu^{2+} ion, thus demonstrating an "anticryptate effect" in which guest binding in the cavity is disfavored rather than favored compared with the free Lewis acid species. One of the most thoroughly investigated binucleating ligands is the monocycle OBISDIEN **4.12**,[194] which also represents the monocyclic parent of **4.93**. Cu^{2+}, Ni^{2+}, Zn^{2+}, and Co^{2+} complexes of **4.12** have been shown to bind a wide variety of anions, including azide,[195] malonate,[196] sulfate,[197] and perchlorate.[198] In all cases anion binding is assumed or proven to occur in a μ-bridging fashion. This was also suggested by molecular modeling undertaken to explain the binding of pyrophosphate^{4-} by the biscopper(II) complex of **4.12**.[199] Around pH 8 the unprotonated complex is the dominating species, exhibiting an extraordinary high 1:1 association constant of 3.1×10^8 M^{-1} with this biologically important anion.

Cavity formation as a prerequisite for anion binding is not restricted to macrocyclic ligands, but can also result when sterically crowded chelating ligands with proper spacing of donor sites are employed. A recent advance aimed at the design of anion hosts using coordinatively bound Lewis acids produced Cu(I)–phosphine complexes of 2:2 stoichiometry, which in the solid state associate ClO_4^- or NO_3^- in μ-bridging fashion.[200] This seems to be a rather stable arrangement, because in acetone solution the anion can be exchanged for organic carboxylates or vanadate VO_3^-, if the latter is used in excess.

4.95

Supplementation of Lewis acid metal complexes by additional binding interactions recruited from the arsenal of conventional anchor functions presents a promising concept for the promotion of guest selectivity. Early attempts in this direction have already been mentioned in the discussion of Zn complexes appended to cyclodextrins and, of course, are involved in Lehn's coreceptor strategies,[190] as well as implicitly participate, for instance, in anion binding of the protonated mononuclear Cu(II) complexes of the bistren macrocycle (ligand of **4.93**; see above).[191] Ultimately, any ligand having functionality in excess of that needed for the basic metal cationic coordination will influence the binding of an incoming guest by virtue of attractive or repulsive interactions and consequently affect the association constant. This is a longstanding notion and closely related to second-sphere coordination,[201] which lacks the direct interaction with the metal center. Though well known to be of fundamental importance, the incorporation of these secondary interactions as an element of host design has recently given another impetus to multitopic receptor development. Strapping an N-methylmesoporphyrin II with an achiral bridge and subsequent treatment with Zn^{2+} yielded enantiomeric zinc–porphyrin complexes, **4.95**, which could be resolved by HPLC.[202] Since backside attack of an anion to the metal is impaired by the blocking N-methyl group, only the front side lined by the strap containing hydrogen-bonding amido groups is open for binding. This constitutes an enforced chiral environment, so that binding of racemic guest species yields diastereomeric complexes. Single extraction experiments of N-acylamino acid anions ($H_2O/CHCl_3$) confirmed a strong enantio preference (up to 90% ee) that could be attributed to the hydrogen bond donor ability of the guest.

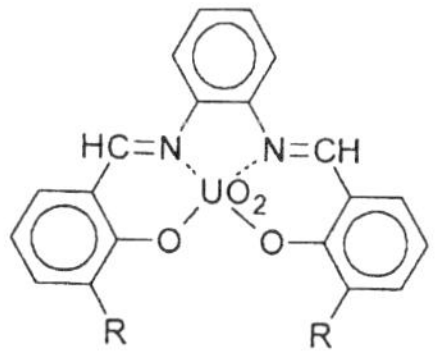

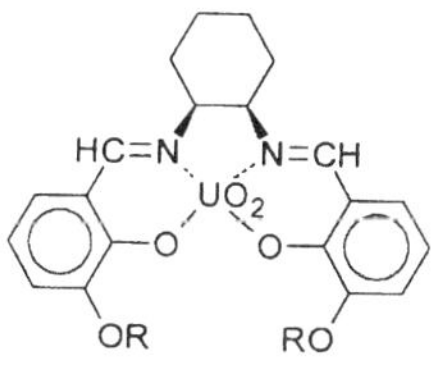

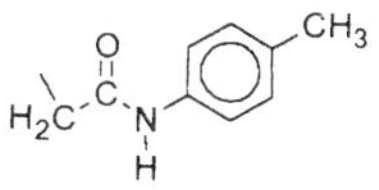

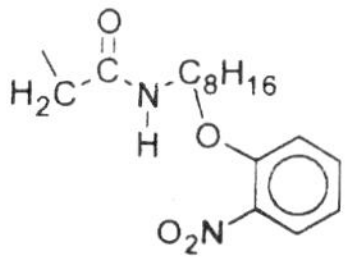

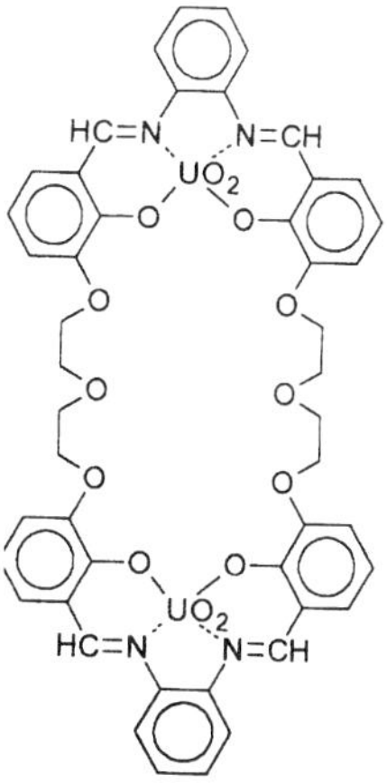

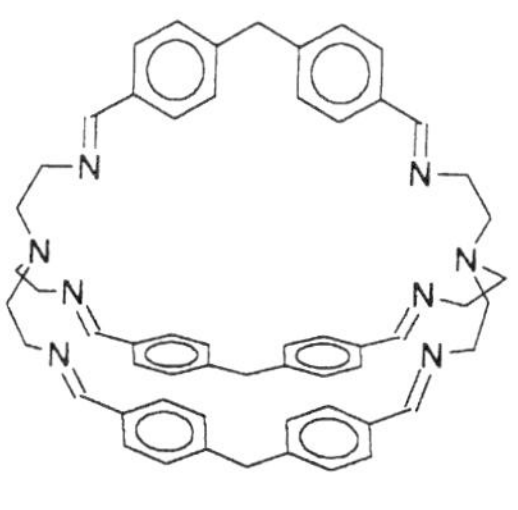

Hydrogen bonding was also exploited as a means to enhance the inherent selectivity of uranyl–salenes **4.96** towards dihydrogen phosphate.[203] Thus the attachment of two additional secondary carbonamide functions to **4.96a** to give **4.97** augments the absolute K_S values for $H_2PO_4^-$ and Cl^- in acetonitrile/DMSO (99:1) by an order of magnitude [$K_S\,(H_2PO_4^-) > 10^5\ M^{-1}$] and sets the discrimination factor to ca. 100. Association appears to be entropically driven, and huge positive entropies of association around $+377\ J\,K^{-1}\,mol^{-1}$ have been reported for **4.96a** binding $H_2PO_4^-$,[203] indicating an extraordinary amount of

4.96a R = H

4.96b R = OCH_3

4.97a

4.97b

4.98

4.99

desolvation of the binding partners, which outmatches the inherently negative entropy of host–guest association. In the crystal dihydrogen phosphate is found to coordinate to the uranyl center as well as to methoxy or amido functions of the ligand via hydrogen bonds. A characteristic feature of all these complexes is the occurrence of dimeric phosphate entities held together by an obviously strong set of two hydrogen bonds. Even small amounts of water in DMSO may disrupt the optimal host–guest configuration, leading to a drastic drop in association of $H_2PO_4^-$ in DMSO/H_2O 9:1 vol % (K_S of 40 M^{-1}).

Nevertheless, the utility of this design of an anion host **4.97b** proved useful in transporting $H_2PO_4^-$ across a supported liquid membrane, demonstrating the advantage of an electroneutral anion receptor.[204] Duplication of the uranyl–salene site gives dinuclear complexes like **4.98** that were shown to be strong receptors for dicarboxylic anions in DMSO, notably fumarate ($K_S > 10^5$ M^{-1}) and terephthalate ($K_S = 1.4 \times 10^4$ M^{-1}).[205] This appears to be the electroneutral version of a more general theme represented by ligand **4.99**, which forms a dinuclear copper(I) cryptate and complexes terephthalate as a polyprotonated host in a highly complementary fashion.[206]

Despite the well-established occurrence of main-group metal cations—in particular Mg^{2+} and Ca^{2+}—in proteins where they serve in binding anionic substrates and structure enforcement or allosteric switching, their use in artificial receptors is very scarce indeed.[207] In an attempt to mimic the coordination environment of Mg^{2+} in some phosphatases, the Kemp acid derivative **4.100** was prepared, which gives a dinuclear magnesium complex on addition of $Mg(NO_3)_2$ in methanol.[208] As revealed by X-ray crystallography, the Mg^{2+} cations bridge the two carboxylates on either side, giving the proper arrangement and spacing to associate another phosphate monoanion (diphenylphosphate) in μ fashion between them. The Mg–Mg distance varies by 0.75 Å in a number of analogous complexes, indicating quite some flexibility in the ligand to accommodate different coordination environments of the Lewis acid sites. Similar adaptation to the changing transition structures coordinated to Mg^{2+} centers might be essential for enzymatic catalysis.

4.100

4.5.4 Anion Hosts Operating by Ion–Dipole Binding

In contrast to the ligation of anions to metal centers, which usually involves contributions from covalent bonding, ion–dipole interactions are primarily electrostatic in nature. Compared to coulombic forces between charged host–guest partners, ion–dipole interactions have the same dependence on the dielectric environment, but are much weaker and have a steeper falloff with distance. However, they are directional and orient the ion and the dipole vector with respect to each other. This is a structure-making property vastly exploited in all biological structures. The prototypical example here is the hydrogen bonds, which by virtue of their accumulative power warrant the defined secondary and tertiary structure of the proteins and nucleic acids. Taken alone an average single hydrogen bond between electronegative atoms can contribute up to about 30–40 kJ mol^{-1} to binding two partners,[209] roughly one-tenth the energy of a typical carbon–carbon or carbon–hydrogen single bond. Stronger hydrogen bonds are known, but their utility and participation in natural and artificial receptors currently is a matter of a vivid controversial debate.[210] In solution the maximal attainable energy from host and guest exclusively interacting via hydrogen bonding is severely attenuated by the dielectric permittivity of the solvent. This applies even to the extent that the attraction between host and guest, doubtlessly existing in less polar solvents, vanishes completely when the solvent is switched to water.[211] In the absence of any competition, however, hydrogen bonding to anions seems to be an almost universal property of matter. Thus, in the gas phase[212] or even in a crystal grown from apolar organic solvents,[213] fluorocrown ethers can associate fluoride anion. As proven by several X-ray crystal structures, the partially fluorinated macrocycle **4.101** hosts F$^-$ in a nesting configuration right in the middle of the folded macrocycle with the help of 4 converging C–H$\cdots$F$^-$ hydrogen bonds.[213] Apparently, the neighboring CF$_2$ groups polarize the CH bond sufficiently to enable this unusual binding motif. In addition, preorganization of the binding groups into a macrocycle is also mandatory to success, because fluoride complexation was not detected with an open-chain analog of **4.101**.

4.101

The special attraction of hydrogen bonding to the construction of abiotic hosts derives from its electroneutrality and from the overwhelmingly rich chemistry for incorporation of appropriate structural elements into molecular frameworks. Electroneutrality, as was mentioned previously, is a very desirable property if applications like membrane transport or potentiometric ion sensing are envisaged. The versatility of construction in combination with the weak but nonspecialized nature of hydrogen-bonding interactions opens a wide arena for receptor design and encompasses almost any class of compounds in organic chemistry. In the absence of interference from competitive H-bond acceptors (frequently this is the major obstacle in experimental design), anion complexation can be detected by even the most primitive H-bond donor hosts. The virtue of the host design correlates with its ability to stand up to competing solvation of both binding partners. Shining examples provided by natural receptors confirm that the goal of complexation of heavily hydrated anionic species solely by the aid of hydrogen bonding is indeed attainable. Sulfate-binding protein, for instance, achieves sequestration of sulfate from aqueous solution with a K_S of 10^6 M^{-1}, discriminating against the very similar hydrogen phosphate by a factor of 10^5. A total of seven highly dedicated hydrogen bonds hold the anion in a deeply buried cavity in the interior of the protein as revealed by the X-ray structure (see Fig. 3.6, p. 74).[214] This primary binding pattern is superimposed on another ion–dipole interaction involving four macrodipoles of α-helices, which converge with their N-terminal head on to the sulfate binding pocket.

Relative binding affinities obtained for a number of oxoanions like carboxylate, phosphate, phosphonate, sulfonate, and isosteric oxostructures like lactone and nitro with the urea derivatives **4.102** and **4.103** revealed the plausible sequence: Complex stability increases with higher charge and greater Brønsted basicity of the guest.[215] Another correlation in the similar system **4.104** that relates complex stability to the pK_a of the H-bond donor has been observed previously by Hamilton.[216] The interdependence though is rather flat in either case.

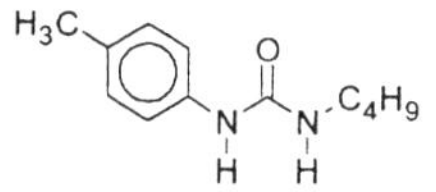

4.102

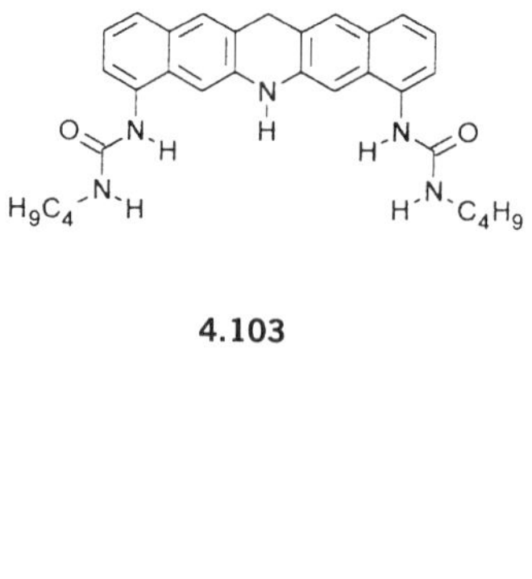

4.103

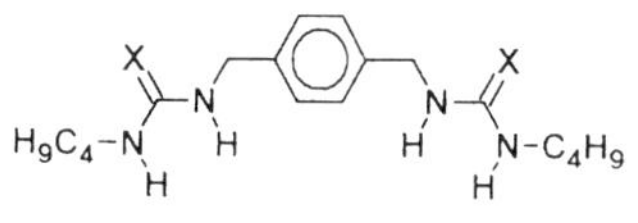

4.104a X = O

4.104b X = S

Much more pronounced is the solvent dependence of host–guest association. Although the general notion holds that K_S decreases in the solvent series CCl$_4$ > CHCl$_3$ > DMSO > DMSO/H$_2$O due to the increasing global polarity, the dissection of the Gibbs free energies of association into their component enthalpy and entropy contributions reveals a nonuniform picture that clearly reflects much more subtle molecular influences. These weak interactions are subject to extensive enthalpy–entropy compensations,[217] which in turn depend heavily on the molecular structures not in direct contact with each other. The pivotal role of small amounts of water in organic solvents in these binding events has been unfolded by Wilcox.[218] The interpretation of binding data is further complicated by multiple equilibria that inevitably accompany the host–guest binding under direct study. Among these are self-dimerization of the host and ion pairing of the anion with countercations, both of which are frequently not considered or discussed, so that one arrives at skewed and overoptimistic projections of the true factors causing host–guest complexation. Some general conclusions, however, appear to be well founded: The accumulation of H-bond donor sites in close proximity to one another seems to be beneficial for two reasons. It enables multiple contacts with the guest, which should add up to improve binding on enthalpic and entropic grounds.[217] Furthermore, the enforced neighborhood yields less efficient solvation of the individual donor sites in the host, resulting in enthalpically more favorable interactions with the guest. Another notion concerns host flexibility. Since hydrogen-bond donor hosts in most cases contain Lewis basic acceptor sites as well, a self-satisfying intramolecular saturation would be deleterious to guest binding. Thus restriction of host flexibility to circumvent the intramolecular compensation pays off in enhanced host–guest complex stability.

Nevertheless, even rather mobile and readily preparable derivatives of the polyamine complexing agent tren, **4.105**, bind chloride, $HClO_4^-$ or $H_2PO_4^-$ in acetonitrile,[219] the latter with the impressive K_S of 1.4×10^4 M^{-1}. Following a very similar concept Morán reported K_S for binding of phenylphosphate dianion to the cyclohexanetricarboxylic amide **4.106** in DMSO as 1.5×10^4.[220] Though very suggestive complex structures can be drawn, no decisive evidence in their favor has been presented yet, so that any interpretation should contain a fair amount of caution. In the same vein, but with an electrochemical detection technique, preferential binding of $H_2PO_4^-$ to the ferrocene derivative **4.107** was observed.[221] Again, the purely qualitative evidence does not justify more speculations about the host–guest structure.

In an attempt to take advantage of an organized array of peptide bond dipoles, Ishida et al. synthesized the cyclic peptide **4.108** composed of dipeptide building blocks containing m-aminobenzoic acid as a rigid, structure-enforcing element.[222] Binding studies with p-nitrophenylphosphate dianion in DMSO monitored by UV–Vis spectroscopy revealed a 1:1 host–guest equilibrium with the extraordinarily high K_S of 1.2×10^6 M^{-1}. Structural variation of the amino acid and the ring size disclosed a strong preference for a cyclohexapeptide structure, while the actual nature of the amino acid side chain is of minor importance. As read from ^{1}H-NMR chemical shifts, rapid host–guest equilibration occurs, affecting all the amide NH resonances, so that the time-averaged symmetry of the host is conserved in the complex. These cyclopeptides may be compared to rigid polycyclic amides directly obtainable in a one-pot reaction (e.g., **4.109**),[223] which can bind linear amino acid derivatives with exceptional selectivity, but so far have not been tested in the complexation of anions.

A different approach to enforce the positional and directional relations among H-bond donors is to attach them to well-known molecular scaffolds.

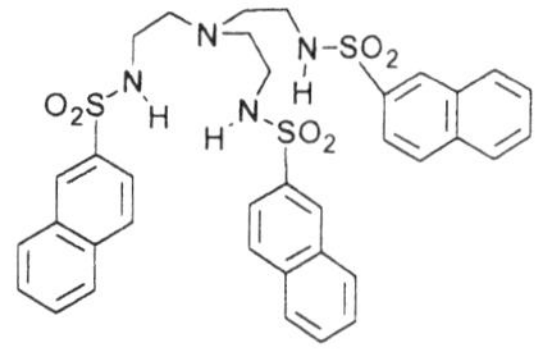

4.105, 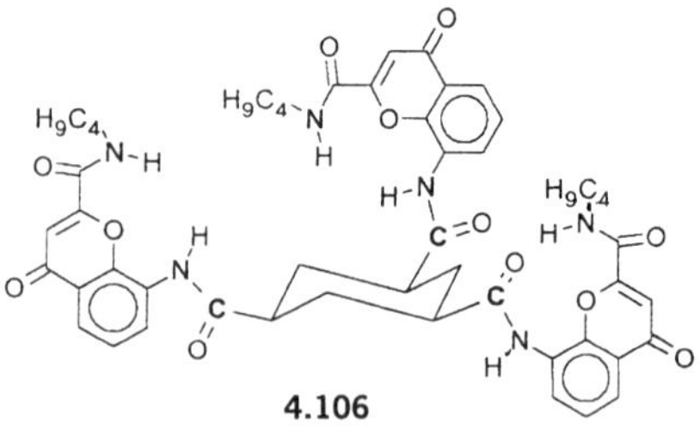 **4.106**, 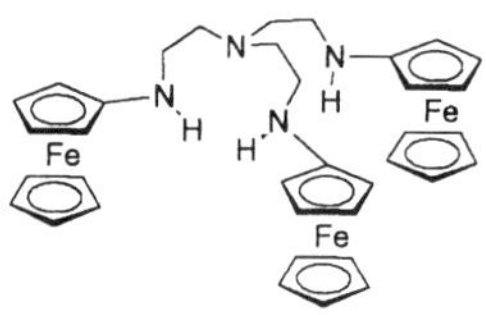 **4.107**, 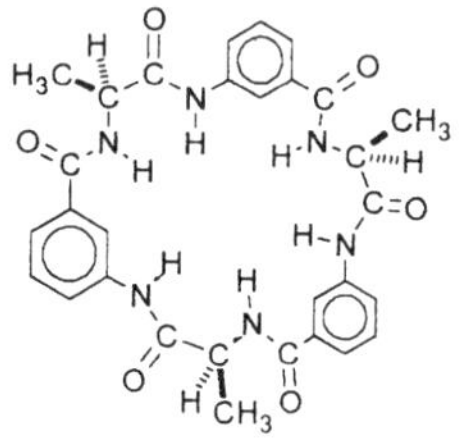 **4.108**, 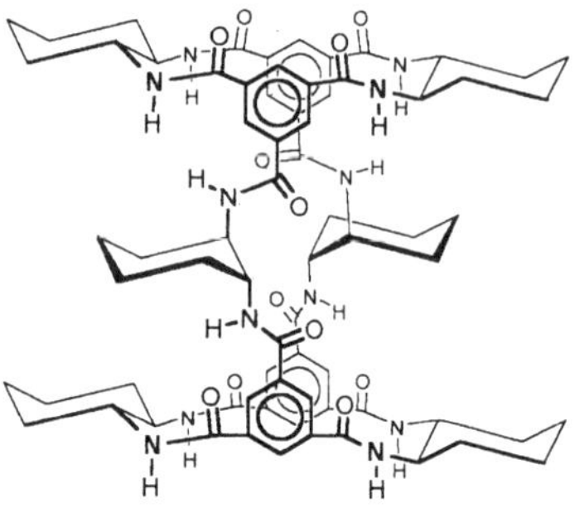 **4.109**

Prominent candidates with a well-proven reputation in cation binding are the calixarenes, which can be functionalized in predetermined positions and may thus provide an ideal platform to investigate systematically spatial and functional consequences of anchor group variation. The attachment of sulfonamide functions to the upper rim of a calix[4]arene yielded host **4.110** capable of binding HSO_4^- selectively over chloride or nitrate.[224] The discrimination against $H_2PO_4^-$ could not be determined, but better complexation of HSO_4^- was likely. In contrast, functionalization of a calix[4]arene[225] or its higher congener calix[6]arene[226] with tethered urea or thiourea moieties to give **4.111** and **4.112**, respectively, did not produce a phosphate host at all. While both calixarenes qualify as rather moderate hosts for the halides in chloroform, **4.112** very

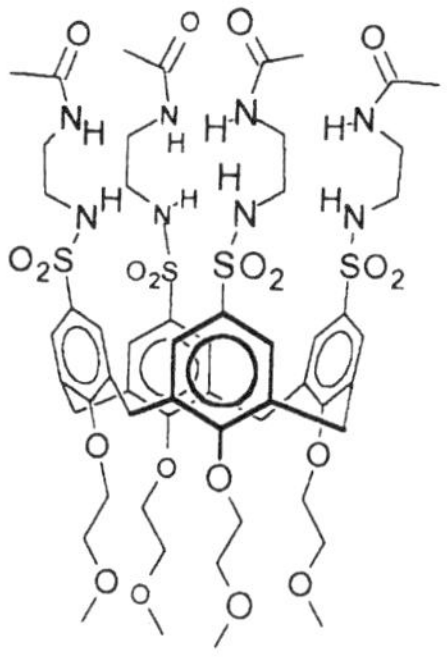

4.110

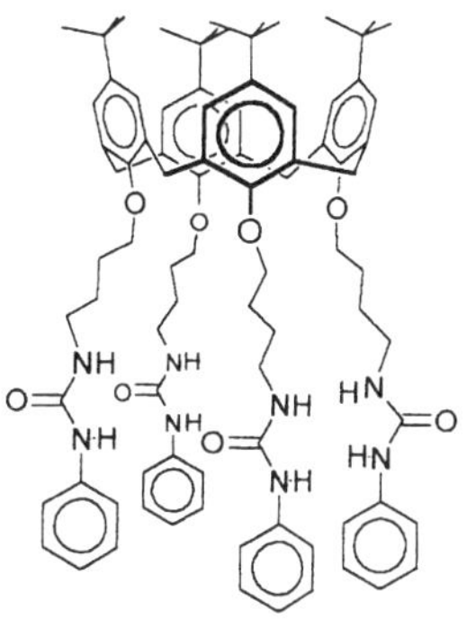

4.111 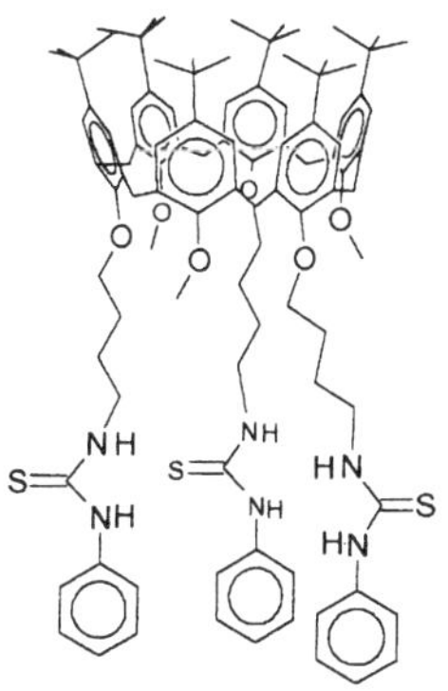 **4.112**

strongly complexes 1,3,5-benzenetricarboxylate ($K_S = 2.0 \times 10^5$ M^{-1} in CDCl$_3$) and exhibits steric recognition, because the 1,2,4- and 1,2,3-regioisomeric benzenetricarboxylates show ten- to hundredfold weaker binding. Apparently, the topological complementary governs this result.

Hydrogen bonding to a first approximation senses points of outstanding electron density, for example, lone electron pairs in the anionic guest. In a fundamentally different approach the bond dipoles of heavier, nonhydrogen elements were used to provide sufficient attraction for the transfer of the negatively charged guest into a molecular cavity in the host. This concept, first advanced by Schmidtchen,[227] requires that strong, but chemically reasonably stable, dipolar bonds are fixed in a configuration that exposes their positive ends toward a binding center. Building on a macrotricyclic tertiary amine, the adduct with borane BH$_3$ **4.113** represents such an example. The four amino–borane bonds converge to the center of the molecular cavity and cannot escape this orientation due to the high connectivity of the macrocyclic framework. The only conceivable interaction mode with the guest is ion–dipole attraction, which is not perturbed by any covalent bonding contributions. In fact the borane–amine **4.113** was shown to complex a number of inorganic ions in chloroform solution exhibiting guest discrimination by size. This and the observation of halide complexes of strict 1:1 stoichiometry by electrospray MS conclusively proved true anion encapsulation by this electroneutral host. A similar binding rationale was deduced in the investigation of guest-binding properties of macrocycle **4.114**.[228] Owing to the stereochemistry in this macrocycle, the sulfur and phosphorus oxide dipoles occupy the same face and thus are roughly aligned in parallel to each other.[229] One hemisphere of the macrocycle, exposing a surface of enhanced positive potential, is therefore open for association of an anion. Indeed, **4.114** forms 1:1 complexes with chloride, bromide, and iodide of $K_S = 40\text{–}60\ M^{-1}$ in chloroform, but binding of fluoride could not be detected. On the opposite side of the macrocycle the presentation of oxide functions well known for their exceptional hydrogen bond acceptor power[230] may simultaneously bind a primary ammonium cation via threefold H bonding. Anion association at the other end of the dipolar bond diminished ammonium cation complexation.

The most simple means to arrive at an anion host of zero net charge is to segregate positive and negative charges while conserving their center of gravity

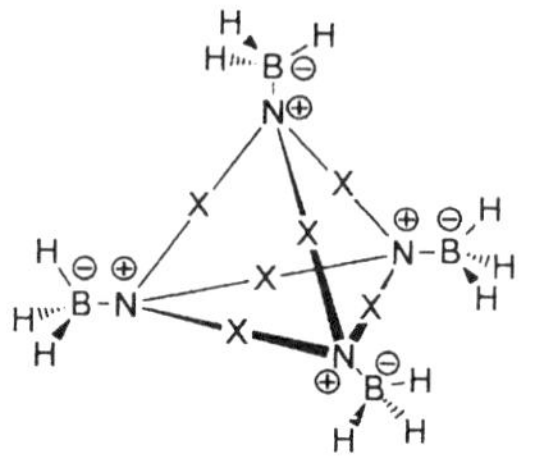

4.113 X = (CH$_2$)$_6$

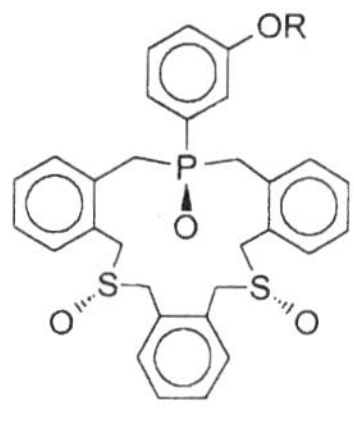

4.114

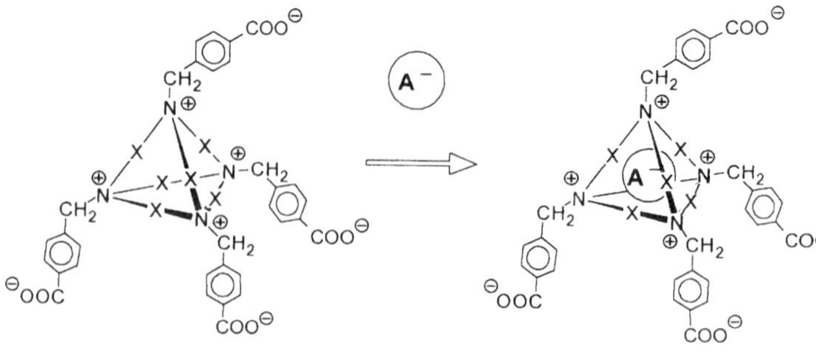

4.115a X = (CH$_2$)$_6$

4.115b X = (CH$_2$)$_8$

4.116

and avoiding their mutual intramolecular contact. In this way zwitterionic structures are formed mimicking the natural hosts (e.g., enzymes) that typically follow the same strategy. The creation of positively charged domains can furnish an anion binding substructure that is embedded into an oppositely charged shell to yield a host of overall electroneutrality. The zwitterionic compound **4.115** is readily available by straightforward alkylation of the macrotricyclic tertiary amine also present as a parent in the borane adduct **4.113**.[231] On the basis of ^{1}H and ^{35}Cl-NMR titration data, **4.115** forms surprisingly stable inclusion complexes with halides and cyanide in water ($K_S = 300\text{–}6000\ M^{-1}$). Dissection of the Gibbs free energy by van't Hoff plots showed bromide and iodide binding as enthalpically driven processes, which are counterbalanced by a negative entropy contribution. This scenario would be expected from an encapsulation of the guest with loss of its solvent shell into the solvent free cage of the host.

4.6 REFERENCES

1. For informative discussion see: (a) J. F. Riordan, *Mol. Cell. Biochem.* **26**, 71 (1979); (b) M. S. Shongwe, C. A. Smith, E. W. Ainscough, H. M. Baker, A. M. Brodie, and E. N. Baker, *Biochemistry* **31**, 4451 (1992); (c) B. J. Calnan, B. Tidor, S. Biancalana, D. Hudson, and A. D. Frankel, *Science* **252**, 1167 (1991).
2. (a) M.-D. Tsai and H. Yan, *Biochemistry* **30**, 6806 (1991); (b) J. P. Abrahams, R. Lutter, G. W. Leslie, and J. E. Walker, *Nature* **370**, 621 (1994); (c) S. Mangani, M. Ferraroni, and P. Orioli, *Inorg. Chem.* **33**, 3421 (1994).
3. (a) J. W. Pflugrath and F. A. Quiocho, *Nature* **314**, 257 (1985); (b) H. Luecke and F. A. Quiocho, *Nature* **347**, 402 (1990); (c) J. J. He and F. A. Quiocho, *Science* **251**, 1479 (1991).
4. R. C. Hayward, *Chem. Soc. Rev.* 285 (1983).
5. J.-L. Pierre and P. Baret, *Bull. Soc. Chim. France* **II**, 367 (1983).
6. E. Kimura, *Top. Curr. Chem.* **128**, 113 (1985).
7. (a) B. Dietrich, in *Inclusion Compounds*, J. L. Atwood, J. E. D. Davies and D. D. MacNichol, Eds., Oxford University Press, Oxford, 1984, Vol. 2, pp. 373–405; (b) B. Dietrich, *Pure Appl. Chem.* **65**, 1457 (1993).
8. (a) F. P. Schmidtchen, *Nachr. Chem. Tech. Lab.* **36**, 8 (1988); (b) F. P. Schmidtchen, *Pure Appl. Chem.* **61**, 1535 (1989).
9. D. N. Reinhoudt, D. M. Rudkevich, and W. Verboom, *Pure Appl. Chem.* **66**, 679 (1994).
10. J. L. Sessler, M. Cyr, H. Furuta, V. Král, T. Mody, T. Morishima, M. Shionoya, and S. Weghorn, *Pure Appl. Chem.* **65**, 393 (1993).
11. (a) E. P. Kyba, R. C. Helgeson, K. Madan, G. W. Gokel, T. L. Tarnowski, S. S. Moore, and D. J. Cram, *J. Am. Chem. Soc.* **99**, 2564 (1977); (b) D. J. Cram, *Angew. Chem. Int. Ed. Engl.* **27**, 1009 (1988).
12. For the debate on stereoelectronics of the carboxylate anion, see: (a) R. Gandour, *Bioorg. Chem.* **10**, 169 (1981); (b) B. M. Tadayoni, J. Huff, and J. Rebek, Jr., *J. Am.*

Chem. Soc. **113**, 2247 (1991); (c) F. H. Allen and A. J. Kirby, *J. Am. Chem. Soc.* **113**, 8829 (1991).

13. (a) R. S. Alexander, Z. F. Kanyo, L. E. Chirlian, and D. W. Christianson, *J. Am. Chem. Soc.* **112**, 933 (1990); (b) Z. F. Kanyo and D. W. Christianson, *J. Biol. Chem.* **266**, 4264 (1991).
14. G. R. J. Thatcher, D. R. Cameron, R. Nagelkerke, and J. Schmitke, *J. Chem. Soc. Chem. Commun.* 386 (1992).
15. I. Marcus, *Chem. Rev.* **88**, 1475 (1988).
16. (a) H. D. B. Jenkins, K. P. Thakur, *J. Chem. Educ.* **56**, 576 (1979); (b) H. Solis-Correa and J. Gomez-Lara, *J. Chem. Educ.* **64**, 942 (1987).
17. (a) G. Boche, *Angew. Chem. Int. Ed. Engl.* **31**, 731 (1992); (b) J. Smid, *Angew. Chem. Int. Ed. Engl.* **11**, 112 (1972); (c) T. E. Hogen-Esch, *Adv. Phys. Org. Chem.* **15**, 153 (1977).
18. S. H. Strauss, *Chem. Rev.* **93**, 927 (1993).
19. (a) M. Bochmann, *Angew. Chem. Int. Ed. Engl.* **31**, 1181 (1992); (b) K. Seppelt, *Angew. Chem. Int. Ed. Engl.* **32**, 1025 (1993).
20. Y. Marcus, *Pure Appl. Chem.* **59**, 1093 (1987).
21. (a) Y. Marcus, *J. Chem. Soc. Faraday Trans. 1* **87**, 2995 (1991); (b) Y. Marcus, *J. Chem. Soc. Faraday Trans. 1* **83**, 339 (1987); (c) Y. Marcus, *J. Chem. Soc. Faraday Trans. 1* **82**, 233 (1986).
22. (a) J. Gao, S. Boudon, and G. Wipff, *J. Am. Chem. Soc.* **113**, 9610 (1991); (b) H. L. Friedman, *Faraday Discuss. Chem. Soc.* **85**, 1 (1988).
23. (a) Y. Marcus, *Ion Solvation*, John Wiley, Chichester, 1985, pp. 45ff; (b) S. Ahrland, *Pure Appl. Chem.* **54**, 1451 (1982).
24. G. Wipff, *J. Coord. Chem.* **27**, 7 (1992).
25. (a) W. L. Jorgensen, *Chemtracts-Org. Chem.* **4**, 91 (1991); (b) W. L. Jorgensen, *Acc. Chem. Res.* **22**, 184 (1989).
26. K. T. Chapman and W. C. Still, *J. Am. Chem. Soc.* **111**, 3075 (1989).
27. D. J. Murphy, *Biochemistry* **34**, 4507 (1995).
28. (a) J. D. Stewart and S. J. Benkovic, *Nature* **375**, 388 (1995); (b) D. M. Epstein, S. J. Benkovic, and P. E. Wright, *Biochemistry* **34**, 11037 (1995).
29. (a) C. S. Wilcox, J. C. Adrian, T. H. Webb, and F. J. Zawacki, *J. Am. Chem. Soc.* **114**, 10189 (1992); (b) C. S. Wilcox, in *Frontiers in Supramolecular Organic Chemistry and Photochemistry*, Schneider, H. J., Dürr, H., Eds., VCH Publishers, Weinheim, 1991, pp. 123–43.
30. H.-J. Buschmann, *Inorg. Chim. Acta* **195**, 51 (1992).
31. (a) R. M. Izatt, J. L. Oscarson, S. E. Gillespie, X. Chen, P. Wang, and G. D. Watt, *Pure Appl. Chem.* **67**, 543 (1995). (b) Y. Inoue, T. Hakushi, Y. Lin, L.-H. Tong, B.-J. Shen, and D.-S. Jin, *J. Am. Chem. Soc.* **115**, 475 (1993); (c) H.-J. Buschmann, E. Cleve, and E. Schollmeyer, *Thermochim. Acta* **207**, 329 (1992).
32. P. M. May, K. Murray, and D. R. Williams, *Talanta* **35**, 825 (1988).
33. C. H. Park and H. E. Simmons, *J. Am. Chem. Soc.* **90**, 2431 (1968).
34. R. A. Bell, G. G. Christof, F. R. Fronzeck, and R. E. Marsh, *Science* **190**, 151 (1975).

35. B. Dietrich, P. Viout, and J.-M. Lehn, *Macrocyclic Chemistry*, VCH Publishers, Weinheim, 1993.
36. (a) P. Knops, N. Sendhoff, H.-B. Mekelburger, and F. Vögtle, *Top. Curr. Chem.* **161**, 1 (1992); (b) N. Kise, H. Oike, E. Okazaki, M. Yoshimoto, and T. Shono, *J. Org. Chem.* **60**, 3980 (1995).
37. H. Stetter and J. Marx, *Liebig's Ann. Chem.* **607**, 59 (1957).
38. (a) I. Labadi, E. Jenei, R. Lahti, and H. Lönnberg, *Acta Chem. Scan.* **45**, 1055 (1991); (b) R. Lahti and H. Lönnberg, *Biochem. J.* **259**, 55 (1989); (c) W. H. Voige and R. I. Elliott, *J. Chem. Educ.* **59**, 257 (1982).
39. B. Dietrich, M. W. Hosseini, J.-M. Lehn, and R. B. Sessions, *Helv. Chim. Acta* **66**, 1262 (1983).
40. B. Dietrich, M. W. Hosseini, J.-M. Lehn, and R. B. Sessions, *J. Am. Chem. Soc.* **103**, 1282 (1981).
41. M. W. Hosseini and J.-M. Lehn, *Helv. Chim. Acta* **70**, 1312 (1987).
42. A. Bencini, A. Bianchi, M. I. Burguete, A. Doménech, E. García-España, S. V. Luis, M. A. Nino, and J. A. Ramirez, *J. Chem. Soc. Perkin Trans. 2* 1445 (1991).
43. A. Bencini, A. Bianchi, I. Burguete, E. García-España, S. V. Luis, and J. A. Ramirez, *J. Am. Chem. Soc.* **114**, 1919 (1992).
44. (a) J. F. Marecek, P. A. Fischer, and C. J. Burrows, *Tetrahedron Lett.* **29**, 6231 (1988); (b) J. F. Marecek and C. J. Burrows, *Tetrahedron Lett.* **27**, 5943 (1986).
45. J. Cullinane, R. I. Gelb, T. N. Margulis, and L. J. Zompa, *J. Am. Chem. Soc.* **104**, 3048 (1982).
46. R. I. Gelb, B. T. Lee, and L. J. Zompa, *J. Am. Chem. Soc.* **107**, 909 (1985).
47. M. S. Searle, M. S. Westwell, and D. H. Williams, *J. Chem. Soc. Perkin Trans. 2* 141 (1995).
48. D. H. Williams, M. S. Searle, J. P. Mackay, U. Gerhard, and R. A. Maplestone, *Proc. Natl. Acad. Sci. USA* **90**, 1172 (1993).
49. A. Bencini, A. Bianchi, P. Paoletti, and P. Paoli, *Pure Appl. Chem.* **65**, 381 (1993).
50. A. Bencini, A. Bianchi, P. Dapporto, E. García-España, M. Micheloni, P. Paoletti, and P. Paoli, *J. Chem. Soc. Chem. Commun.* 753 (1990).
51. A. Bencini, A. Bianchi, P. Dapporto, E. García-España, M. Micheloni, J. A. Ramirez, P. Paoletti, and P. Paoli, *Inorg. Chem.* **31**, 1902 (1992).
52. A. Bianchi, M. Micheloni, P. Orioli, P. Paoletti, and S. Mangani, *Inorg. Chim. Acta* **146**, 153 (1988).
53. (a) M. W. Hosseini, J.-M. Lehn, and M. P. Mertes, *Helv. Chim. Acta* **66**, 2454 (1983); (b) M. W. Hosseini, A. J. Blacker, and J.-M. Lehn, *J. Am. Chem. Soc.* **112**, 3896 (1990); (c) P. G. Yohannes, K. E. Plute, M. P. Mertes, and K. B. Mertes, *Inorg. Chem.* **26**, 1751 (1987).
54. M. W. Hosseini, J.-M. Lehn, L. Maggiora, K. B. Mertes, and M. P. Mertes, *J. Am. Chem. Soc.* **109**, 537 (1987).
55. (a) M. W. Hosseini, J.-M. Lehn, K. C. Jones, K. E. Plute, K. B. Mertes, and M. P. Mertes, *J. Am. Chem. Soc.* **111**, 6330 (1989); (b) M. W. Hosseini, A. J. Blacker, and J.-M. Lehn, *J. Am. Chem. Soc.* **112**, 3896 (1990); (c) M. P. Mertes and K. B. Mertes, *Acc. Chem. Res.* **23**, 413 (1990).

56. (a) M. W. Hosseini and J.-M. Lehn, *Helv. Chim. Acta* **69**, 587 (1986); (b) idem, *J. Am. Chem. Soc.* **104**, 3525 (1982).
57. R. Breslow, R. Rajagopalan, and J. Schwarz, *J. Am. Chem. Soc.* **103**, 2905 (1981).
58. F. Peter, M. Gross, M. W. Hosseini, J.-M. Lehn, and R. B. Sessions, *J. Chem. Soc. Chem. Commun.* 1067 (1981).
59. E. Kimura, *Top. Curr. Chem.* **128**, 113 (1985).
60. (a) E. Kimura, A. Sakonaka, T. Yatsunami, and M. Kodama, *J. Am. Chem. Soc.* **103**, 3041 (1981); (b) E. Kimura, A. Sakonaka, and M. Kodama, *J. Am. Chem. Soc.* **104**, 4984 (1982).
61. (a) E. Kimura, M. Kodama, and T. Yatsunami, *J. Am. Chem. Soc.* **104**, 3182 (1982); (b) Y. Umezawa, M. Kataoka, W. Takami, E. Kimura, T. Koike, and H. Nada, *Anal. Chem.* **60**, 2392 (1988).
62. (a) M. F. Manfrin, N. Sabbatini, L. Moggi, V. Balzani, M. W. Hosseini, and J.-M. Lehn, *J. Chem. Soc. Chem. Commun.* 555 (1984); (b) M. F. Manfrin, L. Moggi, V. Castelvetro, V. Balzani, M. W. Hosseini, and J.-M. Lehn, *J. Am. Chem. Soc.* **107**, 6888 (1985).
63. E. Kimura, Y. Kuramoto, T. Koike, H. Fujioka, and M. Kodama, *J. Org. Chem.* **55**, 42 (1990).
64. (a) J.-M. Lehn, E. Sonveaux, and A. K. Willard, *J. Am. Chem. Soc.* **100**, 4914 (1978); (b) B. Dietrich, J. Guilhem, J.-M. Lehn, C. Pascard, and E. Sonveaux, *Helv. Chim. Acta* **67**, 91 (1984).
65. B. Dietrich, J.-M. Lehn, J. Guilhem, and C. Pascard, *Tetrahedron Lett.* **30**, 4125 (1989).
66. M. W. Hosseini and J.-M. Lehn, *Helv. Chim. Acta* **71**, 749 (1988).
67. (a) E. Graf and J.-M. Lehn, *J. Am. Chem. Soc.* **97**, 5022 (1975); (b) E. Graf and J.-M. Lehn, *Helv. Chim. Acta* **64**, 1040 (1981).
68. E. Graf and J.-M. Lehn, *J. Am. Chem. Soc.* **98**, 6403 (1976).
69. B. Metz, J. M. Rosalky, and R. Weiss, *J. Chem. Soc. Chem. Commun.* 533 (1976).
70. G. Wipff and J.-M. Wurtz, *New J. Chem.* **13**, 807 (1989).
71. B. Owenson, R. D. McElroy, and A. Pohorille, *J. Am. Chem. Soc.* **110**, 6992 (1988); (b) idem. *J. Mol. Struct. (Theochem.)* **179**, 467 (1988).
72. F. P. Schmidtchen, *Angew. Chem.* **89**, 751 (1977); *Angew. Chem. Int. Ed. Engl.* **16**, 720 (1977).
73. F. P. Schmidtchen, *Chem. Ber.* **113**, 864 (1980).
74. (a) F. P. Schmidtchen, *Chem. Ber.* **114**, 597 (1981); (b) K. Ichikawa, A. Yamamoto, and M. A. Hossain, *Chem. Lett.* **12**, 2175 (1993); (c) M. A. Hossain and K. Ichikawa, *Tetrahedron Lett.* **35**, 8393 (1994).
75. F. P. Schmidtchen, *J. Chem. Soc. Chem. Commun.* 1115 (1984).
76. C. A. Bunton, F. Nome, F. H. Quina, and L. S. Romsted, *Acc. Chem. Res.* **24**, 357 (1991).
77. (a) F. P. Schmidtchen, *Angew. Chem.* **93**, 469 (1980); *Angew. Chem. Int. Ed. Engl.* **20**, 466 (1981); (b) F. P. Schmidtchen, *J. Molec. Catal.* **37**, 141 (1986); (c) F. P. Schmidtchen, *Top. Curr. Chem.* **132**, 101 (1986).
78. (a) F. P. Schmidtchen, *Chem. Ber.* **117**, 725 (1984); (b) idem, *ibid.* **117**, 1287 (1984).

79. F. P. Schmidtchen, *J. Molec. Catal.* **38**, 272 (1986).
80. (a) D. S. Kemp, D. D. Cox, and K. G. Paul, *J. Am. Chem. Soc.* **97**, 7312 (1975); (b) S. N. Thorn, R. G. Daniels, M.-T. M. Auditor, and D. Hilvert, *Nature* **373**, 228 (1995).
81. F. P. Schmidtchen, *J. Chem. Soc. Perkin Trans. 2* 135 (1986).
82. F. P. Schmidtchen, *J. Org. Chem.* **51**, 5161 (1986).
83. F. P. Schmidtchen, *Z. Naturforsch.* **42c**, 476 (1987).
84. (a) F. P. Schmidtchen, *J. Am. Chem. Soc.* **108**, 8249 (1986); (b) idem., *Tetrahedron. Lett.* **27**, 1987 (1986).
85. K. Odashima, A. Itai, Y. Iitaka, and K. Koga, *J. Am. Chem. Soc.* **102**, 2504 (1980).
86. (a) I. Tabushi, Y. Kimura, and K. Yamamura, *J. Am. Chem. Soc.* **100**, 1304 (1978); (b) *idem., ibid.* **103**, 6486 (1981).
87. (a) F. Diederich, *Cyclophanes*, Royal Society of Chemistry Publishers, Cambridge, 1991; (b) F. Vögtle, *Supramolekulare Chemie*, B. G. Teubner Publishers, Stuttgart, 1989; (c) R. Breslow, P. J. Duggan, D. Wiedenfeld, and S. T. Waddell, *Tetrahedron Lett.* **36**, 2707 (1995); (d) M. Cesario, J. Guilhem, and C. Pascard, *Supramolec. Chem.* **2**, 331 (1993).
88. (a) H. J. Schneider, R. Kramer, S. Simova, and Y. Schneider, *J. Am. Chem. Soc.* **110**, 6442 (1988); (b) H. J. Schneider, T. Blatter, A. Eliseev, V. Rüdiger, and O. A. Raevsky, *Pure Appl. Chem.* **65**, 2329 (1993); (c) H. J. Schneider, T. Schiestel, and P. Zimmermann, *J. Am. Chem. Soc.* **114**, 7698 (1992).
89. H. J. Schneider, T. Blatter, B. Palm, U. Pfingstag, V. Rüdiger, and I. Theis, *J. Am. Chem. Soc.* **114**, 7704 (1992).
90. H. J. Schneider, T. Blatter, S. Simova, and I. Theis, *J. Chem. Soc. Chem. Commun.* 580 (1989).
91. S. Claude, J.-M. Lehn, and J.-P. Vigneron, *Tetrahedron Lett.* **30**, 941 (1989).
92. S. Claude, J.-M. Lehn, F. Schmidt, and J.-P. Vigneron, *J. Chem. Soc. Chem. Commun.* 1182 (1991).
93. P. Cudic, M. Zinic, V. Tomisic, V. Simeon, J.-P. Vigneron, and J.-M. Lehn, *J. Chem. Soc. Chem. Commun.* 1073 (1995).
94. For analogous approaches see (a) J.-M. Lehn, R. Meric, J.-P. Vigneron, I. Bkouche-Waksman, and C. Pascard, *J. Chem. Soc. Chem. Commun.* 62 (1991); (b) M. Dhaenens, J.-M. Lehn, and J.-P. Vigneron, *J. Chem. Soc. Perkin Trans. 2* 1379 (1993); (c) M.-P. Teulade-Fichou, J.-P. Vigneron, and J.-M. Lehn, *Supramolec. Chem.* **5**, 139 (1995); (d) A. Slama-Schwok, M.-P. Teulade-Fichou, J.-P. Vigneron, E. Taillandier, and J.-M. Lehn, *J. Am. Chem. Soc.* **117**, 6822 (1995).
95. (a) A. J. Blacker, J. Jazwinski, and J.-M. Lehn, *Helv. Chim. Acta* **70**, 1 (1987); (b) P. D. Beer, J. W. Wheeler, A. Griere, C. Moore, and T. Wear, *J. Chem. Soc. Chem. Commun.* 1225 (1992); (c) W. Gender, S. Huenig, and A. Suchy, *Tetrahedron* **42**, 1665 (1986).
96. (a) D. Heyder and J.-M. Lehn, *Tetrahedron Lett.* **27**, 5869 (1986); (b) T. Fujita and J.-M. Lehn, *Tetrahedron Lett.* **29**, 1709 (1988).
97. (a) Y. Murakami, J. Kikuchi, T. Ohno, T. Hirayama, and H. Nishimura, *Chem. Lett.* 1199 (1989); (b) Y. Murakami, J. Kikuchi, T. Ohno, T. Hirayama, Y. Hisaeda, H. Nishimura, J. P. Snyder, and K. Steliou, *J. Am. Chem. Soc.* **113**, 8229 (1991).

98. (a) Y. Murakami, O. Hayashida, T. Ito, and Y. Hisaeda, *Pure Appl. Chem.* **65**, 551 (1993); (b) O. Hayashida, K. Ono, Y. Hisaeda, and Y. Murakami, *Tetrahedron* **51**, 8423 (1995).

99. M. Huser, W. S. Morf, K. Fluri, K. Seiler, P. Schulthess, and W. Simon, *Helv. Chim. Acta* **73**, 1481 (1990).

100. Y. Kuroda, H. Hatakeyama, N. Inakoshi, and H. Ogoshi, *Tetrahedron Lett.* **34**, 8285 (1993).

101. R. F. Pasternak, A. Antebi, B. Ehrlich, and D. Sidney, *J. Molec. Cat.* **23**, 235 (1984).

102. (a) J. L. Sessler, M. Cyr, H. Furuta, V. Kral, T. Mody, T. Morishima, M. Shionoya, and S. Weghorn, *Pure Appl. Chem.* **63**, 393 (1993); (b) J. L. Sessler and A. K. Burrell, *Top. Curr. Chem.* **161**, 177 (1991).

103. B. Franck and A. Nonn, *Angew. Chem. Int. Ed. Engl.* **34**, 1795 (1995).

104. (a) E. Vogel, J. L. Sessler, E. A. Brucker, S. J. Weghorn, M. Kisters, M. Schaefer, and J. Lex, *Angew. Chem. Int. Ed. Engl.* **33**, 2308 (1994); (b) E. Vogel, *Pure Appl. Chem.* **65**, 143 (1993).

105. (a) First described by R. B. Woodward at the conference in Sheffield, UK, 1966, see: V. J. Bauer, D. L. J. Clive, D. Dolphin, J. P. Paine III, F. L. Harris, M. M. King, H. Loder, S.-W. C. Wang, and R. B. Woodward, *J. Am. Chem. Soc.* **105**, 6429 (1983); (b) M. J. Broadhurst, R. Grigg, and A. W. Johnson, *J. Chem. Soc. Perkin Trans. 1*, 2111 (1972).

106. (a) J. L. Sessler, M. J. Cyr, V. Lynch, E. McGhee, and J. A. Ibers, *J. Am. Chem. Soc.* **112**, 2810 (1990); (b) J. L. Sessler, V. Lynch, and M. R. Johnson, *J. Org. Chem.* **52**, 4394 (1987); (c) J. L. Sessler, M. J. Cyr, and A. K. Burrell, *Synlett.* 127 (1991).

107. (a) J. L. Sessler, M. J. Cyr, V. Lynch, E. McGhee, and J. A. Ibers, *J. Am. Chem. Soc.* **112**, 2810 (1990); (b) M. Shionoya, H. Furuta, V. Lynch, A. Harriman, and J. L. Sessler, *J. Am. Chem. Soc.* **114**, 5714 (1992).

108. (a) V. Kral, J. L. Sessler, and H. Furuta, *J. Am. Chem. Soc.* **114**, 8704 (1992); (b) H. Furuta, M. J. Cyr, and J. L. Sessler, *J. Am. Chem. Soc.* **113**, 6677 (1991).

109. B. L. Iverson, R. E. Thomas, V. Kral, and J. L. Sessler, *J. Am. Chem. Soc.* **116**, 2623 (1994).

110. B. L. Iverson, K. Shreder, V. Kral, and J. L. Sessler, *J. Am. Chem. Soc.* **115**, 11022 (1993).

111. V. Kral, A. Andrievsky, and J. L. Sessler, *J. Am. Chem. Soc.* **117**, 2953 (1995).

112. J. L. Sessler, T. D. Mody, D. A. Ford, and V. Lynch, *Angew. Chem. Int. Ed. Engl.* **31**, 452 (1992).

113. H. L. Anderson and J. K. M. Sanders, *J. Chem. Soc. Chem. Commun.* 946 (1992).

114. (a) J. F. Riordan, *Mol. Cell. Biochem.* **26**, 71 (1979); (b) L. Shimoni and J. P. Glusker, *Protein Sci.* **4**, 65 (1995).

115. E. V. Anslyn and C. L. Hannon, The Guanidinium Group: Its Biological Role and Synthetic Analogs, in *Bioorg. Chem. Frontiers, Vol. 3*, H. Dugas, F. P. Schmidtchen, Eds., Springer Publishers, Berlin, 1993, pp. 193–255.

116. (a) A. R. Clarke, T. Atkinson, and J. J. Holbrook, *Trends Biochem. Sci.* **14**, 101 (1989); (b) Y. Inoue, S. Kuramitsu, K. Inoue, H. Kagamiyama, K. Hiromi, S. Tanase, and Y. Morino, *J. Biol. Chem.* **264**, 9673 (1989); (c) P. W. White and J. F. Kirsch, *J. Am. Chem. Soc.* **114**, 3567 (1992).

117. (a) see refs. in P. Chakrabarti, *Int. J. Peptide Protein Res.* **43**, 284 (1994); (b) V. Tsikaris, M. T. Cung, E. Panou-Pomonis, and M. Sakarellos-Daitsiotis, *J. Chem. Soc. Perkin Trans. 2* 1345 (1993); (c) Y. Yokomori and D. J. Hodgson, *Int. J. Peptide Prot. Res.* **31**, 289 (1988).

118. R. L. Wolfenden Andersson, P. M. Cullis, and C. C. B. Southgate, *Biochemistry* **20**, 849 (1981).

119. B. Springs and P. Haake, *Bioorg. Chem.* **6**, 181 (1977).

120. S. Boudon, G. Wipff, and B. Maigret, *J. Phys. Chem.* **94**, 6056 (1990).

121. R. Schwesinger, *Chimica* **39**, 269 (1985).

122. B. Dietrich, T. M. Fyles, J.-M. Lehn, L. G. Pease, and D. L. Fyles, *J. Chem. Soc. Chem. Commun.* 934 (1978).

123. B. Dietrich, D. L. Fyles, T. M. Fyles, and J.-M. Lehn, *Helv. Chim. Acta* **62**, 2763 (1979).

124. (a) D. S. Sigman, A. Mazumder, and D. M. Perrin, *Chem. Rev.* **93**, 2295 (1993); (b) K. N. Dalby, A. J. Kirby, and F. Hollfelder, *Pure Appl. Chem.* **66**, 687 (1994).

125. (a) R. P. Dixon, S. J. Geib, and A. D. Hamilton, *J. Am. Chem. Soc.* **114**, 365 (1992); (b) R. P. Dixon, V. Jubian, and A. D. Hamilton, *J. Am. Chem. Soc.* **114**, 1120 (1992); (c) E. Fan, S. A. V. Arman, S. Kincaid, and A. D. Hamilton, *J. Am. Chem. Soc.* **115**, 369 (1993); (d) R. Gross, G. Dürner, and M. W. Göbel, *Liebig's Ann. Chem.* 49 (1994); (e) R. Gross, J. W. Bats, and M. W. Göbel, *Liebig's Ann. Chem.* 205 (1994).

126. (a) L. S. Flatt, V. Lynch, and E. V. Anslyn, *Tetrahedron Lett.* **33**, 2785 (1992); (b) K. Ariga, E. V. Anslyn, *J. Org. Chem.* **57**, 417 (1992).

127. F. P. Schmidtchen, *Tetrahedron Lett.* **30**, 4493 (1989).

128. (a) R. Breslow and B. Zhang, *J. Am. Chem. Soc.* **116**, 7893 (1994); (b) B. Linkletter and J. Chin, *Angew. Chem. Int. Ed. Engl.* **34**, 472 (1995); (c) S. Amin, J. R. Morrow, C. H. Lake, and M. R. Churchill, *Angew. Chem. Int. Ed. Engl.* **33**, 773 (1994).

129. F. P. Schmidtchen, *Chem. Ber.* **113**, 2175 (1980). For related amidinium salts see (a) F. Heinzer, M. Soukup, and A. Eschenmoser, *Helv. Chim. Acta* **61**, 2851 (1978); (b) M. A. Convery, A. P. Davis, C. J. Dunne, and J. W. Mackinnou, *Tetrahedron Lett.* **36**, 4279 (1995); (c) idem., *J. Chem. Soc. Chem. Commun.* 2557 (1994).

130. A. Echavarren, A. Galan, J. de Mendoza, A. Salmeron, and J.-M. Lehn, *Helv. Chim. Acta* **71**, 685 (1988).

131. (a) F. P. Schmidtchen, *Tetrahedron Lett.* **31**, 2269 (1990); (b) H. Kurzmeier and F. P. Schmidtchen, *J. Org. Chem.* **55**, 3749 (1990).

132. (a) A. Gleich and F. P. Schmidtchen, *Chem. Ber.* **123**, 907 (1990); (b) F. P. Schmidtchen, H. Oswald, and A. Schummer, *Liebig's Ann. Chem.* 539–43 (1991).

133. G. Müller, J. Riede, and F. P. Schmidtchen, *Angew. Chem. Int. Ed. Engl.* **27**, 1516–18 (1988).

134. A. Echavarren, A. Galan, J.-M. Lehn, and J. de Mendoza, *J. Am. Chem. Soc.* **111**, 4994 (1989).

135. A. Gleich, F. P. Schmidtchen, P. Mikulcik, and G. Müller, *J. Chem. Soc. Chem. Commun.* 55 (1990).

136. (a) E. van Aken, H. Wynberg, and F. van Bolhuis, *J. Chem. Soc. Chem. Commun.*

629 (1992); *idem., Acta Chem. Scand.* **47**, 122 (1993); (b) cf. R. Chincilla, C. Najera, and P. Sanchez-Agullo, *Tetrahedron Asym.* **5**, 1393 (1994).

137. P. H. Boyle, M. A. Convery, A. P. Davis, G. D. Hosken, and B. A. Murray, *J. Chem. Soc. Chem. Commun.* 239 (1992).

138. V. Alcazar, J. R. Moran, and J. de Mendoza, *Tetrahedron Lett.* **36**, 3941 (1995).

139. (a) E. J. Corey and M. Ohtani, *Tetrahedron Lett.* **30**, 5227 (1989); (b) P. Molina, M. Alajarin, and A. Vidal, *Tetrahedron* **51**, 5351 (1995); (c) P. Molina, R. Obon, C. Conesa, A. Arques, M. de los Desamparados Velasco, A. L. Llamas-Saiz, and C. Foces-Foces, *Chem. Ber.* **127**, 1641 (1994); (d) P. Molina, M. Alajarin, and A. Vidal, *J. Org. Chem.* **58**, 1687 (1993); (e) J.-L. Chicharro, P. Prados, and J. de Mendoza, *J. Chem. Soc. Chem. Commun.* 1193 (1994).

140. R. S. Hutchins, P. Molina, M. Alajarin, A. Vidal, and L. G. Bachas, *Anal. Chem.* **66**, 3188 (1994).

141. F. P. Schmidtchen, *Tetrahedron Lett.* **30**, 4493 (1989).

142. P. Schiessl and F. P. Schmidtchen, *J. Org. Chem.* **59**, 509 (1994).

143. P. Schiessl and F. P. Schmidtchen, *Tetrahedron Lett.* **34**, 2449 (1993).

144. W. Peschke and F. P. Schmidtchen, *Tetrahedron Lett.* **36**, 5155 (1995).

145. H. Stephan, K. Gloe, P. Schiessl, and F. P. Schmidtchen, *Supramolec. Chem.* **5**, 273 (1995).

146. A. Galan, E. Pueyo, A. Salmeron, and J. de Mendoza, *Tetrahedron Lett.* **32**, 1827 (1991).

147. G. Deslongchamps, A. Galan, J. de Mendoza, and J. Rebek, Jr., *Angew. Chem. Int. Ed. Engl.* **31**, 61 (1992).

148. Y. Kato, M. M. Conn, and J. Rebek, Jr., *J. Am. Chem. Soc.* **116**, 3279 (1994).

149. A. Galan, J. de Mendoza, C. Toiron, M. Bruix, G. Deslongchamps, and J. Rebek, Jr., *J. Am. Chem. Soc.* **113**, 9424 (1991).

150. C. Andreu, A. Galan, K. Kobiro, J. de Mendoza, T. K. Park, J. Rebek, Jr., A. Salmeron, and N. Usman, *J. Am. Chem. Soc.* **116**, 5501 (1994).

151. C. Seel, A. Galan, and J. de Mendoza, *Top. Curr. Chem.* **175**, 101 (1995).

152. A. Galan, C. Andreu, A. M. Echavarren, P. Prados, and J. de Mendoza, *J. Am. Chem. Soc.* **114**, 1511 (1992).

153. F. Gago and J. de Mendoza, in *NATO ASI Series C: Vol. 426*, pp. 79–99 (1994).

154. A. Metzger, K. Gloe, H. Stephan, and F. P. Schmidtchen, *J. Org. Chem.*, **61**, 2051 (1996).

155. P. Vierling and J.-M. Lehn, *Tetrahedron Lett.* **21**, 1323 (1980).

156. (a) J. Szejtli, *Cyclodextrins and their Inclusion Complexes*, Akademiai Kiado Publishers, Budapest, 1982; (b) O. S. Tee, *Adv. Phys. Org. Chem.* **29**, 1 (1994).

157. (a) J. Taraszewska and J. Wojcik, *Supramolec. Chem.* **2**, 337 (1993); (b) I. Sanemasa, M. Fujiki, and T. Deguchi, *Bull. Chem. Soc. Jpn.* **61**, 2663 (1988); (c) R. I. Gelb, L. M. Schwartz, M. Radeos, and D. A. Laufer, *J. Phys. Chem.* **87**, 3349 (1983).

158. (a) K. Fujita, A. Matsunaga, and T. Imoto, *Tetrahedron Lett.* **25**, 5533 (1984); (b) E. van Dienst, B. H. M. Snellnik, I. von Piekartz, M. H. B. Grote Gansey, F. Venema, M. C. Feiters, R. J. M. Nolte, J. F. J. Engbersen, and D. N. Reinhoudt, *J. Org. Chem.* **60**, 6537 (1995).

159. J. Boger and J. R. Knowles, *J. Am. Chem. Soc.* **101**, 7631 (1979).
160. (a) I. Tabushi, Y. Kuroda, and T. Mizutani, *Tetrahedron* **40**, 545 (1984); (b) I. Tabushi, N. Shimizu, T. Sugimoto, M. Shiozuka, and K. Yamamura, *J. Am. Chem. Soc.* **99**, 7100 (1977).
161. I. Willner, Z. Goren, *J. Chem. Soc. Chem. Commun.* 1469 (1983).
162. (a) R. Breslow, *Pure Appl. Chem.* **62**, 1859 (1990); (b) E. V. Anslyn and R. Breslow, *J. Am. Chem. Soc.* **111**, 5972 (1989); (c) R. Breslow, *Acc. Chem. Res.* **24**, 317 (1991).
163. (a) A. V. Eliseev and H.-J. Schneider, *Angew. Chem. Int. Ed. Engl.* **32**, 1331 (1993); (b) *idem., J. Am. Chem. Soc.* **116**, 6081 (1994).
164. P. D. Beer, *Adv. Mater.* **6**, 607 (1994); *idem., Adv. Inorg. Chem.* **39**, 79 (1992).
165. P. D. Beer, D. Hesek, J. Hodakova, and S. E. Stokes, *J. Chem. Soc. Chem. Commun.* 270 (1992).
166. P. D. Beer, Z. Chen, A. J. Goulden, A. Grieve, D. Hesek, F. Szemes, and T. Wear, *J. Chem. Soc. Chem. Commun.* 1269 (1994).
167. P. D. Beer, Z. Chen, M. G. B. Drew, J. Kingston, M. Ogden, and P. Spencer, *J. Chem. Soc. Chem. Commun.* 1046 (1993).
168. X. Yang, Z. Zheng, C. B. Knobler, and M. F. Hawthorne, *J. Am. Chem. Soc.* **115**, 193 (1993).
169. H. E. Katz, *J. Org. Chem.* **50**, 5027 (1985).
170. H. E. Katz, *J. Am. Chem. Soc.* **108**, 7640 (1986).
171. S. Jacobson and R. Pizer, *J. Am. Chem. Soc.* **115**, 11216 (1993).
172. (a) M. T. Reetz, C. M. Niemeyer, and K. Harms, *Angew. Chem. Int. Ed. Engl.* **30**, 1472 (1991); (b) a similar concept was realized employing an aluminium Lewis acidic site: M. T. Reetz, B. M. Johnson, and K. Harms, *Tetrahedron Lett.* **35**, 2525 (1994).
173. M. E. Jung and H. Xia, *Tetrahedron Lett.* **29**, 297 (1988).
174. S. Aoyagi, K. Tanaka, and Y. Takeuchi, *J. Chem. Soc. Perkin Trans. 2*, 1549 (1994).
175. (a) K. Jurkschat, H. G. Kuivila, S. Liu, and J. A. Zubieta, *Organometallics* **8**, 2755 (1989); (b) K. Jurkschat, A. Rühlemann, and A. Tzschach, *J. Organomet. Chem.* **381**, C 53 (1990).
176. (a) M. Newcomb, A. M. Madonik, M. T. Blanda, and J. K. Judice, *Organometallics* **6**, 145 (1987); (b) Y. Azuma, M. Newcomb, *Organometallics* **3**, 9 (1984).
177. M. Newcomb and M. T. Blanda, *Tetrahedron Lett.* **29**, 4261 (1988).
178. M. Newcomb, J. H. Horner, M. T. Blanda, and P. J. Squattrito, *J. Am. Chem. Soc.* **111**, 6294 (1989).
179. M. T. Blanda, J. H. Horner, and M. Newcomb, *J. Org. Chem.* **54**, 4626 (1989).
180. M. T. Blanda and M. Newcomb, *Tetrahedron Lett.* **30**, 3501 (1989).
181. K. Fluri, J. Koudelka, and W. Simon, *Helv. Chim. Acta* **75**, 1012 (1992).
182. (a) N. A. Chaniotakis, K. Jurkschat, and A. Rühlemann, *Anal. Chim. Acta* **282**, 345 (1993); (b) J. K. Tsagatakis and N. A. Chaniotakis, *Helv. Chim. Acta* **77**, 2191 (1994).
183. A. L. Beauchamp, M. J. Olivier, J. D. Wuest, and B. Zacharie, *J. Am. Chem. Soc.* **108**, 73 (1986).
184. (a) V. B. Shur, I. A. Tikhonova, A. I. Yanovsky, Y. T. Struchkov, P. V. Petrovskii,

S. Y. Panov, G. G. Furin, and M. E. Vol'pin, *J. Organomet. Chem.* **418**, C 29 (1991); (b) V. B. Shur, I. A. Tikhonova, F. M. Dolgushin, A. I. Yanovsky, Y. T. Struchkov, A. Y. Volkonsky, E. V. Solodova, S. Y. Panov, P. V. Petrovskii, and M. E. Vol'pin, *J. Organomet. Chem.* **443**, C 19 (1993).

185. (a) X. Yang, C. B. Knobler, and M. F. Hawthorne, *Angew. Chem. Int. Ed. Engl.* **30**, 1507 (1991); (b) idem., *J. Am. Chem. Soc.* **114**, 380 (1992).

186. (a) X. Yang, S. E. Johnson, S. I. Khan, and M. F. Hawthorne, *Angew. Chem. Int. Ed. Engl.* **31**, 893 (1992); (b) X. Yang, C. B. Knobler, Z. Zheng, and M. F. Hawthorne, *J. Am. Chem. Soc.* **116**, 7142 (1994).

187. (a) Z. Zheng, X. Yang, C. B. Knobler, and M. F. Hawthorne, *J. Am. Chem. Soc.* **115**, 5320 (1993); (b) Z. Zheng, C. B. Knobler, and M. F. Hawthorne, *J. Am. Chem. Soc.* **117**, 5105 (1995).

188. X. Yang, Z. Theng, and M. F. Hawthorne, *Pure Appl. Chem.* **66**, 245 (1994).

189. (a) K. Casellato, P. A. Vigato, and M. Vidali, *Coord. Chem. Rev.* **23**, 31 (1977); (b) V. McKee, *Adv. Inorg. Chem.* **40**, 323 (1993); (c) A. Bianchi, A. Bencini, P. Paoletti, and E. García-España, *NATO Asi Ser. Ser. C* **448**, 309 (1994).

190. J.-M. Lehn, in *IUPAC "Frontiers of Chemistry,"* K. J. Laidler, Ed., Pergamon Press, Oxford, 1982, p. 265.

191. R. J. Motekaitis, A. E. Martell, B. Dietrich, and J.-M. Lehn, *Inorg. Chem.* **23**, 1588 (1984).

192. R. J. Motekaitis, A. E. Martell, J.-M. Lehn, and E. I. Watanabe, *Inorg. Chem.* **21**, 4253 (1982).

193. M. W. Hosseini and J.-M. Lehn, *Helv. Chim. Acta* **71**, 749 (1988).

194. R. J. Motekaitis, A. E. Martell, J. P. Lecomte, and J.-M. Lehn, *Inorg. Chem.* **22**, 609 (1983).

195. P. Comarmond, P. Plumere, J.-M. Lehn, Y. Agnus, R. Louis, R. Weiss, O. Kahn, and I. Morgenstern-Badaran, *J. Am. Chem. Soc.* **104**, 6330 (1982).

196. R. J. Motekaitis and A. E. Martell, *Inorg. Chem.* **31**, 5534 (1992).

197. R. J. Motekaitis, W. B. Utley, and A. E. Martell, *Inorg. Chim. Acta* **212**, 15 (1993).

198. P. K. Coughlin and S. J. Lippard, *J. Am. Chem. Soc.* **103**, 3228 (1981).

199. P. E. Jurek, A. E. Martell, R. J. Motekaitis, and R. D. Hancock, *Inorg. Chem.* **34**, 1823 (1995).

200. S. Kitagawa, M. Kondo, S. Kawata, S. Wada, M. Maekawa, and M. Munakata, *Inorg. Chem.* **34**, 1455 (1995).

201. H. M. Colquhoun, J. F. Stoddard, and D. J. Williams, *Angew. Chem. Int. Ed. Engl.* **25**, 487 (1986).

202. K. Konishi, K. Yahora, H. Toshishige, T. Aida, and S. Inoue, *J. Am. Chem. Soc.* **116**, 1337 (1994).

203. (a) D. M. Rudkevich, W. Verboom, Z. Brzozka, M. J. Palys, W. P. V. R. Stauthamer, G. J. van Hummel, S. M. Franken, S. Harkema, J. F. J. Engbersen, and D. N. Reinhoudt, *J. Am. Chem. Soc.* **116**, 4341 (1994); (b) D. M. Rudkevich, W. P. R. V. Stauthamer, W. Verboom, J. F. J. Engbersen, S. Harkema, and D. N. Reinhoudt, *J. Am. Chem. Soc.* **114**, 9671 (1992).

204. H. C. Visser, D. M. Rudkevich, W. Verboom, F. de Jong, and D. N. Reinhoudt, *J. Am. Chem. Soc.* **116**, 11554 (1994).

205. S. M. Lacy, D. M. Rudkevich, W. Verboom, and D. N. Reinhoudt, *J. Chem. Soc. Perkin Trans. 2*, 135 (1995).
206. (a) J. Jazwinski, J.-M. Lehn, D. Lilienbaum, R. Ziessel, J. Guilhem, and C. Pascard, *J. Chem. Soc. Chem. Commun.* 1691 (1987); (b) J.-M. Lehn, R. Méric, J. P. Vigneron, I. Bkouche-Waksman, and C. Pascard, *J. Chem. Soc. Commun.* 62 (1991).
207. C. B. Black, H.-W. Huong, and J. A. Cowan, *Coord. Chem. Rev.* **135**, 165 (1994).
208. J. W. Yun, T. Tanase, L. E. Pence, and S. J. Lippard, *J. Am. Chem. Soc.* **117**, 4407 (1995).
209. J. Macro, J. M. Orza, R. Notario, and J. L. M. Abboud, *J. Am. Chem. Soc.* **116**, 8841 (1994), and refs. cited therein.
210. (a) W. W. Cleland and M. M. Kreevoy, *Science* **264**, 1887 (1994); (b) P. A. Frey, S. A. Whitt, and J. B. Tobin, *Science* **264**, 1927 (1994); (c) A. Warshel, A. Papazyan, and P. A. Kollman, *Science* **269**, 102 (1995); (d) W. W. Cleland and M. M. Kreevoy, *Science* **269**, 103 (1995); (e) P. A. Frey, *Science* **269**, 104 (1995); (f) S. Scheiner and T. Kar, *J. Am. Chem. Soc.* **117**, 6970 (1995).
211. (a) A. R. Fersht, *Trends Biochem. Sci.* **12**, 301 (1987); (b) W. L. Jorgensen, *J. Am. Chem. Soc.* **111**, 3770 (1989).
212. T.-Y. Lin, W.-H. Lin, W. D. Clarke, R. J. Lagow, S. B. Larson, S. H. Simonsen, V. M. Lynch, J. S. Brodbelt, S. D. Maleknia, and C.-C. Liou, *J. Am. Chem. Soc.* **116**, 5172 (1994).
213. W. B. Farnham, D. C. Roe, D. A. Dixon, J. C. Calabrese, and R. L. Harlow, *J. Am. Chem. Soc.* **112**, 7707 (1990).
214. B. L. Jacobson and F. A. Quiocho, *J. Mol. Biol.* **204**, 783 (1988).
215. T. R. Kelly and M. H. Kim, *J. Am. Chem. Soc.* **116**, 7072 (1994).
216. E. Fan, S. A. V. Arman, S. Kincaid, and A. D. Hamilton, *J. Am. Chem. Soc.* **115**, 369 (1993).
217. (a) E. Grunwald and C. Steel, *J. Am. Chem. Soc.* **117**, 5687–92 (1995); (b) D. H. Williams, M. S. Searle, M. S. Westwell, J. P. Mackay, P. Groves, and D. A. Beauregard, *Chemtracts Org. Chem.* **7**, 133–59 (1994); (c) M. C. Chervenak and E. J. Toone, *J. Am. Chem. Soc.* **116**, 10533 (1994); (d) B. R. Peterson, P. Wallimann, D. R. Carcanague, and F. Diederich, *Tetrahedron* **51**, 401 (1995).
218. J. C. Adrian, Jr. and C. S. Wilcox, *J. Am. Chem. Soc.* **114**, 1398 (1992).
219. S. Valiyaveetil, J. F. J. Engbersen, W. Verboom, and D. N. Reinhoudt, *Angew. Chem. Int. Ed. Engl.* **32**, 900 (1993).
220. C. Raposo, N. Pérez, M. Almaraz, M. L. Mussons, M. C. Caballero, and J. R. Morán, *Tetrahedron Lett.* **36**, 3255 (1995).
221. P. D. Beer, Z. Chen, A. J. Goulden, A. Graydon, S. Stokes, and T. Wear, *J. Chem. Soc. Chem. Commun.* 1834 (1993).
222. H. Ishida, M. Suga, K. Donowaki, and K. Ohkugbo, *J. Org. Chem.* **60**, 5374 (1995).
223. (a) S. S. Yoon and W. C. Still, *J. Am. Chem. Soc.* **115**, 823 (1993); (b) S. S. Yoon and W. C. Still, *Tetrahedron Lett.* **35**, 2117 (1994).
224. Y. Morzherin, D. M. Rudkevich, W. Verboom, and D. N. Reinhoudt, *J. Org. Chem.* **58**, 7602 (1993).

225. J. Schreeder, M. Fochi, J. F. J. Engbersen, and D. N. Reinhoudt, *J. Org. Chem.* **59**, 7815 (1994).
226. J. Schreeder, J. F. J. Engbersen, A. Casnati, R. Ungaro, and D. N. Reinhoudt, *J. Org. Chem.* **60**, 6448 (1995).
227. K. Worm, F. P. Schmidtchen, A. Schier, A. Schäfer, and M. Hesse, *Angew. Chem. Int. Ed. Engl.* **33**, 327 (1994).
228. P. B. Savage, S. K. Holmgren, and S. H. Gellman, *J. Am. Chem. Soc.* **116**, 4069 (1994).
229. P. B. Savage, S. K. Holmgren, and S. H. Gellman, *J. Am. Chem. Soc.* **115**, 7900 (1993).
230. (a) M. H. Abraham, P. L. Grellier, D. V. Prior, J. J. Morris, P. J. Taylor, C. Laurence, and M. Berthelot, *Tetrahedron Lett.* **30**, 2571 (1989); (b) M. H. Abraham, *Chem. Soc. Rev.*, 73 (1993).
231. K. Worm, F. P. Schmidtchen, *Angew. Chem. Int. Ed. Engl.* **34**, 65 (1995).

CHAPTER 5

Structural and Topological Aspects of Anion Coordination

JERRY L. ATWOOD and JONATHAN W. STEED

5.1 Introduction
    5.1.1 Historical Perspective
5.2 Protonated Polyamine-Based Receptors
    5.2.1 Acyclic and Monocyclic Hosts
    5.2.2 Polycyclic Hosts
5.3 The Guanidinium Moiety
5.4 Cyclophane Receptors
5.5 Neutral Receptors
    5.5.1 Organic Hosts
    5.5.2 Inorganic Hosts
5.6 Lewis Acid Receptors
    5.6.1 Hosts Containing Boron and Silicon
    5.6.2 Hosts Containing Mercury
    5.6.3 Hosts Containing Tin
5.7 Organometallic Receptors
5.8 Transition Metal Complexes
    5.8.1 Zinc Complexes
    5.8.2 Copper Complexes
5.9 Polymetallic Receptors
    5.9.1 Polyoxometallate Hosts
    5.9.2 Helicates
5.10 Alkalide and Electride Salts
5.11 Anion Complexation in Biochemistry
5.12 Concluding Remarks
5.13 References

Supramolecular Chemistry of Anions, Edited by Antonio Bianchi, Kristin Bowman-James, and Enrique García-España.
ISBN 0-471-18622-8. © 1997 Wiley-VCH, Inc.

5.1 INTRODUCTION

In the chemistry of the life processes anion binding plays a central role, as evidenced by the fact that the substrates of a large majority of characterized enzymes are anionic in nature.[1] In spite of this, the design and synthesis of anion binding hosts has been relatively slow to develop, in contrast to the analogous chemistry of cation receptors. This discrepancy may readily be traced to a number of inherent difficulties in anion binding[2-4]:

1. Anions are relatively large and therefore require receptors of considerably greater size than cations. For example, one of the smallest anions, F^-, is comparable in ionic radius to K^+ (1.36 Å vs. 1.33 Å). Other selected anion radii are shown in Table 5.1.[2]
2. Even simple inorganic anions occur in a range of shapes and geometries, for example, spherical in the case of halides, tetrahedral (PO_4^{3-}, SO_4^{2-}), planar (NO_3^-), linear (SCN^-, N_3^-), as well as more complicated examples as in the case of biologically important oligophosphate anions.[2]
3. In comparison to cations of similar size, anions have high free energies of solvation, and hence, anion hosts must compete more effectively with the surrounding medium (e.g., $\Delta G_{F^-} = -434.3$ kJ mol^{-1}, $\Delta G_{K^+} = -337.2$ kJ mol^{-1}).[5]
4. Many anions only exist in a relatively narrow pH window, which can cause problems, especially in the case of receptors based on polyammonium salts, where the host may not be fully protonated in the pH region in which the anion is present in the desired form.
5. Finally, anions are usually coordinatively saturated, and therefore, other than the strong electrostatic forces provided by positively charged receptors, they bind only via weak forces such as hydrogen bonding and van der Waals interactions.

These problems have been addressed in a wide variety of imaginative and novel ways, and progress in anion complexation has been rapid in recent years[2-4,7-10] to the extent that it has now been described by Lehn as "a full member of the field of supramolecular chemistry."[2b]

In this chapter we focus upon points (1) and (2) above by an examination of the ways in which the sizes and shapes of anionic guests may be utilized in the

TABLE 5.1. Ionic Radii (Å)[2,6]

Anion	Radius	Anion	Radius	Anion	Radius	Anion	Radius
F^-	1.36	OH^-	1.40	NO_2^-	1.55	CO_3^{2-}	1.85
Cl^-	1.81	CN^-	1.82	NO_3^-	1.89	SO_4^{2-}	2.30
Br^-	1.95	IO_3^-	1.82	MnO_4^-	2.40	PO_4^{3-}	2.38
I^-	2.16	Na^-	2.2	Cs^-	3.5		

synthesis of hosts exhibiting a high degree of anion affinity and selectivity. In particular, the results of X-ray crystallographic studies will be examined as well as aspects of solution binding related to discrimination between anionic guests based on structure and shape, as distinct from charge-density considerations.

5.1.1 Historical Perspective

The first report of designed anion hosts, the bicyclic *katapinands* (**5.1**), was published in 1968[11] (cf. the discovery of the aptitude of crown ethers for alkali metal ions in 1967[12,13]). In 1975 the hypothesis that these materials encapsulate halide anions was confirmed by an X-ray crystallographic determination of the structure of chloridekatapinato-*in,in*-1,11-diazabicyclo[9.9.9]nonacosane-*bis*(ammonium)chloride.[14]

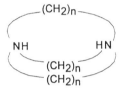

5.1 n = 7-10

Progress in the field lagged until the mid-1970s, however, at which time the structure appeared of chloride ion encapsulated in a macrotricyclic ligand, which became known as the "soccer ball" ligand (**5.2**)[15] Structures of other protonated azamacrocycles capable of binding oxoanions had appeared by this time, but in these structures the anions rested outside the macrocyclic ring.[16] This paucity in structural information reflected both the slow development of the field of anion complexation and the ever-present difficulties associated with the crystallography of supramolecular complexes, namely, poorly diffracting

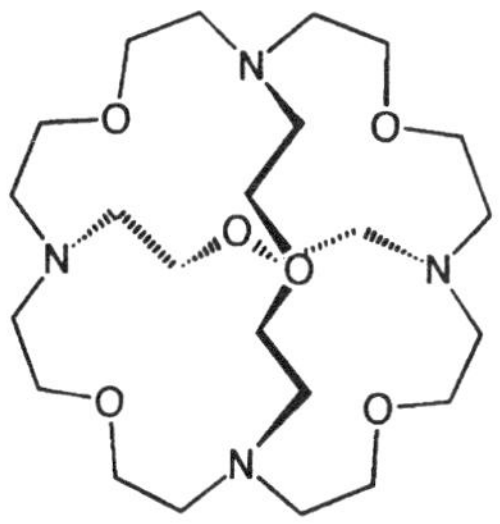

5.2

crystals exhibiting extensive disorder of host, guests, and enclathrated solvent molecules.

A key turning point in the field of supramolecular anion chemistry came in 1984, when Lehn and co-workers reported the structures of four anion cryptates $X^-\cdot\mathbf{BT}\cdot 6H^+$ (**BT** = *bis*-tren, **5.3**), in which the structural complementarity between the hexaprotonated $\mathbf{BT}\cdot 6H^+$ and, particularly, N_3^-, was linked to the high association constant of the *bis*-tren host for the anion.[17] Since that time, the increasing number of anion hosts, coupled with improvements in crystallographic and computational techniques, has resulted in an explosion of structural studies on supramolecular complexes of anions, which will be highlighted in the sections to follow.

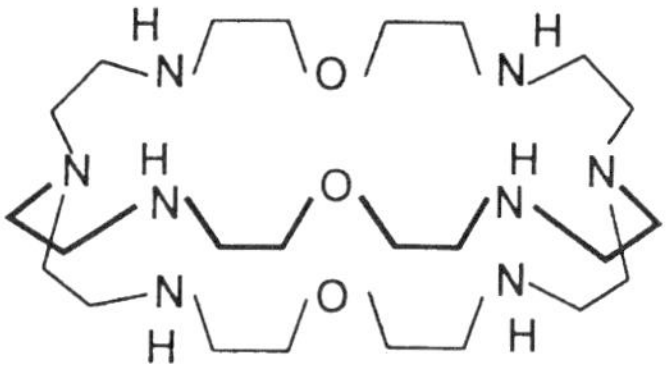

5.3

5.2 PROTONATED POLYAMINE-BASED RECEPTORS

5.2.1 Acyclic and Monocyclic Receptors

By far the most common type of anion binding host is based on numerous substituted forms of the ammonium ion. Alkyl ammonium salts have the advantage of positive charge for anion binding via electrostatic (i.e., nondirectional) forces, polar N–H bonds capable of forming hydrogen-bonding interactions, as well as an extensive degree of synthetic versatility, enabling them to be incorporated into a wide range of multidentate host frameworks of the desired solubility and geometry. Furthermore, polyammonium-based anion hosts are derived, conceptually at least, from analogous cation-binding polyamine ligands. Hence, a change in pH is often sufficient to form an anion binding host from a known amine ligand. To take a very simple example, ethylenediamine ($H_2NCH_2CH_2NH_2$, en) forms numerous Lewis acid–base complexes with a variety of metal ions.[18] Similarly, the ethylenediammonium cation ($H_3NCH_2CH_2NH_3^{2+}$, en·$2H^+$) forms simple salts with anions such as citrate (Fig. 5.1).[19] In contrast to the stabilization of metal complexes by the chelating en ligand, however, the interactions of en·$2H^+$ with anions is not chelating in nature, inasmuch as the central C–C bond adopts a *trans* configuration with the two $-NH_3^+$ moieties binding to different anions. This bridging mode of interaction is readily attributed to the need for the two positively

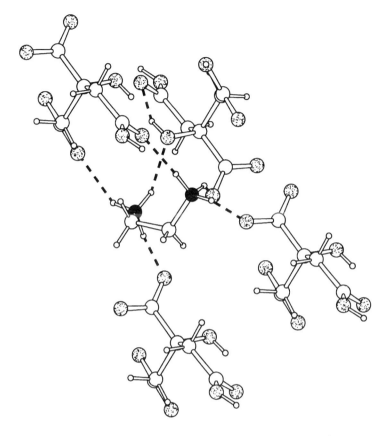

Fig. 5.1. Complex of ethylenediaminium dication ($H_3NCH_2CH_2NH_3^{2+}$) with citrate.

charged centers to move as far away from one another as possible, thus minimizing unfavorable cation–cation interactions.

A similar situation is found in the structure of putrescine diphosphate[20] (a model system for amine–nucleic acid interactions), which divides into layers of $H_2PO_4^-$ anions bridged by protonated putrescine (1,4-diamino-n-butane) cations. In a real biological system (yeast phenylalanine transfer RNA), phosphate residues are found to be enveloped by spermine $NH_2(CH_2)_3NH(CH_2)_4NH(CH_2)_3NH_2$, a polyamine that again adopts a linear nonchelating conformation.[21]

This evidence points towards a markedly decreased tendency of acyclic polyammonium hosts to form chelate complexes with anions as a consequence of the need to separate the cationic nitrogen centers. This tendency is also important in the case of cyclic structures, in which the cationic nitrogen centers cannot move as far away from one another as in the acyclic counterparts. For example, potentiometric titration results demonstrate that hexaprotonated hexacyclen (**5.4**, the nitrogen analog of 18-crown-6) is a strong diprotic acid, existing predominantly as the tetraprotonated form (**5.4**·4H$^+$) in aqueous solution, suggesting that the enforced proximity of six cationic centers is highly energetically unfavorable. Solution complexation experiments indicate that while (**5.4**·4H$^+$) exhibits a greater affinity for NO$_3^-$ over Cl$^-$, Br$^-$, I$^-$, and ClO$_4^-$, it paradoxically forms stronger hydrogen bonds to Cl$^-$. The X-ray crystal structure of **5.4**·2HCl·2HNO$_3$ demonstrates the binding of chloride ions above and below the host ring plane (N $\cdots$ Cl = 3.07–3.28 Å) while nitrate ions are indirectly bound via enclathrated water molecules (Fig. 5.2).[22] In spite of the high acidity of the hexaprotonated form **5.4**·6H$^+$, crystallization of the hydrochloride salt of hexacyclen from concentrated nitric acid yields **5.4**·4HNO$_3$·2HCl. In this case, the nitrate ions are directly hydrogen bonded to the hexacyclen ring in a unidentate fashion (N $\cdots$ ONO$_2$ = 2.77–2.88 Å).[16] In contrast, molecular mechanics calculations (MM2) suggest a bidentate mode of coordination for the binding of carbonate by **5.4**·3H$^+$,[23] indicating that the nature of the hexacyclen–anion interactions is both complex and highly variable. No crystal structure data on the carbonate complex are available. In

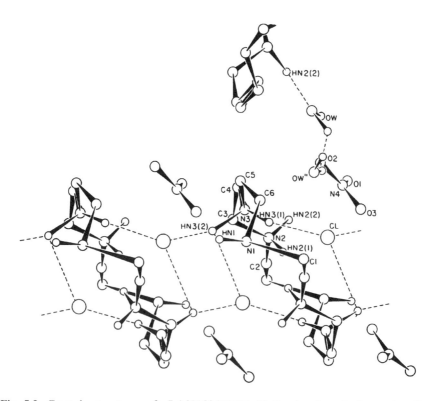

Fig. 5.2. Crystal structure of **5.4**·2HCl·2HNO$_3$·H$_2$O showing hydrogen-bonding interactions of the chloride ions. (Reproduced with permission from Ref. 22).

none of these 18-membered ring structures is anion inclusion within the macrocycle observed crystallographically.

Related work on **5.4** and similar pentaamine macrocycles show that these materials also act as efficient, selective binding agents for syn-type polycarboxylic anions such as those found in the tricarboxylate cycle. This selectivity is attributed to geometrical factors enabling a macrocyclic chelating effect to enhance binding.[24] An analogous material containing amide functionalities has been shown to bind azide in the solid state with the guest anion bridging unsymmetrically across pairs of host cations in the crystal structure (N···N = 2.76–3.07 Å).[25] ^{1}H-NMR measurements of this complex indicate that only weak binding is observed in solution, consistent with the nonincluded position of the guest anion.

Lehn and co-workers have synthesized a variety of larger polyammonium-based macrocycles incorporating greater separations between the nitrogen centers (**5.5–5.7**). These species were all found to exhibit pK_a values around or above 7, and as a consequence, displayed strong anion binding behavior at neutral pH. While no X-ray crystallographic results are available for anion binding, solution studies indicate a number of structural preferences, interpreted in terms of inclusion of anionic guests within the macrocyclic ring. For example, larger anions such as squarate and fumarate form stronger complexes ($\log K_s = 2.9$–3.6) with the larger macrocycle **5.6**·8H$^+$ than with **5.5**·6H$^+$. Similarly, very high stability constants ($\log K_s = 6.0$ and 8.9) are observed for Co(CN)$_6^{3-}$ and Fe(CN)$_6^{4-}$ with the fourfold-symmetric macrocycle **5.6**·8H$^+$. It should be noted, however, that electrostatic effects also play a major role in both the strength and selectivity of anion binding, consistent with the fact that there is usually more than one anion associated with the macrocyclic host.[26]

Protonated forms of the large-ring macrocycle [24]N$_6$O$_2$ (**5.8**) and related compounds have been shown to be active in nucleotide chemistry as dephosphorylation and phosphoryl transfer catalysts,[27,28] including phosphorylation resulting in ATP synthesis.[28] It is possible that in this case the substrate enters the macrocylic cavity, as proposed from the results of molecular modeling studies performed using coordinates obtained from the crystal structure of the hexahydrobromide salt.[27a]

The structural report of the hexahydrochloride salt of **5.8** indicated that halides could indeed be enveloped within a cleft formed by the boat-shaped conformation of the macrocycle (Fig. 5.3A).[29] The hydrochloride salt of the smaller [22]N$_6$, **5.9**, was found, on the other hand, to bind two chloride anions above and below the host plane (Fig. 5.3B). In both cases of chloride binding to **5.8** and **5.9**, N···Cl distances are in the range of 3.02–3.41 Å,[29] although, in the case of **5.8**·6HCl the shortest contacts are not to the centrally located anion, consistent with a lack of size complementarity between the host and Cl$^-$ and guest.[29] Some of the first molecular-dynamics simulations on complex anionic systems were performed for the two chloride complexes with **5.8** and **5.9**[29] These studies indicate that the pocketlike conformation of **5.8**·6H$^+$ is maintained in

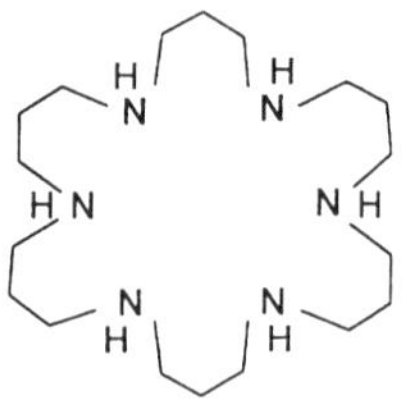

5.5

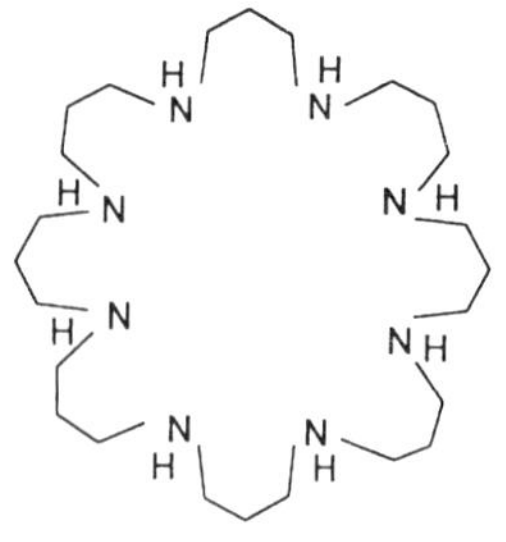

5.6

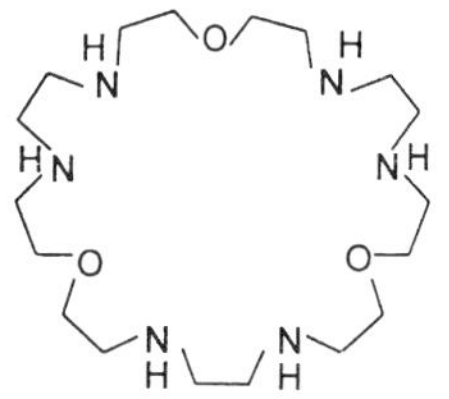

5.7

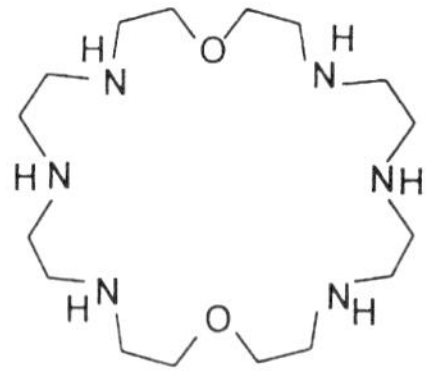

5.8

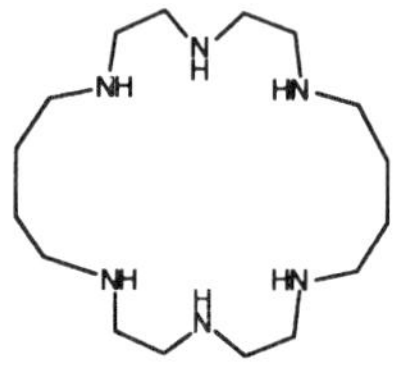

5.9

solution, although ^{35}Cl NMR experiments demonstrate that halide ions are in rapid exchange between the complexed and solvated state.

A crystal structure of the nitrate complex of **5.8** in its tetraprotonated form, **5.8**·4HNO$_3$, has recently been reported, and one of the nitrates was found to be held within the macrocyclic cavity (Fig. 5.4A).[30] Along with the crystal structure of the nitrate complex of **5.8**·4H$^+$, the structure of the smaller 18-membered analog [18]N$_4$O$_2$, **5.10** (Fig 5.4B) was reported. In this case the four nitrates fall above and below the cavity of the rather planar macrocycle. Rather extensive molecular-dynamics simulations have been performed for the nitrate complexes, indicating solvation effects play an important role in conformation changes. Both the chloride and nitrate studies are described in greater detail in Chapter 9.

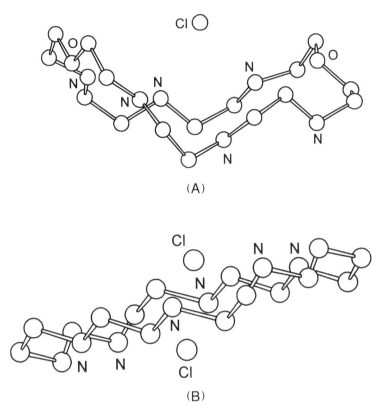

Fig. 5.3. (A) Structure of **5.8**·6HCl showing chloride inside the cleft of the macrocycle. (B) Structure of **5.9** with chlorides above and below the macrocyclic plane.

Crystal structure data for a modified form of **5.8** with a furan heterocycle incorporated into the ring, **5.11**, have been obtained for both the pyrophosphate and oxalate complexes.[31] As seen from the oxalate structure, the macrocycle is flattened, and the anion is clearly outside the cavity (Fig. 5.5A), while the pyrophosphate just sticks into the cavity (Fig. 5.5B).

Consistent with the theme of increasing macrocylic size in order to induce strong complexation by inclusion of anionic guests within the macrocyclic ring, extensive studies with the totally polyaza macrocycles [3·k]aneN$_k$ (**5.12**, $k = 7$–12) as hosts have been performed. These macrocycles, related to hexacyclen, **5.4**, consist of macrocyclic rings of up to 36 atoms.[32–35] Measurements of K_as indicate that, starting from the fully protonated macrocycle, the first three protons are readily lost (e.g., for $k = 12$, log $K_a = 1.0$, 2.3, and 2.65 for the first three protons), again as a consequence of the accumulation of positive charge density on sites in close proximity to one another. In spite of this, the partially protonated macrocycles form a number of interesting "supercomplexes" with coordination complex anions such as Co(CN)$_6^{3-}$,

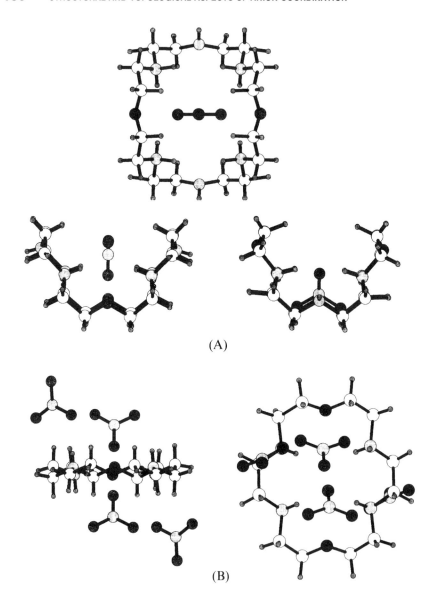

Fig. 5.4. (A) Structure of **5.8**·4HNO$_3$ showing nitrate inside the macrocyclic pocket. (B) Structure of **5.10**·4HNO$_3$ showing nitrates above and below the macrocyclic plane.

Fe(CN)$_6^{4-}$, and PtCl$_6^{2-}$, PdCl$_4^{2-}$, and Pt(CN)$_4^{2-}$. X-ray crystallographic studies demonstrate that in the majority of cases, the anions bridge between the macrocycles, and little anion selectivity is observed, indicating that binding occurs primarily by coulombic attractions as in the structure of **5.12** ($n = 4$) [30]aneN$_{10}$·8H$^+$ with Co(CN)$_6^{3-}$ (Fig. 5.6A).[32]

PROTONATED POLYAMINE-BASED RECEPTORS **157**

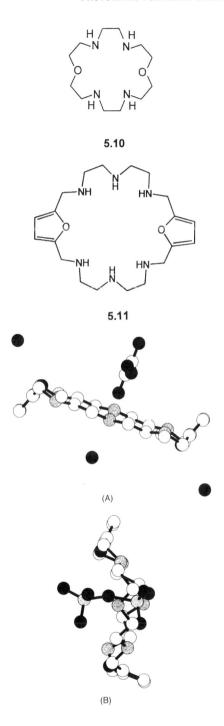

Fig. 5.5. Structure of the (A) oxalate and (B) pyrophosphate complex of **5.11**.

5.12 n = 2 - 7

However, in the case of the $PdCl_4^{2-}$ complex of the fully protonated **5.12** ($n = 4$), [30]aneN$_{10}$·10H$^+$, host, one of the $PdCl_4^{2-}$ anions is included within the macrocyclic cavity. The macrocyclic receptor adopts an "S-shaped" conformation in order to wrap around the guest anion and maximize N–H···Cl hydrogen-bonding interactions, Fig. 5.6B. Solution measurements indicate that $PdCl_4^{2-}$ is inclusively bound by macrocycles with $k = 9$ and greater, with [30]aneN$_{10}$ exhibiting a cavity size most complementary to the dimensions of the complex anion. The included anion is found to exchange only slowly with other complex anions in solution, and the binding is exothermic, with $\Delta H^\circ = -16.3(4)$ kJ mol^{-1} per $PdCl_4^{2-}$ unit (see also Chapter 6).[33]

Binding of $PtCl_2^{6-}$ has also been reported for hosts based on crystalline thiamine, in which the metal complex anions are sandwiched between pairs of host molecules that are hydrogen bonded together to form a large cavity.[36]

Another recent report deals with the complexation of chloride by *tris*(pyrazolyl)hydroborato [HB(3-*t*BpzH)$_3^{2+}$] derivatives. The three protonated nitrogen atoms of the [HB(3-*t*BpzH)$_3^{2+}$] ligand each hydrogen bond to the guest chloride anion in the solid state with N···Cl distances an average of 3.08 Å, indicating reasonably strong bonding and resulting in an unusual trigonal pyramidal geometry at the anion (Fig. 5.7).[37]

More complex, polypyrrole-based macrocycles capable of complexation of anions within the macrocylic ring have been recently synthesized.[38,39] Work by Sessler et al. on the "expanded porphyrin" sapphyrin **5.13** has shown that this molecule, in its diprotonated form, is capable of complexation of fluoride ion. The crystal structure of the mixed fluoride·PF$_6^-$ salt of **5.13**·2H$^+$ (Fig. 5.8A) demonstrates the inclusion of the fluoride ion within a regular array of five N–H···F hydrogen bonds with N···F distances in the range 2.697–2.788 Å.[38a] The deviation of the F$^-$ ion from the plane defined by the five nitrogen atoms is only 0.03 Å, indicating a high degree of size complementarity between the sapphyrin ring and the small fluoride ion. In contrast, the analogous dichloride salt exhibits binding of chloride ions above and below the plane of the macrocyle (Fig. 5.8B),[38b] with relatively long N···Cl contacts in the range 3.12–3.14 Å, an indication that the chloride is too large to fit snugly within the macrocylic ring. Similar results were observed for alkyl phosphate salts.[3b] Consistent with the structural results, the stability constant K_s of the **5.13**·2HF complex is of the order of 10^8 M^{-1}, with a selectivity factor over Cl$^-$ and Br$^-$ in excess of 10^3. Sapphyrins are also capable of binding anions of

PROTONATED POLYAMINE-BASED RECEPTORS 159

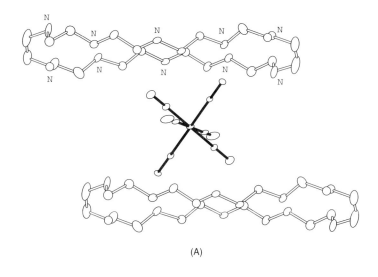

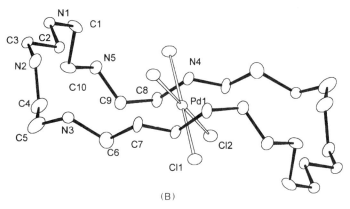

Fig. 5.6. (A) Structure of the octaprotonated **5.12**($n = 4$) showing Co(CN)$_6^{3-}$ bridge. (B) Crystal structure of the encapsulated PdCl$_4^{2-}$ in **5.12** ($n = 4$).

biological interest. The crystal structure of sapphyrin in its dicationic form with monobasic cAMP reveals interaction of a phosphate oxygen with the macrocycle through hydrogen bonds with the two pyrrole hydrogens (Fig. 5.9).[38c] Related structures of sapphyrin with a variety of phosphoric acids have also appeared.[38d–f]

Interestingly, a related anthracene-based, porphyrin-derived macrocycle with a larger cavity does bind a single chloride anion such that it is close to the plane defined by the nitrogen atoms (the chloride anion is 0.794 Å above the plane of the four bonding N atoms, N···Cl = 3.11–3.25 Å).[39]

Finally, another series of novel monocyclic polyamine derivatives, 4n-pyridinium crown-n cations ($n = 4$, **5.14A**; $n = 6$, **5.14B**) has recently been

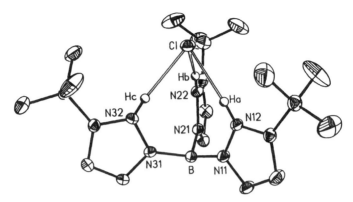

Fig. 5.7. ORTEP diagram of [{η^3-HB(3-Bu^{t}pzH}Cl][AlCl$_4$]. For clarity, [AlCl$_4$]$^-$ and C$_6$H$_6$ of crystallization are not included. (Reproduced with permission from Ref. 37.)

5.13

reported in connection with the complexation of simple anions such as Cl$^-$ and NO$_3^-$ as well as HgI$_4^{2-}$.[40] In the case of both the chloride and nitrate salts of **5.14B**, the macrocycle adopts a pincherlike conformation, with the anion held between two opposing pyridinium rings (Figs. 5.10A and B, respectively). Interestingly, the guest anions do not hydrogen bond to the N–H protons, but are in close contact with the protons of the pyridinium ring. The C···Cl distances range from 3.695 to 3.787 Å, while in the analogous nitrate salt, the C···O contacts are 3.2–3.3 Å. The larger macrocycle **5.14B** has also been crystallographically characterized in the form of the rather complicated salt (24-pyridinium crown-6) [(DMSO)HgI$_3$] [HgI$_4$][Hg$_2$I$_7$]·11DMSO·2H$_2$O.[41] The key feature of this structure involves the inclusion of the [HgI$_4$]$^{2-}$ anion within the cavity of the macrocycle similar to complexes of **5.14B**, with C···I contacts in the range 3.69–3.71 Å (Fig. 5.10C).

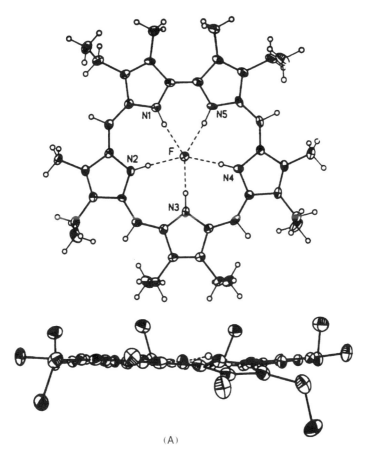

Fig. 5.8. (A) Two views of the structure of sapphyrin, **5.13**·2H$^+$ with fluoride ion. (Reproduced with permission from Ref. 38b.)

5.2.2 Polycyclic Hosts

The original solution work of Simmons and Park, suggesting that the *in*, *in*-form of the bicyclic *katapinands* **5.1** may complex halide anions within the macrocyclic cavity, was confirmed in 1975[14] by the X-ray crystal structure determination of chloride *katapinato-in*,*in*-1,11-diazabicyclo[9.9.9]nonacosane-*bis*(ammonium)chloride (i^+Cli^--[[9.9.9]Cl$^-$]), a host exhibiting significant selectivity for chloride. The structure determination revealed the inclusion of one chloride ion, centrally located within the macrocyclic cavity, with N···Cl hydrogen-bond distances of 3.10(1) Å, suggesting a reasonable degree of size complementarity between the halide and host. The idea of size-based selectivity is borne out by solution work upon similar katapinands. In the case of the smaller macrocycles i^+i^+-[7.7.7] and i^+i^+-[8.8.8] (estimated distance between the faces of NH protons: 1.6 and 2.8 Å, respectively) no binding of Cl$^-$, Br$^-$, or I$^-$ is

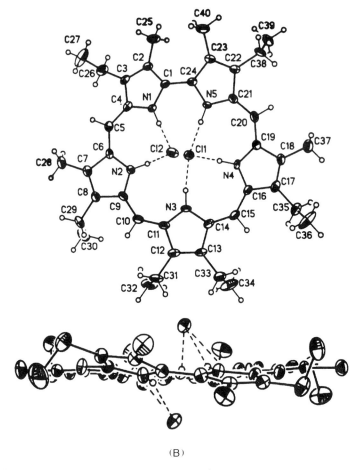

(B)

Fig. 5.8. (B) Two views of sapphyrin, **5.13**·2H$^+$ with chloride showing the chlorides above and below the macrocyclic plane. (Reproduced with permission from Ref. 38b.)

noted. The larger i^+i^+i-[10.10.10] (diameter 4.5 Å, cf. 3.6 Å for i^+i^+ [9.9.9]), however, exhibits a strong affinity for all three anions, although little selectivity is observed.[11]

The macrotricyclic cryptand **5.2** in its tetraprotonated form has been shown to be an excellent host for spherical anions such as halides. Also, unlike many of the smaller polyammonium-type hosts, there is no problem associated with complete protonation of all the nitrogen atoms, since all four are widely separated. Stability constant measurements demonstrate that **5.2** is strongly selective for Cl$^-$ with log $K_s > 4$ in aqueous solution, as opposed to the bromide complex with log $K_s < 1$ (2.75 in methanol/water mixture).[42] It is notable that this value is more than three orders of magnitude larger than the analogous *katapinand* complexes (vide supra).[42] It has been suggested that this selectivity

PROTONATED POLYAMINE-BASED RECEPTORS   **163**

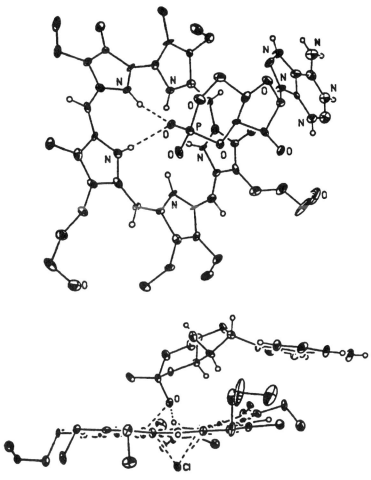

Fig. 5.9. Two views of the structure of sapphyrin, **5.13**·2H$^+$ with monobasic cAMP and chloride ion. (Reproduced with permission from Ref. 38c.)

5.14A n = 4
5.14B n = 6

arises from the presence of a closed and rigid cavity, preorganized for anion binding.[43] The chloride complex of **5.2**·4HCl has been characterized by X-ray crystallography[15] (Fig. 5.11). All four nitrogen atoms adopt the *in* conformation with average N···Cl hydrogen-bonding distances of 3.09 Å, similar to those found for the *katapinate* chloride complex. This compares with an average

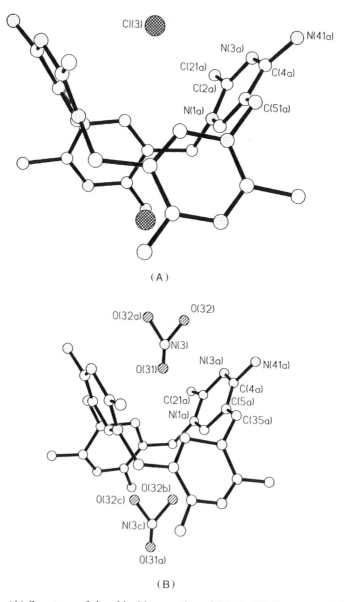

Fig. 5.10. (A) Structure of the chloride complex of **5.14A**. (B) Structure of the nitrate complex of **5.14A**.

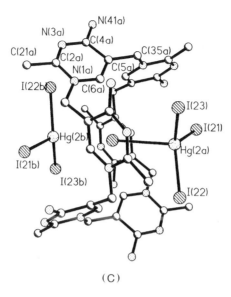

(C)

Fig. 5.10. (C) Structure of the complex of **5.14B** showing coordinated HgI_4^{2-} and HgI_3^- units.

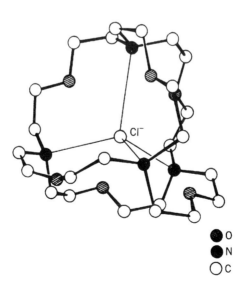

Fig. 5.11. X-ray crystal structure of the macrotricyclic cryptate $[Cl^- \cdot \mathbf{4.12} \cdot 4H^+]$. (Reproduced with permission from Ref. 15.)

distance of 3.25 Å for Cl···O contacts, while in the analogous ammonium cation complex of the unprotonated macrotricycle, the average $NH_4···N$ and $NH_4···O$ distances are almost identical, 3.13 and 3.11 Å.

In almost all the examples mentioned so far, hydrogen bonding has played a central role in stabilizing the host–guest interaction. Schmidtchen, in an effort to evaluate the importance of hydrogen bonding, however, has demonstrated the inclusion of iodide by the macrotricyclic receptor **5.15** solely by a combination of electrostatic and space-filling-type interactions (Fig. 5.12),[44,45] with long N···I contacts averaging 4.5 Å. It is noteworthy that the receptor provides a

5.15

nearly ideal tetrahedral coordination environment about the iodide anion, and the aqueous solution binding constants are still relatively high, although generally less than those obtained for complexes of the soccer ball ligand, **5.2·4H$^+$**. Interestingly, association constants for halide ion complexes of **5.15** are significantly increased in methanol solution, possibly as a consequence of diminished anion–solvent interactions.[45]

In the previous section it was noted that hexacyclen **5.4** may not possess a macrocyclic cavity of sufficient size to encapsulate anionic guests such as Cl$^-$ and NO_3^-. In contrast, the octaaza macrobicyclic analog of hexacyclen (**5.16**) forms the fluoride cryptate F$^-$·**5.16**·6H$^+$ (Fig. 5.13), with incorporation of the fluoride ion within the cryptand.[46] Consistent with the observed high acidity of the hexacyclen hexacation, the cryptand **5.16** carries only six of a possible eight protons, with F···N hydrogen-bonded distances between 2.76 and 2.86 Å. This contrasts to nonbonded F···N contacts in the region of 3.3 Å to the nonprotonated, bridgehead nitrogen atoms, although the macrocycle does retain the characteristic *in,in* conformation maximizing the cavity size. While the fluoride anion is situated within the center of the macrocycle in a quasi-trigonal prismatic coordination environment, the mode of interaction is not dissimilar to that observed for the chloride complex of hexacyclen,[22] with the fluoride anion situated above the plane of each of the eighteen-membered rings.

The majority of the receptor systems mentioned so far have been shown to encapsulate halide guests as a consequence of the symmetrical nature of their binding pockets, which readily adapt to the spherical symmetry of the halide ion. Crystallographic studies of the *bis*-tren macrobicyclic ligand **5.3·6H$^+$**, however, revealed that, in addition to accommodating halide anions, **5.3** is

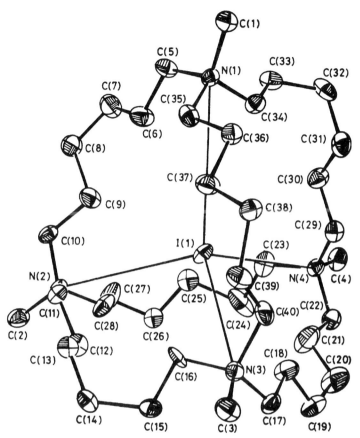

Fig. 5.12. Molecular structure of the complex of iodide with the cation **5.15**$^{4+}$. (Reproduced with permission from Ref. 44.)

also able to encapsulate azide, N_3^-, within the cylindrical macrocylic cavity.[17,47] The X-ray crystal structures of the F^-, Cl^-, and Br^- cryptates of **5.3·6H$^+$** demonstrate the importance of size and electrostatic considerations. In the case of the small, highly electronegative fluoride ion complex, the anion does not seem to be able to bind to all six protons and essentially "slides" toward one side

5.16

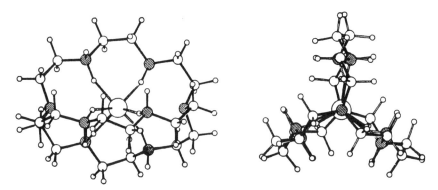

Fig. 5.13. X-ray crystal structure of the fluoride cryptate $F^-\cdot5.16\cdot6H^+$. (Reproduced with permission from Ref. 46a.)

of the cavity, where it "binds" in a tetrahedral fashion with a mean $N\cdots F$ hydrogen bonding distance of 2.72 Å. In contrast, the Cl^- and Br^- cryptates are octahedrally coordinated halide ions, situated almost exactly centrally within the host framework (see Fig. 2.5, p. 49), with $X^-\cdots N$ distances in the range 3.19–3.39 Å (Cl^-) and 3.33–3.47 Å (Br^-). It is noteworthy in the case of both Cl^- and Br^- that the internal hydrogen-bonded distances within the cryptand host are longer by up to 0.15 Å compared to those for the other anions in the lattice. This increased distance is possibly a result of the need for considerable deformation of the cavity in order to accommodate the two halides, with the effect being less obvious in the case of the larger bromide ion. In the structure of the azide cryptate $N_3^-\cdot5.3\cdot6H^+$, azide ion can be considered to be a linear or cylindrical, "ditopic" substrate, exceptionally suited for a snug and well-matched fit with the bicyclic receptor (Fig. 5.14). The $azide\cdots N_{macrocycle}$ hydrogen-bonded contacts are symmetrical and fall in the range 2.81–3.02 Å. Solution stability constant measurements indicate that the macrocycle is indeed selective for the azide anion with $\log K_s = 4.30$. Interestingly, the F^- cryptate is also very stable ($\log K_s = 4.19$) in comparison to the Cl^-, Br^-, and I^- salts ($\log K_s = 3.0, 2.6$, and 2.15, respectively), a reflection of the strong electrostatic and hydrogen-bonding interactions.

Large stability constants (up to $\log K_s = 10.30$ in the case of $P_2O_7^{4-}$) with bis-tren, 5.3, are also observed for a number of polyanions, especially polyphosphates, including ATP, ADP, and other nucleotides.[17] These results can be rationalized in terms of a large electrostatic component to the binding. Thus the macrocyclic receptor is selective for anions of high charge density such as F^- and for multiply charged ions. These findings must to a great degree arise as a consequence of the fact that only one of up to six anions may be included within the molecular cavity at any time, and thus there is no basis for assuming any size or shape selectivity. As noted above, the high preference for N_3^- binding in comparison to halides other than F^- and other monovalent anions such as nitrate (factor of ca. 100) represents strong evidence for structural selectivity

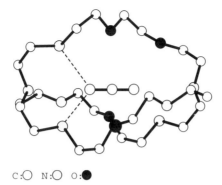

Fig. 5.14. Azide inclusion within the cylindrical *bis*-tren macrocycle **5.2·6H⁺**. (Reproduced with permission from Ref. 2a.)

Indeed, the observed selectivity sequence ClO_4^-, Cl^-, $I^- < CH_3CO_2^-$, $Br^- < HCO_2^- < NO_3^-$, $NO_2^- < N_3^-$ parallels neither the sequence of hydration energies nor the lyotropic series (based on the effectiveness of different salts in bringing about the precipitaton of proteins);[48] but rather arises as a consequence of topological discrimination by the protonated host molecule.[9]

The binding of larger, linear anions by *bis*-tren–type macrocycles has been extended to the cyclophane receptor **5.17·6H⁺**. Solution measurements at pH

5.17

5.5 indicate that this host exhibits an impressive degree of structural selectivity within a series of linear α,ω-dicarboxylate anions $O_2C(CH_2)_nCO_2^{2-}$, with adipate ($n = 4$) being bound much more strongly than either shorter or longer homologues. Also notable is the extremely strong binding of the terphthalate dianion ($O_2CC_6H_4CO_2^{2-}$, tph) arising from both electrostatic and hydrophobic interactions, as well as a high degree of structural complementarity. The X-ray crystal structure of tph·**5.17·6H⁺** (Fig. 5.15) demonstrates that with the inclusion of one of the three thp anions within the macrocyclic cavity, there is also a significant structural change in the macrocycle itself upon

170   STRUCTURAL AND TOPOLOGICAL ASPECTS OF ANION COORDINATION

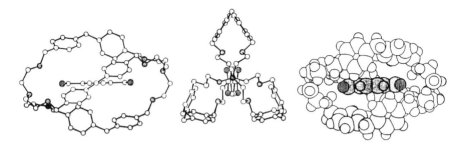

Fig. 5.15. X-ray crystal structure of the terphthalate dianion cryptate tph·**5.17**·6H$^+$. (Reproduced with permission from Ref. 49.)

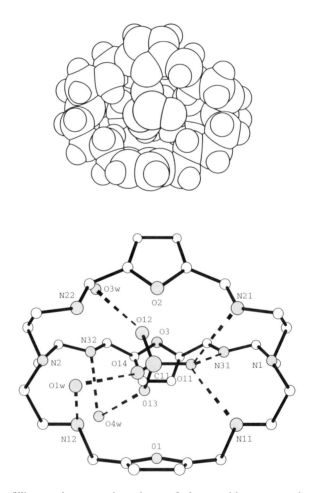

Fig. 5.16. Space-filling and perspective views of the perchlorate complex of **5.18**. (Reproduced with permission from Ref. 50.)

complexation, in comparison to the structure of the free (nonprotonated) host. This change involves the movement of the two bridgehead nitrogen atoms approximately 1.8 Å toward one another in order to maximize interactions with the guest anion. The hydrogen-bonding coordination sphere of both macrocycle and substrate is completed by a number of water molecules, and it is noteworthy that one of the nonincluded tph anions interacts with the host only via one of these waters.[49]

In related structures of polyatomic anionic species, compound **5.11** has been structurally characterized as both the perchlorate (Fig. 5.16) and SiF_6^{2-} salts, and both were found to be included in the macrocyclic cavity.[50] This is the first example of crystallographically characterized polyatomic inorganic anionic cryptates other than the azido complex with *bis*-tren, **5.3**, to date. Again both electrostatic and hydrogen-bonding effects play major roles in complex stabilization. Interestingly, in the accompanying attempt to isolate the encapsulated BF_4^- complex, a mixed BF_4^-/SiF_6^{2-} salt was obtained, in which the SiF_6^{2-} was found to be encapsulated within the bicyclic cavity. The silicon contamination was attributed to the reaction of HBF_4 on the glass reaction vessel.[50] An outcome of these structures was the illustration of the flexibility of the cryptand itself. For example, the length of the cavity for the perchlorate complex was found to be 9.33 Å, while in the dicopper complex $[Cu_2 \cdot 5.3(OH)]^{3+}$, the length was considerably contracted at 8.05 Å.[51] The width of the cavity was not found to change significantly, however (Fig. 5.16).

5.3 THE GUANIDINIUM MOIETY

The guanidinium ion ($CN_3H_6^+$) **5.18** and its many derivatives have been widely studied in the context of anion binding. In particular, the guanidinium ions contained in arginine residues of enzymes play a major role in the binding of anionic substrates and in maintaining protein tertiary structure (Sect. 5.10). Most important, the guanidinium moiety exhibits a particularly high pK_a (13.5 for the parent ion), resulting in its protonation over a wide pH range during which it retains its positive charge and hydrogen-bond donor properties.[2]

The X-ray crystal structure determination of a number of simple guanidinium salts (e.g., methylguanidinium dihydrogenphosphate) clearly demonstrates the existence of bidentate ionic hydrogen bonds as shown in **5.18**, with N···O contacts in the region of 2.87 Å (Fig. 5.17).[52,53]

Early work on the guanidinium group by Lehn and co-workers resulted in the synthesis of a number of guanidinium-containing macrocycles with two or three guanidinium moieties separated by various spacer groups.[2,54] No crystallographic results were reported for these compounds, and pH-metric titration results with PO_4^{3-} indicated only weak binding in aqueous solution (log K_s = 1.7–2.4) with little macrocyclic effect being observed. The weakness of the anion binding was ascribed to the delocalized nature of the positive charge on the guanidinium fragment, resulting in fewer regions of high positive charge density in comparison with related ammonium-containing macrocycles. In general, observed anion selectivities are attributable simply to electrostatic interactions. Selectivity patterns between monovalent anions reveals that those such as carboxylate, however, which are capable of engaging in bidentate hydrogen-bonding interaction of the type shown in Fig. 5.17, are bound more strongly than halides.[54]

Despite these relatively unencouraging results with guanidinium-based receptors, the biological role of the guanidinium group has recently prompted a significant amount of work in the design and synthesis of hosts capable of catalyzing biological reactions and of the transport of phosphate and polyphosphate anions.[55] In particular, in relatively noncompetitive solvent media, the *bis*(acylguanidinium) host **5.19**, with a *bis*(bidentate) binding related to that shown in Fig. 5.18 proposed, has been shown to accelerate the rate of phosphodiester cleavage by factors of up to 10^3, resulting in association constants in the region of 5×10^4 M^{-1}.[55b]

In an attempt to achieve anion binding in a protic medium, Schmidtchen et al. have developed a number of acyclic, bifunctional hosts such as **5.20**.

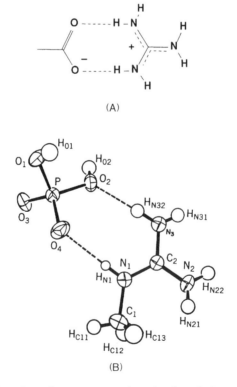

Fig. 5.17. Perspective view of one asymmetric unit of methylguanidinium dihydrogen phosphate. (Reproduced with permission from Ref. 53.)

Molecular modeling studies suggest that **5.20A** adopts an extended conformation and, indeed, shows no significant affinity for monoanions such as iodide or acetate.[56] However, NMR titration results in deuterated methanol solution with a wide variety of dicarboxylate anions indicate that **5.20** possesses a significant affinity and selectivity for malonate ($K_s = 16,500$ M^{-1}) over both longer- and shorter-chain analogs, suggesting a significant degree of size compatibility with that substrate. Binding is not dominated by simple geometrical considerations, however. A similarly large association constant ($K_s = 14,500$ M^{-1}) is

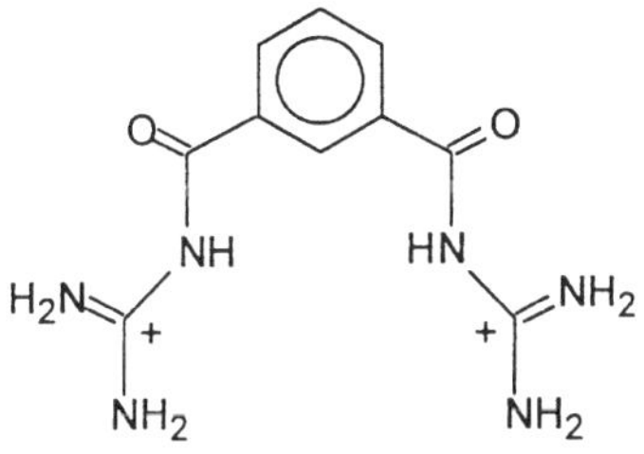

5.19

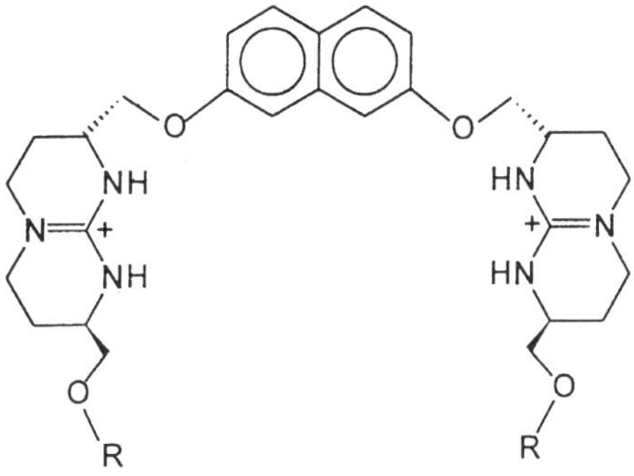

5.20A R = Si(C$_6$H$_6$)$_2$(*t*-C$_4$H$_9$)
5.20B R = H

found for nitro-isophthalate. This is compared to the unnitrated analog, which exhibits a much smaller association constant ($K_s = 6060$ M^{-1}), indicating the influence of effects such as stacking and charge transfer. Related work with *bis*(guanidinium)-, *bis*(urea)-, and *bis*(thiourea)-based hosts incorporating *p*-xylyl spacer units has demonstrated binding constants of up to 50,000 M^{-1} in the relatively competitive solvent DMSO, although addition of water results in a dramatic decrease in association.[57] More interestingly, the related host system **5.20B** is capable of binding the highly solvated HPO$_4^{2-}$ anion in aqueous solution ($K_s = 970$ M^{-1}), as well as biologically relevant materials such as 5'-AMP ($K_s = 9330$ M^{-1}).[55c] Comparison of these results with work in MeOH

174 STRUCTURAL AND TOPOLOGICAL ASPECTS OF ANION COORDINATION

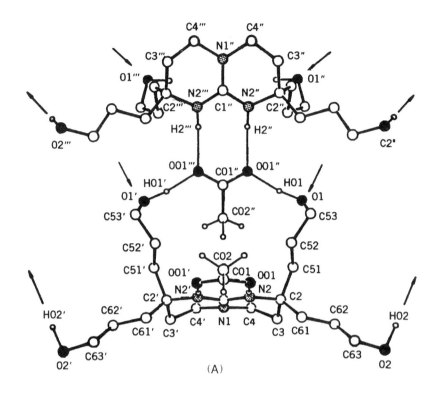

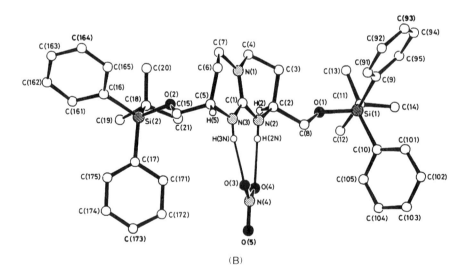

Fig. 5.18. (A) Section of the crystal structure of **5.21A** with acetate. (Reproduced with permission from Ref. 55d.) (B) Molecular structure of **5.21B** with nitrate. (Reproduced with permission from Ref. 55e.)

and DMSO, however, highlights the strong electrostatic contribution to the binding.

Unfortunately, given the complexity of receptors such as **5.20**, no crystallographic results are currently available. The simpler hosts **5.21A** and **B** have been structurally characterized, however, as their acetate and nitrate salts, respectively (Fig. 5.18A and B).[55d,e] In the case of **5.21A**, the results of the X-ray structure determination clearly indicate bidentate hydrogen bonding of the acetate anion to the guanidinium NH functionalities [N···O = 2.850(5) Å] (Fig. 5.18A) similar to that shown in Fig. 5.15. Additional stabilization comes from interactions with the pendant hydroxyl functionalities with hydrogen-bonded O···O contacts of 2.733(5) Å.[55d] Host **5.21B** contains two chiral centers. In the optically resolved form, **5.21B** is capable of enantiodifferentiation of racemic mixtures of chiral carboxylic acids such as N-acetyl-D,L-alanine, as determined by NMR in acetonitrile solution. The X-ray crystal structure of the nitrate salt of **5.21B** demonstrates the chiral nature of the molecular cleft[55e] with the nitrate ion fitting only poorly within the large cavity (Fig. 5.18B). Similar chiral recognition has also been reported in related species by Lehn et al.[58]

Finally, bidentate *bis*(guanidinium) hosts reported by Anslyn et al. have been shown crystallographically to bind dibenzyl phosphate in a *bis*(bidentate) fashion, incorporating two interactions to one guanidinium as shown in Fig. 5.19A, while a chloride ion is bound to the other guanidinium.[59] Solution studies in 15% H_2O/DMSO give binding constants of ca. 500 M^{-1}. The corresponding dianionic phosphomonoester phenyl phosphate shows a two-point hydrogen–bonding network with both guanidiniums holding the phosphate. Hydrogen-bonding N···O distances are in the range 2.658(7)–2.868(6) Å.

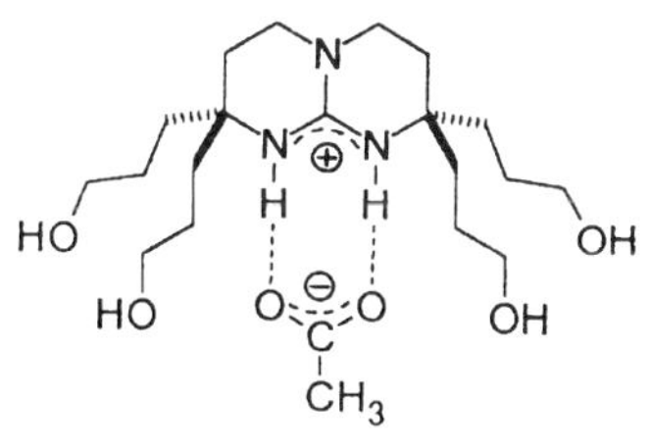

5.21A

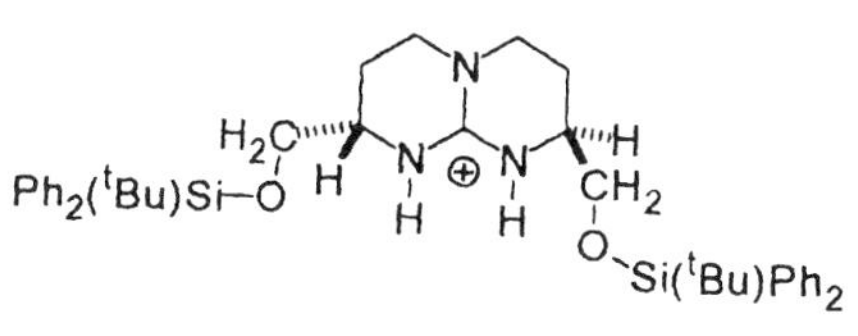

5.21B

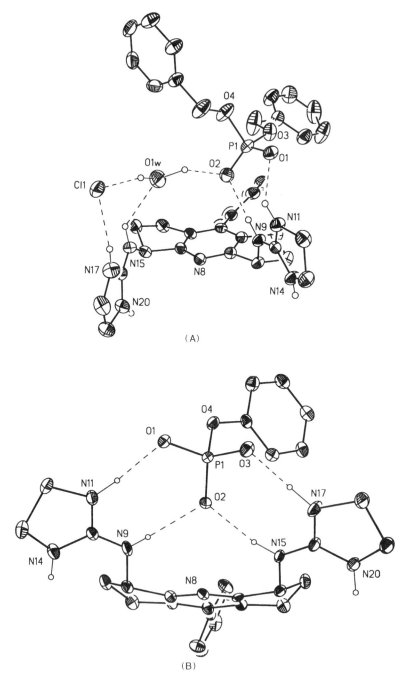

Fig. 5.19. (A) Structure of a *bis*-guanidinium salt with chloride and dibenzyl phosphate. (Reproduced with permission from Ref. 59.) (B) Structure of the same receptor with phenyl phosphate. (Reproduced with permission from Ref. 59.)

5.4 CYCLOPHANE RECEPTORS

The broad definition of a cyclophane as a molecule incorporating any bridged aromatic ring[60] results in some difficulty in classification of cyclophane-type anion hosts. Mention has already been made of the cyclophane macrocycle **5.18**. Extension of the *katapinand* complexes **5.1** by addition of rigid aromatic spacer groups gives **5.22**, which, in its diprotonated form, can include one or even two bromide or iodide ions within its large, macrocyclic cavity.[61] This contrasts to complexes of type **5.1**, in which inclusion of only a single halide ion is observed as a consequence of the smaller cavity size. A related octaaza cryptand, **5.23**, in its tetraprotonated form has been shown by X-ray crystallography partially to include two trifluoromethane sulfonate anions (Fig. 5.20), each via two hydrogen-bonding interactions to the sulfonate head group, $SO_3 \cdots N = 2.90–2.98$ Å. Slightly shorter hydrogen-bonding distances are observed to extracavity $CF_3SO_3^-$ anions, which bridge between pairs of host molecules.[62]

A number of interesting square-cavity heterocyclophanes have been synthesized containing not only nitrogen but also iodinium, sulfonium, and even transition metal bridges.[63–66] In particular, the closely related hosts **5.24A** and **B** have been shown by fluorescence spectroscopy to include the 1-anilinonaphthalene-8-sulfonate anion with K_s up to 1.6×10^3 M^{-1}.[53] More recently a related series of "molecular boxes" have been synthesized,[64–66] incorporating four biphenyl or bipyridyl spacer units. Some preliminary work has shown that these materials may complex neutral molecules in solution,[64] but their ability to function as anion complexones is as yet unknown.[66]

Preliminary reports of other smaller cyclophane-type receptors suggest that anions are bound only outside the cavity. However, solution measurements indicate the possibility of exchange of both internally and externally bound anions.[67]

5.5 NEUTRAL RECEPTORS

Inevitably, in the case of cationic receptors, the binding of guest anions must compete with that of other counterions. Hence, association constants represent more a measure of the effectiveness with which a target guest anion competes with the other anions also present, rather than being an absolute measure of the magnitude of binding. It has been pointed out that neutral receptors do not suffer from this drawback and, furthermore, have potentially greater anion selectivity, since they do not rely upon nondirectional electrostatic forces to achieve anion coordination.

5.5.1 Organic Hosts

A very simple modification of the parent amine of the macrotricyclic cation **5.15** by formation of its borane adduct is sufficient to produce the neutral

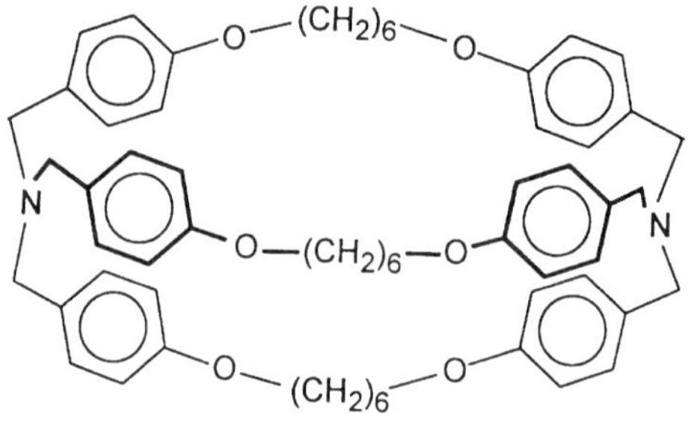

178 STRUCTURAL AND TOPOLOGICAL ASPECTS OF ANION COORDINATION

5.22

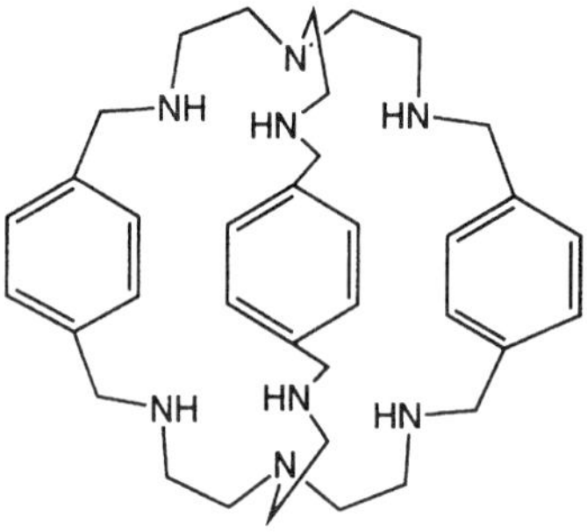

5.23

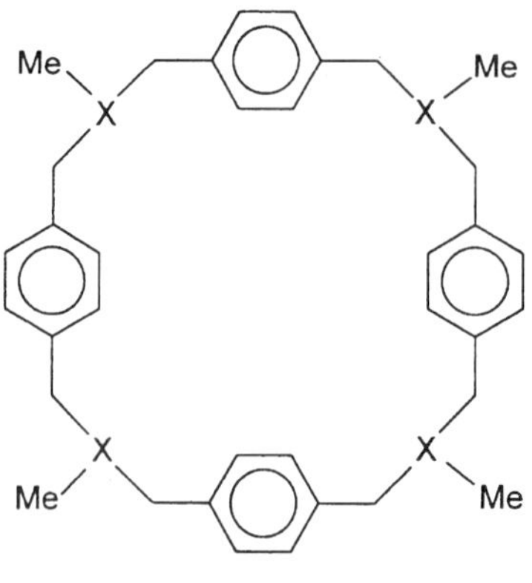

5.24

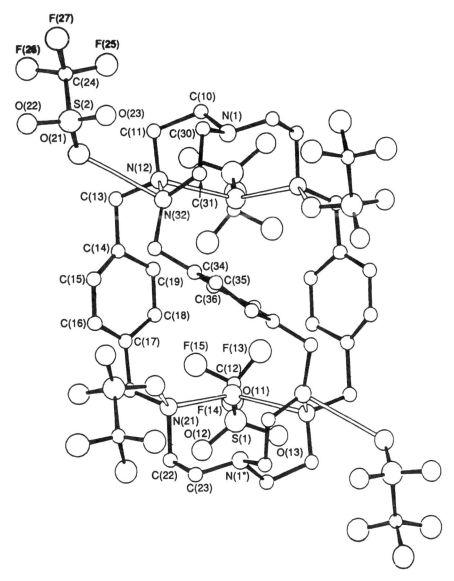

Fig. 5.20. General view of the crystal structure of **5.23** with trifluoromethane sulfonate. (Reproduced with permission from Ref. 62.)

zwitterionic host **5.25**.[68] Complex **5.25** has been characterized by X-ray crystallography (in the absence of guest anion) and adopts a related conformation to that observed for the iodide complex I⁻·**5.15**.[45] However, the cavity is significantly distorted from tetrahedral symmetry for the neutral receptor, N···N distances for **5.15** = 7.10–7.52 Å, cf. **5.25** 6.92–8.10 Å.[68] Molecular modeling calculations indicate that **5.25** is capable of forming 1:1 inclusion complexes with small anions such as Br$^-$ and CN$^-$ (calculated $\Delta H_f = -16$ and -26 kcal mol^{-1}, respectively), although ^{1}H-NMR titrations indicate weak binding of Br$^-$ with $K_{\text{diss}} = 80 \times 10^{-3}$ M.

In addition to the possibility of ion-dipole-type interactions such as found for **5.25**, it is possible to design neutral receptors with either hydrogen bond donor–acceptor and/or Lewis acid character. Work by Reinhoudt et al. has shown that extremely simple hosts such as **5.26**,[69] containing both hydrogen-bond donor and acceptor functionalities, may act as remarkably effective anion hosts, mimicking the extensive array of hydrogen-bonding interactions found in natural anion binding proteins (Sect. 5.10).[70–73] Ligand **5.26** and a number of related compounds were all found to be selective for H$_2$PO$_4^-$ over HSO$_4^-$ and Cl$^-$ with binding constants up to 14,200 M^{-1}.[69] This affinity is attributed to

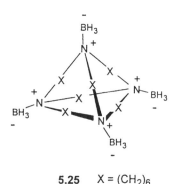

5.25   X = (CH$_2$)$_6$

the high electrophilicity of the sulfonamide NH group and preorganization of the binding sites by π-stacking interactions involving the naphthyl groups. Extension of this work to a range of calix[4]arenesulfonamides **5.27** resulted in binding constants of up to 103,400 M^{-1} and significant selectivity for HSO$_4^-$ over chloride and nitrate. This striking result is attributed to the combination of a hydrophobic calixarene cavity and an array of approximately preorganized amide functionalities, which functionally mimics protein anion binding environments.[70–73]

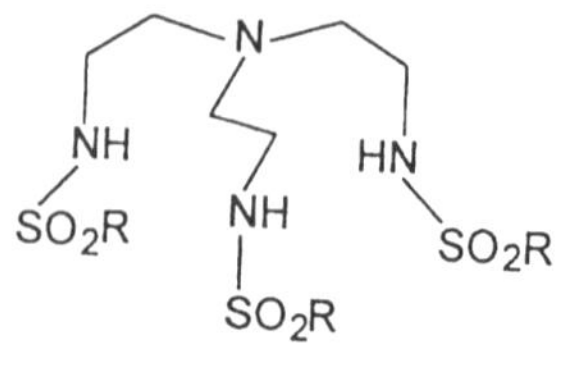

5.26

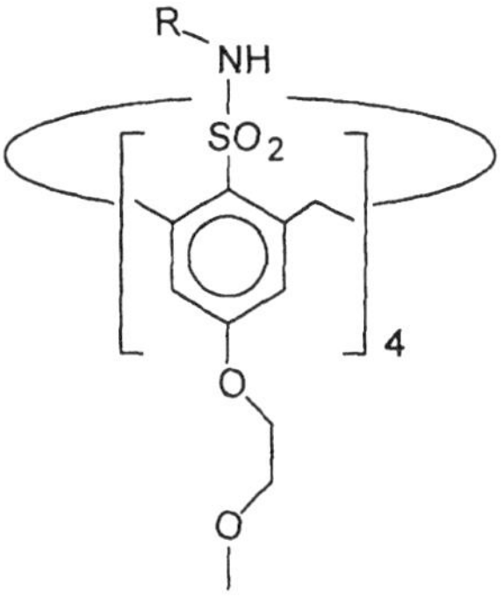

5.27
R = CH$_2$CH$_2$NHC(O)Me

5.5.2 Inorganic Hosts

Macrocycles such as the calixarenes have become increasingly topical in recent years as a basic framework for the design of various biomimetic host systems.[74,75] In addition to compounds such as **5.27**, a number of inorganic calixarene-based systems have also been reported in which a Lewis acid metal center is coupled with the hydrophobic nature of the calixarene cavity. The uranyl calixarene complex **5.28** with a salene "handle" is an example of a bifunctional receptor for both uranyl ion as well as anionic species. The uranyl complex has been found to be selective for H$_2$PO$_4^-$, although binding constants are only in the region of 400 M^{-1}.[76] It is suggested that anion binding occurs via coordination of the phosphate anion to the oxophilic uranium center and is stabilized by additional hydrogen bonding interactions to the amide functionalities. In such a case, the anion is likely to be significantly removed from the calixarene cavity. Indeed, a similar host strategy involving uranyl salene complexes containing pendant benzo-15-crown-5 moieties results in the simultaneous complexation of K$^+$ and H$_2$PO$_4^-$ with association constants in the region of 1000 M^{-1}.[77] Further studies involving X-ray crystallography, multinuclear NMR, cyclic voltammetry, and conductometry upon a wide range of similar hosts such as **5.29** indicate that the anionic guests do indeed coordinate to the uranium center with H$_2$PO$_4 \cdots$ U distances in the region of 2.28(2) Å. Additional stabilization is gained by phosphate$\cdots$amide and phosphate$\cdots$acetoxy hydrogen-bonding interactions, with O$\cdots$N and O$\cdots$O distances of 2.79(2) and 2.84(2) Å, respectively, suggesting strong interactions (Fig. 5.21).[78] In addition, in the case of **5.29**, a second phosphate anion is also bound in the solid state via hydrogen-bonding interactions. It is likely that this guest is more weakly held in solution. This multiple recognition, in which the phosphate guest anion acts both as a hydrogen-bond donor and acceptor, as well as a ligand for the uranium center, is reflected in the high value for K_s (measured by

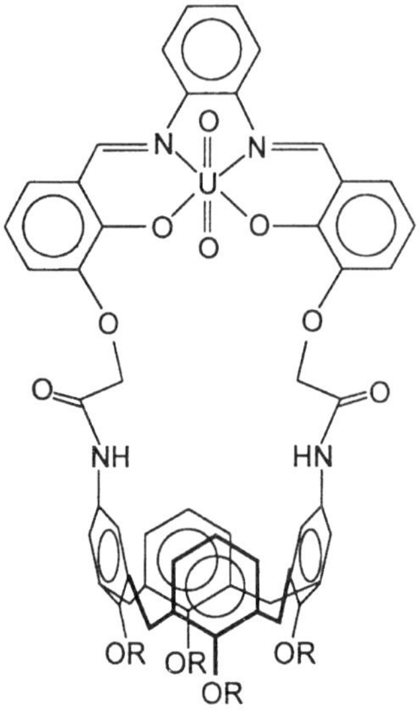

5.28 R = n-Pr, CH$_2$C(O)OEt

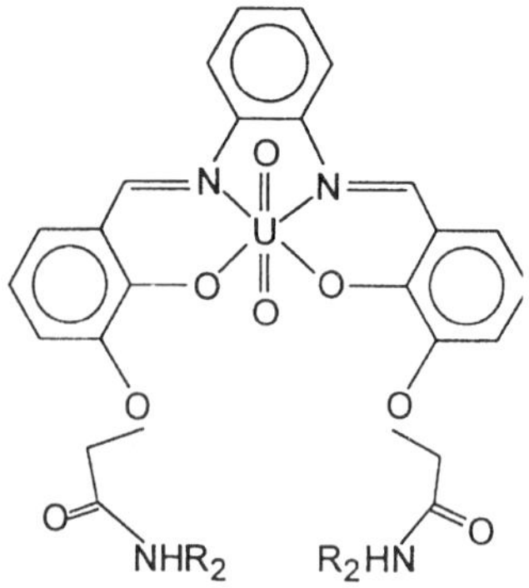

5.29 R = p-MeC$_6$H$_4$

conductometry and NMR), which is in excess of 10^5 in acetonitrile solution and drops to 1.5×10^3 in DMSO, presumably as a consequence of greater anion solvation.[78]

A remarkable example of halide anion recognition based solely upon size considerations has recently been reported by Puddephatt et al.[79,80] Reaction of the phosphonite calix[4]arene derivative **5.30** (L) with silver or copper salts gives the respective tetranuclear metal halide-bridged complexes [LM$_4$(–Cl)$_4$($_n$–Cl)]$^-$ (M = Cu, n = 3: M = Ag, n = 4), in which a guest chloride engages in a multidentate face-bridging interaction with the square plane containing the metal centers. In the case of the copper complex (Fig. 5.22A), the X-ray crystal structure reveals a triply bridging coordination mode with bound Cu–Cl distances ranging from 2.467(5) to 2.548(6) Å. In the case of the silver complex, the presence of the larger metal ion results in a much more symmetrical quadruply bridging coordination mode with Ag–Cl distances 2.69(1)–2.76(1) Å (Fig. 5.22B).[79,80] Fascinatingly, the bridging chloride ligand in the copper complex may be selectively replaced by iodide to give a compound related to that observed in the silver case, with a symmetrically bridging iodide guest anion (Cu···I = 2.75–2.85 Å).[81] In each case, the guest anion is not within the plane of the metal centers, but resides in the center of the host cavity, with a unique trigonal or square pyramidal coordination. It is clear that both coordination number and geometry are significantly more malleable in the case of anions than cations.

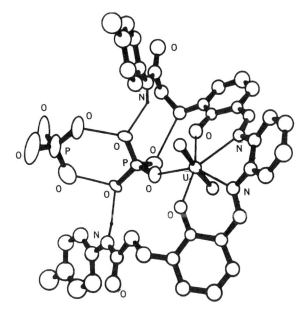

Fig. 5.21. Binding of two $H_2PO_4^-$ anions by the uranyl salene host **5.29**. (Reproduced with permission from Ref. 78.)

5.30 R = CH_2CH_2Ph

5.6 LEWIS ACID RECEPTORS

Even before the discovery of the *katapinands*, work on bidentate Lewis acid hosts suggested the possibility of chelation of anionic species by acyclic boron-containing ligands such as $BF_2CH_2CH_2BF_2$.[82] In many ways multidentate

184 STRUCTURAL AND TOPOLOGICAL ASPECTS OF ANION COORDINATION

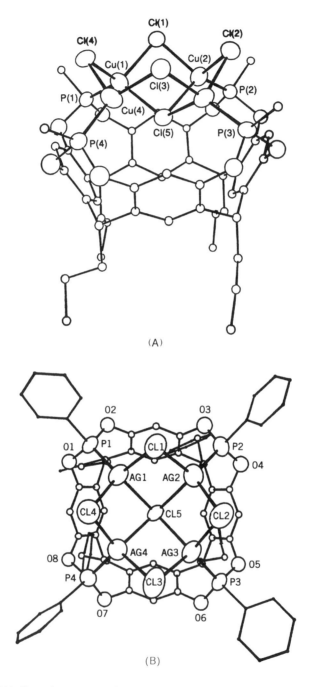

Fig. 5.22. (A) Crystal structure of the copper derivative of the phosphonito calixarene **5.30**. (Reproduced with permission from Ref. 80.) (B) Crystal structure of the silver complex of **5.30**. (Reproduced with permission from Ref. 79.)

LEWIS ACID RECEPTORS

Lewis acid hosts may be regarded as the anion binding equivalent of the crown ethers, and cyclic as well as acyclic compounds of B, Sn, Si, and Hg have been extensively studied. Much of the following chemistry has been reviewed by Katz in 1991,[7] and the interested reader is referred to this comprehensive summary as the primary source for further reading. In this section important structural results will be summarized and more recent work covered.

5.6.1 Hosts Containing Boron and Silicon

One of the simplest neutral anion hosts so far examined is the so-called hydride sponge[83] (1,8-naphthalenediyl*bis*(dimethylborane), **5.31**, cf. proton sponge[84]) which contains two strongly Lewis acidic $-BMe_2$ functionalities preorganized in such a fashion to be able to chelate an anionic or Lewis basic guest. Compound **5.31** readily and irreversibly abstracts a single hydride anion from a range of hydride sources, and the resulting complex is stable even in the presence of moderately strong acids such as acetic acid and $HNEt_3Cl$, although it is slowly decomposed by trityl cation. The X-ray crystal structure of **5.31**·H$^-$ (Fig. 5.23) shows the chelation of the hydride anion by the two $-BMe_2$ moieties, with refined B–H bond distances of 1.20(5) and 1.49(5) Å.[83] It is unclear whether the

5.31

hydride anion is genuinely situated asymmetrically between the boron atoms since related hydride systems exhibit symmetrical bridging modes.[85] Compound **5.31** also forms stable anion complexes with F$^-$ and OH$^-$ but not with Cl$^-$. This relative instability is presumably a consequence of the larger ionic radius of the chloride ion, which gives a poor fit between the two boron binding sites, as well as the generally weak Lewis acidic nature of **5.31**.

In contrast, the more strongly Lewis acidic $-BCl_2$ analog of **5.31**[86] is capable of binding chloride. The crystal structure of this material reveals that the guest chloride lies out of the plane of the naphthalene ring, again consistent with the larger size of the chloride anion. The chloride is asymmetrically situated between the two boron atoms, with B$\cdots$Cl distances of 1.66–1.94 Å (two disordered orientations) and B–Cl–B angles of 102° and 110°.

Related work on mixed borane–silyl compounds of type **5.32** also results in anion binding, as evidenced by the X-ray crystal structure of the fluoride complex **5.32**·F$^-$, in which the silicon expands its coordination number to five, albeit with one long bond to fluorine, 2.714(7) Å, and shorter bonds to the four carbon atoms averaging 1.87 Å. The B$\cdots$F distance is much shorter at

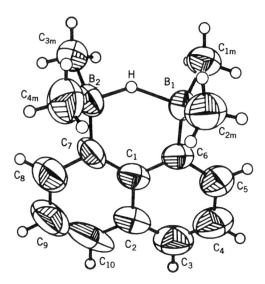

Fig. 5.23. Crystal structure of the hydride complex of 1,8-naphthalenediyl-*bis*-(dimethyl-borane), **5.31**·H$^-$. (Reproduced with permission from Ref. 83.)

1.475(6) Å. It is clear that the relatively low Lewis acidity of tetravalent silicon results in only a weak interaction with F$^-$, thought to stabilize the complex relative to the unhindered dimethylnaphthylborane kinetically, but not thermodynamically, as a result of steric crowding.[7,87] Interestingly, these results contrast with the more closely five-coordinate silicon centers found for the

5.32

disilane fluoride complex [K$^+$·18-crown-6][o-C$_6$H$_4$(SiPhF$_2$)$_2$F$^-$] **5.33**, in which the bridging Si···F distances are 1.898(4) and 2.065(4) Å compared with terminal Si···F bonds of about 1.60–1.65 Å (Fig. 5.24).[88] Larger Lewis basic guests may be accommodated by the use of longer rigid spacer units as in 1,8-anthracenediethynyl*bis*(catechol boronate).[89]

Simultaneous complexation of anions and cations, similar to that reported for 15-crown-5 uranyl salene complexes (Sect. 5.5.2),[77] has been observed in the 21-membered ring boronate crown ether **5.34**. Ligand **5.34** is capable of dissolving stoichiometric amounts of KF in dichloromethane at room temperature to give the KF adduct shown in Fig. 5.25.[90] The fluoride ion is bound to the

LEWIS ACID RECEPTORS **187**

[Structures 5.33A and 5.33B shown with F⁻ arrow between them]

5.33A → **5.33B**

boron atom out of the plane of the crown ether, which contains the K⁺ ion. Ligand **5.34** fails to dissolve either KCl or KBr as a consequence of the weaker nature of the B–X bonds. KI and KSCN are dissolved by **5.34** but without complexation of the boron atom. The stabilization of the K⁺ ion by the crown ether moiety is apparently sufficient in these cases.[90]

A recent report gives the results of theoretical (AM1 molecular orbital) treatments of boron analogs of the katapinand series as well as macrotricyclic compounds related to **5.15**. As with the ammonium-based host molecules, the macrotricyclic hosts containing four boron atoms exhibited a greater degree of

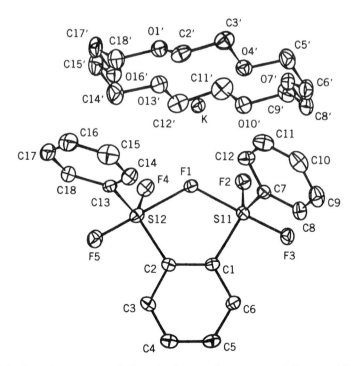

Fig. 5.24. Crystal structure of the mixed potassium crown disilyl fluoride complex **5.33**·F⁻. (Reproduced with permission from Ref. 88.)

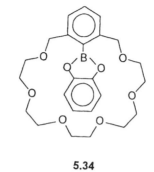

5.34

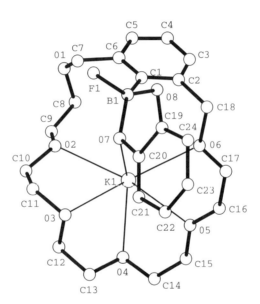

Fig. 5.25. Crystal structure of the KF adduct of **5.34**. (Reproduced with permission from Ref. 90.)

anion specificity as a consequence of the rigidity of their binding sites. In all cases, size-selective complexation of halide anions was observed with accompanying decreases in B···B distances and partial rehybridization (sp^2 to sp^3).[91]

5.6.2 Hosts Containing Mercury

A large number of receptors containing mercury have been reported based upon the *o*-phenylenedimercurial moiety **5.35**.[92] Interaction of **5.35A** with halide ions results in the formation of complexes soluble in nonpolar media under conditions in which the free hosts are insoluble, while monomercurials disproportionate. Solution ^{199}Hg-NMR measurements indicate that 1:1 complexes with

5.35 A: X = Cl
B: X = Br
C: X = I
D: X = CF$_3$CO$_2$

chloride are favored in solution, while the 2:1 host–guest complex crystallizes (Fig. 5.26).[92a] Primary bonds to the mercury centers adopt the usual linear arrangement, Hg–Cl = 2.93 Å, while the guest anion as well as the secondary Hg···Cl interactions form longer bonds at 3.17 Å.[92a] This chemistry has been extended to give macrocyclic o-phenylenedimercurial-based host compounds by reaction of polymeric mercuric oxo species with various α,ω-dicarboxylic acids. Tractable products were only obtained from perfluoroglutaric acid, which gave the tetradentate macrocycle **5.36** in ca. 80% yield. An X-ray crystal structure determination of the *bis*(thf) adduct of this macrocycle revealed an approximately planar ring of dimension about 12 × 7 Å, with the thf guests each chelated by two mercury atoms. No structural results relating to the anion-binding properties of these macrocycles have yet been reported.[92b]

More recent work has yielded another class of carborane-based mercury-containing macrocycles **5.37** and **5.38** related to crown ethers such as 12-crown-4 and 9-crown-3. Complex **5.37** associates with one or two moles of lithium halide, depending on conditions. The X-ray crystal structure of the chloride salt is shown in Fig. 5.27.[93] The guest chloride anion interacts equally with all four

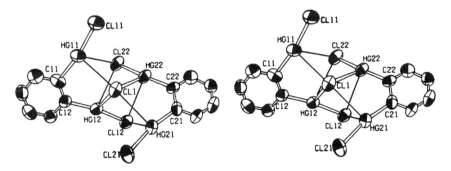

Fig. 5.26. Crystal structure of the chloride complex of the o-diphenylenedimercurial receptor **5.35A**. (Reproduced with permission from Ref. 92a.)

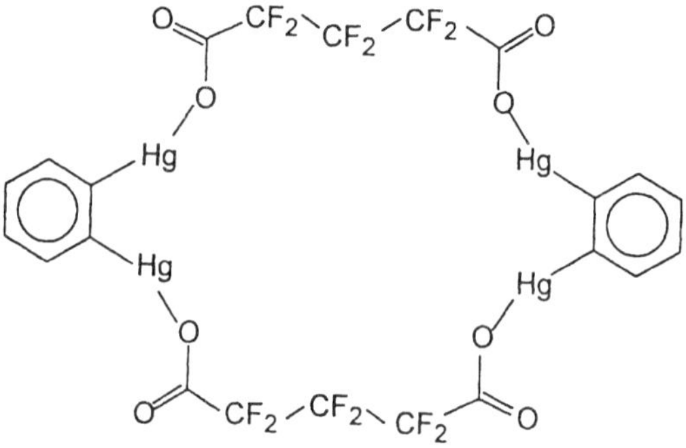

5.36

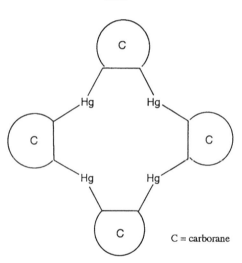

C = carborane

5.37

C = carborane

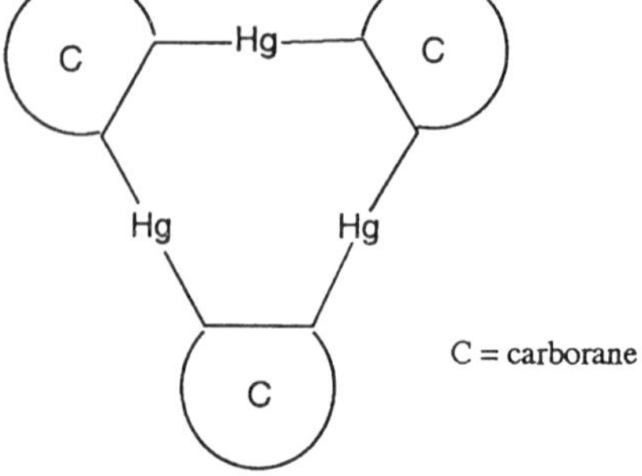

5.38

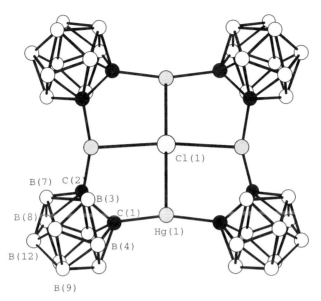

Fig. 5.27. Crystal structure of the chloride complex of **5.37**. (Reproduced with permission from Ref. 93a.)

electrophilic mercury centers with long Hg···Cl "bonds" of 2.944(2) Å (cf. 2.93 Å for **5.35A**). The anion is situated almost in the plane of the four mercury atoms, which deviate from linear coordination with a C–Hg–C angle of 162.0(3)°, suggesting a distortion of the mercury centers toward the guest anion. In the analogous bromide complex the larger ionic radius of the bromide ion results in its being displaced some 0.96 Å out of the plane containing the mercury atoms. The average Hg–Br contact in this structure is 3.063(5) Å.

A 1:2 structure is formed between **5.37** and iodide, with iodide ions 1.96 Å above and below the plane of the mercury atoms. The Hg–I distances are not equal, but fall into two distinct categories with both iodide ions equidistant from two opposite mercury centers (3.31 Å) and closer to a third (3.28 Å) than the fourth (3.77 Å).[93]

The X-ray crystal structures of the related, though less complex, mercury-containing macrocycles **5.39** and **5.40** have also recently been reported.[94,95] Compound **5.39** may form either 1:1 complexes with Br⁻ or I⁻ or a 3:2 complex with Cl⁻. In the case of the bromo derivative, crystallographic results reveal an infinite chain of alternating bromide and **5.39** with each halide bridging between six mercury atoms, Hg···Br = 3.07–3.39 Å (Fig. 5.28). It is postulated that the related 3:2 chloride complex exhibits a similar, though finite, layered structure. The related pentameric species **5.40** forms 1:2 complexes with Cl⁻ and Br⁻ with halide ions laying above and below the plane containing the mercury ions, Hg···Cl = 3.09–3.39 Å. It is symptomatic of the poor fit of the halide ions within these cyclic host systems, and of the large number of interactions

STRUCTURAL AND TOPOLOGICAL ASPECTS OF ANION COORDINATION

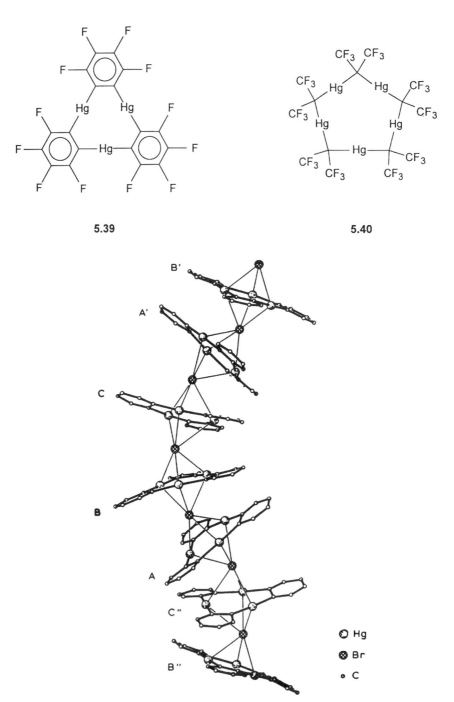

Fig. 5.28. The fragment of infinite polyanionic chain of **5.39** with bromide ion. (Reproduced with permission from Ref. 94.)

involved, that these distances are markedly longer than those observed in the LiBr bridged chain complex $\{[Hg(CH_2P(O)Ph_2)_2]BrLi\}_n$ in which the bromide ion bridges between two mercury centers in a nonlinear fashion with Hg···Br distances of 2.981(1) and 3.187(1) Å. Also, interestingly, the C–Hg–C angle is almost linear in this case, 174.7(4)°, because of the absence of macrocyclic ring strain.[96]

5.6.3 Hosts Containing Tin

Work by Newcomb et al. has resulted in the synthesis of a number of macroyclic and macrobicyclic host molecules somewhat related to the *katapinands*,[11] containing two Lewis acidic chloro-substituted tin centers separated by macrocyclic cavities of varying dimensions.[97] In the case of the monocyclic compound, acetonitrile solution ^{119}Sn-NMR measurements indicate that the halide affinity is relatively weak (K_s up to ca. 1000 M^{-1}), presumably as a consequence of the relatively flexible nature of the ligands. In contrast, the macrobicyclic analogues **5.41**, in which the conformation at tin and the size of the macrocyclic cavity are more rigidly enforced, demonstrate a marked anion affinity and selectivity.[98] In particular, complexes **5.41B** (n = 8) can bind chloride markedly more strongly than either the smaller- or longer-chain analogs, while **5.41A** (n = 6) is selective for fluoride.

$$Cl-Sn \begin{array}{c} -(CH_2)_n- \\ -(CH_2)_n- \\ -(CH_2)_n- \end{array} Sn-Cl$$

5.41 n = 6, 8, 10, 12

The binding of halide ions within the cavity of these macrocycles has been confirmed for **5.41A·F$^-$** and **5.41B·Cl$^-$** by X-ray crystal structure determinations (Fig. 5.29).[98b] Notably, in both cases the macrocycles adopt an outward conformation of the chloride substituent on the tin center (in contrast to the *in* form observed for the nitrogen atom in **5.1**[14]) simply as a consequence of the ability of the tin center to adopt a five-coordinate geometry in anion binding. In the case of **5.41A**, the fluoride anion adopts an approximately symmetrical bridging mode with relatively short Sn–F distances of 2.12(4) and 2.28(4) Å. In comparison, the chloride ion is situated asymmetrically in the larger macrocycle **5.41B**, apparently frozen between dissociation from one tin center and binding to the other, with Sn···Cl distances 2.610(5) and 3.388(5) Å. This may be compared with Sn···Cl distances to the chloride substituents outside the cavity of 2.745(5) Å for the trigonal bipyramidal metal center, 2.415(5) Å for the tetrahedrally coordinated tin and 2.373(3) Å in the free macrobicycle. Solution ^{119}Sn-NMR measurements indicate that a rapid equilibrium exists in solution with the chloride ion hopping between the two metal centers even at $-100°$C.[98]

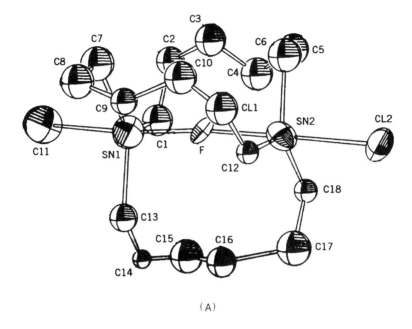

(A)

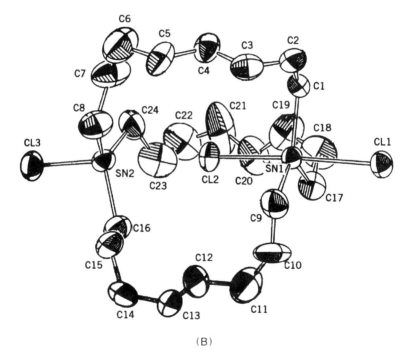

(B)

Fig. 5.29. (A) ORTEP drawing of **5.41**·F$^-$. (Reproduced with permission from Ref. 98b.) (B) ORTEP drawing of **5.41B**·Cl$^-$. (Reproduced with permission from Ref. 98b.)

The X-ray crystal structures of a number of other tin-containing macrobicycles have been reported. The crystallographically determined cavity sizes are in good agreement with those calculated by molecular mechanics methods.[99]

5.7 ORGANOMETALLIC RECEPTORS

Extensive work by Beer et al. has resulted in the synthesis of a range of anion-binding hosts containing electrochemically active cations *tris*(bipyridyl)-ruthenium(II),[100] cobalticinium,[101–105] and ferrocenyl[102,106] moieties, coupled with additional recognition sites such as crown ethers,[105,106] calixarenes,[100–102] and, particularly, amide functionalities.[100–104] A number of these materials are capable of recognition of both inorganic phosphate[100,102,106] and biologically relevant polyphosphates,[106] as well as halides and HSO_4^-. In each case, anion recognition is guided by electrostatic attraction to the transition metal ion and hydrogen-bonding interactions with the amide functionality, as well as hydrophobic and topological considerations. Unfortunately, relatively few X-ray crystal structure determinations have appeared, and little is known about the exact nature of the synergy between these diverse interactions. The X-ray crystal structure of the cobalticinium complex **5.42** does, however, establish the inclusion of two molecules of acetonitrile between the cobalticinium binding sites. These solvent guest molecules do not interact with the amide functionalities.[101] Also, related work using resolved, chiral cobalticinium complexes has resulted in a series of receptors capable of responding to the chirality of guests such as camphor-10-sulfonate.[107]

Recently, our group has taken a slightly different approach to the use of calixarenes as anion hosts.[108–112] Direct attachment of cationic transition metal fragments to the outside of the calixarene bowl results in a cationic binding

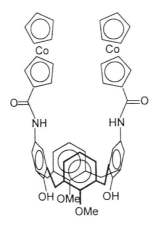

5.42

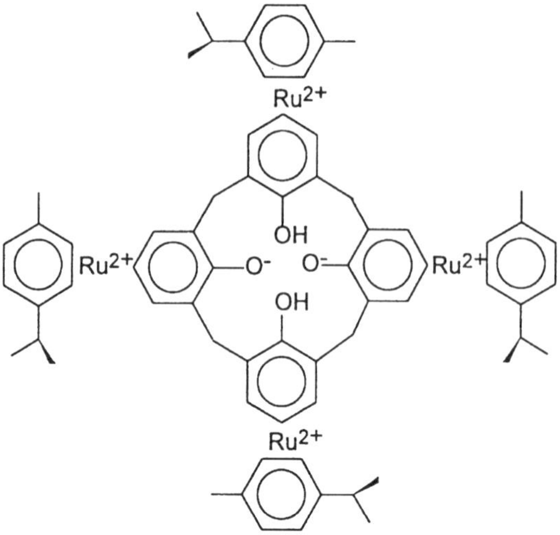

5.43

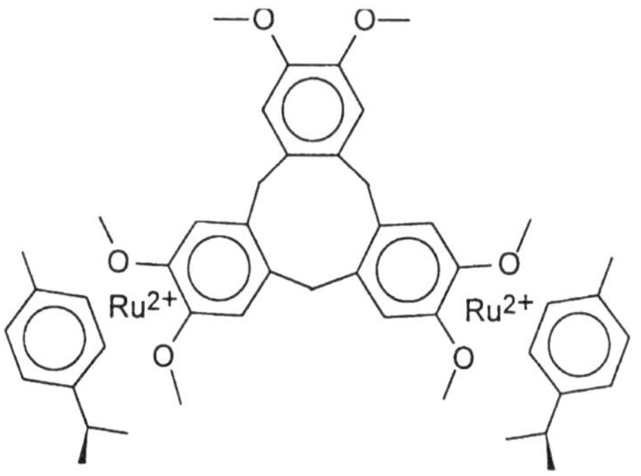

5.44

pocket of very specific size and shape into which anions are very deeply included, **5.43** and **5.44**. The X-ray crystal structure of the BF_4^- salt of the water-soluble "bear trap" **5.43** demonstrates the suitability of the calix[4]arene cavity for small tetrahedral anions (Fig. 5.30A). Strong electrostatic attractions exist between the fluorine atoms of the anion and the ring carbon atoms, with F···C contacts as short as 2.85 Å. For comparison, BF_4^-···C distances are generally longer than 3.10 Å in noncavity systems.[108] The tetrafluoroborate anion may be displaced by treatment with Bu_4NI to give the analogous iodide salt (Fig. 5.30B), which exhibits I···C contacts of ca. 3.7 Å, consistent with the large ionic radius of iodide. Interestingly, the iodide anion is situated some

0.11 Å lower in the cavity than the boron atom in the BF_4^- salt, consistent with the larger size of the tetrafluoroborate anion. Formation of such organometallic hosts is extremely general, and a number of crystallographic results have also been obtained with *p-t*-butylcalix[5]arene[110] as well as related macrocycles such as cyclotriveratrylene (CTV).[111,112] Indeed, the bimetallic CTV-based host **5.44** exhibits a significant selectivity for ReO_4^-, presumably as a consequence of the wide, shallow nature of the CTV cavity.[112]

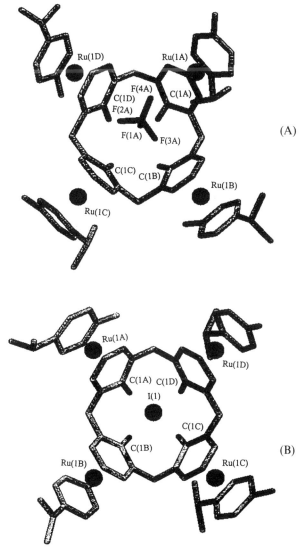

Fig. 5.30. Crystal structure of the organometallic "bear trap" host: (A) with included BF_4^- anion and (B) with iodide.

5.8 TRANSITION METAL COMPLEXES

An obvious possibility in the design of anion complexation agents is the incorporation of a metal center with which the anion may form a coordinate bond. Such "bond" formation is by definition outside the realm of supramolecular chemistry; however, a few examples that touch on the border are included in the discussion below. This section deals with a number of other metal complexes capable of "multipoint" molecular recognition, based on binding anionic ligands to a metal center as well as the formation of other stabilizing interactions such as hydrogen bonds.

5.8.1 Zinc Complexes

Reaction of $Zn(ClO_4)_2 \cdot 6H_2O$ with acridine-pendant cyclen affords the Zn(II) host molecule **5.45**.[113] The combination of π-stacking interactions with the acridine functionality, hydrogen bonding with the NH protons of the cyclen ring, and coordinative interaction with the metal center enables **5.45** to recognize and bind thymidine and its analogs from a mixture of nucleosides in aqueous solution at physiological pH. Selectivity for thymidine analogues is dependent on the fact that these substrates possess an imide functionality, enabling them to bind to the Zn(II) center. Stability constant measurements in aqueous buffer suggest a high affinity of both **5.45** and its acridine-free analog for thymidine-related nucleoesides with log K values in the range 5.7–7.2. The X-ray crystal structure of the complex formed between **5.45** and deprotonated methylthymine (Fig. 5.31) demonstrates the remarkably tight binding of the methylthymine anion, with a short $Zn-N_{thym}$ bond of 1.987(4) Å [cf. 2.04(1) Å to the neutral substrate guanine].[113] In addition, two of the three cyclen NH groups form hydrogen bonds with the methylthymine carbonyl functionalities either directly, 2.881(5) Å, or via a water molecule. Evidence for π-stacking interactions comes from the face-to-face stacking of the thymine and acridine moieties (dihedral angle 2.5°) with nonbonded C···C separations in the range 3.285–3.419 Å.[113]

A related receptor based on a chiral, resolved Zn(II) porphyrin complex has

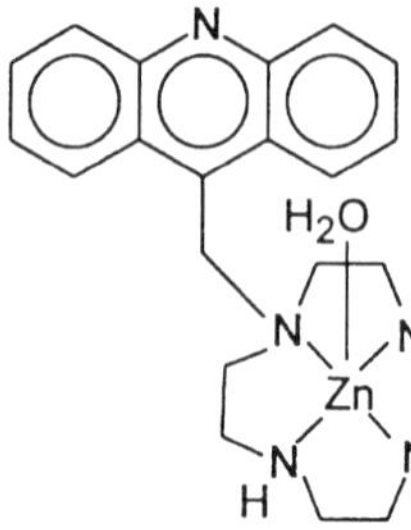

5.45

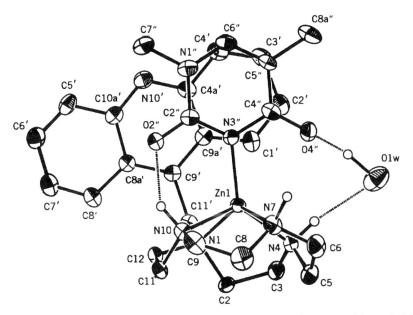

Fig. 5.31. Crystal structure of the zinc(II) cyclen complex **5.45** with methylthymine anion. (Reproduced with permission from Ref. 113.)

been shown to differentiate enantiomers of a range of substituted amino acids with selectivity of up to 96% for L-benzyloxycarbonylvalinate. Again the combination of metal coordination and hydrogen-bonding interactions are responsible for the observed selectivity.[114]

5.8.2 Copper Complexes

Reaction of copper perchlorate with a silver complex of the 30-membered macrocyclic Schiff base ligand **5.46** (*L*) gives the dicopper host $Cu_2L(ClO_4)_4 \cdot 3H_2O$. In the presence of anions such as N_3^- and OH^-, the perchlorate anions are replaced by both terminal and bridging azide or hydroxide to give a five-coordinate square pyramidal dicopper "cascade" complex. The X-ray crystal structure of the azide derivative has been determined and shows that longer bonds are formed to the singly bridging azide anion (2.25 Å) than to the terminal ligands (1.94 Å). The anion "zigzags" between the two metal centers to give a Cu···Cu distance of 6.02 Å (Fig. 5.32).[115]

A related dicopper complex of the 24-membered macrocycle **5.47** binds four azide ions to give a neutral species containing two bridging and two terminal ions.[116] In this instance, the $Cu-N_{bridge}$ distance, averaging 2.00 Å, is significantly shorter than that observed for the analogous copper complex of **5.46** and may reflect the doubly bridged nature of the complex as well as the smaller size of the macrocycle. The Cu···Cu separation is 5.145 Å (Fig. 5.33).

5.46

Another related dicopper species of a 24-membered cryptand has also been shown to bind the imidazolate anion in a singly bridging fashion, Cu–N$_{imid}$ = 1.946 Å.[117] Work by Martell et al. has shown that a wide range of related complexes may be formed in solution with various anions, including inorganic phosphates, often with remarkably high stability constants.[118]

5.9 POLYMETALLIC RECEPTORS

5.9.1 Polyoxometallate Hosts

An additional class of anion host is based upon the reaction of various vanadate salts with alkyl phosphates in the presence of halides. Formation of a wide variety of oxovanadium cagelike cluster species has been observed, many of

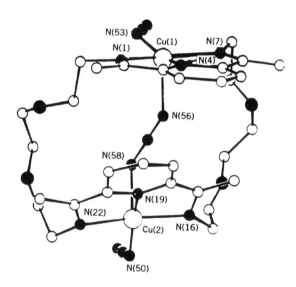

Fig. 5.32. Crystal structure of the dicopper complex of **5.46** with azide anion. (Reproduced with permission from Ref. 115.)

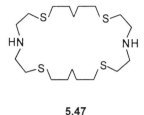

5.47

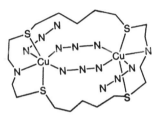

Fig. 5.33. Schematic drawing of the binding of azide ion in the dicopper complex of 5.47.

which incorporate anions or both anions and cations within large, spherical cavities. Reaction of t-BuPO$_3$H with [PPh$_4$][VO$_2$Cl$_2$] gives the chloride inclusion species [(V^{V}O)$_5$(V^{IV}O)(t-BuPO$_3$)$_8$Cl]. The X-ray structure of this material (Fig. 5.34) reveals the fact that the single halide anion is encapsulated within the cagelike oxovanadium system, and indeed may well act as a template for cluster formation.[119]

A related reaction results in the formation of the *bis*(ammonium chloride) inclusion species [2NH$_4^+$, 2Cl$^-$ · V$_{14}$O$_{22}$(OH)$_4$(H$_2$O)$_2$(C$_6$H$_5$PO$_3$)$_8$]$^{6-}$ in which a nanometer-wide cavity contains a square planar array of two ammonium cations and two chloride anions. The ammonium cation is located at the center of a near-planar coronand fragment analogous to the hypothetical 16-crown-8 macrocycle with NH$_4$···O distances of 2.99–3.66 Å. The chloride anions are situated between 3.32 and 3.63 Å to the vanadium centers (Fig. 5.35).[120]

This type of compound is not limited to vanadium, but is observed for other polyoxometallates, for example, the polyoxotungstate [NaP$_5$W$_{30}$O$_{110}$]$^{14-}$, which contains five internal phosphate groups.[121]

5.9.2 Helicates

With the rising era of self-assembly, new possibilities begin to unfold for anion recognition. The self-assembly of ligand strands binding metal ions can generate double and triple helical arrays, as exemplified by the cyclic bipyridine helicate

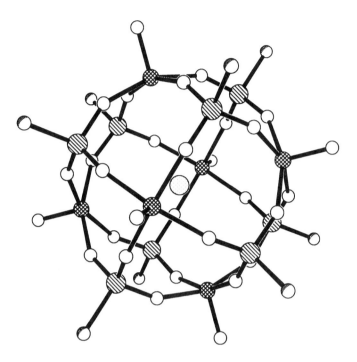

Fig. 5.34. Structure of the chloride inclusion complex of a polyoxovanadate. (Reproduced with permission from Ref 119a.)

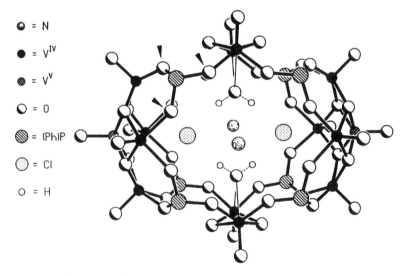

Fig. 5.35. The *bis*(ammonium chloride) inclusion into a polyoxovanadate core. (Reproduced with permission from Ref. 120a.)

of Lehn, Hasenknopf, and co-workers.[122] This pentameric complex helicate, formed by the reaction of a tris-bipyridine ligand and $FeCl_2$, results in a double helical torus carrying a +9 charge (Fig. 5.36). The outer diameter of the helicate is ca. 22 Å, but an inner cavity with a 1.75 Å radius is also formed, ideally suited for incorporation of a chloride ion (ionic radius = 1.8 Å) (Fig. 5.37). The formation of this elegant receptor may have broader implications with respect to a combinatorial approach to the recognition process. In such an instance, the component complexes would assemble in the presence of a given substrate (in this case chloride ion) in a fashion that serves to sequestor or bind that substrate in the most effective manner.

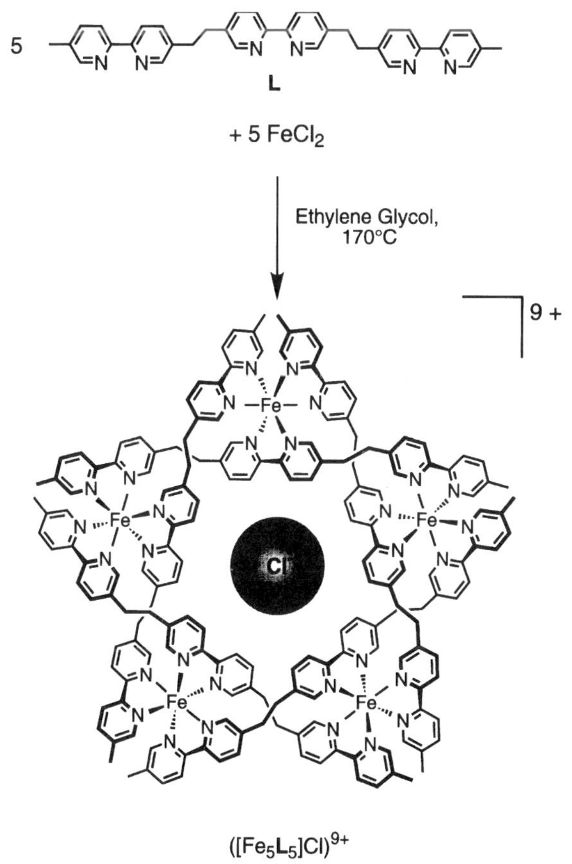

Fig. 5.36. Reaction of $FeCl_2$ with the tris-bipyridine ligand showing the formation of the circular double helicate. (Reproduced with permission from Ref. 122.)

204 STRUCTURAL AND TOPOLOGICAL ASPECTS OF ANION COORDINATION

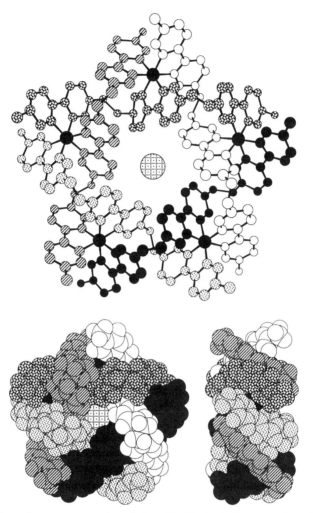

Fig. 5.37. Crystal structure results of the chloride inclusion complex of the circular double helicate shown formed in Fig. 5.36. (Reproduced with permission from Ref. 122.)

5.10 ALKALIDE AND ELECTRIDE SALTS

One of the primary motivations behind much of the early work on crown ethers and related cation hosts was the observation that such ligands may dissolve small quantities of alkali metals such as sodium and potassium to give deep blue solutions containing solvated electrons. In 1983, the first crystalline electride complex, $Cs^+(18\text{-crown-6})_2 \cdot e^-$, **5.48**, in which a naked electron acts as the counteranion to the more conventional cesium *bis*(crown ether) cation, was isolated.[123,124] Since that time, extensive work by Dye et al. has resulted in X-ray crystal structure determinations of four highly air, moisture, and thermally sensitive electride complexes[125] as well as a number of closely related alkalide

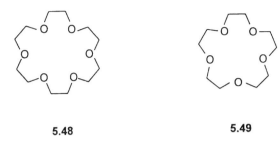

5.48   **5.49**

salts containing the unusual anions Na⁻, K⁻, Rb⁻, and Cs⁻. In each of these complexes it is an alkali metal cation that is bound within the crown ether or cryptand host molecules while the anion resides in specific sites in the lattice. However, the strong dependence of the properties of the electrides upon the geometry and connectivity of these lattice sites justifies their treatment in this discussion.

Complex **5.48** has been shown to be a localized electride by optical absorption spectroscopy with the electrons trapped in potential wells 0.9 eV below the conduction band.[126] As a consequence, the dc powder conductivity of this material is low (ca. $10^{-10} \, \omega^{-1} \, cm^{-1}$ at 200 K[125]), while magnetic susceptibility studies show a Curie–Weiss Law dependence with $\Theta = -1.5$ K, indicating very weak antiferromagnetic coupling between electrons. Interestingly, however, if extreme care is taken to keep the highly thermally sensitive complex at very low temperatures, rather stronger antiferromagnetic coupling is observed equivalent to thermal energies at 30–50 K. These properties may be rationalized from the X-ray crystal structure of this material,[127] which suggests the electrons to be localized within nearly spherical, otherwise empty cavities of radius ca. 2.4 Å, formed by eight encircling crown sandwiched Cs⁺ cations. Cs–O bond lengths are in the expected region of 3.3–3.5 Å. The shortest electron–electron distances are 8.68 Å, precluding significant interelectron interactions. It has been suggested[128] that the curious drop in antiferromagnetic coupling upon warming and recooling the sample arises from an irreversible phase transition involving the disordering of the atoms of the crown ether ring, which results in the closing of a narrow communication channel between each anionic site. The related electride complex $Cs^+(15\text{-crown-}5)_2 \cdot e^-$ **5.49** has also been structurally characterized.[129] As with **5.48**, the electron is apparently localized in a lattice cavity ca. 2.35 Å wide, with electron–electron distances upwards of 8.597 Å. Complex **5.49** does exhibit some degree of antiferromagnetic coupling, but again is only a very poor conductor.

There is some question as to whether the electron in these materials is actually located within the lattice cavities or is more localized on the alkali metal cation, since the "free" electron cannot be directly observed by diffraction methods.[130] Compelling evidence for the former interpretation comes from the X-ray crystal structures of the sodide (Na⁻) analogs of **5.48** and **5.49** as well as the kalide (K⁻) analogue of **5.49**, which demonstrates the inclusion of the alkalide anion in the lattice cavity in a fashion closely related to that suggested for the electride complexes.[126,131,132] Also a recent theoretical study on **5.49** has demonstrated that the electron is delocalized over the lattice cavity as a consequence of the need to minimize its kinetic energy in spite of the fact that the calculated potential is repulsive in this region.[133]

A more interesting electride is formed from the interaction of potassium metal with cryptand[2.2.2] (C222) in dimethyl ether at $-50°C$. In contrast to **5.48** and **5.49**, this material, $K^+(C222) \cdot e^-$, **5.50**,[134] demonstrates a high positive temperature coefficient of conductivity, with dc conductivity ranging from $9.7\omega^{-1}$ cm^{-1} at 100 K to about $10\omega^{-1}$ cm^{-1} at 200 K, a factor of at least a million times that of **5.48** and **5.49**.[126,135] The X-ray crystal structure of **5.50** demonstrates an interconnecting series of dumbbell-shaped cavities, each occupied by a pair of electrons separated by only 5.3 Å, with distances of 7.8

5.50

and 8.4 Å to electrons in adjacent cavities. The complex adopts a diamagnetic electronic configuration as the temperature approaches 0 K, but magnetic susceptibility increases with temperature, suggesting electron pair dissociation or the population of a triplet state. These results suggest only weak electron pairing and a significant degree of delocalization, although the compound is not actually metallic, but has an activation energy of 0.02 eV. The structures of the analogous sodide, kalide, rubidide, and ceside complexes $M^+(C222) \cdot M^-$ (M = Na, K, Rb, Cs) have also been determined and are closely related to that of **5.50**, with pairs of alkalide anions in close contact within the dumbbell-shaped cavities with a $K \cdots K^-$ distance of 4.90 Å rising to 6.38 Å for the cesium analog,[6,135] suggesting the presence of the unlikely dimers M_2^{2-}. In the case of the cesium complex, the resultant perturbation from spherical symmetry is sufficient to mix 5d character into the ground-state 6s wavefunction resulting in substantial spin-orbit coupling and the loss of the solid-state ^{133}Cs-NMR signal.

The X-ray crystal structure of the mixed crown electride $[Cs^+(15\text{-crown-}5)(18\text{-crown-}6) \cdot e^-]_6(18\text{-crown-}6)$ has been reported.[125] Even though it might be assumed that this material should be closely related to the parent compounds **5.48** and **5.49**, it exhibits radically different properties. Six mixed-sandwich Cs^+ complexes pack in a double three-bladed propeller-type array around a central 18-crown-6 ligand (occupied 25% of the time by a further Cs^+ ion, which contributes extra electrons) to give a puckered hexagon of six inter-

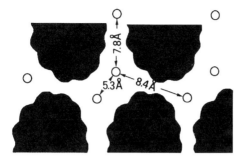

Fig. 5.38. Schematic diagram of the channel structure of $K^+(2.2.2.)\cdot e^-$ showing probable electron locations.

connected cavities containing the electron anions (ca. 8 Å interelectron separation) with further links of about 10 Å to adjoining electron sites (Fig. 5.38). The temperature-dependent dc conductivity of this material ranges up to 22.2(7) Ω^{-1} cm^{-1} while the weak, field-independent paramagnetism of the sample is indicative of strong electron coupling, consistent with diamagnetic six-electron rings weakly coupled together.

Finally, it is noteworthy that hosts such as crown ether and cryptand complexes are not the only ones capable of formation of electride salts. Inorganic materials such as sodalite may also admit extraframework anions to form complexes of the form $M_4[e^-](AlSiO_4)_3$ (M = monovalent metal ion) in which the electron is situated within the sodalite framework in cavities about 6.5 Å in diameter.[136]

5.11 ANION COMPLEXATION IN BIOCHEMISTRY

Anionic substrates are of great importance in biological systems, and a great deal of work has been expended upon their interactions with biological host molecules such as enzymes and transport proteins. Unfortunately, it is not possible to present a full treatise here, and a more in-depth treatment is offered in Chapter 3. Nonetheless, a number of recent results relating to structural aspects of anion binding in biological systems are of interest within the context of this discussion.

Several recent X-ray crystal structure determinations[70–73,137] have demonstrated that biochemical anion binding is a result of multiple hydrogen bonding interactions and a size-selective fit process. In particular, the 2.5 Å resolution structure of the complex of Yersina protein tyrosine phosphatase with tungstate anion shows the sequestration of the WO_4^{2-} guest within a network of twelve hydrogen bonds ranging in length from 2.7 to 3.4 Å. Notably four of these hydrogen bonds arise from the guanidinium moieties of arginine residues within the enzyme, highlighting the importance of these functionalities in biological systems. The structure of the tungstate complex differs from that of the native protein in that a ten-residue loop has moved an average of 3.3 Å in order to encapsulate the bound anion. If such a conformational change is also observed on binding a phosphotyrosine substrate, the structures give an excellent insight into the mechanism of proton transfer during the proposed hydrolysis steps undergone by the enzyme–substrate complex.[137]

The structure of the sulfate binding protein (SBP) of *Salmonella typhimurium* has also recently been reported.[72,73] In this case, the sulfate anion lies completely isolated from both solvent and counterions or salt bridges and is stabilized solely by seven hydrogen bonds between 2.67 and 2.84 Å in length. Interestingly, the related phosphate binding protein (PBP) is highly selective for HPO_4^{2-} and $H_2PO_4^-$ over SO_4^{2-} in spite of the similar shapes and charges of the substrates. As with SBP, the HPO_4^{2-} guest in PBP is sequestered some 8 Å below the protein surface and is held in place by a total of twelve hydrogen-bonding interactions of between 2.63 and 2.92 Å in length, again without the occurrence of any salt bridges. The crucial difference between SBP and PBP is the presence in the latter of a carboxylate side chain capable of acting as a hydrogen-bond acceptor, and hence substrate recognition depends primarily upon the state of protonation of the oxyanion.[70,71]

The biological recognition of phosphate and sulfate has also been the subject of a survey[138] of the Cambridge Crystallographic Database. This study reveals that the average hydrogen bonding $S=O-H$ angle of 127.9° is some 9° larger than in the analogous phosphate interaction. Also sulfonyl hydrogen bonding interactions are much more densely clustered than phosphoryl analogues. The sulfate also tends toward a nearly eclipsed geometry, in contrast to the phosphoryl preference for gauche interactions.

5.12 CONCLUDING REMARKS

For reasons such as high anion solvation energy, coordinative saturation, and large size as outlined in Section 5.1, it is clear that anion binding occurs much less readily than in related cation systems. Host–anion interactions are, to a large extent, dominated by nondirectional electrostatic forces, with small chelate and macrocyclic effects as well as considerations based on host geometry, size and degree of preorganization playing a distinctly secondary role.

It is tempting to try to attribute to anions a preferential coordination geometry analogous to that so well established for various metal cations. In many cases, simple anions such as the halides exist in approximately tetrahedral or octahedral environments, but it is clear from the diversity of examples reviewed herein that anion coordination geometry is highly flexible and may be adjusted to fit the properties of the various host systems.

It is clear that, in spite of dramatic recent advances and the enormous variety of anion hosts synthesized to date, the problem of achieving strong, selective anion complexation has not yet been solved and will doubtless be the subject of a great deal of rich and innovative chemistry in the future.

Acknowledgments

We thank the American Association for the Advancement of Science, the American Chemical Society, Gordon and Breach, Oxford University Press, Pergamon, the Royal Society of Chemistry, and VCH for permission to reproduce copyrighted material.

5.13 REFERENCES

1. (a) L. G. Lange III, J. F. Riordan, and B. L. Vallèe, *Biochem.* **13**, 4361 (1974); (b) F. P. Schmidtchen, *Nachr. Chem. Tech. Lab.* **36**, 8 (1988).
2. (a) B. Dietrich, in *Inclusion Compounds*, J. L. Atwood, J. E. D. Davies, and D. D. MacNicol, Eds., Oxford University Press Publishers, Oxford, 1984, Vol. 2, ch.10, pp. 373–405; (b) B. Dietrich, *Pure Appl. Chem.* **7**, 1457 (1993).
3. (a) M. W. Hosseini and J.-M. Lehn, *Helv. Chim. Acta* **70**, 1312 (1987); (b) J. L. Sessler, H. Furata, and V. Král, *Supramol. Chem.* **1**, 209 (1993).
4. D. Kaufmann, A. Otten, *Angew. Chem. Int. Ed. Engl.* **33**, 1832 (1994).
5. S. Goldman and R. G. Bates, *J. Am. Chem. Soc.* **94**, 1476 (1972).
6. (a) F. J. Tehan, B. L. Barnett, and J. L. Dye, *J. Am. Chem. Soc.* **96**, 7203 (1974); (b) R.-H. Huang, D. L. Ward, M. E. Kuchenmeister, and J. L. Dye, *J. Am. Chem. Soc.* **109**, 5561 (1987).
7. H. E. Katz, in *Inclusion Compounds*, J. L. Atwood, J. E. D. Davies, and D. D. MacNicol Eds., Oxford University Press Publishers, Oxford, 1991, Vol. 4, ch. 9, pp. 391–405.
8. R. M. Izatt, K. Pawlak, J. S. Bradshaw, and R. L. Breuning, *Chem. Rev.* **91**, 1721 (1991).
9. F. Vögtle, H. Sieger, and W. M. Müller, *Top. Curr. Chem.* **98**, 107 (1981).
10. J.-L. Pierre and P. Baret, *Bull. Soc. Chim. Fr.*, 367 (1983).
11. C. H. Park and H. E. Simmons, *J. Am. Chem. Soc.* **90**, 2431 (1968).
12. C. J. Pedersen, *J. Am. Chem. Soc.* **89**, 7017 (1967).
13. C. J. Pedersen, *Angew. Chem. Int. Ed. Engl.* **27**, 1021 (1988).
14. R. A. Bell, G. G. Christoph, F. R. Fonczek, and R. E. Marsh, *Science* **190**, 151 (1975).
15. B. Metz, J. M. Rosalky, and R. Weiss, *J. Chem. Soc., Chem. Commun.* 533 (1976).
16. T. N. Margulis and L. J. Zompa, *Acta Crystallogr., Sect. B* **37**, 1428 (1981).
17. B. Dietrich, J. Guihem, J.-M. Lehn, C. Pascard, and E. Sonveaux, *Helv. Chim. Acta* **67**, 91 (1984).
18. D. A. House, in *Comprehensive Coordination Chemistry*, G. Wilkinson, R. D. Gillard, and J. A. McCleverty, Eds., Pergamon Press, Oxford, 1987, Vol. 2, ch. 13, pp. 30–34.

19. N. Gavrushenko, H. L. Carrell, W. C. Stallings, and J. P. Glusker, *Acta Crystallogr., Sect. B* **33**, 3936 (1977).
20. N. H. Woo, N. C. Seeman, and A. Rich, *Biopolymers* **18**, 539 (1979).
21. G. J. Quigley, M. M. Teeter, and A. Rich, *Proc. Natl. Acad. Sci. USA* **75**, 64 (1978).
22. J. Cullinane, R. I. Gelb, T. N. Margulis, and L. J. Zompa, *J. Am. Chem. Soc.* **104**, 3048 (1982).
23. M. A. Santos and M. G. B. Drew, *J. Chem. Soc., Faraday Trans.* **87**, 1321 (1991).
24. (a) E. Kimura, A. Sakonaka, T. Yatsunami, and M. Kodama, *J. Am. Chem. Soc.* **103**, 3041 (1981); (b) E. Kimura, A. Sakonata, and M. Kodama, *J. Am. Chem. Soc.* **104**, 4984 (1982).
25. E. Kimura, H. Anan, T. Koike, and M. Shiro, *J. Org. Chem.* **54**, 3998 (1989).
26. B. Dietrich, M. W. Hosseini, J.-M. Lehn, and R. B. Sessions, *J. Am. Chem. Soc.* **103**, 1282 (1981).
27. (a) M. P. Mertes and K. B. Mertes, *Acc. Chem. Res.* **23**, 413 (1990); (b) M. W. Hosseini, J.-M. Lehn, and M. P. Mertes, *Helv. Chim. Acta* **68**, 818 (1985); (c) M. W. Hosseini, J.-M. Lehn, L. Maggiora, K. B. Mertes, and M. P. Mertes, *J. Am. Chem. Soc.* **109**, 537 (1987); (d) P. G. Yohannes, M. P. Mertes, and K. B. Mertes, *Inorg. Chem.* **26**, 1751 (1987); (e) M. W. Hosseini, J.-M. Lehn, K. C. Jones, K. E. Plute, K. B. Mertes, and M. P. Mertes, *J. Am. Chem. Soc.* **111**, 6330 (1989); (f) L. Qian, Z. Sun, J. Gao, B. Movassagh, L. Morales, and K. B. Mertes, *J. Coord. Chem.* **23**, 155 (1991); (g) A. Bencini, A. Bianchi, E. García-España, E. Scott, L. Morales, V. Wang, M. P. Mertes, and K. B. Mertes, *Bioorg. Chem.* **20**, 8 (1992).
28. (a) M. W. Hosseini and J.-M. Lehn, *J. Chem. Soc., Chem. Commun.* 397 (1988); (b) M. W. Hosseini and J.-M. Lehn, *Helv. Chim. Acta* **70**, 1312 (1987); (c) H. Fenniri and J.-M. Lehn, *J. Chem. Soc., Chem. Commun.* 1819 (1993).
29. S. Boudon, A. Decian, J. Fischer, M. W. Hosseini, J.-M. Lehn, and G. Wipff, *J. Coord. Chem.* **23**, 113 (1991).
30. G. Papoyan, K. Gu, J. Wiorkiewicz-Kuczera, K. Kuczera, and K. Bowman-James, *J. Am. Chem Soc.* **118**, 1354 (1996).
31. Q. Lu, R. J. Motekaitis, J. J. Reibenspies, and A. E. Martell, *Inorg. Chem.* **34**, 4958–64 (1995).
32. (a) A. Bianchi, S. Mangani, M. Micheloni, P. Orioli, and P. Paoletti, *Chem. Comm.* 729 (1987); (b) A. Bianchi, M. Micheloni, and P. Paoletti, *Pure Appl. Chem.* **60**, 525 (1988).
33. (a) A. Bencini, A. Bianchi, M. Micheloni, P. Paoletti, P. Dapporto, P. Paoli, and E. García-España, *J. Incl. Phenom.* **12**, 291 (1992); (b) A. Bencini, A. Bianchi, P. Dapporto, E. Garcia-España, M. Micheloni, P. Paoletti, and P. Paoli, *J. Chem. Soc., Chem. Commun.* 753 (1990).
34. A. Bencini, A. Bianchi, P. Dapporto, E. García-España, M. Micheloni, J. A. Ramirez, P. Paoletti, and P. Paoli, *Inorg. Chem.* **31**,1902 (1992).
35. (a) A. Bianchi, E. García-España, S. Mangani, M. Micheloni, P. Orioli, and P. Paoletti, *J. Chem. Soc., Chem. Commun.* 729 (1987); (b) A. Bencini, A. Bianchi, E. García-España, M. Giusti, S. Mangani, M. Micheloni, P. Orioli, and P. Paoletti, *Inorg. Chem.* **26**, 3902 (1987).
36. K. Aoki, T. Tokuno, K. Takagi, Y. Hirose, I.-H. Suh, A. O. Adeyemo, and G. N. Williams, *Inorg. Chim. Acta* **210**, 17 (1993).

37. A. Looney, G. Parkin, and A. L. Rheingold, *Inorg. Chem.* **30**, 3099 (1991).
38. (a) J. L. Sessler, M. J. Cyr, V. Lynch, E. McGhee, and J. A. Ibers, *J. Am. Chem. Soc.* **112**, 2810 (1990); (b) M. Shionoya, H. Furuta, V. Lynch, A. Harriman, and J. L. Sessler, *J. Am. Chem. Soc.* **114**, 5714 (1992); (c) B. L. Iverson, K. Shreder, V. Kral, P. Sansom, V. Lynch, and J. L. Sessler, *J. Am. Chem. Soc.* **118**, 1608 (1996); (d) J. L. Sessler, H. Furuta, and V. Kral, *Supramol. Chem.* **1**, 209 (1993); (e) V. K. Kral, K. Shreder, H. Furuta, V. Lynch, and J. L. Sessler, *J. Am. Chem. Soc.* **118**, 1595 (1996); (f) J. L. Sessler, M. Cyr, H. Furuta, V. Kral, T. Mody, T. Morishimsa, M. Shionoya, and S. Weghorn, *Pure & Appl. Chem.* **65**, 393 (1993).
39. (a) J. L. Sessler, T. D. Mody, D. A. Ford, and V. Lynch, *Angew. Chem. Int. Ed. Engl.* **31**, 452 (1992); (b) *idem.*, *Angew. Chem.* **104**, 461 (1992).
40. R. E. Cramer, V. Fermin, E. Kuwabara, R. Kirkup, M. Selman, K. Akoi, A. Adeyemo, and H. Yamazaki, *J. Am. Chem. Soc.* **113**, 7033 (1991).
41. R. E. Cramer and M. J. J. Carrié, *Inorg. Chem.* **29**, 3902 (1990).
42. E. Graf and J.-M. Lehn, *J. Am. Chem. Soc.* **98**, 6403 (1976).
43. (a) D. J. Cram and K. N. Trueblood, *Top. Curr. Chem.* **98**, 43 (1981); (b) D. J. Cram, *Angew. Chem. Int. Ed. Engl.* **25**, 1039 (1986).
44. F. P. Schmidtchen and G. Müller, *J. Chem. Soc., Chem. Commun.* 1115 (1981).
45. F. P. Schmidtchen, *Chem. Ber.* **114**, 597 (1981).
46. (a) B. Dietrich, J.-M. Lehn, J. Guilhem, and C. Pascard, *Tetrahedron Lett.* **30**, 4125 (1989); (b) B. Dietrich, B. Dilworth, J.-M. Lehn, J. P. Souchez, M. Cesario, J. Guilhem, and C. Pascard, *Helv. Chem. Acta* **79**, 569 (1996).
47. J.-M. Lehn, E. Sonveaux, and A. K. Willard, *J. Am. Chem. Soc.* **100**, 4914 (1978).
48. W. P. Jencks, *Catalysis in Chemistry and Enzymology*, McGraw Hill Publishers, New York, 1969, p. 358.
49. J.-M. Lehn, R. Méric, J.-P. Vigneron, I. Bkouche-Waksman, and C. Pascard, *J. Chem. Soc., Chem. Commun.* 62 (1991).
50. G. Morgan, V. McKee, and J. Nelson, *J. Chem. Soc., Chem. Commun.* 1649 (1995).
51. C. J. Harding, V. McKee, J. Nelson, and Q. Lu, *J. Chem. Soc. Chem. Commun.*, 1768 (1993).
52. (a) J. M. Adams and R. W. H. Small, *Acta. Crystallogr., Sect. B* **30**, 2191 (1974); (b) J. M. Adams and R. G. Pritchard, *Acta Crystallogr., Sect. B* **32**, 2438 (1976); (c) J. M. Adams and R. W. H. Small, *Acta Crystallogr., Sect. B* **32**, 832 (1976).
53. F. A. Cotton, V. W. Day, E. E. Hazen, Jr., and S. Larsen, *J. Am. Chem. Soc.* **95**, 4834 (1973).
54. B. Dietrich, D. L. Fyles, T. M. Fyles, and J.-M. Lehn, *Helv. Chim. Acta.* **62**, 2763 (1979).
55. (a) R. P. Dixon, S. J. Geib, and A. D. Hamilton, *J. Am. Chem. Soc.* **114**, 365 (1992); (b) V. Jubian, R. P. Dixon, and A. D. Hamilton, *J. Am. Chem. Soc.* **114**, 1120 (1992); (c) P. Schießl and F. P. Schmidchen, *J. Org. Chem.* **59**, 509 (1994); (d) G. Müller, J. Riede, and F. P. Schmidtchen, *Angew. Chem. Int. Ed. Engl.* **27**, 1516 (1988); (e) A. Gleich, F. P. Schmidtchen, P. Mikulcik, and G. Müller, *J. Chem. Soc., Chem. Commun.* 55 (1990).
56. P. Schießl and F. P. Schmidtchen, *Tetrahedron Lett.* **34**, 2449 (1993).

57. E. Fan, S. A. Van Arman, S. Kincaid, and A. D. Hamilton, *J. Am. Chem. Soc.* **115**, 369 (1993).
58. A. Echavarren, A. Galan, J.-M. Lehn, and J. de Mendoza, *J. Am. Chem. Soc.* **111**, 4994 (1989).
59. D. M. Kneeland, K. Ariga, V. M. Lynch, C.-Y Huang, and E. V. Anslyn, *J. Am. Chem. Soc.* **115**, 11042 (1993).
60. V. Boekelheide, *Acc. Chem. Res.* **13**, 65 (1980).
61. (a) N. Wester and F. Vogtle, *Chem. Ber.* **113**, 1487 (1980); (b) L. Rossa and F. Vogtle, *Liebig's Ann. Chem.*, 459 (1981).
62. M. G. B. Drew, J. Hunter, D. J. Marrs, J. Nelson, and C. Harding, *J Chem. Soc., Dalton Trans.*, 3235 (1992).
63. (a) I. Tabushi, H. Sasaki, and Y. Kuroda, *J. Am. Chem. Soc.* **98**, 5727 (1976); (b) I. Tabushi, Y. Kuroda, and Y. Kimura, *Tetrahedron Lett.*, 3327 (1976).
64. M. Fujita, J. Yazaki, and K. Ogura, *J. Am. Chem. Soc.* **112**, 5645 (1990).
65. (a) P. J. Stang and D. H. Cao, *J. Am. Chem. Soc.* **116**, 4981 (1994); (b) P. J. Stang and K. Chen, *J. Am. Chem. Soc.* **117**, 1667 (1995).
66. R. Baum, *Chem. Eng. News* **73**(7), 37 (1995).
67. D. Heyer and J.-M. Lehn, *Tetrahedron Lett.* **27**, 5869 (1986).
68. K. Worm, F. P. Schmidtchen, A. Schier, A. Schafer, and M. Hesse, *Angew. Chem. Int. Ed. Engl.* **33**, 327 (1994).
69. S. Valiyaveettil, J. J. F. Engbersen, W. Verboom, and D. N. Reinhoudt, *Angew. Chem. Int. Ed. Engl.* **32**, 900 (1993).
70. J. J. He and F. A. Quiocho, *Science* **251**, 1497 (1991).
71. H. Luecke and F. A. Quiocho, *Nature* **347**, 402 (1990).
72. J. W. Pflugrath and F. A. Quiocho, *Nature* **314**, 257 (1985).
73. J. W. Pflugrath and F. A. Quiocho, *J. Mol. Biol.* **200**, 163 (1988).
74. C. D. Gutsche, *Calixarenes*, J. F. Stoddart, Ed., Royal Society of Chemistry, Cambridge, 1989.
75. J. Vincens and V. Bohmer, Eds., *Calixarenes: A Versatile Class of Macrocyclic Compounds*, Kluwer Publishers, Dordrecht, 1991.
76. D. M. Rudkevich, W. Verboom, and D. N. Reinhoudt, *J. Org. Chem.* **59**, 3683 (1994).
77. (a) D. M. Rudkevich, Z. Brzozka, M. Palys, H. C. Visser, W. Verboom, and D. N. Reinhoudt, *Angew. Chem.* **106**, 480 (1994); (b) idem., *Angew. Chem. Int. Ed. Engl.* **33**, 467 (1994).
78. D. M. Rudkevich, W. Verboom, Z. Brzozka, M. J. Palys, W. P. R. V. Stauthamer, G. J. van Hummel, S. M. Franken, S. Harkema, J. F. J. Engbersen, and D. N. Reinhoudt, *J. Am. Chem. Soc.* **116**, 4341 (1994).
79. W. Xu, J. J. Vittal, and R. J. Puddephatt, *J. Am. Chem. Soc.* **115**, 6456 (1993).
80. W. Xu, J. P. Rourke, J. J. Vittal, and R. J. Puddephatt, *J. Chem. Soc., Chem. Commun.*, 145 (1993).
81. We thank Prof. R. J. Puddephatt for permission to quote these results prior to publication.
82. D. F. Shriver and M. J. Biallas, *J. Am. Chem. Soc.* **89**, 1078 (1967).

83. H. E. Katz, *J. Org. Chem.* **50**, 5027 (1985).
84. R. W. Alder, P. S. Bowman, W. R. S. Steel, and D. R. Winterman, *J. Chem. Soc., Chem. Commun.*, 723 (1968).
85. D. J. Saturnino, M. Yamauchi, W. R. Clayton, R. W. Nelson, and S. G. Shore, *J. Am. Chem. Soc.* **97**, 6063 (1975).
86. H. E. Katz, *Organometallics* **6**, 1134 (1987).
87. H. E. Katz, *J. Am. Chem. Soc.* **108**, 7640 (1986).
88. K. Tamao, T. Hayashi, Y. Ito, and M. Shiro, *J. Am. Chem. Soc.* **112**, 2422 (1990).
89. H. E. Katz, *J. Org. Chem.* **54**, 2179 (1989).
90. (a) M. T. Reetz, C. M. Niemeyer, and K. Harms, *Angew. Chem.* **103**, 1515 (1991); (b) *idem., Angew. Chem. Int. Ed. Engl.* **30**, 1427 (1991).
91. S. Jacobson and R. Pizer, *J. Am. Chem. Soc.* **115**, 11216 (1993).
92. (a) A. L. Beauchamp, M. J. Oliver, J. D. Wuest, and B. Zacharie, *J. Am. Chem. Soc.* **108**, 73 (1986); (b) J. D. Wuest and B. Zacharie, *J. Am. Chem. Soc.* **109**, 4714 (1987).
93. (a) X. Yang, C. B. Knobler, Z. Zheng, and M. F. Hawthorne, *J. Am. Chem. Soc.* **116**, 7142 (1994); (b) M. F. Hawthorne, X. Yang, and Z. Zheng, *Pure Appl. Chem.* **66**, 245 (1994).
94. V. B. Shur, I. A. Tikhonova, A. I. Yanovsky, Yu. T. Struchkov, P. V. Petrovskii, S. Yu. Panov, G. G. Furin, and M. E. Vol'pin, *J. Organomet. Chem.* **418**, C29 (1991).
95. V. B. Shur, I. A. Tikhonova, F. M. Dolgushin, A. I. Yanovsky, Yu. T. Struchkov, A. Yu. Volkonsky, E. V. Solodova, S. Yu. Panov, P. V. Petrovskii, and M. E. Vol'pin, *J. Organomet. Chem.* **443**, C19 (1993).
96. J. P. Fackler, Jr. and R. A. Kresinski, *Organometallics* **10**, 3392 (1991).
97. M. Newcomb, A. M. Madonik, M. T. Blanda, and J. K. Judice, *Organometallics* **6**, 145 (1987).
98. (a) M. Newcomb, J. H. Horner, and M. T. Blanda, *J. Am. Chem. Soc.* **109**, 7878 (1987); (b) M. Newcomb, J. H. Horner, M. T. Blanda, and P. J. Squattrito, *J. Am. Chem. Soc.* **111**, 6294 (1989).
99. J. H. Horner, P. J. Squatritto, N. McGuire, J. P. Riebenspies, and M. Newcomb, *Organometallics* **10**, 1741 (1991).
100. P. D. Beer, Z. Chen, A. J. Goulden, A. Grieve, D. Hesek, F. Szemes, and T. Wear, *J. Chem. Soc., Chem. Commun.*, 1269 (1994).
101. P. D. Beer, M. G. B. Drew, C. Hazelwood, D. Hesek, J. Hodacova, and S. E. Stokes, *J. Chem. Soc., Chem. Commun.*, 229 (1993).
102. P. D. Beer, Z. Chen, A. J. Goulden, A. Graydon, S. E. Stokes, and T. Wear, *J. Chem. Soc., Chem. Commun.*, 1834 (1993).
103. P. D. Beer, D. Hesek, J. Hodacova, and S. E. Stokes, *J. Chem. Soc. Chem. Commun.*, 270 (1992).
104. P. D. Beer, C. Hazelwood, D. Hesek, J. Hodacova, and S. E. Stokes, *J. Chem. Soc., Dalton Trans.*, 1327 (1993).
105. P. D. Beer and A. R. Graydon, *J. Organomet. Chem.* **466**, 241 (1994).
106. P. D. Beer, Z. Chen, M. G. B. Drew, J. Kingston, M. Ogden, and P. Spencer, *J. Chem. Soc. Chem. Commun.*, 1046 (1993).

107. M. Uno, N. Komatsuzaki, K. Shirai, and S. Takahashi, *J. Organomet. Chem.* **462**, 343 (1993).
108. (a) J. W. Steed, R. K. Juneja, and J. L. Atwood, *Angew. Chem.* **106**, 2571 (1994); (b) *idem., Angew. Chem. Int. Ed. Engl.* **33**, 2456 (1994).
109. J. W. Steed and J. L. Atwood, unpublished results.
110. J. W. Steed, C. P. Johnson, R. K. Juneja, R. S. Burkhalter, and J. L. Atwood, *Supramol. Chem.* **6**, 235 (1996).
111. J. W. Steed, P. C. Junk, J. L. Atwood, M. J. Barnes, C. L. Raston, and R. S. Burkhalter, *J. Am. Chem. Soc.* **116**, 10346 (1994).
112. A. B. Mitchell, J. W. Steed, K. T. Holman, M. M. Halihan, J. Montgomery, S. S. Jurisson, J. L. Atwood, and R. S. Burkhalter, *J. Am. Chem. Soc.* **118**, 9567 (1996).
113. M. Shionoya, T. Ikeda, E. Kumura, and M. Shiro, *J. Am. Chem. Soc.* **116**, 3848 (1994).
114. K. Konishi, K. Yahara, H. Toshishige, T. Aida, and S. Inoue, *J. Am. Chem. Soc.* **116**, 1337 (1994).
115. M. G. B. Drew, M. McCann, and S. M. Nelson, *J. Chem. Soc., Chem. Commun.*, 481 (1979).
116. Y. Agnus, R. Louis, and R. Weiss, *J. Am. Chem. Soc.* **101**, 3381 (1979).
117. P. K. Coughlin, J. C. Dewan, S. J. Lippard, E. Watanabe, and J.-M. Lehn, *J. Am. Chem. Soc.* **101**, 265 (1979).
118. R. J. Motekaitis and A. E. Martell, *Inorg. Chem.* **31**, 5534 (1992).
119. J. Salta, Q. Chen, Y.-D. Chang, and J. Zubieta, *Angew. Chem. Int. Ed. Engl.* **33**, 757 (1994); (b) *idem., Angew. Chem.* **106**, 781 (1994).
120. (a) A. Müller, K. Hovemeier, and R. Rohlfing, *Angew. Chem. Int. Ed. Engl.* **31**, 1192 (1992); (b) *idem., Angew. Chem.* **104**, 1214 (1992).
121. M. H. Alizadeh, S. P. Harmalker, Y. Jeannin, Y, J. Martin-Frere, and M. T. Pope, *J. Am. Chem. Soc.* **107**, 2662 (1985).
122. B. Hasenknopf, J.-M. Lehn, B. O. Kneisel, G. Baum, and D. Fenske, *Angew. Chem., Int. Ed. Engl.* **35**, 1838 (1996).
123. A. Ellaboudy, J. L. Dye, and P. B. Smith, *J. Am. Chem. Soc.* **105**, 6490 (1983).
124. A. Ellaboudy and J. L. Dye, *Chem. Br.* **20**, 210 (1984).
125. M. J. Wagner, R.-H. Huang, J. L. Eglin, and J. L. Dye, *Nature* **368**, 726 (1994).
126. J. L. Dye and R.-H. Huang, *Chem. Br.* **26**, 239 (1990).
127. S. B. Dawes, D. L. Ward, R.-H. Huang, and J. L. Dye, *J. Am. Chem. Soc.* **108**, 3534 (1986).
128. J. L. Dye, *Nature* **365**, 10 (1993).
129. D. L. Ward, R.-H. Huang, M. E. Kuchenmeister, and J. L. Dye, *Acta Crystallogr., Sect. C* **46**, 1831 (1990).
130. S. Golden and T. R. Tuttle, Jr., *Phys. Rev. B* **45**, 13913 (1992).
131. D. L. Ward, R.-H. Huang, and J. L. Dye, *Acta Crystallogr., Sect. C* **46**, 1838 (1990).
132. S. B. Dawes, D. L. Ward, O. Fussa-Rydel, R.-H. Huang, and J. L. Dye, *Inorg. Chem.* **28**, 2132 (1989).
133. D. J. Singh, H. Krakauer, C. Haas, and W. E. Pickett, *Nature* **365**, 39 (1993).
134. (a) D. L. Ward, R.-H. Huang, and J. L. Dye, *Acta Crystallogr., Sect. C* **44**, 1374

(1988); (b) R.-H. Huang, M. K. Faber, K. J. Moeggenborg, D. L. Ward, and J. L. Dye, *Nature* **331**, 599 (1988).
135. J. L. Dye, *Science* **247**, 663 (1990).
136. T. M. Nenoff, W. T. A. Harrison, T. E. Gier, N. L. Keder, C. M. Zaremba, V. I. Srdanov, J. M. Nicol, and G. D. Stucky, *Inorg. Chem.* **33**, 2472 (1994).
137. J. A. Stuckey, H. L. Schubert, E. B. Fauman, Z.-Y. Zhang, J. E. Dixon, and M. A. Saper, *Nature* **370**, 571 (1994).
138. Z. F. Kanyo and D. W. Christianson, *J. Biol. Chem.* **266**, 4264 (1991).

CHAPTER 6

Thermodynamics of Anion Complexation

ANTONIO BIANCHI and ENRIQUE GARCÍA-ESPAÑA

6.1 Introduction
6.2 A Simple Electrostatic Model for Ion Pairing
6.3 Anion Solvation
    6.3.1 Properties of Solvents
    6.3.2 Free Energies of Anion Solvation
6.4 Energetics of Noncovalent Interactions
    6.4.1 Electrostatics
    6.4.2 Effects of Induced Dipoles
    6.4.3 Hydrogen Bonding
    6.4.4 Hydrophobic Effect
    6.4.5 Stacking Effect
6.5 Covalently Bonded Anion Complexes
6.6 A Cautionary Word on Selectivity
6.7 Appendix. Methods for the Determination of Thermodynamic Parameters
    6.7.1 Introduction
    6.7.2 Methods
6.8 References

6.1 INTRODUCTION

Supramolecular coordination of simple as well as polyatomic anions has developed as a result of the synthesis of highly structured receptors, forwarding the knowledge of anion chemistry in both abiotic and biological systems. As demonstrated by the many crystal structures of anion complexes (see Chapter 5), the interaction between partners occurs via a large number of intermolecular

---

Supramolecular Chemistry of Anions, Edited by Antonio Bianchi, Kristin Bowman-James, and Enrique García-España.
ISBN 0-471-18622-8.   © 1997 Wiley-VCH, Inc.

218 THERMODYNAMICS OF ANION COMPLEXATION

contacts differing from each other in the nature and strength of the forces involved. Electrostatic attraction, hydrogen bonding, van der Waals and dispersion interactions, π-stacking and solvophobic effects make the principal contributions to the stability of anion complexes in solution. In addition, with the growing field of anion binding based on noncationic, and Lewis-acid-type receptors, covalency now also needs to be considered. With the exception of, at most, the latter receptors, the coordination geometries of anions are largely flexible, being determined by forces that are not directional, or strictly directional. For this reason, as already commented on in the discussion of the topological aspects of anion coordination (Chapter 5), criteria for strong and selective anion binding have not yet developed in as detailed a form as for metal cation recognition. Nevertheless, there is an increasing number of anion complexes showing very high stability, comparable with the stability of many coordination compounds of metal cations. It is clear that the driving force for most of the anion complexation reactions in solution is, to a large extent, determined by electrostatic attraction.

6.2 A SIMPLE ELECTROSTATIC MODEL FOR ION PAIRING

Although the correct method for correlating the structural and electronic properties of reacting species with the energetic characteristics of the anion binding process in solution should consider both reactants and products, respectively, as a whole, including solvating molecules, the mathematical treatments of such models are quite laborious and, time consuming, and, to date, they are not generally applicable. On the other hand, the use of simplified models can furnish a few general rules that, although not rigorous, are of immediate application and can be of great help in achieving an understanding of the real systems at the molecular level.[1-6]

To this purpose let us consider each binding site of both anions and receptors as separate entities and reduce these entities to hard spheres with embedded point charges.

The electric work to be gained per mole for bringing two similar gaseous ions of opposite charges $e\nu^+$ and $e\lambda^-$ (e = electronic charge, ν and λ = number of positive and negative charges, respectively) from infinity to distance d is given by

$$W_{el} = (Ne^2\nu\lambda)/d \quad (1)$$

where N is Avogadro's constant. The free energy change for this process in the gas phase is

$$\Delta G_g^o = -T\,\Delta S_t^o - W_{el} \quad (2)$$

ΔS_t° being the difference in translational entropy of reactants and products. Since W_{el} in the gas phase is independent of temperature, differentiation of $\Delta G_g^\circ (\partial \Delta G^\circ / \partial T = -\Delta S^\circ)$ gives

$$\Delta S_g^\circ = \Delta S_t^\circ \tag{3}$$

and according to the Gibbs–Helmholtz equation

$$\Delta H_g^\circ = \Delta G_g^\circ + T \Delta S_g^\circ = -W_{el} \tag{4}$$

Assuming that d is the sum of the radii (about 2 Å) of two singly charged spherical partners in a ion pair, Eq. (1) gives a value of about 200 kcal mol^{-1} for the electric work W_{el}. Consequently, ΔH_g° is strongly negative and largely predominant over the entropy loss ($T \Delta S_g^\circ$ about -10 kcal mol^{-1} at 298 K) produced by association of the ions. Therefore, the free energy change for ion pairing in vacuum is mainly determined by the electrostatic energy, and the process is largely exothermic.

In solution, the interaction between ions is weakened by the dielectric properties of solvents. For many years the Bjerrum theory of ion pairs has been used to interpret the behavior of electrolyte solutions.[7] Many other theories were developed later to account for the deviation of experimental results from theoretical expectations.[8–13] All the proposed equations predict an almost linear dependence of the free energy change for ion pairing from the product, $e^2 \nu \lambda$, of cationic and anionic charges.

Let us simply consider the previous process occurring in a structureless homogeneous medium of dielectric constant ε. The ions immersed in this dielectric continuum do not affect the bulk solvent properties even near the surface of the ions. The electric work

$$W_{el} = (Ne^2 \nu \lambda)/d\varepsilon \tag{5}$$

is now temperature dependent as far as ε varies with temperature

$$\frac{\partial W_{el}}{\partial T} = \frac{-(Ne^2 \nu \lambda)}{d\varepsilon^2} \left(\frac{\partial \varepsilon}{\partial T}\right) = -W_{el} \left(\frac{1}{\varepsilon}\right)\left(\frac{\partial \varepsilon}{\partial T}\right) \tag{6}$$

The dielectric shielding of charges is more pronounced at low temperature, where the reduced thermal motion disturbs the orientation of solvent molecules in the electric field to a lesser extent. For instance, the dielectric constant of water is 88.0, 78.5, and 55.3 at 273, 298, and 373 K, respectively, and $\partial \varepsilon / \partial T = -0.361$. This means that at higher temperatures a stronger association is expected owing to an enhanced electrostatic attraction.

The thermodynamic functions (2), (3), and (4) thus become

$$\Delta G_{aq}^o = -T\,\Delta S_t^o - W_{el} \tag{7}$$

$$\Delta S_{aq}^o = \Delta S_t^o - W_{el}\left(\frac{1}{\varepsilon}\right)\left(\frac{\partial \varepsilon}{\partial T}\right) \tag{8}$$

$$\Delta H_{aq}^o = -W_{el}\left[1 + \left(\frac{T}{\varepsilon}\right)\left(\frac{\partial \varepsilon}{\partial T}\right)\right] \tag{9}$$

The electrostatic energy is now strongly reduced by the dielectric shielding of charges and the ion-pairing process is much less exothermic, or endothermic. In water at 298 K, for instance, the quantity within square brackets in Eq. (9) is negative, $[1 + 298/78.5(-0.361)] = -0.370$, making positive ΔH_{aq}^o. On the other hand, the quantity $W_{el}(1/\varepsilon)(\partial \varepsilon/\partial T)$ is also negative, and $\Delta S_{aq}^o > \Delta S_t^o$, accounting for a greater mobility of the solvent molecules around the approaching ions caused by charge neutralization. Therefore, in a structureless homogeneous solvent the processes involved with ion association are mainly promoted by the increase of translational entropy.

Although Eqs. (7)–(9) offer a correct qualitative interpretation of ion pairing in solution, they do not furnish accurate quantitative evaluations of the relevant thermodynamic functions.

A considerable improvement is obtained by treating the solvent as a structureless but nonhomogeneous medium. Actually the solvent molecules display different behavior in the bulk of the solution and in the vicinity of the metal ions. The electric field generated by a pair of oppositely charged ions increases with decreasing distances between them. When the two ions are very close to each other, the electric field becomes so enormous that the solvent molecules in the vicinity are strictly oriented and dielectric saturation occurs. These phenomena determine a dielectric deshielding of the charges, namely, the dielectric constant, having ε value in the bulk of the solution, is rapidly reduced to an effective ε_e in proximity of the ions. The solvent molecules constrained within these electric fields are less influenced by the thermal motion than in the bulk, and therefore ε_e presents a less pronounced decrease with increasing temperature than the macroscopic dielectric constant $\varepsilon(-\partial \varepsilon_e/\partial T < -\partial \varepsilon/\partial T)$.

In conclusion, considering the solvent as a structureless medium, association of spherical ions in solution is expected to be promoted by a positive dominant entropic term, the enthalpic contribution being almost negligible.

The simplified model we have developed so far is limited to low concentrations where the concept of ion association is unambiguous, while at higher concentrations the formation of ion clusters occurs through association of three or more ions. However, even in dilute solution, this model is able to account for a limited number of cases in which the interacting ions have spherical symmetry and remain practically unsolvated in the particular solvent, and there is negligible charge redistribution within the solute and solvent.

When the actual structures of the ion partners, as well as of the solvent, in the association process are considered, a number of short-range contributions deriving from dipole–dipole, dipole–induced dipole, and induced dipole–induced dipole interactions (London dispersion interactions, hydrogen bonding, stacking effects, and hydrophobic forces) also have to be considered. Moreover, multipolar interactions deriving from charge distributions that are more complicated than dipoles can contribute to the stability of anion complexes. All these short-range forces can produce favorable enthalpic contributions to anion association. Therefore, anion complexation is often an exothermic process, in contrast to the expectations for ion pairing based on the simple electrostatic model.

6.3 ANION SOLVATION

It is to be stressed here that, along with the charge-shielding effect due to macroscopic and local dielectric properties of solvent, the molecular properties and the real structure of the solvent itself also play a fundamental role in determining the nature and the strength of anion-receptor interaction.

6.3.1 Properties of Solvents

The principal physical and empirical parameters that may be useful in understanding the solvating properties of the most common solvents are listed in Table 6.1.

The dielectric constant and dipole moment of a solvent indicate how strongly it may interact with a solute via electrostatic forces. Furthermore, molecular polarizability is strictly connected with the ability of solvent molecules to give rise to dispersion interactions. Dipolar aprotic solvents are much easily polarized than protic solvents. Both electrostatic and dispersion interactions produce strong solvation of polarizable anions like I^-, SCN^-, and ClO_4^- in dipolar aprotic solvents.

The enthalpy change for the binding of a donor (D) to $SbCl_5$ in 1,2-dichloroethane as solvent

$$D + SbCl_5 \rightleftharpoons D \cdot SbCl_5, \quad (-\Delta H°(\text{kcal mol}^{-1}) = \text{DN})$$

has been considered by Gutmann and Wychera as a measure of the ability of solvent molecules to donate electron pairs.[17,18] The donor number (DN) of a large variety of solvents has been directly determined by calorimetric measurements, while those for many other solvents have been inferred by indirect methods.

In the case of anion solvation the acceptor properties of solvents are of greater interest.[18] An empirical parameter related to the ability of a solvent molecule to accept electron pairs has been proposed based on the ^{31}P-NMR

TABLE 6.1. Physical and Empirical Parameters of Most Common Solvents[14–21]; Values Determined at 298 K Unless Otherwise Reported

Solvent	Dielectric constant	Dipole moment	Molecular polarizability (10^{-24} cm^3; Å^3)	Donor number	Acceptor number	$E_T(30)$
H$_2$O	78.54	1.85	1.45	18	54.8	63.1
MeOH	32.63	1.70	3.32	19	41.3	55.5
EtOH	24.30	1.69	5.11	20	37.1	51.9
PrOH	20.1	1.68	6.74		34	50.7
NH$_3$ (liquid)	16.9	1.47	2.26	59		
Et$_3$N	2.42	0.66	13.1	61		33.3
Py	12.3	2.19	9.5	33.1	14.2	40.2
CHCl$_3$	4.81 (293 K)	1.01	9.5		23.1	39.1
Me$_2$SO	48.9	3.96		29.8	19.3	45.1
Me$_2$CO	20.7	2.88	6.39	17.0	12.5	42.2
THF	6.4	1.63		20	8.0	37.4
Et$_2$O	4.33 (293 K)	1.15	10.2	19.2	3.9	34.6
Dioxane	2.21	0	8.60	14.8	10.8	36.0
DMF	36.7	3.82	7.81	26.6	16.0	43.9
MeCN	37.5	3.92	4.48	14.1	19.3	46.0
MeNO$_2$	35.8	3.46	7.37	2.7	20.5	46.2
C$_6$H$_5$NO$_2$	34.8	4.21	14.7	4.4	14.8	41.9
C$_6$H$_6$	2.28 (293 K)	0	10.32	0.1	8.2	34.5
Hexane	1.89 (293 K)	0	11.9	0.0	0.0	30.9

chemical shift of triethylphosphine in the considered solvent.[19,20]

$$Et_3P=O \rightarrow solvent$$

The acceptor solvent attacking at the oxygen atoms of Et$_3$PO causes a decrease of the electron density on the phosphorus atom, which is manifested by a down-field shift of the ^{31}P-NMR signal. A scale of acceptor numbers (AN) was constructed by arbitrarily assigning the reference values 0 and 100 to the ^{31}P chemical shifts of Et$_3$PO in pure hexane and of the Et$_3$PO → SbCl$_5$ adduct in 1,2-dichloroethane, respectively. As can be seen from Table 6.1, protic solvents show more marked acceptor properties than polar aprotic and apolar solvents. In contrast with physical parameters, the acceptor number allows the interpretation of many solvent-dependent phenomena. For example, the high reactivity of anions in diethyl ether, such as in electrophilic substitution reactions, would not be expected considering its dipole moment, but it is in evident agreement with its very weak acceptor properties. On the other hand, benzene, which has been considered an inert solvent owing to its apolarity, manifests chemical activity in accordance with its acceptor number.[21]

$E_T(30)$ is another widely used empirical parameter to express solvent polarities.[21] This parameter is the lowest-energy transition of the pyridinium phenol betain indicator (kcal mol^{-1}) in the presence of a given solvent. Such spectroscopic solvent polarity indexes, which are linearly correlated with the acceptor numbers, provide a measure of the ability of solvents to stabilize anions.[45]

Solvation by protic solvents takes great advantage of the formation of strong hydrogen bonds with appropriate anions containing small, negatively charged atoms, such as F, Cl, O, N. However, the presence of these atoms does not lead necessarily to hydrogen bonding as far as charge dispersal by electron-withdrawing groups, as in ClO_4^-, SCN^-, and picrate, reduces significantly the hydrogen-bond acceptor character of the anion.[22]

Solvents with high dipole moments or hydrogen-bonding ability are more organized than apolar solvents. Water, in particular, has a well-organized structure.[23-26] Owing to their strong hydrogen-bond donor and acceptor properties, water molecules in the liquid phase are constrained within "icelike" dynamically structured clusters, which are formed in such a way as to achieve a maximum number of hydrogen bonds. Consequently, the binding forces between different clusters are weaker than those inside the clusters. These clusters, characterized by different sizes, have a limited half-life period of about 10^{-11} s. Dissolved electrolytes can enhance or break this structure. Ions such as F^-, OH^-, H_3O^+, and NH_4^+ are structure makers; they fit the water structure by hydrogen bonding, enhancing its stability though their charges. In contrast, large ions like Br^- and I^- are structure breakers.[27,28] A particularly illustrative behavior is shown by large ions with centrally localized charge such as tetraalkylammonium cations. While Me_4N^+ is a structure breaker, Pr_4N^+ and Bu_4N^+ are, surprisingly, structure makers, Et_4N^+ remaining on the borderline. This has been explained by considering that, as the length of the hydrocarbon chains increases, the cluster structures of water are more and more deeply enforced between these hydrophobic portions, which protect the cluster from disruption.[29,30] On the other hand, large anions in which the charge is distributed on the surface (BF_4^-, $PdCl_4^{2-}$, ClO_4^-, etc.) are structure breakers.

Upon tight association, the separate identities of the reacting species disappear in forming the complex, and the final solvent structure depends on the specific properties of the complex itself. Consequently, the solvent can undergo drastic changes towards both higher or lower organization of its structure during complexation reactions, giving rise to favorable or unfavorable contributions. Among all possible combinations, the association of anions and receptors, both having marked structure-breaking properties, generally leads to highly favorable solvent reorganization.

For this reason complexation by apolar receptors of anions containing large apolar regions in their structures is strongly enhanced in polar solvents characterized by highly cohesive interactions (hydrophobic effect).[31,32] Such solvent molecules interact more favorably with bulk solvent molecules than with the apolar surfaces and cavities of anions and receptors, and consequently, the release of solvating molecules into the bulk during complexation promotes the process. At the same time the apolar surfaces of the interacting partners bind more strongly to each other than to the polar solvent molecules. In the case of tight association this overall redistribution of binding interactions gives rise to favorable (negative) enthalpy changes and generally unfavorable entropic contributions due to the loss of degrees of freedom by the interacting partners during the association process.[33]

Hence water is the most appropriate solvent for anion complexation reactions promoted by hydrophobic interactions. On the contrary, association occurring via electrostatic attraction and extensive hydrogen-bond formation is weakened by solvents with marked dielectric properties and hydrogen-bonding ability, like water, in which the interacting partners extinguish their charges and saturate their hydrogen-bonding valences.

6.3.2 Free Energies of Anion Solvation

Anion complexation in solution is generally accompanied by partial or complete desolvation of the interacting partners. Earlier studies[34–37] on the formation of ion pairs in solution led to the division of ion pairs in two broad categories, depending on the structure of the solvation shell around the ionic aggregates: (1) contact ion pairs when the cation and the anion touch each other inside a common solvation sphere, (2) solvent-separated ion pairs when the solvation spheres of both cation and anion overlap. In the light of more recent investigations[38–43] this subdivision appears to be a simplification representing two limiting structures for ion pairs in solution compared to their real structure, especially when the interacting species are highly structured. At least, in the latter cases, ion pairs might be better regarded as mixtures of interionic isomers that, because of the actual structure of the ions and the nature of the solvent, differ in interionic tightness and geometry and whose equilibrium composition varies with the solvent. Accordingly, the concept of "average interionic geometry" has been introduced.[40]

As a consequence of the nonspherical structures of most ions, the effective mean interionic distance in a tight ion pair is generally shorter than that predicted assuming a hard-sphere model. This is observed even in the formation of ion pairs between cations and anions, such as R_4N^+ and BPh_4^-, that are frequently considered as spherical, or roughly spherical, since they approach each other by alignment of two of their triangular faces. In the case of small anions such as halides or BH_4^-, penetration into the voids between the chains of the ammonium cation may occur.[42] The size of the resulting penetrated ion pair is much smaller than that of the sum of the individual ions.

An extreme case is found in anions included in cryptand cavities where all solvent molecules are removed from the anion solvation sphere, and the host, commonly a cationic species, can be forced to contract upon itself to accommodate the anionic guest.[44]

In water, the free energy change for anion solvation, ΔG_h^o (free energy of hydration), is defined by the equation $A^{n-}(g) + H_2O(l) \leftrightarrows A(aq)^{n-}$. ΔG_h^o values for the most common anions are reported in Table 1.1 (p. 6). The free energy of hydration of such anions is mostly determined by electrostatic forces,

TABLE 6.2. Free Energy Changes (kJ mol^{-1}) for the Addition of Liquid Water Molecules to Gaseous Anions (X^-(g) + nH$_2$O(l) ⇌ [XH$_2$O)$_n$]$^-$(g)) at 298 K^a

	F$^-$	Cl$^-$	Br$^-$	I$^-$	OH$^-$	ClO$_4^-$	NO$_3^-$	CN$^-$
$-\Delta G_{0,1}$	66.9	25	20	14	61.9			
$-\Delta G_{0,2}$	104	43.9	35	23	97.8			
$-\Delta G_{0,3}$	127	53.6	43.1	27	121			
$-\Delta G_{0,4}$	146	58.6	46.4	28	135			
$-\Delta G_{0,5}$	159	61.9	48.1		144			
$-\Delta G_h^o$	436	311	285	247	403	178	270	310

^{a} Values from Refs. 45–47 corrected to a standard state for water of liquid water at 298 K.

increasing with decreasing ionic radius and increasing charge, and it is also sensitive to the hydrogen-bond acceptor properties of anions like F$^-$, OH$^-$, and carboxylate.

A different approach to anion solvation, which is more helpful in understanding anion complexation, is to consider the hydration process as a stepwise addition of water molecules. According to the equation A^-(g) + nH$_2$O(l) ⇌ [A(H$_2$O)$_n$]$^-$(g), the binding of the first few water molecules to halides and OH$^-$ anions follows the same pattern of total hydration free energies (Table 6.2), indicating that the specific properties of anions towards hydration are manifested in the first part of the process.

After the addition of a small number of water molecules (3–4), the $\Delta G_{0,n}$ values tend to level off, and the successive solvation is almost independent of the specific anion. The difference between the free energy changes for the assembling of the first sphere and for complete hydration is approximately constant (250 ± 30 kJ mol^{-1}). It is worth noting that the energy gained in the formation of the first sphere of hydration comprises a considerable portion of the total hydration energy in the case of F$^-$ and OH$^-$ (greater than $\Delta G_h^o/3$), while it decreases with decreasing hydrogen-bond acceptor properties and increasing dimensions and polarizability (Table 6.3) of anions.

It is clear that anion solvation plays a greater role in anion binding in solution when the complexation reaction involves the replacement of solvent molecules from the first solvation sphere, that is, formation of contact ion pairs. On the other hand, formation of anion complexes with typical structure

TABLE 6.3. Polarizability of Anions[15]

	r (Å)	Polarizability (cm^3 mol^{-1})		r (Å)	Polarizability (cm^3 mol^{-1})
F$^-$	1.36	2.60	NO$_3^-$	1.89	11.01
Cl$^-$	1.81	9.06	CN$^-$	1.90	8.34
Br$^-$	1.95	12.66	OH$^-$	1.40	5.15
I$^-$	2.16	19.22	SCN$^-$		16.54
ClO$_4^-$	2.36	13.24			

TABLE 6.4. Free Energy of Transfer $[\Delta G^\circ_{tr}(\text{kJ mol}^{-1})]$ of Anions from Water to Methanol, Acetonitrile, Dimethylsulfoxide, Dimethylformamide, Nitromethane, Propylene Carbonate, and Pyridine at 298 K[48,49]

	H$_2$O to MeOH	H$_2$O to MeCN	H$_2$O to DMSO	H$_2$O to DMF	H$_2$O to MeNO$_2$	H$_2$O to PC	H$_2$O to Py
Cl$^-$	13.7	41.7	39.4	45.9	37.7	39.4	34.2
Br$^-$	11.4	31.4	26.3	31.7	29.0	30.3	20.5
I$^-$	7.4	19.4	9.1	20.2	18.8	16.8	19.4
SCN$^-$	5.7	12.6	9.1				20.0
ClO$_4^-$	5.7	4.0	−1.1	7.0	4.7	−3.0	16.0
CH$_3$COO$^-$	16.1	56.9	60.3				
CF$_3$SO$_3^-$	−16.0	−23.4					−14.3
Pic$^-$	−4.6	−2.3		−7.0		−6.0	−4.6
BPh$_4^-$	−23.4	−33.1	−37.1	−38.4	−32.6	−36.2	−37.7

of solvent-separated anion pairs occurs at smaller energetic cost for anion desolvation, but are stabilized by weaker electrostatic attraction.

In nonaqueous solutions simple inorganic anions are less solvated than in water, as demonstrated by the free energy of transfer from water to organic solvents listed in Table 6.4.

The weakening of solvation as evidenced by the larger positive values of ΔG°_{tr} is much more apparent for aprotic than for protic solvent (i.e., alcohols) and decreases with ion polarizability. For instance, anions with higher charge density like Cl$^-$, Br$^-$, and acetate are more affected by transfer from water than larger, more polarizable anions like I$^-$ and ClO$_4^-$. Contrarily, bulky organic anions (BPh$_4^-$, picrate) show stronger solvation in organic, especially aprotic, solvents, as seen by the large negative values of ΔG°_{tr}.

The free energies of solvation are of great help in the understanding of many association reactions, in contrast with expectations from simple electrostatic considerations. For instance, the association of electrolytes in protic solvents appears to increase with ionic size, namely, with decreasing charge density. This is illustrated by the equilibrium constants reported in Table 6.5 for the association of tetraalkylammonium cations in water and in alcohols with halide anions,[50-55] which are among the most representative partners in anion complexation. In the gas phase the association constants for a given cation are expected to increase in the order Cl$^-$ > Br$^-$ > I$^-$; however, in protic solvents a reversed trend is observed. This behavior is easily rationalized by considering that the release of solvent molecules occurring upon complexation is accompanied by the corresponding loss of free energy of solvation, which is larger for the smaller anions (Tables 1.1 and 6.4). On the other hand, in solvents of weaker anion-solvating properties, including acetonitrile, nitrobenzene, and acetone, the "gaslike" conditions are approached, and the stability constants vary as predicted by the elementary electrostatic theory (Cl$^-$ > Br$^-$ > I$^-$).

TABLE 6.5. Equilibrium Constants Determined for Association of Tetraalkylammonium Cations with Halide Anions in Different Solvents at 298 K[50-55]

	H_2O	MeOH	EtOH	PrOH	BuOH	$C_6H_5NO_2$	MeCN	DMSO	Me_2CO
Me_4NCl	0.8	7	122	456	2270		1.74		
Me_4NBr	1.3	14	146	638	2110			0.65	
Me_4NI	1.5	18						0.34	
Et_4NCl						80			
Et_4NBr	1.7	10	99	373	1330	62	0.67	0.56	330
Et_4NI	3.3		133	466	1410	29			155
Pr_4NCl									
Pr_4NBr	2.3	6	78	270	920				335
Pr_4NI	3.8	17	120	391	1160				174
Bu_4NCl	1.8		39	149	620				430
Bu_4NBr	2.8	3	75	266	869	56	1.23		285
Bu_4NI	5.1	16	123	415	1180	27	1.26		155

6.4 ENERGETICS OF NONCOVALENT INTERACTIONS

An empirical role frequently observed in molecular recognition is the additivity, in terms of free energy contributions, of noncovalent interactions.[56] Conscious of this role, synthetic chemists have obtained a variety of target supramolecular assemblies by controlling the association mechanisms through the choice of the structures and the functional groups of the reacting species. In this section we describe some examples of anion coordination in which it is possible to evidence the effects of the specific contributions. In several cases the thermodynamic data available allow an estimation of such contributions. Similar correlation between structural features and energetics of the interactions allows the design of new receptors to be based on quantitative provisions.

6.4.1 Electrostatics

Electrostatic attraction is the driving force in anion complexation by positively charged receptors, furnishing in many cases the dominant contribution to the organization and the activities of anions, not only in inorganic environments but also in the very complex structures of biological systems.[57]

The role of electrostatics in the supramolecular chemistry of anions is not limited to attractive forces, since repulsion between positive functionalities in the receptor framework can determine the shape of the receptor itself and the spatial disposition of the charged binding sites. When charges of the same sign are built on a molecule, the molecular framework tends to expand in order to minimize the electrostatic repulsion. Such an expansion more greatly affects flexible frameworks than rigid ones. This phenomenon is particularly evident in macrocyclic and, to a much lesser extent, in macropolycyclic ligands such as **6.1–6.11**, in which a large number of positive charges can be gathered together

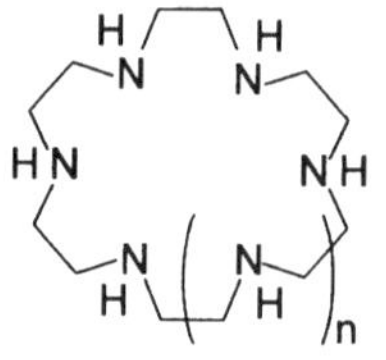

n = 1 - 7   **6.1-6.7**

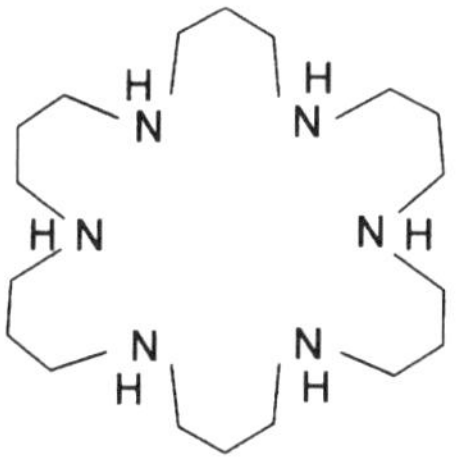

6.8

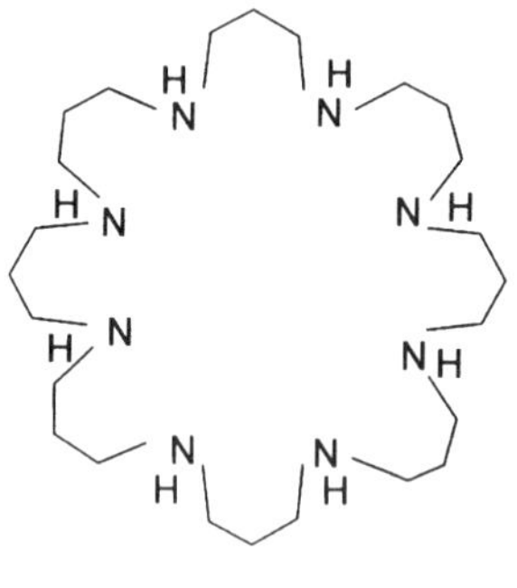

6.9

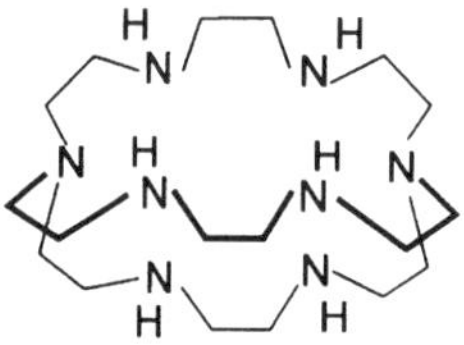

6.10

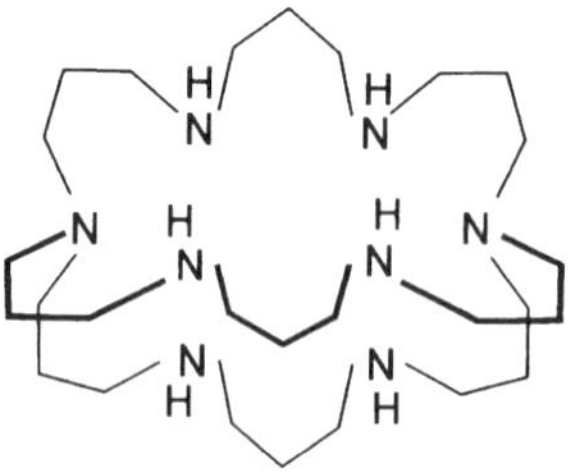

6.11

at close distance by protonation of the amine groups.[58-59] The effect of extensive protonation on the molecular conformation of similar ligands has been shown by the crystal structures of partly or fully protonated **6.1**,[70,71] **6.2**[72] **6.5**,[73-77] and **6.10**.[78]

A further consequence of the repulsion between positive charges in polyammonium macrocycles and macropolycycles is the acquisition of the *out* configuration by the ammonium groups, providing an additional expansion of the ligand cavity.

Equilibria between stereoisomeric species differing in the *in–out* configuration of the ammonium groups appeared to play a fundamental role in the

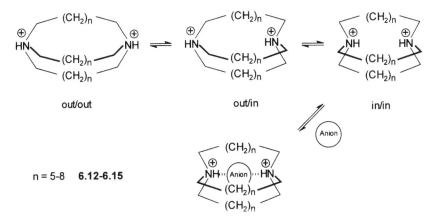

n = 5-8   **6.12–6.15**

Fig. 6.1. Equilibria between stereoisomeric species of diprotonated katapinand macrobicycles. Anion inclusion.

inclusion of halide anions by katapinand molecules **6.12–6.15** in acidic solutions (Fig. 6.1).[79,80] While protonation of such molecules favor the *out–out* conformation, Cl^- encapsulation into the diprotonated species arises from prior ligand rearrangement to the *in–in* conformation, which allows the formation of two $N-H\cdots Cl^-$ hydrogen bonds.[81]

The interconversion from *in* to *out* conformations requires a certain accumulation of charge on the cyclic frameworks, as demonstrated, for instance, by the structures of the diprotonated tetraazacycloalkane cyclam $H_2(\mathbf{6.16})^{2+}$ (all *in* configurations)[82] and of its tetraprotonated analog, $H_4(\mathbf{6.16})^{4+}$ (all *out* configuration),[83] as shown in Fig. 6.2, and deduced from thermodynamics of protonation of the same ligand in aqueous solution.[84]

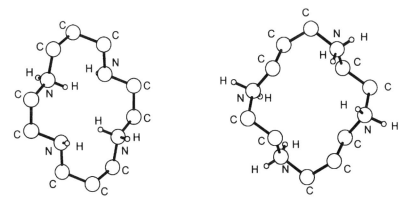

Fig. 6.2. Molecular structures of $H_2(\mathbf{6.16})^{2+}$ and $H_4(\mathbf{6.16})^{4+}$ cations.

Mixed *in–out* conformations as well as intermediate conformations are commonly observed in anion complexes of polyammonium receptors.[70–77] Preorganization of protonated receptors in *in* or *out* conformations can facilitate or reduce anion complexation depending on the structure of the receptor prior to or after complexation and on the location of the anion inside or outside the cavity.

The location of the protonated amine groups, the binding sites for anions, is also controlled by electrostatics, since the H^+ ions tend to bind to the amine groups, which allow their location as far as possible from each other. Of course, this tendency can cooperate with or oppose the propensity of H^+ ions to bind to amine groups of higher basicity. For this reason, while the protonation patterns in aqueous solution of the polyazamacrocycles **6.1–6.9**, containing equivalent secondary amine groups, appear to be simply controlled by electrostatic repulsion,[58–67,85] localization of H^+ ions on the secondary nitrogens is found for the hexaprotonated forms of **6.10** and **6.11**[68,69] owing to the reduced basicity of the tertiary bridgehead nitrogens. Redistribution of the protonated sites is possible, however, upon interaction with the anionic guest.

Conversely, fixed binding sites are present in anion receptors containing quaternary ammonium groups such as **6.17** and **6.18**.[86] These ligands, which are unable to form hydrogen bonds, allows anion complexation to be studied independently of configurational and protonation equilibria at the nitrogen atoms. In addition, due to the high connectivity of their molecular skeleton, they are resistant to deformation, rendering the cavity almost unable to adjust to changes in anion size. Owing to these characteristics, the stability of the anion complexes formed by similar ligands is predominantly determined by coulombic interactions. The tetrahedral arrangement of the four quaternary ammonium centers gives these ligands an appropriate symmetry for the inclusion of spherical anions. The equilibrium constants for the formation in aqueous solution of the halide complex of **6.17** and **6.18** revealed two different sequences of stability, $Cl^- < I^- < Br^-$ for **6.17** and $Cl^- < Br^- \leq I^-$ for the larger **6.18**,

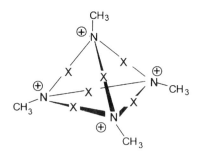

X = $(CH_2)_6$ **6.17**

X = $(CH_2)_8$ **6.18**

evidencing selectivity in the encapsulation of spherical anions based on the mutual dimensions of anion and ligand cavity.[86–88]

As observed for almost all the ligands used in anion coordination studies, these molecules also inevitably suffer from competitive binding to the counter-anions. This problem has been overcome, at least in part, by transforming **6.17** and **6.18** into the zwitterionic host molecules **6.19** and **6.20** (Fig. 6.3), although intramolecular association between neighboring carboxylates and quaternary ammonium groups seems to be occurring.[89] The free energy changes determined for the interaction of **6.19** with anions in D_2O revealed a general enhancement of complex stability with respect to **6.17**,[86–88] as expected, and a sequence of stability $Cl^- < Br^- < I^-$. Dominating negative enthalpy changes contribute to the stability of the Br^- and I^- complexes, the entropic terms being unfavorable (Fig. 6.3).

Selectivity in electrostatic coordination of an anion can be achieved by spatial matching of charges. An illustrative example is given by the interaction of α,ω-dicarboxylate anions with quaternarized diammonium cations (Fig. 6.4).[90] The best fit between the length of the chain $[-(CH_2)_n-]$ connecting the two $-COO^-$ groups and the separation of the ammonium centers produces the most stable complexes. Several ditopic hosts have been synthesized in order to achieve this recognition pattern.[44,67,69,91–94] Spatial charge matching with

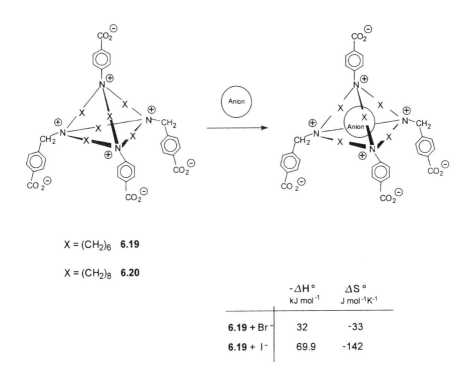

X = $(CH_2)_6$ **6.19**

X = $(CH_2)_8$ **6.20**

	$-\Delta H°$ kJ mol^{-1}	$\Delta S°$ J mol^{-1}K^{-1}
6.19 + Br$^-$	32	-33
6.19 + I$^-$	69.9	-142

Fig. 6.3. Anion inclusion by zwitterionic host molecules **6.19**, **6.20**.

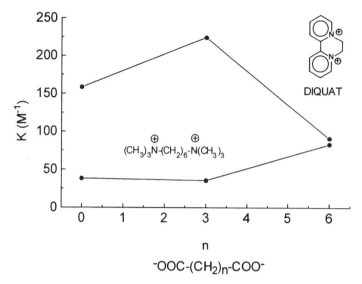

Fig. 6.4. Graphical representation of the stability constants for the complexes formed by quaternarized diammonium cations with differently sized α,ω-carboxylate anions.

anions of more complicated two-dimensional and three-dimensional geometries has also been accomplished by the design of appropriate receptors.[65,95,96] In all these cases other forces in addition to the predominant electrostatic attraction contribute to the stability of the anion complexes.

Apart from considerations of the geometrical complementarity of charge distribution on anions and receptors, the strength of the interaction between anions and charged ligands in solution is mostly determined by the overall charges or by local charge densities, on the interacting species, accordingly to provisions based on simple electrostatic models.

Polyazamacrocycles **6.1–6.7** (ligands with $n = 6$–12 in the [3k]aneN$_k$ series) offer useful examples in this respect.[72,74,77,97,98] Along this series of ligands the dimensions of the cyclic frameworks are increased by successive additions of –NH–CH$_2$–CH$_2$– fragments, so that the effects of both increasing overall charge and decreasing charge density on anion complexation can be analyzed. For a given anion the complex stability increases with the charge of the ligand and, for given ligand protonation degrees, with the anion charge, while for a given anion and ligand protonation degree the complex stability decreases with increasing ligand dimensions, namely, with decreasing charge density on the receptor. However, in the case of Co(CN)$_6^{3-}$, PO$_4^{3-}$, and Pt(CN)$_4^{2-}$, inversion of this trend is observed from **6.5** ([30]aneN$_{10}$) to **6.6** ([33]aneN$_{11}$) (see Chapter 4, Fig. 4.2), the enhanced stability of the complexes of **6.6** being considered evidence of inclusive coordination of the anionic guests.[72,77,99] The same inference was also drawn by Pina et al. on the basis of the quantum yield measured for the CN$^-$ photoaquation reaction of Co(CN)$_6^{3-}$ in the presence of these polyammonium receptors.[100]

In contrast to these stability arguments, molecular mechanics calculations on the isolated $H_{10}(6.5)^{10+}$ showed that the most stable conformation of the decacharged receptor, which is almost circular, possesses appropriate dimensions for the inclusion of all the above-mentioned complex anions.[77] Actually, the inclusion of $PdCl_4^{2-}$ into $H_{10}(6.5)^{10+}$ was shown by the crystal structure of $[(PdCl_4) \subset H_{10}(6.5)](PdCl_4)_2Cl_4$ (Fig. 6.5).[75,76] Microcalorimetric measurements in very acidic solutions evidenced that the formation of stable complexes between $PdCl_4^{2-}$ and the fully protonated forms of **6.1–6.6** is enthalpically favored ($-\Delta H^\circ = 6.3, 6.3, 6.7, 12.1, 16.3, 13.0 \, kJ\,mol^{-1}$ for $H_6(6.1)^{6+}$, $H_7(6.2)^{7+}$, $H_8(6.3)^{8+}$, $H_9(6.4)^{9+}$, $H_{10}(6.5)^{10+}$, and $H_{11}(6.6)^{11+}$, respectively).[75,76,99] The enthalpy of reaction remains constant for the three smaller receptors and then rises to a maximum for $H_{10}(6.5)^{10+}$. The increase has been attributed to a greater degree of penetration into the cavity until, with $H_{10}(6.5)^{10+}$, the anion is fully encapsulated. While reactions of $PdCl_4^{2-}$ with **6.1**, **6.2** are fast, those with the larger macrocycles of the series **6.3–6.6** are much slower. This slowness, which is maximum for **6.5**, can be ascribed to the reorganization undergone by the macrocycles to include the anion. Hydrogen bonding observed in the solid state[75,76] between the ammonium groups and chloride atoms in $PdCl_4^{2-}$ is expected to play a minor but nevertheless synergistic role.

Schneider et al. carried out a great deal of work in an attempt to quantify the electrostatic contribution to the overall free energy change associated with the formation of a large number of anion complexes in water.[56,101] Many charged groups were considered, including, among others, $R_2NH_2^+$, R_4N^+, positively charged pyridinium nitrogens, R_4P^+, $R-CO_2^-$, $R-SO_3^-$, $RO-PO_3H^-$, $RO-PO_3^{2-}$, and phenolates. Although these groups present evident differences in size, polarizability, and charge localization, a plot of the determined electrostatic terms versus the number of possible ionic interactions, or salt bridges, resulted in a surprisingly linear correlation furnishing a value of $-5 \pm 1 \, kJ\,mol^{-1}$ for the electrostatic contribution to the formation of a single salt bridge in water.[56,101] The unique requirement needed for the observed role to be followed is the formation of strainless complexes with contact ion pairs. The observed steady increase of the overall electrostatic contribution with the number of salt bridges implies that only the interaction between neighboring groups of opposite charge produces significant anion–receptor attraction.

Electrostatic attraction between ions and permanent dipoles is weaker than ion–ion interactions. Consequently ion–dipole effects in solution are easily hindered by competitive interaction with solvent molecules, especially in polar solvents where alignment of solvent dipoles with the dipolar functionalities extinguishes, or reduces, their binding ability. Obviously a minimum separation of charges is even more important for ion–dipole than for ion–ion interactions, and, furthermore, ion–dipole interaction is also dependent on orientation.

Hydrogen bonding is surely the principal phenomenon where ion–dipole attractions manifest their large contribution in the binding of anionic species. Further factors participate in this type of bond, however, and for this reason, hydrogen bonds will be considered separately from pure ion–dipole interactions.

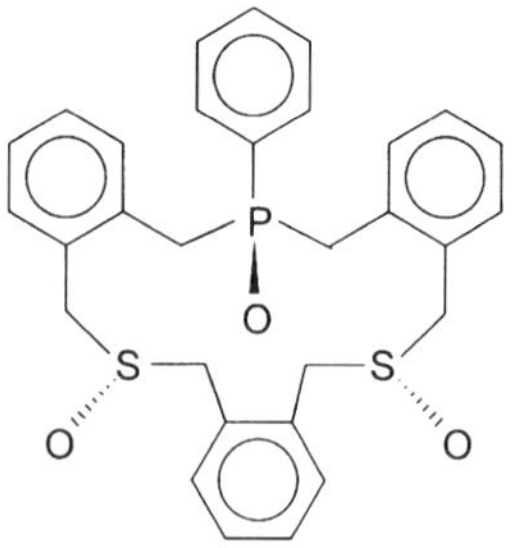

6.21

According to this choice, to date the number of anion complexation reactions in solution in which pure anion–dipole interactions are the driving forces is probably limited to the few following cases:

1. ^{1}H-NMR studies, performed in solution of 2% CD_3OD in $CDCl_3$ (v/v) containing the macrocyclic phosphine oxide disulfoxide ligands **6.21** in the presence of various anionic substrates, proved the formation of complexes with halides, HSO_4^-, and $H_2PO_4^-$ anions.[102] The crystal structure of **6.21** showed that the three oxygen atoms of the ligand are oriented on the same side of the macrocycle mean plane, allowing the positive lobes of the P=O and S=O dipoles to be suitably arranged for interaction with anions (Fig. 6.5).[103] Attraction of anions to this cluster of positive termini of strong dipoles is almost the unique force promoting and stabilizing anion complexation by **6.21**. The trend of affinity observed for these anions, $Cl^- \approx Br^- > I^- > HSO_4^- > H_2PO_4^- > F^-$, can be rationalized in terms of anion/macrocycle mutual dimensions, higher solvation of F^-, HSO_4^-, and $H_2PO_4^-$ by methanol, and the different polarizability of the anions. In the case of the best match, occurring between Cl^- (Br^-) and **6.21**, ^{1}H-NMR analysis based on two different resonances gave formation constants of 70 and 60 M^{-1} ($\Delta G° \approx -10\,kJ\,mol^{-1}$),[101] corresponding to about $-3.3\,kJ\,mol^{-1}$ contribution for a single anion–dipole interaction, in good agreement with a mean value ($\approx -3\,kJ\,mol^{-1}$) deduced for other anion–dipole bindings in chloroform.[56]

2. A macrotricyclic uncharged host (**6.22**) possessing dipolar binding sites for anions has been synthesized by Worm and Schmidtchen et al. by borane–amine adduction.[104] Inclusion of Cl^-, Br^-, I^-, and CN^- into **6.22** was followed by ^{1}H-NMR spectroscopy in 20% $CD_2Cl_2/CDCl_3$ (v/v) solution, yielding the stability sequence $Br^- > Cl^- > I^- > CN^-$. The free energy change ($-6.3\,kJ\,mol^{-1}$ at 297 K) determined for the complexation of Br^- is smaller by about $10\,kJ\,mol^{-1}$ than that obtained for the analogous ligand **6.20**,[87,88] in agreement with the formation of ion–dipole instead of ion–ion interactions.

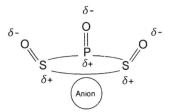

Fig. 6.5. Anion–dipole attraction. Convergent orientation of P=O and S=O dipoles in **6.21** promotes the complexation of anions.

X = (CH₂)₆   **6.22**

More commonly, interactions between anions and permanent dipoles furnish minor contributions to the stability of anion complexes and start being operative when the solvent is released from the interacting functions and the complex is almost assembled. In many cases the dipole, or the dipoles, are located in a region of the anion structure while the net charges reside on the receptor. Interesting examples are given by the interaction of anion-bearing aromatic moieties with positively charged ligands. Benzene, for instance, presents a quadrupole moment whose direction lies on the C_6 axis (Fig. 6.6) and consequently can attract positive charges along this axis and negative charges on the molecular plane. Figure 6.7 illustrates the free energy changes for the inclusion of similarly sized aromatic and aliphatic anions into the quaternary tetraazaparacyclophane **6.23**. The

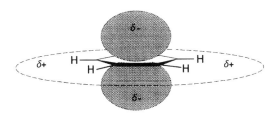

Fig. 6.6. Schematic charge distribution in benzene.

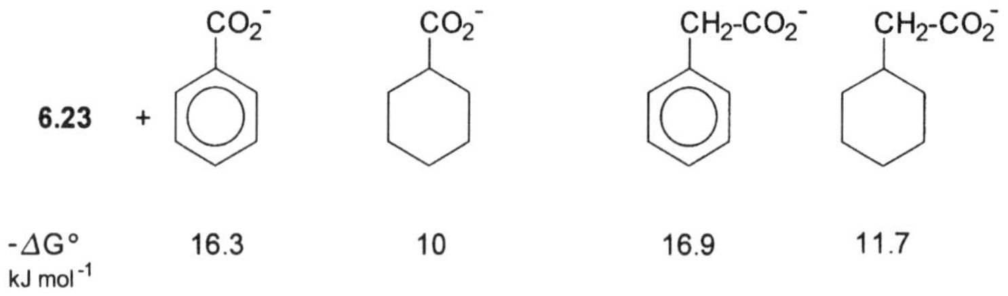

236 THERMODYNAMICS OF ANION COMPLEXATION

6.23 + [benzoate: CO$_2^-$ on benzene] [cyclohexanecarboxylate: CO$_2^-$ on cyclohexane] [phenylacetate: CH$_2$-CO$_2^-$ on benzene] [cyclohexylacetate: CH$_2$-CO$_2^-$ on cyclohexane]

$-\Delta G°$ kJ mol^{-1}: 16.3   10   16.9   11.7

Fig. 6.7. Free energy changes for the inclusion of similarly sized aromatic and aliphatic anions into the quaternary tetraazaparacyclophane **6.23**.

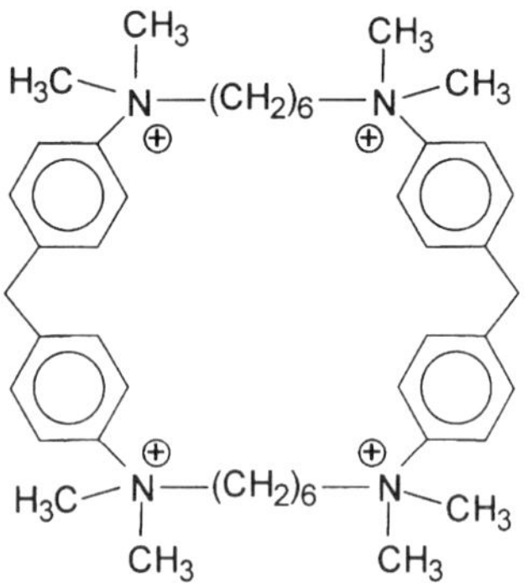

6.23

equilibrium data obtained in D_2O/CD_3OD 80:20 (v/v) at 298 K indicate that the aromatic anionic guests are more strongly retained inside the host cavity than aliphatic ones, the difference in stability (4–6 kJ mol^{-1}) being mostly ascribable to the interaction of the ammonium groups with the partial negative charge residing on the benzene π system.[105]

6.4.2 Effects of Induced Dipoles

Charged particles have the general properties of polarizing the neighboring species; nevertheless, anions, which are atoms or molecules with enriched electron clouds, can be polarized by the neighboring species. For this reason we can expect van der Waals and London dispersion interactions to accompany all anion complexation reactions. For example, the crystal structure of the I$^-$ complex with the ligand **6.17** (see discussion on X-ray structure in Chapter 5, Fig. 5.12) revealed significant contacts between the encapsulated anion and hydrogen atoms of the aliphatic chains.[89] I$^-$ presents high polarizability (Table 6.3), and therefore its spherical distribution of charge is expected to be significantly modified when subjected to the tetrahedral arrangement of positive charges imposed by **6.17**. Further modification of the charge distribution is

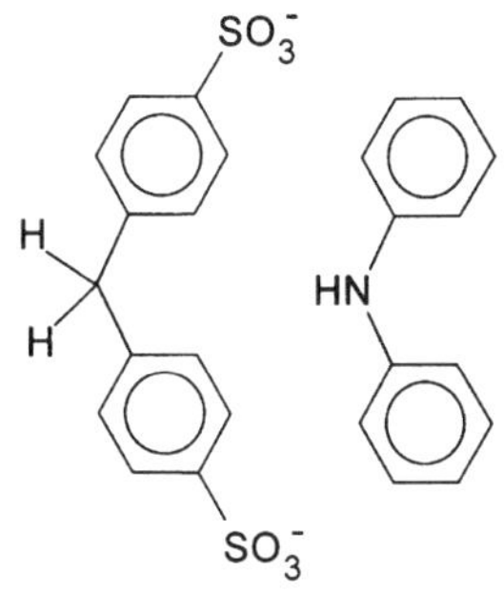

Fig. 6.8. Anion–induced dipole attraction. $-SO_3^-$ groups induce dipoles in the lobes of the electron clouds of diphenylamine.

expected as a consequence of the interaction with the aliphatic chains of the ligand. Attractive interactions between the anion charge and the dipoles induced in the ligand, and between the latter and the resulting polar distribution of charge on I^-, contribute to the stability of the $[I^- \subset (\mathbf{6.17})]$ complex. This stability is, however, mostly determined by charge–charge electrostatic attraction.

Actually, there is an intrinsic difficulty, principally connected with the presence of net charges on anions, and commonly also on receptors, in distinguishing between the effects of induced dipoles and the dominant charge–charge attraction, and hence, in quantifying contributions deriving from charge–induced dipole and induced dipole–induced dipole interactions in anion coordination. Effectively, there is no special need for separating these contributions since dispersion forces have a small effect in the binding of anionic groups. However, multiple dispersion interactions become evident in the hydrophobic effect observed in tight binding of apolar surfaces in water.[33] It deserves to be mentioned here that there is evidence that anions can induce dipoles in the lobes of the π-electron clouds of aromatic systems, giving rise to attractive interaction instead of the expected repulsion. An example is offered by the complex of diphenylamine in Fig. 6.8.[56] A study performed by Schneider et al. on singly charged anionic groups allowed an estimate of an energetic contribution of ca. $-1.5\,\text{kJ}\,\text{mol}^{-1}$ per single anion–arene interaction.

6.4.3 Hydrogen Bonding

Hydrogen bonding is largely determined by $^{\delta-}X-H^{\delta+}\cdots Y^{\delta-}$ electrostatic attraction involving an H atom bound to an electronegative atom X and another electronegative atom Y acting as a base, although significant contributions are also furnished by charge-transfer, dispersive, and covalent forces. F, O, and N are very efficient donors ($X-H$) and acceptors (Y) of hydrogen bonds, while Cl, Br, P, and S present similar although reduced properties. In a few cases C–H can also act as donor in hydrogen bonding. This is observed when the carbon atom is bound to electronegative groups as in $H-CCl_3$ or is involved in triple bonds as in H–CN or $RC \equiv C–H$. It has been recently reported that $–CH_2–$ groups attached to fluorinated groups in perfluoro macrocycles also constitute the donor sites of hydrogen bonds for the coordination of anions such as F^- and O_2^-.[106]

A qualitative justification of this bonding interaction is provided by viewing the formation of XHY adducts as a partial, or total, proton transfer from $X–H$ to Y. From this point of view the basicity of the X^- anion (i.e., the pK_a of the corresponding $X–H$ molecule) can be considered a measure of the anion propensity to form hydrogen bonds ($PO_4^{3-} > R–CO_2^- > ClO_4^-$ and $F^- > Cl^-$). Increasing the extent of the proton transfer, the complexed system shifts from a neutral $^{\delta-}X–H^{\delta+}\cdots Y^{\delta-}$ species to a hydrogen-bonded ion pair $X^-\cdots{^+}H–Y$, the second arrangement of bonding interactions being favored by a greater basicity of Y with respect to X^-.

A quantitative interpretation of this process, which describes also a mechanisms of anion coordination, has been recently furnished by ab initio molecular orbital calculations correlating the bonding strength in the XHY system with the hydrogen bond $X–Y$ distance and the dielectric nature of the medium.[107] Using as an example the formic acid/methyleneimine systems, HCOOH/NHCH$_2$, the computed proton transfer potentials indicate that at relatively long separation R (3.5 Å) between the hydroxylic oxygen and the imine nitrogen, in the absence of any external influence (dielectric constant $\varepsilon = 1$), the neutral complex $HCOOH\cdots NHCH_2$ is about 40 kJ mol^{-1} more stable than the separated species, while the $HCOO^-\cdots{^+}HNHCH_2$ ion pair is some 200 kJ mol^{-1} less stable, due to the difficulty in attaining the charge separation intrinsic to the latter. By progressive increasing of the dielectric constant, the ion pair is stabilized much more than the neutral complex, becoming more stable for relatively larger ε values ($\varepsilon > 4$), while the electronic barrier to proton transfer decreases, vanishing for very short ($R = 2.5$ Å) hydrogen bonds.

In the absence of structural restraints, the strongest such complexes tend to form very short hydrogen bond distances, some less than 2.5 Å, giving rise in several cases to systems where the proton oscillates around the center of the bond.[108] A classic example is given by HF_2^-, where the $F\cdots H\cdots F$ bond is symmetrical. In such cases the hydrogen-bond interactions involve covalent forces, which necessarily require the three center–two electron bond scheme, producing particularly stable adducts. However, as already discussed in Chapter 4, the participation of very strong hydrogen bonds in general substrate–receptor pairing in solution, including anion binding, seems to be rare, and it is a matter of controversy.[107,109–111]

As a matter of fact, although hydrogen bonds are responsible for strong association processes in the gas phase, the strength of such interactions is attenuated by the medium and depends on the nature of the medium. For instance, while the enthalpy values derived by thermochemical measurements or theoretical approaches change for the formation of HF_2^- according to $HF + F^- \rightleftharpoons HF_2^-$, ranging from -150 to -250 kJ mol^{-1} in the gas phase,[112] a positive ΔH° value of about 4 kJ mol^{-1} is obtained in aqueous solution, even if a favorable entropic term promotes the formation of such a species ($-\Delta G^\circ < 3.5\,\text{kJ}\,\text{mol}^{-1}$ at 298 K).[113] On the other hand, a lower attenuation of hydrogen-bond interactions is observed in apolar solvents like CCl_4. A systematic analysis of hydrogen-bond associations, spanning six orders of magnitude in K_S, has recently revealed that a single hydrogen bond between O and/or N atoms of different partners can provide up to about $30\,\text{kJ}\,\text{mol}^{-1}$ in the gas phase at 295 K in terms of free energy change, while a lowering of 30–40% in free energy change for association reactions with $\log K_S > 1$ and much greater attenuation for less stable adducts is found in CCl_4 solutions.[114]

Nevertheless, even small additions of solvents with larger polarity or higher hydrogen-bonding ability produce drastic attenuation of the complex stability.[56] Decreasing aggregation is commonly observed along the solvent series $CCl_4 > C_6H_6 > CHCl_3 > CH_2CCl_2 > \text{dioxane} \gg H_2O$,[115] although this trend is not totally valid owing to rather common enthalpy–entropy compensation effects, which have been mainly attributed to the real structures of the interacting partners and of the solvent. This aspect has been addressed earlier in this book (see Chapter 4), and therefore it will not be further discussed here.

Using the association data currently available, we can say that, approaching optimal conditions, a hydrogen bond in solution can be expected to furnish a contribution not greater than $20\text{--}30\,\text{kJ}\,\text{mol}^{-1}$ to the free energy of association, namely, less than one-tenth the energy of a common carbon–carbon or carbon–hydrogen bond. Only in rare cases, as in phosphate and sulfate binding proteins, which incorporate such anions into isolated clefts, are hydrogen bonds evidently more stable. At such locations the anions are away from the receptor surface (see Chapter 3) and protected from solvent and counterions,[116] where the average dielectric constant is estimated to be in the 2–4 range.[117] In spite of these particularly favorable conditions, the noticeable stability of anion complexes with such natural receptors is achieved via the formation of a considerable number of hydrogen bonds.

Thermodynamic data for association via hydrogen bonding are not generally available. Consequently the strength of these kind of bonds is commonly discussed in terms of bond distances determined by X-ray or neutron diffraction or by physical properties, such as vibrational changes in infrared and Raman spectra, and chemical shifts in ^{1}H-NMR spectroscopy, all of which are connected, within certain limits, with the strength of the interaction. For example, when the proton of X–H participates in hydrogen bonding, the ν_{X-H} stretching frequencies are depressed, the bands becomes broader, the X–H wagging modes increase in frequency, and the ^{1}H-NMR resonances are downfield shifted with respect to the non-hydrogen-bonded X–H species.

Considering that the negative charge in hydrogen-bond acceptors is concentrated along the lone pairs, a proper question arises: Do hydrogen bonds form preferentially along the direction of the lone pairs? In other words: Is the stability of the adducts connected with the structural features of hydrogen bonds? The answer to this question is rather controversial because the geometrical characteristics of hydrogen bonds showed by crystal structures are often determined by packing and steric effects, which produce a large variety of intermediate orientations. Nevertheless, there is clear evidence, from many crystal structures showing $NH\cdots O=C$ interactions, that hydrogen bonding occurs in, or near, the directions of the conventionally viewed oxygen sp^2 lone pairs.[118] Namely, hydrogen bonds of type II in Fig. 6.9 seem to be more stable than type I. The preferred approach along the lone-pair direction is also in good agreement with the formation of rather common three-centered hydrogen bonds like III, but it doesn't correlate well with the rare four-centered structures IV.

Further evidence is that the directional influence of the lone pairs becomes more pronounced as the $H\cdots O$ distance is increased. In other words, shorter (stronger) bonds take less advantage than longer bonds from orientation along the direction of the lone pairs. An explanation for this behavior could be found in the increasing charge polarization that the $X-H$ donor induces in the acceptor Y as the $X-H\cdots Y$ distance is reduced.[119]

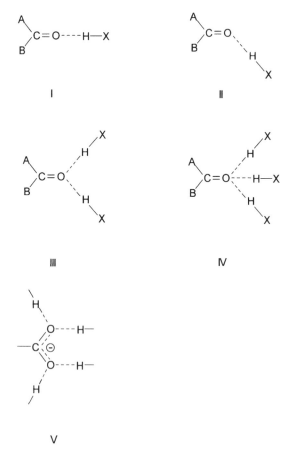

Fig. 6.9. Directionality of $N-H\cdots O=C$ hydrogen bonds.

Hydrogen bonds involving C=O groups in carboxylate anions present some differences with respect to those formed by un-ionized carbonyl acceptors, mainly as a consequence of the net negative charge present on the carboxylate, which induces closer hydrogen-bond contacts (also the directional influence of the lone pairs is correlated with the hydrogen-bond distance) and of the larger propension of carboxylate oxygens to form three-centered bonds, leading carboxylates to be commonly involved in the hydrogen-bond pattern V in Fig. 6.9. These characteristics are also determinant in making carboxylate anions stronger acceptors than other compounds containing C=O groups like amides, ketones, and carboxylic acids, respectively.[120]

The strength of the hydrogen-bond interactions also depends on the nature of the donor species. Table 6.6 lists the mean H$\cdots$O distances in N–H$\cdots$O=C(carboxylate) bonds obtained from crystal structures of six categories of donors: uncharged >N–H donors, charged trigonal >N$^+$–H donors, the NH$_4^+$ ion, and mono-, di-, and trisubstituited ammonium ions (RNH$_3^+$, R_2NH$_2^+$, R_3NH$^+$).[120] The strongest hydrogen bonds seem to be formed by R_3NH$^+$, the H$\cdots$O distances increasing along the series R_3NH$^+$ < R_2NH$_2^+$ < RNH$_3^+$ < >N$^+$–H < NH$_4^+$ < >N–H. This trend has been explained considering the synergy of hydrogen bonding and electrostatic attraction between ammonium groups and anionic species, and the steric hindrance produced by multiple hydrogen bonding around low-substituted ammonium ions. It is generally observed that bonds involving a single acceptor are shorter than those involving multiple acceptors.

In the case of sp^3 lone pairs, it appears that the directionality of hydrogen bonds is less important than when involving sp^2 lone pairs.[120] Charge-density studies revealed that the deformation density of the lone pairs in carbonyl groups is usually resolved into two separated maxima, corresponding approximately to the direction of the sp^2 lone pairs. Conversely, in the case of hydroxyl groups, including water, the deformation density of the lone pairs is generally found as a broad peak extending over a large part of the sp^3 lone-pair region.[121] Nevertheless, these results are contradicted by spectroscopic studies on the formation of hydrogen fluoride adducts, in the gas phase, with oxirane, oxetane,

TABLE 6.6. Mean H$\cdots$O Distances of N–H$\cdots$O=C (Carboxylate) Bonds

Donor	Distance (Å)
>N–H	1.928(19)[a]
>N$^+$–H	1.869(28)
NH$_4^+$	1.886(18)
RNH$_3^+$	1.841(8)
R_2NH$_2^+$	1.796(14)
R_3NH$^+$	1.722(25)

[a] Numbers in parentheses are standard deviations on the last significant figures.

and water, which evidence an energetic preference for O···H–F hydrogen bonding in approximately the directions of the oxygen sp^3 lone pairs.[122]

The above structural considerations, although only partly correlated with the energetics of hydrogen bonding, offer essential information for taking advantage of hydrogen bonding in the design of specific receptors for all kinds of substrates, including anions. To this purpose, the number of hydrogen bonds, the complementarity of donor and acceptor functionalities, and charge–charge attraction are the principal parameters to be optimized. A few significant examples of striking anion recognition by synthetic receptors involving hydrogen-bond networks are depicted in Fig. 6.10.[123–125]

In order to mimic natural anion binding through hydrogen bonds, Reinhoudt and co-workers developed some synthetic ligands bearing sulfonamide binding groups (Fig. 6.11). The sulfonamide receptors derived from TREN [tris(aminoethyl)amine] **6.24** and **6.25** revealed a remarkable selectivity in CH_3CN for $H_2PO_4^-$ over HSO_4^- and Cl^- ($H_2PO_4^- > Cl^- > HSO_4$),[126] while the calix[4]arene derivatives **6.26–6.28** demonstrated preferential binding of HSO_4^- over $H_2PO_4^-$, Cl^-, and NO_3^- in $CDCl_3$ (Fig. 6.11).[127]

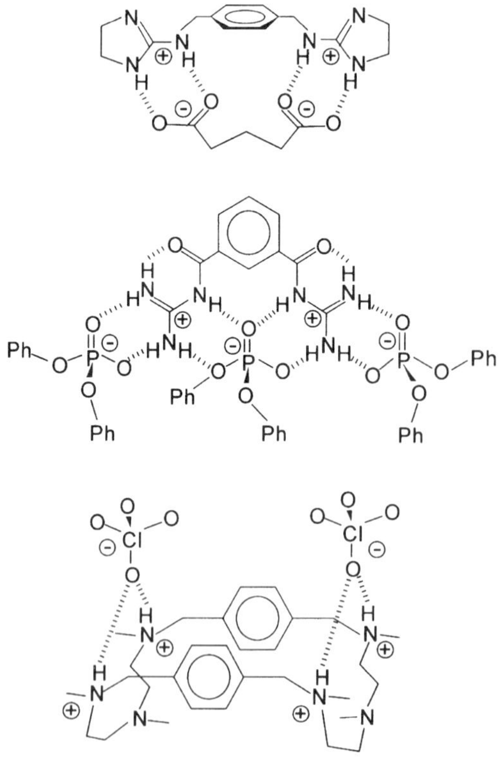

Fig. 6.10. Examples of anion binding by synthetic receptors involving multiple hydrogen bonding.

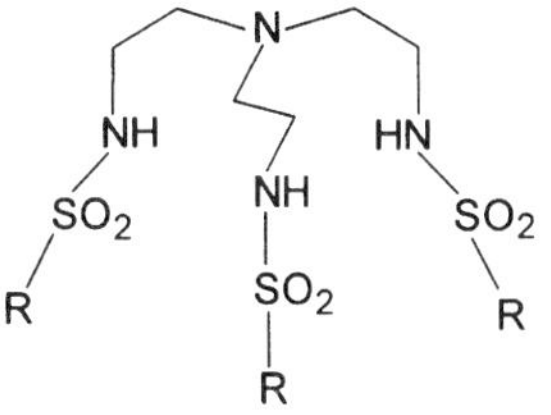

Association constants K_S (±5%) (M^{-1}, CH$_3$CN, 298 K)

	H$_2$PO$_4^-$	HSO$_4^-$	Cl$^-$
6.24	3500	79	540
6.25	14200	38	1600

6.24 R = –⟨C$_6$H$_4$⟩–CH$_3$

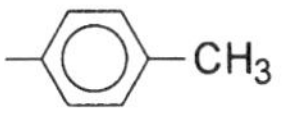

6.25 R = naphthyl

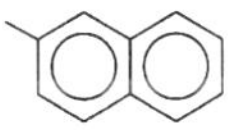

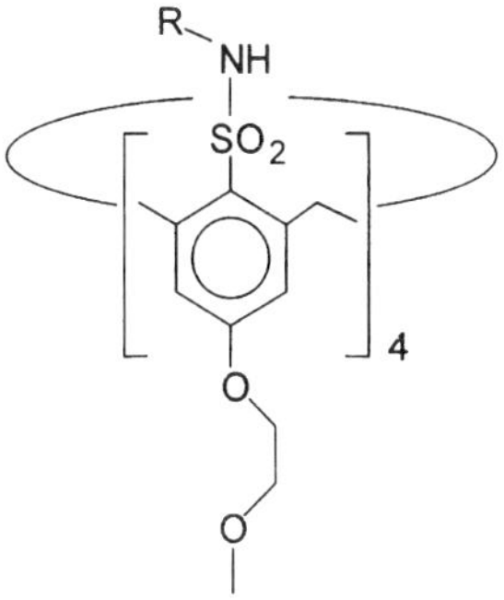

Association constants K_S (±5%) (M^{-1}, CDCl$_3$, 298 K)

	H$_2$PO$_4^-$	HSO$_4^-$	Cl$^-$	NO$_3^-$
6.26	350	970	360	240
6.27	<10	134	72	43
6.28		103400	1250	513

6.26 R = n-Pr

6.27 R = tert-Bu

6.28 R = CH$_2$CH$_2$NHC(O)CH$_3$

Fig. 6.11. Anion coordination by synthetic ligands bearing sulfonamide groups.

Calix[4]arene derivatives with urea residues anchored to the lower rim are effective in the coordination of halide anions; the association constants up to 7.1×10^3 M in CDCl$_3$, with stability trend Cl$^-$ > Br$^-$ > I$^-$,[128] confirm that the urea moiety is a good hydrogen-bond donor, as had been already shown by Hamilton et al.[123,129] and Rebek et al.[130] in the binding of carboxylate anions.

In most cases, anion binding in aqueous solution, as well as in other protic solvents, is largely promoted by the electrostatic attraction exerted by cationic receptors, while the formation of hydrogen bonds between the interacting partners represents a supplemental reinforcement of the complex structure. This reinforcement can be decisive in the formation of anion complexes, as demonstrated by the tetracharged macrocyclic cation **6.29**. As a consequence of nitrogen quaternarization, which prevents the formation of hydrogen bonds, **6.29** is unable to interact appreciably even with highly charged anions such as ATP^{4-} or $Co(CN)_6^{3-}$, both in aqueous solution[131] and in the solid state.[132]

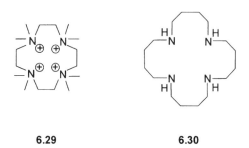

6.29 **6.30**

In contrast with this behavior, the tetraprotonated form of the larger tetraazamacrocycle **6.30**, which has a lower charge density but can still form hydrogen bonds, produces stable complexes in aqueous solution with ATP^{4-}, $Fe(CN)_6^{4-}$, and even with the poor hydrogen-bond acceptor $Co(CN)_6^{3-}$ ($H_4\mathbf{6.30}^{4+} + A^{n-} = (H_4\mathbf{6.30}A)^{(4-n)+}$, $\log K = 3.81, 3.62, 2.38$ for $A^{n-} = ATP^{4-}$, $Fe(CN)_6^{4-}$, and $Co(CN)_6^{3-}$, respectively).[131,132]

Microcalorimetric measurements demonstrated that the binding of $Fe(CN)_6^{4-}$ and $Co(CN)_6^{3-}$ by $H_4\mathbf{6.30}^{4+}$ is promoted by both enthalpic and entropic contributions ($-\Delta H° = 4.6\,kJ\,mol^{-1}$, $T\,\Delta S° = 16\,kJ\,mol^{-1}$ for $Fe(CN)_6^{4-}$ and $-\Delta H° = 10.7\,kJ\,mol^{-1}$, $T\,\Delta S° = 2.9\,kJ\,mol^{-1}$ for $Co(CN)_6^{3-}$).[132] The greater stability observed for the complex with the more charged $Fe(CN)_6^{4-}$ anion is mainly due to a large entropic term, the enthalpic contribution being small and less favorable than that for the weaker complex $[(H_4\mathbf{6.30})Co(CN)_6]^+$. The complete charge neutralization occurring upon formation of the former complex and the consequent desolvation effect can explain this very favorable entropy change. On the other hand, the favorable enthalpic contribution originates from two opposite main terms: an endothermic effect due to the removal of water molecules from the interacting ions and an exothermic effect produced by the formation of salt bridges and hydrogen bonds ($-C\mathord{=}N\cdots H\mathord{-}N^+$) between the cyanide groups and the hydrogen atoms of the ammonium centers.

Another example of anion binding in aqueous solution where hydrogen bonding plays a crucial role is given by the similarly structured ligands **6.31** and **6.32**. As shown by potentiometric and ^{31}P-NMR measurements, the polyprotonated forms of **6.31** give rise to the formation of stable complexes with the anionic species of ATP and ADP ($\log K_S$ values varying in the ranges 3.3–4.8 and 3.0–3.8 for ATP and ADP, respectively), while similar nucleotide complexes are not formed by **6.32**.[125] Such behavior can be rationalized on the basis of the structural information furnished by two crystal structures containing the $H_4(\mathbf{6.31})^{4+}$ and $H_5(\mathbf{6.32})^{5+}$ cations (Fig. 6.12), which show a convergent orientation of the four acidic protons in $H_4(\mathbf{6.31})^{4+}$, in contrast with the

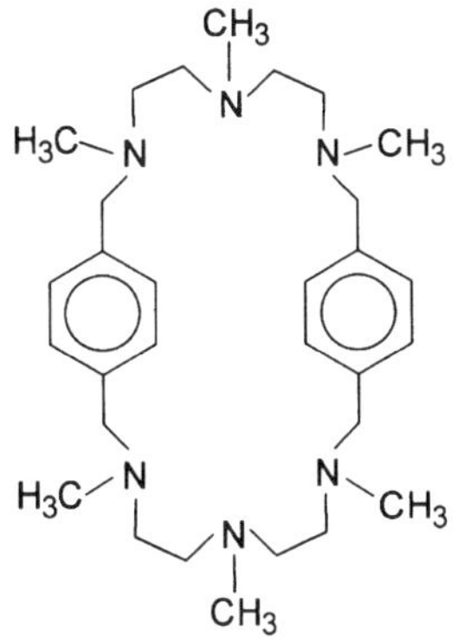

6.31

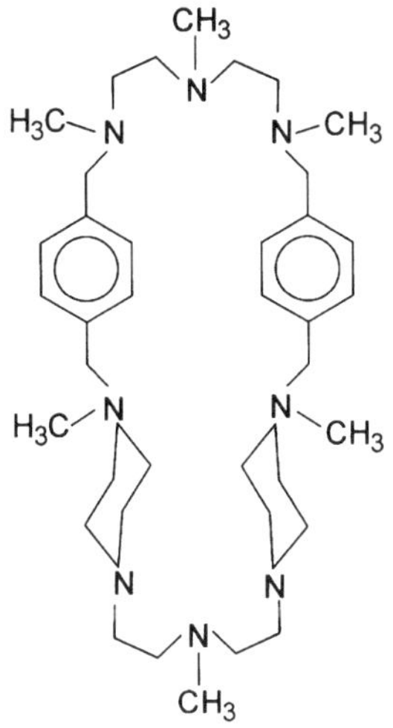

6.32

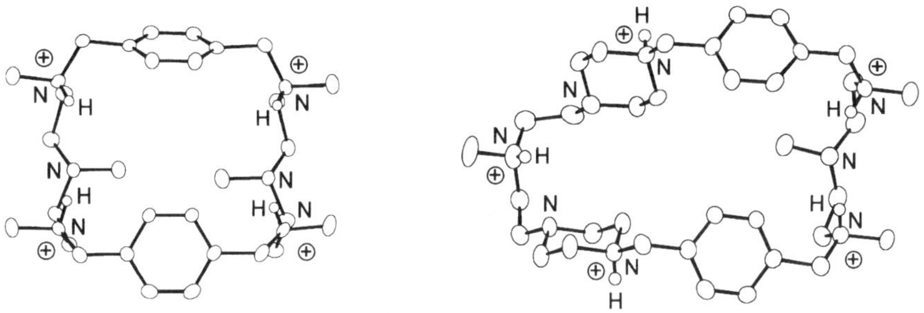

$H_4(\mathbf{6.31})^{4+}$ $H_5(\mathbf{6.32})^{5+}$

Fig. 6.12. Molecular structures of the $H_4(\mathbf{6.31})^{4+}$ and $H_5(\mathbf{6.32})^{5+}$ polyammonium macrocycles.

divergent array of the five acidic protons in $H_5(\mathbf{6.32})^{5+}$.[125] It seems reasonable that, due to the rigidity of **6.32**, this divergent orientation is maintained in solution, thus preventing $H_5(\mathbf{6.32})^{5+}$ from forming a sufficient number of hydrogen bond-mediated binding interactions with the nucleotides.

In this sense, the criterion of spatial charge matching must take into account the hydrogen-bond anchorage occurring between cationic and anionic functionalities. Since from the historic work[91] by Lehn on the recognition of dicarboxylate anion by ditopic polyammonium receptors **6.33**, **6.34** (Fig. 6.13), the spatial matching of charges and hydrogen

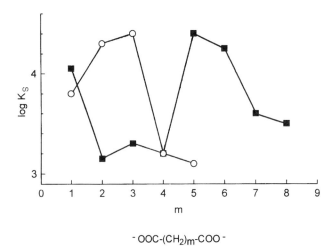

Fig. 6.13. Graphical representation of the stability constants (log K_S) of the complexes formed by the polyammonium macrocycles $H_6(\mathbf{6.33})^{6+}$ and $H_6(\mathbf{6.34})^{6+}$ with the dicarboxylates $-O_2C-(CH_2)_m-CO_2^-$ as a function of chain length m. Reproduced by permission from Ref. 91.

bonds has defined, and is still defining, a large number of selective processes of anion binding in solution. In this context the binding of a variety of simple inorganic anions,[44,68,70,71,79–81,133–142] carboxylates and polycarboxylates,[44,67,65,92,93,123,129,130,143–153] phosphate and nucleotidic anions (ATP, ADP, AMP, etc.), and phosphoesters,[44,65,72,92,99,131,146,150,152–170] as well as complex anions that give rise to second-sphere species ("supercomplexes"),[65,73–77,95,97–99,146,152,167,168,170–174] among others, has been analyzed. In many cases, the well-defined stereochemistry of binding contacts around the anions and the considerable stability of anion adducts inevitably recall the distinctive features of the classical "coordination chemistry" involving metal ions, thus promoting for similarly structured formation of ion pairs (anion adducts) the more descriptive expression "anion coordination chemistry."

6.4.4 Hydrophobic Effect

Tight complexation of small apolar guests by apolar hosts in aqueous solution, or other polar solvents, is accompanied by favorable enthalpy changes ($\Delta H^\circ < 0$) and unfavorable entropic contribution ($T\Delta S^\circ < 0$), in contrast to the classical hydrophobic effect ($\Delta H^\circ \approx 0$, $T\Delta S^\circ > 0$) observed in the formation of membranes and micelles.[29,175] An explanation of this behavior resides mostly in the desolvation of the interacting surfaces and in the tightness of the binding of these surfaces through London dispersion forces.

These observations promote the idea that the association strength could be proportional to the contact area. Actually, an almost linear correlation between the "maximal overlap area" between interacting partners orientated in a plane-to-plane manner and the free energy changes of complexation was evidenced by the work of Connors et al. on the association of planar aromatic molecules in water, furnishing an estimation of about $-0.4\,\text{kJ}\,\text{mol}^{-1}$ per Å^2 of arene–arene interacting area.[176] The correlation did not reveal a dependence of the association free energy changes from specific characteristics of the substrates, such as the presence of charges.

Similar features have also been evidenced by the complexation of a large variety of substrates, including anionic species, into cyclodextrins in water.[177] With this kind of ligand the overlap area is strictly connected with the cavity penetration by the substrate. In the case of anions the negatively charged groups are not involved in binding to the internal surface of cyclodextrins but tend to remain outside the cavity, in contact with the solvent, controlling the orientation of the included hydrophobic moieties. This is clearly shown in the complexation of naphthalene sulfonate and carboxylate anions by β-cyclodextrin.[177a] The thermodynamic data reported in Table 6.7 demonstrate that the complexation reaction with β-cyclodextrin is almost insensitive to the type of anionic group, but it is strongly influenced by the site of the anion in the naphthalene structure. Some reduction in stability seems to occur in the complexes of β-cyclodextrin as a consequence of the increasing number of anionic substituents, which add hydrophilicity to the guests, although charge effects are largely of minor importance compared to sterical effects. As can be

TABLE 6.7. Thermodynamic Parameters ($\text{kJ}\,\text{mol}^{-1}$) for the Inclusion Complex Formation of Naphthalene Anionic Derivatives with β-cyclodextrins

Guest	$-\Delta G^\circ$	$-\Delta H^\circ$	$T\Delta S^\circ$
1-naphthalenesulfonate	4.64	1.49	3.15
2-naphthalenesulfonate	7.33	7.01	0.32
2,6-naphthalenedisulfonate	4.49	2.79	1.70
2,7-naphthalenedisulfonate	3.33	6.75	−3.42
2,3,6-naphthalenetrisulfonate	3.03	3.09	−0.06
4-amino-1-naphthalenesulfonate	2.32	2.38	0.06
1-naphthaleneacetate	5.93	1.11	4.82

seen from the table, the inclusion complexation reactions of β-cyclodextrins are exclusively exothermic and mostly enthalpy-driven. Of these substrates only the 1-substituted naphthalenes give rise to entropy-driven reactions.

In spite of the great importance of hydrophobic effects in supramolecular chemistry, they are predestinated to be effective only in the presence of apolar, not charged, substrates, and, therefore, they appear in anion coordination chemistry when the substrates contain hydrophobic regions and not by virtue of their anionic nature. Nevertheless, when weakly solvated anionic groups are sequestered into hydrophobic cavities, in water or in other particularly polar media, contributions from favorable hydrophobic effects could be envisaged.

6.4.5 Stacking Effect

A large effort has been made in defining the individual contributions, including van der Waals, charge-transfer interactions, and hydrophobic effects, that are involved in the so-called "stacking effect." The fact that charge-transfer, or more generally electron donor–acceptor, forces, originate from the principal attraction between electron-rich and electron-poor partners further suggests that electrostatic forces are also at the origin of such interactions.

Nevertheless, there is an intrinsic difficulty in quantifying these individual contributions, since the boundaries between them are neither well defined nor generally distinguishable. Moreover, as already observed for van der Waals interactions and hydrophobic effects, each one of these forces, taken alone, produces small contributions to the stability of the stacked complexes. This applies also for charge-transfer forces: Even considering very strong interactions like that occurring between the electron-rich hexamethylbenzene and the electron-poor 1,3,5-trinitrobenzene, the free energy change of association in CCl_4 at 298 K is only $\Delta G^\circ = -4.6 \, \text{kJ} \, \text{mol}^{-1}$.[178] For these reasons stacking is more commonly discussed as an overall effect.

Anions, by virtue of their negative charge, are inevitably electron-rich species, and therefore they are open to stacking effects. However, such binding interactions are mostly recognized when lipophilic portions of the anionic species match lipophilic functionalities of the receptors.

It has recently been shown that the polyprotonated hexaazametacyclophane **6.35** forms stable complexes with nucleotidic anions derived from ATP, ADP, and AMP in aqueous solution.[166] While the phosphate chain is a good electrostatic binding point and hydrogen-bond acceptor, ^{1}H-NMR spectra provided unambiguous evidence that the nucleoside part operates as an adequate site for stacking interactions with the aromatic moiety of the macrocycle. A comparison with the similarly sized hexaazacycloalkane **6.8**[65,156] shows that while the presence of the benzene ring in **6.35** reduces its binding ability toward ATP and ADP, probably due to a mismatch of polyphosphate groups and ammonium centers as a result of the bulky ligand functionality, in the case of the smaller AMP having suitable dimensions for **6.35**, the metacyclophane forms more stable complexes. The increase in stability of about $4.6 \, \text{kJ} \, \text{mol}^{-1}$ for

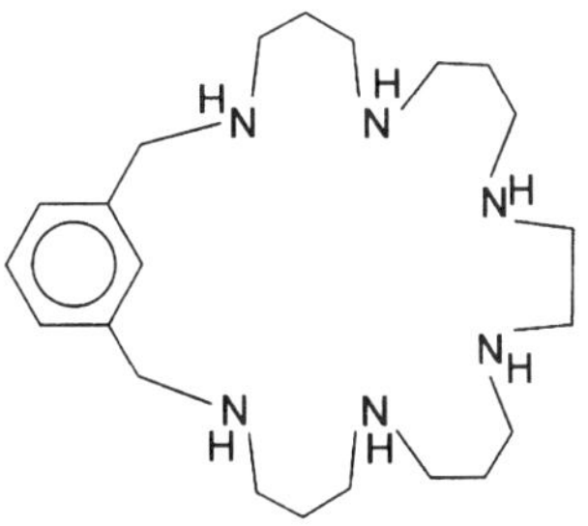

6.35

AMP binding by the fully protonated ligand can be considered an estimate of the π-stacking contribution.

Nucleotidic anions are particularly adequate for taking advantage of stacking effects in the association with host molecules containing aromatic parts. Accordingly, macrobicyclic and acyclic *bis*-intercaland receptor molecules **6.36-6.43** containing two flat acridine residues positioned at a distance suitable for

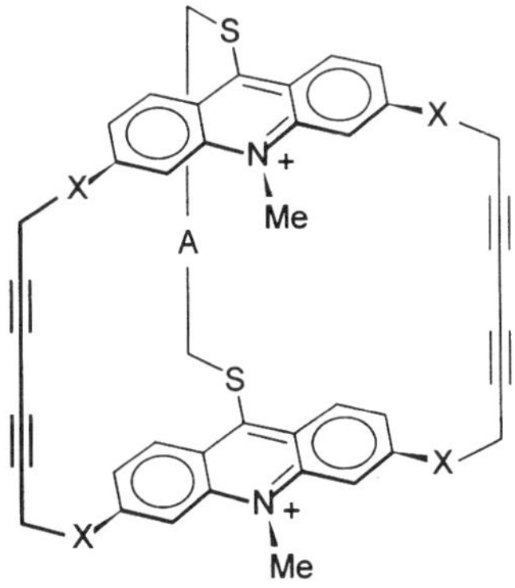

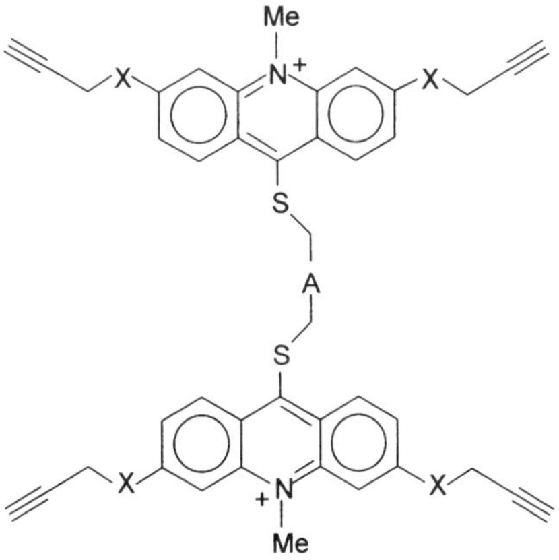

6.36 X = O, A = (CH$_2$)$_4$

6.37 X = O, A = CH$_2$OCH$_2$

6.38 X = O, A = CH$_2$OCH$_2$CH$_2$OCH$_2$

6.39 X = NH, A = CH2OCH2

6.40 X = O, A = (CH$_2$)$_4$

6.41 X = O, A = CH$_2$OCH$_2$

6.42 X = O, A = CH$_2$OCH$_2$CH$_2$OCH$_2$

6.43 X = NH, A = CH2OCH2

the intercalation of nucleoside moieties were tailored and tested in the binding of various nucleotides and some other planar anions in aqueous solution.[150] Although **6.36–6.43** contain two positively charged ammonium centers, electrostatic effects appear to play only a minor role, as demonstrated by the similarity of the binding constants determined at 293 K for AMP^{2-} (log K_S = 4.08), ADP^{3-} (log K_S = 3.84), and ATP^{4-} (log K_S = 3.91) with **6.37**, or for the formation of **6.37** complexes with benzene derivatives bearing one to four carboxylate functions (log K_S = 3.45 for benzoate to log K_S = 3.88 for 1,2,4,5-benzene tetracarboxylate). On the other hand, the complexes exhibit a marked increase in stability with the size of the flat substrates as one goes from one to two or three aromatic rings, as is seen, for instance, in the binding of the equally charged terephthalate (log K_S = 3.54), 2,6-naphthalene dicarboxylate (log K_S = 4.25), and 2,6-anthraquinone disulfonate (log K_S = 5.60) with **6.36**. These features indicate that the binding of such substrates by receptors **6.36–6.43** is dominated by stacking interactions that produce a mean stabilization of about $5 \pm 2 \, \text{kJ} \, \text{mol}^{-1}$ per additional benzene ring.

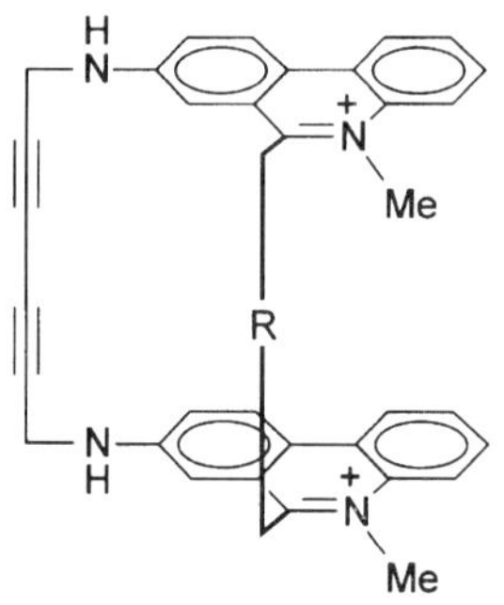

6.44 R = (CH$_2$)$_4$
6.45 R = (CH$_2$)$_6$
6.46 R = p-C$_6$H$_4$

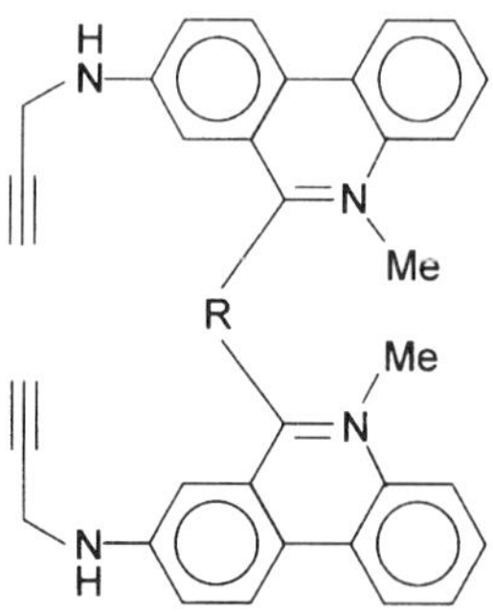

6.47 R = (CH$_2$)$_6$

The importance of π-stacking effects in nucleotidic anions binding in water is even more evident in the case of *bis*-intercaland receptors such as **6.44–6.47** containing two phenanthridinium residues.[170] The equilibrium constants of the complexes formed by these ligands with nucleotides, ranging from 10^5 to 10^6 M^{-1}, are the highest measured to date for receptors that bind only the nucleic base part of a nucleotide, and are strictly independent of host charge. For example, the equilibrium constants for the binding of the differently charged ATP^{4-}, ADP^{3-}, and AMP^{2-} by the *bis*-intercaland receptor **6.44** are log K_S = 5.80, 5.65, and 5.38, respectively, while that determined under the same experimental conditions for the electrically neutral adenosine is log K_S = 5.62.

6.5 COVALENTLY BONDED ANION COMPLEXES

The general strategy so far outlined in the construction of receptor molecules for anions consists in providing a stable environment to the guest species via noncovalent, or essentially noncovalent, intermolecular interactions by the assembly of electron-deficient functionalities. Using a similar criterion, ligands with several electrophilic centers have recently been obtained, making recourse to atoms such as tin,[179–189] mercury,[190–198] silicon,[199,200] boron,[201–204] aluminum,[205] and germanium.[206] Unlike the former receptors, the new Lewis acid ligands bind anions by directional, coordinate covalent bonds, and complementary orbital interactions provide the stabilization of the supramolecular species formed. Hence, the stability of the anion complexes with these electrophilic ligands is strictly controlled by the preorganization of the electron-deficient centers, which determines the number of covalent bonds as well as bond distances and directionalities. An interesting consequence of this is that, in contrast with the other anion receptors mostly based on adirectional forces, size selectivity in anion coordination by multidentate Lewis acid ligands is not based on the ionic radius of the bound anion, as demonstrated by the fact that anions can fit into cavities that are smaller than themselves. This is particularly evident for cagelike ligands such as the organoboron macrocycles **6.48–6.58** for which the binding energies for the complexation of H^-, F^-, Cl^-, and O^{2-} have been computed by means of molecular-orbital (AM1) calculations.[204] Partial sp^2–sp^3 boron rehybridation occurs upon anion complexation. The strength of the anion–host interaction can also be described by such changes in the boron geometric parameters.[204]

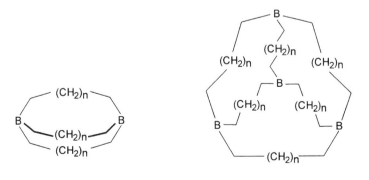

n = 1-8 **6.48-6.55**   n = 0-2 **6.56-6.58**

```
   Cl         (CH₂)n        Cl                    (CH₂)n
    \       /       \      /              /                \
     Sn              Sn            X—Sn                    Sn—X
    /       \       /      \              \    (CH₂)n      /
   Cl         (CH₂)n        Cl             \             /
                                             (CH₂)n
```

n = 6 **6.59**   X = Cl   X = Br

n = 8 **6.60**   n = 6   **6.61**   n = 8   **6.66**

                                         n = 7   **6.62**   n = 10 **6.67**

                                         n = 8   **6.63**   n = 12 **6.68**

                                         n = 10 **6.64**

                                         n = 12 **6.65**

Experimental equilibrium data for the formation of anion complexes with similar Lewis acid ligands are rather rare and to date limited to tin-containing receptors. Association constants were also pursued for the formation of halide complexes with organomercury compounds, but their determination was prevented by very high complex stability or insufficient complex solubility.[193,197]

The equilibrium constants for the formation of halide complexes with the macrocyclic and macrobicyclic Lewis acid ligands containing tin **6.59–6.68** have been have been reported by Newcomb et al.[180–182,184,185] ^{119}Sn-NMR spectra recorded in acetonitrile at 295–298 K revealed the formation of chloride complexes of **6.59** and **6.60** with 1:1 and 1:2 ligand-to-halide stoichiometry, providing estimates of the equilibrium constants.[180] Comparison of these equilibrium data for the addition of the first and the second Cl$^-$ anion to the macrocyclic ditin ligands **6.59** ($K_{S1} = 814$, $K_{S2} = 863$ M^{-1}) and **6.60** ($K_{S1} = 684$, $K_{S2} = 556$ M^{-1}) with the analogous data for the acyclic model compound BuCl$_2$Sn–(CH$_2$)$_{10}$–SnCl$_2$Bu **6.69** ($K_{S1} = 427$, $K_{S2} = 527$ M^{-1}) and the monotin species dibutyltin dichloride Bu$_2$SnCl$_2$ **6.70** ($K_S = 204$ M^{-1}) reveals some increase in stability for the complexes of the cyclic **6.59** as well as cooperativity of the two tin atoms in the coordination to Cl$^-$ anion. These binding tendencies were confirmed by direct competition experiments.[180]

Cooperative effects are of great importance in the coordination of halide anions by ditin Lewis acid ligands with more rigid macrocyclic structure. Strong binding of fluoride by **6.61** was observed from 223 to 303 K in CDCl$_3$ and C$_2$D$_2$Cl$_4$ solutions with equilibrium constants on the order of 1–2×10^4 M^{-1}.[182] Despite the lack of good models, it appears that **6.61** binds fluoride more strongly than would be expected for simple acyclic ligands, the increased stability being ascribed to the simultaneous coordination of the acidic tin atoms to F$^-$ inside the cavity, as already observed in the crystal structure of the (F·**6.61**)$^-$ complex.[182]

Size selectivity is evidenced in the binding of Cl$^-$ and Br$^-$ by **6.61–6.68** in CDCl$_3$ at 293 K, the maximum stability being observed for the chloride complex of **6.63** ($K_S = 17$ M^{-1}) and the bromide complex of **6.67** ($K_S = 1.4$ M^{-1}).[182,185] There is a sharp decrease of the equilibrium constants when the dimensions of the ligand cavity become too small to accommodate the anion. On the other hand, oversized ligands produce only a small decrease of the equilibrium constant with respect to the host molecules that provide the best fit. These features together with the fact that even best-fit complexes are not more stable than complexes with simple trialkyl halides strongly suggest that these macrocyclic tin ligands bind Cl$^-$ and Br$^-$ with only one Lewis acid site inside the cavity. Such binding characteristics are present in the crystal structure of (Cl**6.63**)$^-$ complex.[182] ^{119}Sn spectra of this complex recorded at low temperature suggest a dynamic process of "chloride jump" from one tin to the other with an activation energy of 22 kJ mol^{-1}, some 10 kJ mol^{-1} higher than the energetic barrier for a similar hopping process of fluoride in its complex with **6.61**.

Stronger binding of chloride has been achieved by means of the tricyclic tetratin receptor **6.71**, which form a 1:1 complex ($K_S = 500$ M^{-1} in CDCl$_4$ at 292 K).[184] The binding behavior of **6.71** is unique in comparison to that seen with previous macrocyclic ligands, in that the exchange of free host and complex by dissociation of the bound chloride from the cavity tends to be a slow process. This has been ascribed to cooperative effects of the acceptor sites in the binding of chloride, which produce an increase in stability of about 8 kJ mol^{-1} over the analogous complex of the bicyclic ditin ligand **6.63**.

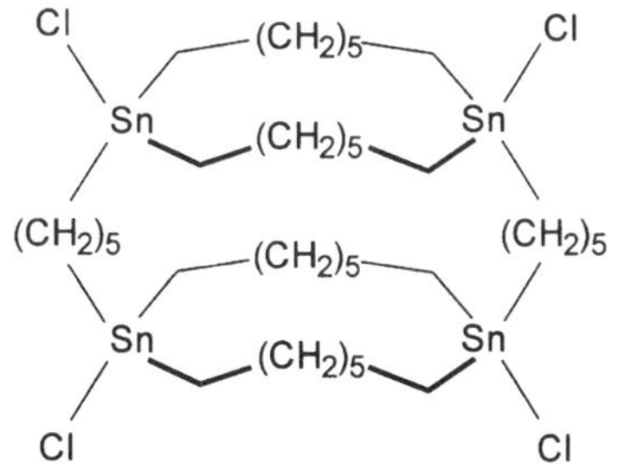

6.71

6.6 A CAUTIONARY WORD ON SELECTIVITY

Thermodynamic or kinetic discrimination among different anions is a major goal in the design of host species. As will be seen in Chapter 10, thermodynamic selectivity is required for most of the actual or envisaged applications this chemistry has or may have. Therefore, it is interesting to make some remarks about the real meaning that selectivity has and how to use this old concept within the framework of anion recognition. Usually quotients of stepwise stability constants are used to express selectivity ratios. For instance, one of the pioneering works in the field was the study of Lehn's group on the interaction between the macrobicyclic receptor bistren (**6.72**) and a series of mono- and polyvalent anions.[207] The constants reported for the interaction of halide anions F^-, Cl^-, Br^-, and I^- with the hexa- and pentaprotonated forms of (**6.72**) ($A^- + H_n(\mathbf{6.72})^{n+} \rightleftharpoons AH_n(\mathbf{6.72})^{(n-1)+}$) in 0.1 mol dm^{-3} NaTsO to yield complexes of 1:1 receptor:anion stoichiometries were: $\log K_S = 4.10$ (F^-), 3.00 (Cl^-), 2.60 (Br^-), 2.15 (I^-) for $H_6(\mathbf{6.72})^{6+}$ and $\log K_S = 3.20$ (F^-), 1.95 (Cl^-), 1.60 (Br^-), 1.55 (I^-) for $H_5(\mathbf{6.72})^{5+}$. The larger stability found for the F^- complexes was ascribed to its smaller size and consequently larger charge density and propensity to form hydrogen bonds. These constants lead, for instance, to selectivity ratios for the formation of halide complexes by the hexaprotonated receptor F^-/Cl^- 12.5, F^-/Br^- 31.7, and F^-/I^- 89.2, as expressed by the quotient of stability constants.

The question underlying these stability constant ratios is whether they are sufficiently large to permit selective coordination of F^- over the other halide anions or not. In this sense, distribution diagrams provide clear pictures of the situation. As seen in Figs. 6.14a and 6.14b, the percentage of formation of the fluoride adducts is much larger than that of the chloride analogues according to the higher stability constants obtained for the former anion. The results of mixed distribution diagrams of the system F^-–Cl^-–**6.72** (Fig. 6.14c), calculated for equimolar amounts of all the reactants, are very illustrative. In this diagram it can be seen that, as could be expected according to the stability constants, F^- complexes are predominant throughout the whole pH range where complexation occurs.

6.72

However, this simple picture does not hold for more complicated systems, particularly those in which both receptor and anion are involved in multiple protonation equilibria. Therefore, for these systems a criterion balancing the differences in basicity is required to establish concentration selectivity patterns. The most straightforward way to do this is to calculate the distribution of complexed species as a function of pH for the mixed systems anion 1–anion 2–

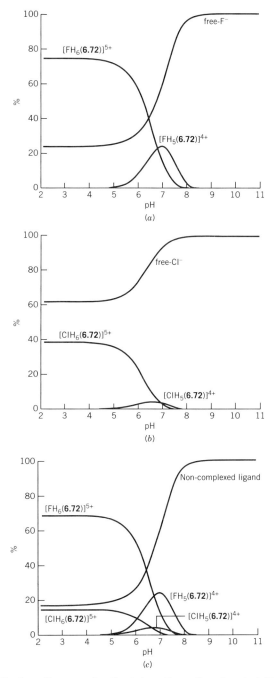

Fig. 6.14. Distribution diagrams for the interaction of protonated (**6.72**) and: (a) F^-, (b) Cl^-. (c) The distribution diagram for the mixed system (**6.72**)–F^-–Cl^-. All diagrams have been plotted for 1×10^{-3} mol dm^{-3} concentrations in all the reactants.

receptor and to represent their overall percentages of formation. This method, which allows for attributing selectivity over the entire pH range, does not require any assumption on the location of the protons in the interacting species, which is a common source of erroneous interpretation of such a parameter. Examples of such systems are, for instance, those regarding the interaction of phosphate type or carboxylate anions with polyazamacrocycles. To illustrate this method, let us present some of these examples.

In 1992 the interaction of 1,4,7,10,13,16,19-heptaazacyclohenicosane (**6.2**) with a series of polycarboxylate anions with different topologies and numbers of functional groups was reported.[148,152] In Tables 6.8 and 6.9 are presented the cumulative and stepwise basicity constants for some of these hosts as well as those of their adduct species with **6.2**. Basicity constants of **6.2** are included in the footnote of Table 6.9. These data offer the opportunity to get a closer look at selectivity criteria, since all four guests display different basicities (Table 6.8) and, even more, two of them are dicarboxylic acids, while the other two are tricarboxylic ones.

The usual practice in speaking about thermodynamic selectivity is to compare the stepwise association constants for the different systems. The stepwise association constants for the equilibria $A + H_rL \rightleftharpoons H_rLA$ (see Table 6.9, Fig. 6.15), derived from the cumulative association constants taking into account the protonation constants of **6.2**, could erroneously suggest that, at least for protonation degrees greater than 4, c,t-TMCTC complexes should predominate over those of both 1,2-BDC and 1,3-BDC.

However, the use of the type of distribution diagrams mentioned above (see Fig. 6.16) allows one to establish that for equimolar amounts of 1,2-BDC, 1,3-BDC, c,t-TMCT, and **6.2**, 1,2-BDC and 1,3-BDC complexes clearly prevail over those of c,t-TMCTC throughout the entire pH range studied. Therefore,

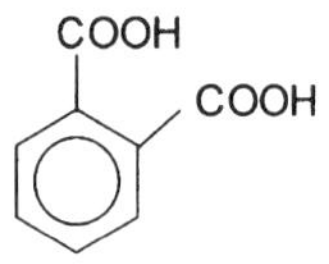

1,2-BDC

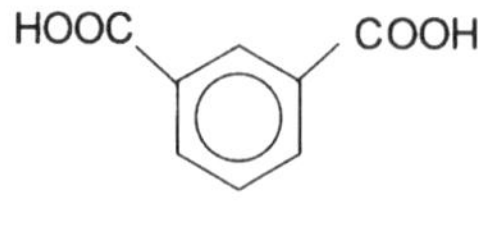

1,3-BDC

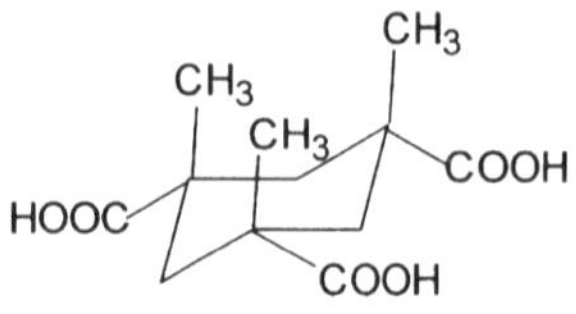

c,c-TMCTC

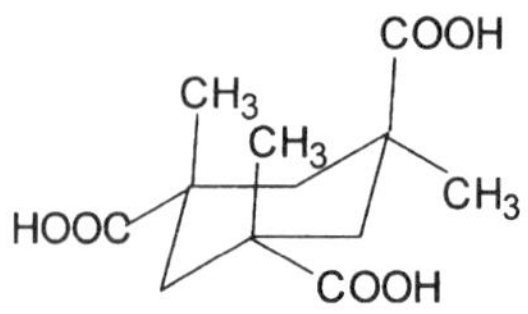

c,t-TMCTC

TABLE 6.8. Logarithms of the Cumulative Basicity Constants for Polycarboxylic Acids 1,2-Benzenedicarboxylic (1,2-BDC), 1,3-Benzenedicarboxylic (1,3-BDC), cis,cis-1,3,5-Trimethyl-1,3,5-Benzenetricarboxylic (c,c-TMCTC) and cis,trans-1,3,5-Trimethyl-1,3,5-Benzenetricarboxylic (c,t-TMCTC) Determined at 298 K in 0.15 mol dm^{-3} NaClO$_4$

Reaction	1,2-BDC	1,3-BDC	c,c-TMCTC	c,t-TMCTC
$A + H \rightleftharpoons A$[a]	4.916(2)[b]	4.206(2)	7.289(5)	6.921(8)
$2H + A \rightleftharpoons H_2A$	7.678(2)	7.444(2)	13.47(1)	11.64(1)
$3H + A \rightleftharpoons H_3A$			17.07(2)	15.46(2)
$HA + H \rightleftharpoons H_2A$	2.76	3.24	6.18	4.72
$H_2A + H \rightleftharpoons H_3A$			3.60	3.82

[a] Charges omitted for clarity.
[b] All constants taken from Ref. 152.

TABLE 6.9. Logarithms of the Cumulative and Stepwise Stability Constants for the Interaction of (5:1) (L) with Anions (A) Derived from 1,2-Benzenedicarboxylic (1,2-BDC), 1,3-Benzenedicarboxylic (1,3-BDC), cis,cis-1,3,5-Trimethyl-1,3,5-Benzenetricarboxylic (c,c-TMCTC), and cis,trans-1,3,5-Trimethyl-1,3,5-Benzenetricarboxylic (c,t-TMCTC) Acids Determined at 298 K in 0.15 mol dm^{-3} NaClO$_4$

Reaction	1,2-BDC	1,3-BDC	c,c-TMCTC	c,t-TMCTC
$A + L + 3H \rightleftharpoons AH_3L$[a,b]	30.95(2)	30.66(4)	31.05(7)	30.23(4)
$A + L + 4H \rightleftharpoons AH_4L$	38.58(2)	38.00(2)	39.32(4)	38.06(3)
$A + L + 5H \rightleftharpoons AH_5L$	44.03(2)	42.90(4)	46.46(5)	44.46(4)
$A + L + 6H \rightleftharpoons AH_6L$	48.20(2)	46.73(4)	52.83(4)	49.61(5)
$A + L + 7H \rightleftharpoons AH_7L$	51.42(4)	50.30(4)	57.63(4)	53.60(7)
$A + L + 8H \rightleftharpoons AH_8L$	53.64(3)	52.77(4)	61.23(5)	57.61(7)
$A + L + 9H \rightleftharpoons AH_9L$			64.31(5)	59.90(1)
$A + H_3L \rightleftharpoons AH_3L$	3.1	3.0	3.4	2.5
$A + H_4L \rightleftharpoons AH_4L$	4.5	3.9	5.2	4.0
$A + H_5L \rightleftharpoons AH_5L$	6.2	5.1	8.6	6.6
$A + H_6L \rightleftharpoons AH_6L$	8.2	6.8	12.9	9.7
$A + H_7L \rightleftharpoons AH_7L$	9.5	8.4	15.7	11.7

[a] Charges omitted for clarity.
[b] The stepwise protonation constants of **6.2** determined at 298 K in 0.15 mol dm^{-3} NaClO$_4$ are: log K_1 = 9.76, log K_2 = 9.28, log K_3 = 8.63, log K_4 = 6.42, log K_5 = 3.73, log K_6 = 2.13, log K_7 = 2.0. Taken from Ref. 208.

unless protonation degrees of the host and guest species at a given pH are exactly known, stepwise stability constants cannot be directly used to derive selectivity trends since the differences in basicity are not well contemplated in these constants, and thus the proper constants to be applied would be the conditional ones at a given pH.

Another system that highlights the usefulness of this type of selectivity analysis is the one formed by **6.1**, its methylated derivative Me$_4$[18]aneN$_6$ (**6.73**), and ATP. Methylation of **6.1** yields an important reduction in basicity,

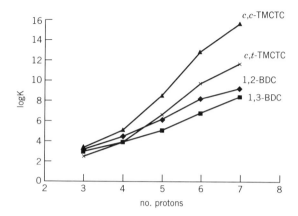

Fig. 6.15. Plot of the logarithms of the stepwise stability constants (log K_S; $A^{a-} + H_rL^{(r-a)+} \leftrightarrows (AH_rL)^{(r-a)+}$) for the interaction of anions derived from tri- and dicarboxylic acids c,c-TMCTC, c,t-TMCTC, 1,2-BDC, 1,3-BDC, and protonated forms of the macrocycle **6.2**.

and therefore also significantly alters its binding strength towards anions. The distribution diagram for the mixed system **6.1–6.73**–ATP (Fig. 6.17) calculated for equimolar amounts shows changes in selectivity with the pH.

ATP-Me$_4$[18]aneN$_6$ complexed species slightly predominate throughout the pH range 6.5–8.2, while for higher and lower values the reverse situation occurs. The pH values 6.5 and 8.2 correspond to points at which both receptors display

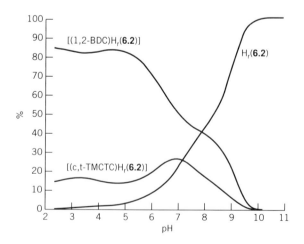

Fig. 6.16. Distribution diagram for mixed system (**6.2**)–1,2-BDC–c,t-TMCT. Overall formation percentages of complexed species are represented versus pH. The diagram has been calculated for 1×10^{-3} mol dm^{-3} initial concentration in all the reactants.

APPENDIX: METHODS FOR THE DETERMINATION OF THERMODYNAMIC PARAMETERS

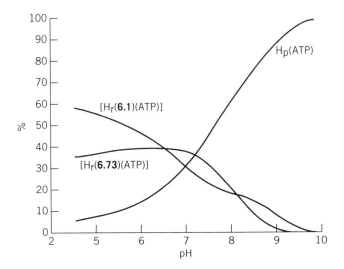

6.73

the same affinity for ATP, and, by analogy with the terminology adopted in other representations, they might be called isoselectivity points. This a clear example of how in many systems selectivity can be modulated by different factors, one of them being the pH.

6.7 APPENDIX: METHODS FOR THE DETERMINATION OF THERMODYNAMIC PARAMETERS

6.7.1 Introduction

Anion complexation leads, most of the time, to significant changes in some of the physical chemical properties of solutions. A crucial, but also initial point, in the investigation in solution of these systems is to identify which are the different

Fig. 6.17. Distribution diagram for mixed system **(6.1)**–**(6.73)**–ATP. Overall formation percentages of complexed species are represented versus pH. The diagram has been calculated for 1×10^{-3} mol dm^{-3} initial concentration in all the reactants.

species formed (speciation) as well as to determine their stability constants. To accomplish this, measurements of any physical chemical property whose magnitude significantly changes in the complexation event may be used. Therefore, in principle, the same general procedures employed for determining stability constants of metal ion complexes can be applied. These techniques have been extensively reviewed and described in several publications.[45,209-211] A point that deserves again to be emphasized is that the parameter being measured must fulfill the following series of conditions which are valid for any equilibrium study:[210]

(1) The equation correlating the extent of anion complexation and the parameter being measured must be known.

(2) The parameter has to change significantly over the experimental error.

(3) Measurements have to be performed in a concentration range where complex formation is significant but not complete.

A large amount of equilibrium data regarding anion complexation has been obtained in aqueous solution by potentiometry, and most concretely by pH-metry.[85,212] Thus, we will devote the major part of the following section to describe the capabilities and also the cautions to be taken when this technique is to be applied within the framework of anion binding. Another technique of high interest in this field is multinuclear NMR, since in some favorable cases it allows for determining the stability constants and additionally provides information at the molecular level, sites of interaction, conformation of receptor and guest species, etc. This is probably the most used technique in nonaqueous solution. Other different spectroscopic methods (IR, UV–vis) can also be used for going beyond the limits of potentiometry, which will be surveyed later. CV, polarography, and photochemistry are techniques that have also found room within the context of supramolecular chemistry and that may be used to obtain valuable information on stability data. The reader is referred to Chapters 7 and 8 for the foundations of these techniques and the information they can offer. Other methods of interest are based on extraction and ion-exchange techniques (see Chapter 2).

It should be noted that it is always advisable to analyze each system using more than one single technique. Different techniques can provide complementary information, and, on the other hand, this combined analysis may help to overcome the intrinsic difficulties associated with the frequently cumbersome systems found in anion recognition.

Finally, some words will be devoted to direct determination of heats of reaction. Breaking down ΔG° values into their ΔH° and ΔS° components provides a masterpiece of information necessary to interpret any kind of complexation event. In spite of this, not many ΔH° values have been reported,[85,212] and very often the enthalpy values published have been calculated through the dependence of the stability constants with temperature (Van't

Hoff equation), which, as is well known, may give results with high uncertainties. Therefore, obtaining accurate enthalpy terms by direct microcalorimetric measurements will surely constitute an aspect of research in anion coordination of increasing interest in the next years.

6.7.2 Methods

Potentiometry. The stability constants for a complexation reaction between an anion (A^{a-}) and a receptor (L^{p+}) in solution, as given by Eq. 10, can be readily calculated when the concentration at equilibrium of one of the intervening species is exactly known. This concentration is then related to the concentration of the other reactants and to the equilibrium constants through the mass balance equations.

$$pA^{a-} + qL^{b+} \rightleftharpoons (A_pL_q)^{(qb-pa)+} \tag{10}$$

This measurement can be made by means of electrodes that respond selectively to a determined reagent. Although the number of metal ion-selective electrodes available has grown considerably in recent years, the number of anion-selective electrodes is much smaller (see Chapter 10). However, in recent years several halide anion-selective electrodes have been successfully used for the determination of formation constants of some metal complexes.[213] Among them, the F^-- selective electrode is the one most widely used. Cl^-, Br^-, and I^- electrodes are also suitable in some instances for equilibrium studies. Nevertheless, experts usually advise of the convenience of coupling them, when possible, with a glass electrode in order to have measurements of both free concentrations of the anion under study and of hydrogen ions.[210]

As pointed out, there is little doubt that pH-metry is the technique most widely used for the determination of stability constants in aqueous solution. In order to apply this method, the consumption or liberation of protons in the complexation process has to be related to the concentration of complex species formed. This means than an additional reactant should be added to Eq. 10 to yield

$$pA^{a-} + qL^{b+} + rH^+ \rightleftharpoons (A_pH_rL_q)^{(qb+r-pa)+} \tag{11}$$

Usually ligands and anions themselves may also capture protons in a stepwise way, and therefore the following equilibria are to be considered

$$L + jH^+ \rightleftharpoons H_jL^{j+} \tag{12}$$

$$A^{a-} + wH^+ \rightleftharpoons (H_wA)^{(w-a)+} \tag{13}$$

Despite the apparent difficulties that the introduction of a new mass balance equation may yield, it contributes to a considerable gain in accuracy in the analysis since the concentration of hydrogen can be very precisely measured.

Nowadays, virtually all the pH-metric studies are carried out with the use of the glass electrode.[214] In the investigation of a particular anion–receptor system the usual strategy consists of determining, in first place, the basicity constants of the ligand and anion separately. Once these constants are exactly known, a solution in which all three reactants, anion, ligand, and protons, are present in known amounts is titrated with an alkali or a mineral acid, and the variation in the free hydrogen ion concentration is followed. Actually, these measurements are carried out most of the time in fully automated equipment controlled by personal computers. Finally, the constants relative to adduct formation are generally derived by computer analysis. To this purpose there are several very good, proven programs. For the interested reader, a more detailed description of the method can be found in different textbooks and reviews.[209–211,215]

Here, however, we will focus our discussion more on presenting the virtues and possible weaknesses of this method when used in anion coordination than to describe the method itself. In this respect it is interesting to note that most of the studies have been performed in water, since this medium is the one of greatest biological relevance. Another point to be noted is that excess background electrolyte should be added in the potentiometric titrations in order to keep the ionic strength constant. These considerations can provide, if disregarded, the main source of errors in the determination of stability constants.

All chemicals used, including the solvent itself, have to be appropriately purified in order to avoid protic impurities that may hamper the measurements. For instance, water used in the measurements and in the preparation of the solutions has to be distilled (or better doubly distilled), and it is advisable to cool it in an inert gas atmosphere in order to prevent CO_2 uptake. Titrant concentrations should be carefully determined, and, in this respect, there are several well-established analytical procedures. Even small amounts of impurities present in the background electrolyte may seriously affect the experiments due to its high concentration with respect to the other reactants.[216] Thus care has to be taken in checking the purity of the background electrolyte. With respect to the host and guest themselves, they have to be analytical-grade reagents, or, if synthetic, they must fulfill the usual quality controls. Obviously, these comments on purity should be applied independently of the method used for the determination of stability constants, even though sometimes the technique might not be so sensitive to impurities like potentiometry or microcalorimetry. Also of interest is that the glass electrode has to be properly calibrated for avoiding systematic errors in recording the concentration of hydrogen ions. A number of procedures have been described; among them Gran's method[217] is one of the most extended and useful.

The selection of the salt to be used as background electrolyte is of great importance. As the electrolyte itself contains anions prone to interact with the guest species, mainly taking into account that its concentration is in a large excess with respect to the host and guest species, stability constants derived from these studies are always conditional and refer to a particular concentration and to a given ionic medium. Therefore, changes in the salt used as ionic medium and in its concentration yield higher variations in the values of the stability constants than those usually reported for the formation of metal ion complexes. This observation has to be kept in mind when comparing stability data from distinct sources. In this sense, some authors have analyzed the influence of ionic strength changes in the stability constants of some systems.

Another challenge these systems offer is the computer treatment of the data and the consequent speciation and derivation of stability constants. Often the systems are complicated with the high number of overlapping equilibria involved. This leads to a high correlation between the different stability constants, and thus poor determination of one of them may considerably alter the constants of all the other species present in the model. Measurements for each system have to be repeated several times, and statistical treatments like the F-test[218] may be critical to selecting the right system.

As a consequence of this and, although potentiometry, when applicable, is by far the most accurate technique to obtain reliable information on stability constants, it is convenient to contrast the results using other independent techniques. For instance, CV, photochemistry (see Chapters 7 and 8), UV-Vis, and NMR have shown adequate applications in a number of systems.

Solid-state structures of anion complexes are a main target in any research. They always provide valuable information; however, within this field, due to the importance that electrostatic and weak intermolecular forces have in the stabilization of the adduct species (vide supra), very often the solution structures do not correspond exactly with those found in the solid state. Sometimes even changes in the stoichiometries of the adducts are observed. Nevertheless, the importance of solid-state structural information must be stressed, particularly in this research field, where the number of structures is still limited.

NMR. Nuclear magnetic resonance probably is, with potentiometry, the technique most widely used for the determination of anion binding constants.[219] On the other hand, NMR has surely provided the most detailed structural information in solution for these systems.

The experimentally measurable parameters in an NMR experiment (chemical shifts, band widths, coupling constants, etc.) depend on the NMR-sensitive nuclei and their chemical environments as well as on the exchange reactions rates occurring between them. Two situations can be differentiated as a function of the relationship between the time scale of the chemical exchange and that of the NMR experiment: slow and fast exchange.

In the former, separated groups of lines are obtained for the free and complexed species, while in the latter one averaged signals corresponding to the free and complexed species are found in the spectra. In the first case integration of the NMR signals would permit the amount of the different species present in equilibria to be deduced and, therefore to provide an estimate of the stoichiometries and stability constants.

In the second case the observed chemical shift, δ, for a particular nucleus is the average of the chemical shifts, δ_i, of that nucleus in the different species present, weighted according to their fractional populations α_i:

$$\delta = \sum_i \alpha_i \delta_i$$

where

$$\alpha_i = x_i C_i / T_X$$

T_X being the total concentration of the reagent containing the nucleus under consideration, x_i the stoichiometric coefficient of reagent X in the ith species, and C_i the concentration at equilibrium of such species. Since α_i depends on the equilibrium constants, δ is a function of δ_i, and the above expressions may be used to compute equilibrium constants.

Recently P. Gans, A. Sabatini, A. Vacca, et al. have reported on the new computer program HypNMR, which permits one to obtain stability constants in multiequilibria systems by using NMR data.[220] Also, in this paper, the authors provide detailed insights on the preparation of solutions, isotopic corrections, ionic strength influence, etc. Another computer program recently reported that can be applied to the fast-exchange situation is EQNMR,[221] although it seems that so far it has just been tested with computer-generated sets of data.

Optical Methods. Very often anion coordination is accompanied by changes in the electromagnetic spectra of the anion and/or receptor in the 10^{-8}–10^{-4} m wavelength region (UV, vis, IR).

All these techniques relate the spectral properties of solutions with the concentration of optically active species present in the equilibrium through the Beer–Lambert–Bouguer law

$$A_p = \sum_i \varepsilon_{ip} l C_i$$

where A_p and ε_{ip} are the absorbance of the solution and the molar absorption coefficient of the ith species at the wavelength denoted by the index p, C_i is the concentration of the ith species, and l is the optical path.

In this way absorption data may be related to concentrations and equilibrium constants through the mass balance equations.

Spectrophotometric methods can be of great help in complementing the other methods of quantifying anion complexation and may prove superior in some studies since they offer a range of advantages: (1) The interaction need not be coupled to a deprotonation equilibrium; (2) investigations at low and high pH are possible; (3) no restriction to aqueous solution applies; (4) structural information about the adducts may be gained from the absorption spectra of the adducts.

There exist different computer programs that deal with spectrophotometric data alone or that may treat conjointly sets of spectrophotometric and potentiometric data.[211,215,216] Factor analysis and evolving factor analysis have recently been revealed as important tools for data treatment, since, in principle, they do not require a previous knowledge of the model system.[222]

Very often, the formation of anion complexes yields very small changes in the spectral properties. Such a situation is found, for example, in the formation of outer-sphere anion complexes between polycyanometallate anions and polyammonium receptors in which spectral changes result only from changes in the solvation shells. However, if enough care on the reproducibility of the results is taken and an appropriate informatic package for the data analysis is available, satisfactory results may be achieved.[223]

Microcalorimetry. There are two ways in which the enthalpy change accompanying a chemical process can be measured.[224] One is to measure directly, by means of a calorimeter, the amount of heat involved in the reaction. The other way consists of determining the equilibrium constants at various temperatures, and applying the Van't Hoff isochore to derive the value of ΔH°. Once ΔH° is known, the entropy term ΔS° can be obtained from the relationship $\Delta G^\circ = \Delta H^\circ - T \Delta S^\circ$. The standard free energy change is related to the equilibrium constant by the expression $\Delta G^\circ = -RT \ln K_S$.

From these two methods, the second, as indicated above, may lead to high uncertainties if not used with great care. Since there is a logarithmic correlation between stability constants and ΔH°, the propagation of the experimental errors in the determination of equilibrium constants on the enthalpy values is exponential

$$\delta(R \log K_S)/\delta(1/T) = -\Delta H^\circ$$

and

$$\log K_S = -\Delta H^\circ/2.303 RT + \text{const}$$

Therefore, the temperature range explored should be as wide as possible, and the equilibrium constants used have to be of great accuracy. However, the temperature range of study is limited by the experimental conditions, solvent boiling and freezing point, etc; furthermore, the assumed constancy of ΔH° with temperature is less likely to occur the larger the temperature range used.

Thus the dependence of ΔC_p on temperature should be known to correct the enthalpy for its temperature variation, $\delta(\Delta H°)/\delta T = \Delta C_p$.

The direct method is, by far, much more accurate provided adequate instrumentation (calorimeter) is used. A calorimeter is an instrument that allows an unknown amount of heat to be determined provided a reference amount of heat is known. The heat involved in a chemical process is usually measured by recording temperature variations. Different microcalorimetic devices have been designed in order to cover fast and relatively slow reactions.

As pointed out above, the information derived from calorimetric studies is of great help to evaluate properly the factors contributing to anion complex formation.

6.8 REFERENCES

1. *Interaction in Ionic Solution*, Discussions of the Faraday Society, No. 24, The Aberdeen University Press, Aberdeen, 1958.
2. C. W. Davies, *Ion Association*, Butterworths, London, 1962.
3. J. Burgess, *Ions in Solution: Basic Principles of Chemical Interactions*, Ellis Horwood, Chichester, 1988.
4. S. A. Rice, *J. Chem. Phys.* **15**, 875 (1947).
5. G. Schwarzenbach, *Pure and Appl. Chem.* **XX**, 306 (1970).
6. G. Schwarzenbach, in *Proceedings of the Summer School on Stability Constants*, P. Paoletti, R. Barbucci, and L. Fabbrizzi, Eds., Editore Scuola Universitaria, Florence, 1974, pp. 151–81.
7. K. Bjerrum, *Danske Vidensk Selsk Mat. fys Medd.* **7**, no. 9 (1926).
8. J. T. Denison and J. B. Ramsey, *J. Am. Chem. Soc.* **77**, 2615 (1955).
9. W. R. Gilkerson, *J. Chem. Phys.* **25**, 1199 (1956).
10. W. R. Gilkerson, *J. Chem. Phys.* **74**, 746 (1970).
11. R. M. Fuoss, *J. Am. Chem. Soc.* **80**, 5059 (1958).
12. W. Ebeling, *Z. Physik. Chem. (Leipzig)* **238**, 400 (1968).
13. H. Yokoyama and H. Yamatera, *Bull. Chem. Soc. Jpn.* **48**, 1770 (1975).
14. *Handbook of Chemistry and Physics*, 67th Edition, R. C. Weast, Ed., CRC Press, Boca Raton, Florida, 1987.
15. Landolt–Bornstein, *Atoms and Ions*. Vol. 1, Part 1, Springer-Verlag, Berlin, 1950.
16. J. Appequist, J. R. Carl, and K.-K. Fung, *J. Am. Chem. Soc.* **94**, 2953 (1972).
17. V. Gutmann and E. Wychera, *Inorg. Nucl. Chem. Lett.* **2**, 257 (1966).
18. V. Gutmann, *The Donor–Acceptor Approach to Molecular Interactions*, Plenum, New York, 1978.
19. U. Mayer, V. Gutmann, and W. Gerger, *Mh. Chem.* **106**, 1235 (1975).
20. V. Gutmann, *Electrochim. Acta.* **21**, 661 (1976).
21. C. Reichardt, *Solvents and Solvent Effects in Organic Chemistry*, Verlag Chemie, Weinheim, Basel, 1988.

22. A. J. Parker, *Chem. Rev.* **69**, 1 (1969).
23. H. F. Franks and M. W. Evans, *J. Chem. Phys.* **13**, 507 (1945).
24. H. F. Franks and W. Y. Wen, *Disc. Faraday Soc.* **24**, 133 (1957).
25. D. Eisemberg and W. Kauzmann. *The Structure and Properties of Water*, Oxford University Press, London, 1969.
26. H. F. Franks and D. J. G. Ives, *Quart. Rev.* **20**, 1 (1966).
27. W. G. Breck and J. Lin, *Trans. Faraday Soc.* **61**, 223 (1965).
28. Y. Marcus and A. S. Kertes. *Ion Exchange and Solvent Extraction of Metal Complexes*, Wiley, London, 1969.
29. A. Ben-Naim, *J. Phys. Chem.* **69**, 1922 (1965); A. Ben-Naim, *Hydrophobic Interactions*, Plenum, New York, 1980.
30. W. P. A. Luck, *Topics in Current Chem.* **64**, 115 (1976).
31. I. Tabushi, Y. Kiyosuke, T. Sugimoto, and K. Yamamura, *J. Am. Chem. Soc.* **100**, 916 (1978).
32. I. Tabushi and T. Mizutani, *Tetrahedron* **43**, 1439 (1987).
33. A comprehensive discussion of enthalpically controlled tight molecular complexation in water is given in: F. Diederich, *Cyclophanes*, The Royal Society of Chemistry, Cambridge, 1991.
34. E. Grunwald, *Anal. Chem.* **26**, 1696 (1954).
35. S. Winstein, E. Clippinger, A. H. Fainberg, and G. C. Robinson, *J. Am. Chem. Soc.* **76**, 2597 (1954).
36. R. M. Fuoss and H. Sadek, *J. Am. Chem. Soc.* **76**, 5897 (1954).
37. R. M. Fuoss and H. Sadek, *J. Am. Chem. Soc.* **76**, 5905 (1954).
38. T. C. Pochapsky and P. M. Stone, *J. Am. Chem. Soc.* **112**, 6714 (1990).
39. T. C. Pochapsky, P. M. Stone, and S. Pochapsky, *J. Am. Chem. Soc.* **113**, 1460 (1991).
40. M. K. Begum and E. Grunwald, *J. Am. Chem. Soc.* **112**, 5104 (1990).
41. A. P. Abbott and D. J. Schiffrin, *J. Chem. Soc. Faraday Trans.* **86**, 1449 (1990).
42. A. P. Abbott and D. J. Schiffrin, *J. Chem. Soc. Faraday Trans.* **86**, 1453 (1990).
43. X. Yang, A. Zaitsev, B. Sauerwein, S. Murphy, and G. B. Schuster, *J. Am. Chem. Soc.* **114**, 793 (1992).
44. B. Dietrich, J. Guilhem, J.-M. Lehn, C. Pascard, and E. Sonveaux, *Helv. Chim. Acta* **67**, 91 (1984).
45. B. G. Cox and H. Schneider, *Coordination and Transport of Macrocyclic Compounds in Solution*, Elsevier, Amsterdam, 1992.
46. M. Arshadi, R. Yamdagni, and P. Kebarle, *J. Phys. Chem.* **74**, 1475 (1970).
47. M. Arshadi and P. Kebarle, *J. Phys. Chem.* **74**, 1483 (1970).
48. M. Johnsson and I. Persson, *Inorg. Chim. Acta* **127**, 15 (1987).
49. B. G. Cox and W. E. Waghorne, *Chem. Soc. Rev.*, **9**, 381 (1980).
50. E. Högfeldt. *Stability Constants of Metal–Ion Complexes. Part A: Inorganic Ligands*, Pergamon, Oxford, 1983.
51. *Stability Constant Database*, IUPAC and Academic Software, 1993.
52. R. L. Kay, C. Zawoyski, and D. F. Evans, *J. Phys. Chem.* **69**, 4208 (1965).

53. D. F. Evans and P. Gardam, *J. Phys. Chem.* **72**, 3280 (1968).
54. D.F. Evans and P. Gardam, *J. Phys. Chem.* **73**, 158 (1969).
55. D. F. Evans, J. Thomas, J. A. Nadas, and S. M. A. Matesich, *J. Phys. Chem.* **75**, 1714 (1971).
56. (a) H.-J. Schneider, *Angew. Chem. Int. Ed. Engl.* **30**, 1417 (1991); (b) H.-J. Schneider, T. Schiestel, and P. Zimmermann, *J. Am. Chem. Soc.* **114**, 7698 (1992).
57. S.-C. Tam and R. J. P. Williams, *Structure and Bonding, Berlin* **63**, 102 (1985).
58. A. Bencini, A. Bianchi, M. Micheloni, P. Paoletti, E. García-España, and M. A. Nino, *J. Chem. Soc., Dalton Trans.*, 1171 (1991).
59. A. Bianchi, M. Micheloni, and P. Paoletti, *Inorg. Chem.* **24**, 3702 (1985).
60. A. Bencini, A. Bianchi, P. Dapporto, E. García-España, M. Micheloni, and P. Paoletti, *Inorg. Chem.* **28**, 1188 (1989).
61. A. Bianchi, S. Mangani, M. Micheloni, V. Nannini, P. Orioli, P. Paoletti, and B. Seghi, *Inorg. Chem.* **24**, 1182 (1985).
62. A. Bencini, A. Bianchi, E. García-España, M. Giusti, M. Micheloni, and P. Paoletti, *Inorg. Chem.* **26**, 681 (1987).
63. A. Bencini, A. Bianchi, E. García-España, M. Giusti, S. Mangani, M. Micheloni, P. Orioli, and P. Paoletti, *Inorg. Chem.* **26**, 1243 (1987).
64. A. Bencini, A. Bianchi, E. García-España, M. Micheloni, and P. Paoletti, *Inorg. Chem.* **27**, 176 (1988).
65. B. Dietrich, M. W. Hosseini, J.-M. Lehn, and R. B. Sessions, *J. Am. Chem. Soc.* **103**, 1282 (1981).
66. B. Dietrich, M. W. Hosseini, J.-M. Lehn, and R. B. Sessions, *Helv. Chim. Acta* **66**, 1262 (1983).
67. M. W. Hosseini and J.-M. Lehn, *Helv. Chim. Acta* **69**, 587 (1986).
68. S. D. Reilly, G. R. K. Khalsa, D. K. Ford, J. R. Brainard, B. P. Hay, and P. H. Smith, *Inorg. Chem.* **34**, 569 (1995).
69. M. W. Hosseini and J.-M. Lehn, *Helv. Chim. Acta* **71**, 749 (1988).
70. T. N. Margulis and L. J. Zompa, *Acta Crystallogr., Sect. B* **337**, 1426 (1981).
71. J. Cullinane, R. I. Gelb, T. N. Margulis, and L. J. Zompa, *J. Am. Chem. Soc.* **104**, 3048 (1982).
72. A. Bencini, A. Bianchi, E. García-España, E. C. Scott, L. Morales, B. Wang, T. Deffo, F. Takusagawa, M. P. Mertes, K. B. Mertes, and P. Paoletti, *Bioorg. Chem.* **20**, 8 (1992).
73. A. Bianchi, E. García-España, S. Mangani, M. Micheloni, P. Orioli, and P. Paoletti, *J. Chem. Soc., Chem. Commun.*, 729 (1987).
74. A. Bencini, A. Bianchi, E. García-España, M. Giusti, S. Mangani, M. Micheloni, P. Orioli, and P. Paoletti, *Inorg. Chem.* **26**, 3902 (1987).
75. A. Bencini, A. Bianchi, P. Dapporto, E. García-España, M. Micheloni, P. Paoletti, and P. Paoli, *J. Chem. Soc., Chem. Commun.*, 753 (1990).
76. A. Bencini, A. Bianchi, M. Micheloni, P. Paoletti, P. Dapporto, P. Paoli, and E. García-España, *J. Inclusion Phen. Mol. Recognition in Chem.* **12**, 291 (1992).
77. A. Bencini, A. Bianchi, P. Dapporto, E. García-España, M. Micheloni, J. A. Ramírez, P. Paoletti, and P. Paoli, *Inorg. Chem.* **31**, 1902 (1992).

78. B. Dietrich, J.-M. Lehn, J. Guilhem, and C. Pascard, *Tetrahedron Lett.* **30**, 4125 (1989).
79. H. E. Simmons and C. H. Park, *J. Am. Chem. Soc.* **90**, 2428 (1968).
80. C. H. Park and H. E. Simmons, *J. Am. Chem. Soc.* **90**, 2429 (1968).
81. R. A. Bell, G. G. Christoph, F. R. Fronczck, and R. E. Marsh, *Science* **190**, 151 (1975).
82. C. Nave and M. R. Truter, *J. Chem. Soc.*, 2351 (1974).
83. M. Studer, A. Riesen, and T. A. Kaden, *Helv. Chim. Acta* **72**, 1253 (1989).
84. M. Micheloni, P. Paoletti, and A. Vacca, *J. Chem. Soc., Perkin Trans. II*, 945 (1978).
85. A. Bianchi, M. Micheloni, and P. Paoletti, *Coord. Chem. Rev.* **110**, 17 (1991).
86. F. P. Schmidtchen, *Angew. Chem. Int. Ed. Eng.* **16**, 720 (1977).
87. F. P. Schmidtchen, *Chem. Ber.* **114**, 597 (1981).
88. F. P. Schmidtchen and G. Müller, *J. Chem. Soc., Chem. Commun.*, 1115 (1984).
89. K. Worm and F. P. Schmidtchen, *Angew. Chem. Int. Ed. Engl.* **34**, 65 (1995).
90. S.-C. Tam and R. J. P. Williams, *J. Chem. Soc., Faraday Trans.* **80**, 2255 (1984).
91. M. W. Hosseini and J.-M. Lehn, *J. Am. Chem. Soc.* **104**, 3525 (1982).
92. M. Dhaenens, J.-M. Lehn, and J.-P. Vigneron, *J. Chem. Soc., Perkin Trans. 2*, 1379 (1993).
93. J.-M. Lehn, R. Méric, J.-P. Vigneron, I. Bkouche-Waksman, and C. Pascard, *J. Chem. Soc., Chem. Commun.*, 62 (1991).
94. F. P. Schmidtchen, *J. Am. Chem. Soc.* **108**, 8249 (1986).
95. M. F. Manfrin, L. Moggi, V. Castelvetro, V. Balzani, M. W. Hosseini, and J.-M. Lehn, *J. Am. Chem. Soc.* **107**, 6888 (1985).
96. K. B. Mertes and J. M. Lehn. "Multidentate Macrocyclic and Macropolicyclic Ligand," in *Comprehensive Coordination Chemistry*, Pergamon Press, Oxford, 1987.
97. E. García-España, M. Micheloni, P. Paoletti, and A. Bianchi, *Inorg. Chim. Acta* **102**, L9 (1985).
98. J. Aragó, A. Bencini, A. Bianchi, A. Doménech, and E. García-España, *J. Chem. Soc., Dalton Trans.*, 319 (1992).
99. A. Bencini, A. Bianchi, P. Paoletti, and P. Paoli, *Pure and Appl. Chem.* **65**, 381 (1993).
100. F. Pina, L. Moggi, M. F. Manfrin, V. Balzani, M. W. Hosseinim, and J.-M. Lehn, *Gazz. Chim. Ital.* **119**, 65 (1989).
101. H.-J. Schneider and I. Theis, *Angew. Chem. Int. Ed. Eng.* **28**, 753 (1989).
102. P. B. Savage, S. K. Holmgren, and S. H. Gellman, *J. Am. Chem. Soc.* **116**, 4069 (1994).
103. P. B. Savage, S. K. Holmgren, and S. H. Gellman, *J. Am. Chem. Soc.* **115**, 7900 (1993).
104. K. Worm, F. P. Schmidtchen, A. Schier, A. Schäfer, and M. Hesse, *Angew. Chem. Int. Ed. Engl.* **33**, 327 (1994).
105. H.-J. Schneider, T. Blatter, S. Simova, and I. Theis, *J. Chem. Soc., Chem. Commun.*, 580 (1989).

106. W. B. Farnham, D. C. Ror, D. A. Dixon, J. C. Calabrese, and R. L. Harlow, *J. Am. Chem. Soc.* **112**, 7707 (1990); T.-Y. Lin, W.-H. Lin, W. D. Clark, R. J. Lagow, S. B. Larson, S. H. Simonsen, V. M. Lynch, J. S. Brodbelt, S. D. Maleknia, and C.-C. Liou, *J. Am. Chem. Soc.* **116**, 5172 (1994).

107. S. Scheiner and T. Kar, *J. Am. Chem. Soc.* **117**, 6970 (1995).

108. J. Emsley, *Chem. Soc. Rev.*, 91 (1980) and references therein.

109. P. A. Frey, S. A. Whitt, and J. B. Tobin, *Science* **264**, 1927 (1994); P. A. Frey, *Science* **269**, 104 (1995).

110. W. W. Cleland and M. M. Kreevoy, *Science* **264**, 1887 (1994); ibid. **269**, 103 (1995).

111. A. Warshel, A. Papazyan, and P. A. Kollman, *Science* **269**, 102 (1995).

112. J. H. Clark, J. Emsley, D. J. Jones, and R. E. Overill, *J. Chem. Soc., Dalton Trans.*, 1219 (1981) and references therein.

113. R. M. Smith and A. E. Martell, *Critical Stability Constants Database*, NIST project, 1993.

114. J. Marco, J. M. Orza, R. Notario, and J.-L. M. Abboud, *J. Am. Chem. Soc.* **116**, 8841 (1994) and references therein.

115. W. L. Jorgensen, *J. Am. Chem. Soc.* **111**, 3770 (1989) and Ref. 3 therein.

116. H. Luecke and F. A. Quiocho, *Nature* **347**, 402 (1990); J. J. He and F. A. Quiocho, *Science* **251**, 1479 (1991), and references therein.

117. M. K. Gilson and B. H. Honig, *Biopolymers* **25**, 2097 (1986).

118. R. Taylor, O. Kennard, and W. Versichel, *J. Am. Chem. Soc.* **105**, 5761 (1983).

119. G. A. Jeffrey and S. Takagi, *Acc. Chem. Res.* **11**, 264 (1978); P. Schuster, in *The Hydrogen Bond*, P. Schuster, G. Zundel, C. Sandorf Eds., Elsevier, Amsterdam, 1976, Vol. 1, Chapter 2; C. Ceccarelli, G. A. Jeffrey, and R. Taylor, *J. Mol. Struct.* **70**, 255 (1981); P. A. Kollman, in *Modern Theoretical Chemistry*, H. F. Schaefer, Ed., Plenum, New York, 1977, Vol. 4, Chapter 3.

120. R. Taylor and O. Kennard, *Acc. Chem. Res.* **17**, 320 (1984).

121. I. Olovsson, *Croat. Chem. Acta* **55**, 171 (1982).

122. D. J. Millen, *Croat. Chem. Acta* **55**, 133 (1982).

123. E. Fan, S. A. Van Arman, S. Kincaid, and A. D. Hamilton, *J. Am. Chem. Soc.* **115**, 369 (1993).

124. R. P. Dixon, S. J. Geib, and A. D. Hamilton, *J. Am. Chem. Soc.* **114**, 365 (1992).

125. C. Bazzicalupi, A. Bencini, A. Bianchi, V. Fusi, C. Giorgi, P. Paoletti, A. Stefani, and B. Valtancoli, *J. Chem. Soc. Perkin Trans. 2*, 275 (1995); C. Bazzicalupi, A. Bencini, A. Bianchi, V. Fusi, C. Giorgi, A. Granchi, P. Paoletti, and Barbara Valtancoli, *J. Chem. Soc., Perkin Trans. 2*, 775 (1997).

126. S. Valiyaveettil, J. F. J. Engbersen, W. Verboom, and D. N. Reinhoudt, *Angew. Chem. Int. Ed. Engl.* **32**, 900 (1993).

127. Y. Morzherin, D. M. Rudkevich, W. Verboom, and D. N. Reinhoudt, *J. Org. Chem.* **58**, 7602 (1993).

128. J. Scheerder, M. Fochi, J. F. J. Engbersen, and D. N. Reihoudt, *J. Org. Chem.* **59**, 7815 (1994).

129. J. S. Albert and A. D. Hamilton, *Tetrahedron Lett.* **34**, 7363 (1993).

130. B. C. Hamann, N. R. Branda, and J. Rebek, Jr., *Tetrahedron Lett.* **34**, 6837 (1993).

131. A. Bianchi, M. Micheloni, and P. Paoletti, *Inorg. Chim. Acta* **151**, 269 (1988).

132. A. Bianchi, M. Micheloni, P. Orioli, P. Paoletti, and S. Mangani, *Inorg. Chim. Acta* **146**, 153 (1988).

133. E. Kimura and A. Sakonaka, *J. Am. Chem. Soc.* **104**, 4984 (1982).

134. R. J. Motekaitis, A. E. Martell, B. Dietrich, and J.-M. Lehn, *Inorg. Chem.* **23**, 1588 (1984).

135. R. I. Gelb, B. T. Lee, and L. J. Zompa, *J. Am. Chem. Soc.* **107**, 909 (1985).

136. R. I. Gelb, L. M. Schwartz, and L. J. Zompa, *Inorg. Chem.* **25**, 1527 (1986).

137. R. J. Motekaitis, A. E. Martell, and I. Murase, *Inorg. Chem.* **25**, 938 (1986).

138. M. W. Hosseini, J.-P. Kintzinger, J.-M. Lehn, and A. Zahidi, *Helv. Chim. Acta* **2**, 1078 (1989).

139. J. S. Alper, R. I. Gelb, and M. H. Schwartz, *J. Incl. Phenom. Mol. Recognition in Chem.* **11**, 333 (1991); *Ibid.* **11**, 349 (1991).

140. R. I. Gelb, J. S. Alper, and M. H. Schwartz, *J. Phys. Org. Chem.* **5**, 443 (1992).

141. M. Shionoya, H. Furuta, V. Lynch, A. Harriman, and J. L. Sessler, *J. Am. Chem. Soc.* **114**, 5714 (1993).

142. J. L. Sessler, M. Cyr, H. Furuta, V. Král, T. Mody, T. Morishima, M. Shionoya, and S. Weghorn, *Pure Appl. Chem.* **65**, 393 (1993) and references therein.

143. E. Kimura, A. Sakonaka, T. Yatsunami, and M. Kodama, *J. Am. Chem. Soc.* **103**, 3041 (1981).

144. T. Yatsunami, A. Sakonaka, and E. Kimura, *Anal. Chem.* **53**, 477 (1981).

145. A. Echavarren, A. Galán, J.-M. Lehn, and J. de Mendoza, *J. Am. Chem. Soc.* **111**, 4994 (1989).

146. E. Kimura, Y. Kuramoto, T. Koike, H. Fujioka, and M. Kodama, *J. Org. Chem.* **55**, 42 (1990).

147. J. Suh, Y. Cho, and K. J. Lee, *J. Am. Chem. Soc.* **113**, 4198 (1991).

148. A. Bencini, A. Bianchi, M. I. Burguete, E. García-España, S. V. Luis, and J. A. Ramírez, *J. Am. Chem. Soc.* **114**, 1919 (1992).

149. J. S. Albert and A. D. Hamilton, *Tetrahedron Lett.* **34**, 7363 (1993).

150. S. Claude, J.-M. Lehn, F. Schmidt, and J.-P. Vigneron, *J. Chem. Soc., Chem. Commun.*, 1182 (1991).

151. P. D. Beer, M. G. B. Drew, C. Hazlewood, D. Hesek, J. Hodacava, and S. E. Stokes, *J. Chem. Soc., Chem. Commun.*, 229 (1993).

152. A. Bencini, A. Bianchi, M. I. Burguete, P. Dapporto, A. Doménech, E. García-España, S. V. Luis, P. Paoli, and J. A. Ramírez, *J. Chem. Soc., Perkin Trans. 2*, 569 (1994).

153. Q. Lu, R. J. Motekaitis, J. J. Reibenspies, and A. E. Martell, *Inorg. Chem.* **34**, 4958 (1995).

154. I. Tabushi, J. Imuta, N. Seko, and Y. Kobuke, *J. Am. Chem. Soc.* **100**, 6287 (1978).

155. B. Dietrich, T. M. Fyles, J.-M. Lehn, L. G. Pease, and D. L. Fyles, *J. Chem. Soc., Chem. Commun.*, 934 (1978).

156. M. W. Hosseini and J.-M. Lehn, *Helv. Chim. Acta* **70**, 1312 (1987).

157. M. W. Hosseini, A. J. Blacker, and J.-M. Lehn, *J. Am. Chem. Soc.* **112**, 3896 (1990).

158. H. Furuka, D. Magda, and J. L. Sessler, *J. Am. Chem. Soc.* **113**, 978 (1991).
159. D. Y. Sasaki, K. Kurihara, and T. Kunitake, *J. Am. Chem. Soc.* **113**, 9685 (1991).
160. V. Jubian, R. P. Dixon, and A. D. Hamilton, *J. Am. Chem. Soc.* **114**, 1120 (1992).
161. H.-J. Schneider, T. Blatter, B. Palm, U. Pfingstag, V. Rüdiger, and I. Theis, *J. Am. Chem. Soc.* **114**, 7704 (1992).
162. A. V. Eliseev and H.-J. Schneider, *Angew. Chem. Int. Ed. Engl.* **32**, 1331 (1993).
163. B. L. Iverson, K. Shreder, V. Král, and J. L. Sessler, *J. Am. Chem. Soc.* **115**, 11022 (1993).
164. R. Motekaitis and A. E. Martell, *Inor. Chem.* **33**, 1032 (1994).
165. A. V. Eliseev and H.-J. Schneider, *J. Am. Chem. Soc.* **116**, 6081 (1994).
166. J. A. Aguilar, E. García-España, J. A. Guerrero, S. V. Luis, J. M. Linares, J. F. Miravet, J. A. Ramírez, and C. Soriano, *J. Chem. Soc., Chem. Commun.*, 2237 (1995).
167. A. Andrés, J. Aragó, A. Bencini, A. Bianchi, A. Doménech, V. Fusi, E. García-España, P. Paoletti, and J. A. Ramírez, *Inorg. Chem.* **32**, 3418 (1993).
168. A. Bianchi, A. Doménech, E. García-España, and S. V. Luis, *Anal. Chem.* **65**, 3137 (1993).
169. A. Andrés, C. Bazzicalupi, A. Bencini, A. Bianchi, V. Fusi, E. García-España, C. Giorgi, N. Nardi, P. Paoletti, J. A. Ramírez, and B. Valtancoli, *J. Chem. Soc. Perkin Trans. 2*, 2367 (1994).
170. P. Cudic, M. Zinic, V. Tomisic, V. Simeon, J.-P. Vigneron, and J.-M. Lehn, *J. Chem. Soc., Chem. Commun.*, 1073 (1995).
171. F. Peter, M. Gross, M. W. Hosseini, J.-M. Lehn, and R. B. Sessions, *J. Chem. Soc., Chem. Commun.*, 1067 (1981); F. Peter, M. Gross, M. W. Hosseini, and J.-M. Lehn, *J. Electroanal. Chem.* **144**, 279 (1983).
172. A. Bianchi, M. Micheloni, P. Orioli, P. Paoletti, and S. Mangani, *Inorg. Chim. Acta* **146**, 153 (1988).
173. A. Bianchi, M. Micheloni, and P. Paoletti, *Pure Appl. Chem.* **60**, 525 (1988).
174. M. A. Bernardo, A. J. Parola, F. Pina, E. García-España, V. Marcelino, S. V. Luis, and J. F. Miravet, *J. Chem. Soc. Dalton Trans.*, 993 (1995).
175. C. Tanford, *The Hydrophobic Effect: Formation of Micelles and Biological Membranes*, 2nd ed., Wiley, New York, 1980.
176. J. L. Cohen and K. A. Connors, *J. Pharm. Sci.* **59**, 1271 (1970); J. L. Connors and S. Sun, *J. Am. Chem. Soc.* **93**, 7239 (1971); J. L. Connors, M. H. Infeld, and B. J. Kline, *J. Am. Chem. Soc.* **91**, 3597 and 3697 (1971).
177. (a) Y. Inoue, T. Hakushi, Y. Liu, L.-H. Tong, B.-J. Shen, and D.-S. Jin, *J. Am. Chem. Soc.* **115**, 475 (1993); (b) references therein.
178. R. Foster, *Organic Charge-Transfer Complexes*, Academic Press, London, 1969.
179. Y. Azuma and M. Newcomb, *Organometallics* **3**, 9 (1984).
180. M. Newcomb, A. M. Madonik, M. T. Blanda, and J. K. Judice, *Organometallics* **6**, 145 (1987).
181. M. Newcomb and M. T. Blanda, *Tetrahedron Lett.* **29**, 4261 (1988).

182. M. Newcomb, J. H. Horner, M. T. Blanda, and P. J. Squattrito, *J. Am. Chem. Soc.* **111**, 6294 (1989).
183. K. Jurkschat, H. G. Kuivila, S. Liu, and J. A. Zubieta, *Organometallics* **8**, 2755 (1989).
184. M. T. Blanda and M. Newcomb, *Tetrahedron Lett.* **30**, 3501 (1989).
185. M. T. Blanda, J. H. Horner, and M. Newcomb, *J. Org. Chem.* **54**, 4626 (1989).
186. K. Jurkschat, A. Rühlemann, and A. Tzschach, *J. Organomet. Chem.* **381**, C53 (1990).
187. K. Fluri, J. Koudelka, and W. Simon, *Helv. Chim. Acta* **75**, 1012 (1992).
188. N. A. Chaniotakis, K. Jurkschat, and A. Rühlemann, *Anal. Chim. Acta* **282**, 345 (1993).
189. J. K. Tsagatakis and N. A. Chaniotakis, *Helv. Chim. Acta* **77**, 2191 (1994).
190. A. L. Beauchamp, M. J. Olivier, J. D. Wuest, and B. Zacharie, *J. Am. Chem. Soc.* **108**, 73 (1986).
191. V. B. Shur, I. A. Tikhonova, A. I. Yanovsky, Y. T. Struchkov, and P. V. Petrovskii, *J. Organomet. Chem.* **418**, C29 (1991).
192. X. Yang, C. B. Knobler, and M. F. Hawthorne, *Angew. Chem. Int. Ed. Engl.* **30**, 1507 (1991).
193. X. Yang, C. B. Knobler, and M. F. Hawthorne, *J. Am. Chem. Soc.* **114**, 380 (1992).
194. X. Yang, S. E. Johnson, S. I. Khan, and M. F. Hawthorne, *Angew. Chem. Int. Ed. Engl.* **31**, 893 (1992).
195. Z. Zheng, X. Yang, C. B. Knobler, and M. F. Hawthorne, *J. Am. Chem. Soc.* **115**, 5320 (1993).
196. V. B. Shur, I. A. Tikhonova, F. M. Dolgushin, A. I. Yanovsky, Yu. T. Struchkov, A. Yu. Volkonsky, E. V. Solodova, S. Yu. Panov, P. V. Petrovskii, and M. E. Vol'pin, *J. Organomet. Chem.* **443**, C19 (1993).
197. X. Yang, C. B. Knobler, Z. Zheng, and M. F. Hawthorne, *J. Am. Chem. Soc.* **116**, 7142 (1994).
198. Z. Zheng, C. B. Knobler, and M. F. Hawthorne, *J. Am. Chem. Soc.* **117**, 5105 (1995).
199. H. E. Katz, *J. Am. Chem. Soc.* **108**, 7640 (1986).
200. M. E. Jung and H. Xia, *Tetrahedron Lett.* **29**, 297 (1988).
201. H. E. Katz, *J. Org. Chem.* **50**, 5027 (1985).
202. M. T. Reetz, C. M. Niemeyer, and K. Harms, *Angew. Chem. Int. Ed. Engl.* **30**, 1472 (1991).
203. M. T. Reetz, C. M. Niemeyer, and K. Harms, *Angew. Chem. Int. Ed. Engl.* **30**, 1474 (1991).
204. S. Jacobson and R. Pizer, *J. Am. Chem. Soc.* **115**, 11216 (1993).
205. M. T. Reetz, B. M. Johnson, and K. Harms, *Tetrahedron Lett.* **35**, 2525 (1994).
206. S. Aoyagi, K. Tanaka, and Y. Takeuchi, *J. Chem. Soc. Perkin Trans. 2*, 1549 (1994).
207. B. Dietrich, J. Guilhem, J.-M. Lehn, C. Pascard, and E. Souveaux, *Helv. Chim. Acta* **67**, 91 (1984).

208. A. Bencini, A. Bianchi, P. Dapporto, E. García-España, M. Micheloni, and P. Paoletti, *Inorg. Chem.* **28**, 1188 (1989).

209. A. E. Martell and R. J. Moteikatis, *Determination and Use of Stability Complexes*, VCH, 1988; A. Braibanti, G. Ostacoli, P. Paoletti, L. D. Pettit, and S. Sanmartano, *Pure. Appl. Chem.* **59**, 1721 (1987); P. W. Linder, in *Handbook of Metal–Ligand Interactions in Biological Fluids*, G. Berthon, Ed., Marcel Dekker, New York—Basel—Hong Kong, 1995; F. R. Harley, C. Burgess, and R. M. Alcock, *Solution Equilibria*, Ellis Horwood Ltd, Chichester, 1980; A. J. Bard, and L. R. Faulkner, *Electrochemical Methods*, Wiley, New York, 1980; H. Rossotti, *The Study of Ionic Equilibria*, Longman, London and New York, 1978.

210. M. T. Beck and I. Nagypal, *Chemistry of Complex Equilibria*, Akedémiai Kiadó, Budapest, 1990.

211. D. J. Legget, Ed., *Computational Methods for the Determination of Stability Constants*, Plenum Press, New York, 1985; P. Gans, *Data Fitting in the Chemical Sciences*, Wiley, Chichester, 1992.

212. R. M. Izaat, K. Pawlak, J. S. Bradshaw, and R. L.Bruening, *Chem. Rev.* **91**, 1721 (1991); R. M. Izaat, J. S. Bradshaw, K. Pawlak, R. L. Bruening, and B. J. Tarbet, *Chem. Rev.* **92**, 1261 (1992).

213. Some examples of the use of anion-selective electrodes can be found in Chapter 10 and in: A. M. Bond and G. Hefter, *Inorg. Chem.* **9**, 1021 (1970); L. Newman, P. Klotz, A. Mukherji, S. Fedberg, and L. Newman, *Inorg. Chem.* **10**, 740 (1971); S. Gilford, W. Cherry, J. Jecmen, and M. Reaudnour, *Inorg. Chem.* **13**, 1474 (1974); J. W. Bixler and A. M. Bond, *Inorg. Chem.* **17**, 3684 (1978); S. Y. Shetty and R. M. Sathe, *J. Inorg. Nucl. Chem.* **39**, 1837 (1977); J. W. Bixler and T. M. Larson, *J. Inorg. Nucl. Chem.* **36**, 224 (1974); K. Burger, S. Stafford, H. J. Haupt, and G. Huber, *Inorg. Chim. Acta* **11**, 207 (1974).

214. P. W. Linder, R. G. Torrington, and D. P. Williams, *Analysis Using Glass Electrodes*, Open University Educational Enterprise, 1984.

215. For programs dealing with computation of equilibrium constants, see Ref. 211. Some selected references are: L. G. Sillén, *Acta. Chem. Scand.* **18**, 1085 (1964); R. J. Moteikatis and A. E. Martell, *Canad. J. Chem.* **60**, 168 (1982); P. Gans, A. Sabatini, and A. Vacca, *J. Chem. Soc., Dalton Trans.*, 1195 (1985); A. Sabatini, A. Vacca, and P. Gans, *Coord. Chem. Revs.* **120**, 389 (1992); L. Zekany and I. Nagypal, Ref. 211, p. 291.

216. M. Micheloni, P. M. May, and D. R. Williams, *J. Inorg. Nucl. Chem.* **40**, 1029 (1978); M. Micheloni, A. Sabatini, and A. Vacca, *Inorg. Chim. Acta* **25**, 41 (1977).

217. G. Gran, *Analyst (London)* **77**, 661 (1952); F. J. C. Rossotti and H. Rossotti, *J. Chem. Educ.* **42**, 375 (1965).

218. J. E. Freund, *Mathematical Statistics*, Prentice Hall, New York, p. 199, 1962.

219. K. A. Connors, *Binding Constants, The Measurements of Molecular Complex Stability*, John Wiley & Sons, New York, Chichester, Brisbane, Toronto, Singapore, 1987.

220. C. Frassineti, S. Ghelli, P. Gans, A. Sabatini, M. S. Moruzzi, and A. Vacca, *Anal. Biochem.* **231**, 374 (1995).

221. M. J. Hynes, *J. Chem. Soc., Dalton Trans.*, 311 (1993).

222. J. J. Kankare, *Anal. Chem.* **42**, 375 (1970); H. Gamp, M. Maeder, C. J. Meyer, and A. Zuberbühler, *Talanta* **32**, 1133 (1985); H. Gamp, M. Maeder, C. J. Meyer, and A. Zuberbühler, *Talanta* **33**, 943 (1986).
223. A. Bayada, G. A. Lawrance, and M. Maeder, *Supramol. Chem.* **3**, 261 (1994); A. Bayada, G. A. Lawrance, M. Maeder, M. Martínez, B. W. Skelton, and A. H. White, *Inorg. Chim. Acta* **227**, 71 (1994).
224. A. Bianchi and P. Paoletti, in *Handbook of Metal–Ligand Interactions in Biological Fluids*, G. Berthon, Ed., Marcel Dekker, New York—Basel—Hong Kong, 1995.

CHAPTER 7

Electrochemical Aspects of Anion Chemistry

ANTONIO DOMÉNECH CARBÓ

7.1 Introduction
7.2 The Nature of Electrode Processes
    7.2.1 Thermodynamics of Electrode Processes. Electrode Potentials
    7.2.2 Reversibility and Irreversibility. Kinetics of Electrode Processes
    7.2.3 Coupled Chemical Reactions
7.3 Review of Selected Electrochemical Techniques
    7.3.1 Chronoamperometry
    7.3.2 Cyclic Voltammetry
    7.3.3 Polarography
7.4 Electrochemical Pattern of Anion Coordination
    7.4.1 Simple Systems
    7.4.2 Systems Involving Coupled Chemical Reactions
7.5 Electrochemical Analysis of Anion Coordination Equilibria
    7.5.1 Analysis of Coordination Equilibria from Electrode Potential Data
    7.5.2 Analysis of Coordination Equilibria from Current Data
    7.5.3 Analysis of Adduct Formation in Unstable Oxidation States
7.6 Electrochemical Analysis of Coordination Equilibria for Nonelectroactive Anions: Competitive Methods
    7.6.1 Effects of Competing Complexation with Metal Ions
    7.6.2 Effect of Competing Complexation with Anionic Substrates
7.7 The Significance of Electrochemical Data
    7.7.1 Thermochemical and Structural Considerations
    7.7.2 Solvent and Supporting Electrolyte Effects
7.8 Other Issues
7.9 References

Supramolecular Chemistry of Anions, Edited by Antonio Bianchi, Kristin Bowman-James, and Enrique García-España.
ISBN 0-471-18622-8. © 1997 Wiley-VCH, Inc.

7.1 INTRODUCTION

Electrochemical methods have traditionally occupied a prominent place for analyzing coordination equilibria in solution.[1-5] Electrochemical studies have proven very useful in obtaining not only thermodynamic data, but also in providing structural, kinetic, and mechanistic information concerning reaction mechanisms and electrocatalysis.[6] Studies on electrochemically generated free radicals,[7-9] biosynthetic reaction pathways,[10,11] and preparation of electrochemical sensors[12] have been among the electrochemical techniques exploited. Among other issues, preparation of chemically modified electrodes, electrochemical treatment of surfaces, electrosynthesis of new products, and combined spectroelectrochemical methods are of interest in various research fields.[13]

In the context of anion coordination chemistry,[14] electrochemical methods are closely related with the design of chemical sensors, modeling electron transfer in biological systems, and producing redox catalysts.[15]

In the following we shall be concerned with a reexamination of the more general electrochemical topics dealing with supramolecular anion chemistry. This study is specifically devoted to processes involving an electron transfer between an electrode and electroactive anions that undergo chemical transformations. In the most simple case, this can be represented as:

$$A^{m-} + ne^- \leftrightarrow A^{(m+n)-} \tag{1}$$

As shown by the Born–Haber cycle depicted in Fig. 7.1, the free enthalpy for the electrode process (Eq. 1) is related to the free enthalpies of anion solvation and electron transfer in the gas phase. Therefore, not only structural changes into isolated anions, but also solvent effects, are pertinent for properly elucidating the electrochemical data for electrode processes.

It should be noted that electron-transfer processes can be described as heterogeneous reactions occurring only at the electrode–electrolyte interface.[6] The rate of the overall electrode process depends on the rates of mass transfer from the bulk solution to the electrode surface (migration, convection,

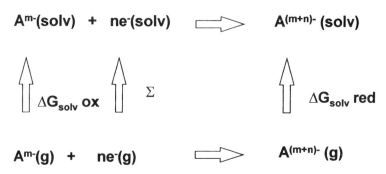

Fig. 7.1. Thermochemical cycle for an electrode process.

diffusion), electron transfer at the electrode surface, chemical reactions preceding or following the electron transfer, and processes such as adsorption/desorption, crystallization, among others. Accordingly, the electrode process can be described from the chemical kinetics methodology in terms of electron-transfer rate constants.

7.2 THE NATURE OF ELECTRODE PROCESSES

7.2.1 Thermodynamics of Electrode Processes. Electrode Potentials

Let us consider the case of an anionic electroactive species that undergoes a reversible n-electron transfer as represented by Eq. 1. The electrode potential, E_A, is then given by the Nernst equation:

$$E_A = E_A^O + \frac{RT}{nF} \ln \frac{a_{ox}}{a_{rd}} \qquad (2)$$

where a represents the thermodynamic activity of the oxidized and reduced species. E_A is the actual electrode potential, and E_A^O is the standard electrode potential, measured when all species are at unity activity. Usually, activities are replaced by concentrations, and the Nernst equation becomes

$$E_A = E_A^{O'} + \frac{RT}{nF} \ln \frac{[A^{m-}]}{[A^{(m+n)-}]} \qquad (3)$$

where $E_A^{O'}$ is the formal potential of the $A^{m-}/A^{(m+n)-}$ couple, obtained for concentrations equal to unity. The formal potential incorporates the standard potential and the activity coefficients of the species involved at the electrode process being affected, in particular, by the nature and concentration of the supporting electrolyte.[6]

As is well known, the free enthalpy changes associated to the electrode process are related to the electrode potentials by the Nernst equation:

$$\Delta G = -nFE \qquad (4)$$

The entropy and enthalpy changes in the cell reaction can be expressed in terms of the temperature dependence of the electrode potential at a constant pressure:

$$\Delta S = -\left(\frac{\delta G}{\delta T}\right)_P = nF\left(\frac{\delta E}{\delta T}\right)_P \qquad (5)$$

$$\Delta H = \Delta G + T\Delta S = nF\left[T\left(\frac{\delta E}{\delta T}\right)_P - E\right] \qquad (6)$$

Thermochemical cycles involving electrode potentials are of interest in several fields of chemistry, including studies dealing with properties of unstable reaction intermediates,[16] biosynthetic reactions,[17,18] and multinuclear clusters,[19] among others.[20–22]

For our purposes, the formation of stable adducts between both the oxidized and reduced forms of the A species and a receptor L can be represented as,

$$A^{m-} + xL = AL_x^{m-} \tag{7}$$

$$A^{(m+n)-} + yL = AL_y^{(m+n)-} \tag{8}$$

Upon addition of a large excess of receptor, the electrode process may be described as a direct reduction/oxidation of the coordinated forms:

$$AL_x^{m-} + (y-x)L + ne^- = AL_y^{(m+n)-} \tag{9}$$

Figure 7.2 presents a thermochemical cycle for relating the electrode potentials and stability constants from which the Nernst equation can be written as:

$$E_{AL} = E_{AL}^{O'} + \frac{RT}{nF} \ln \frac{[AL_x^{m-}][L]^{y-x}}{[AL_y^{(m+n)-}]} \tag{10}$$

where $E_{AL}^{O'}$ is the formal potential of the $AL_x^{m-}/AL_y^{(m+n)-}$ couple.

In cation coordination electrochemistry it is frequently assumed that electrode processes formally affect the uncomplexed ions after dissociation of the coordinated species. Then the shift in the electrode potential is given by:

$$\Delta E = E_{AL} - E_A = \frac{RT}{nF} \ln \frac{\beta_{\text{rd}}}{\beta_{\text{ox}}} + (y-x) \frac{RT}{nF} \ln c_L \tag{11}$$

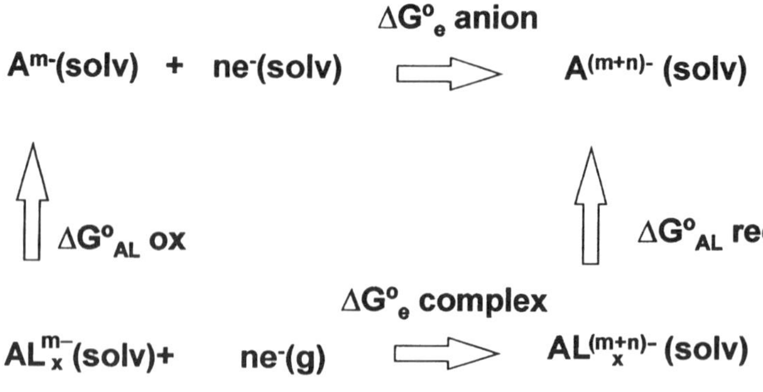

$$\begin{array}{ccc}
& \Delta G^\circ_e \text{ anion} & \\
A^{m-}(\text{solv}) + ne^-(\text{solv}) & \Longrightarrow & A^{(m+n)-}(\text{solv}) \\
\\
\Big\Updownarrow \Delta G^\circ_{AL} \text{ ox} & & \Big\Updownarrow \Delta G^\circ_{AL} \text{ red} \\
\\
& \Delta G^\circ_e \text{ complex} & \\
AL_x^{m-}(\text{solv}) + ne^-(g) & \Longrightarrow & AL_x^{(m+n)-}(\text{solv})
\end{array}$$

Fig. 7.2. Thermochemical cycle for electrode processes involving free and complexed anionic couples.

where $\beta_{ox} = [AL_x^{m-}]/[A][L]^x$ and $\beta_{rd} = [AL_y^{(m+n)-}]/[A][L]^y$ are the stability constants of the complexes formed in both oxidation states. Strictly, these are conditional formation constants, which depend upon the conditions of the medium. Usually, the changes in formal potentials will be basically the same as those of standard potentials, and the ratio of the conditional formation constants can be taken as the ratio of the thermodynamic stability constants.

7.2.2 Reversibility and Irreversibility. Kinetics of Electrode Processes

As previously noted, the electrode process can be regarded as a heterogeneous reaction taking place at the electrode surface. Then a given energy barrier must be associated to the electron transfer through the electrode–electrolyte interface. For a simple one-step electron transfer under linear diffusion at a planar electrode, the current potential characteristic is[6]

$$i = nFAk^O\{c_{ox} \exp[-\alpha n(E - E^{O'})] - c_{rd} \exp[-(1 - \alpha n(E - E^{O'})]\} \quad (12)$$

in which c_{ox} and c_{rd} represent, respectively, the concentration of the oxidized and reduced forms at the electrode surface, n the overall number of electrons involved in the electrode process, and A the electrode area; k^O is the standard (or intrinsic) rate constant, which represents the kinetics of the heterogeneous electron-transfer process, and α is the transfer coefficient, which is a nondimensional parameter whose values vary from zero to unity. In brief, it arises because only a fraction of the energy furnished by the applied potential lowers the activation energy barrier.

Since most electrode processes are mechanisms of several steps, it is convenient to replace the overall n value of the electrode process by the number of electrons involved in the rate-determining step, n_a. Then the current–potential characteristic becomes[6]:

$$i = nFAk^O\{c_{ox} \exp[-\alpha n_a(E - E^{O'})] - c_{rd} \exp[-(1 - \alpha n_a(E - E^{O'})]\} \quad (13)$$

Kinetic data are closely related to the electrochemical reversibility of the electrode process. One can discern between reversible, quasireversible,[23] and totally irreversible electrode processes depending on the values of the heterogeneous electron-transfer rate constant. Notice that irreversible systems can provide only kinetic information; thermodynamic results such as formal potentials and free energies are not available. It should be noted that electrochemical reversibility is not synonymous with chemical reversibility. This last term is applied when the same net electrode process is observed upon current reversal, whereas, operationally, the term *electrochemical reversibility* is applied to electrode processes that follow the Nernst equation.[6]

7.2.3 Coupled Chemical Reactions

Electrode processes are frequently accompanied by coupled chemical reactions that precede parallel, and/or follow the electron-transfer step. Each transient electrochemical technique provides a set of diagnostic criteria for distinguishing between reversible, quasireversible, and irreversible electron transfers as well as different processes involving coupled chemical reactions to multistep electron transfer.[24,25] However, in many cases, it is not easy to discern between irreversibility and homogeneous coupled chemical reactions or surface phenomena. In addition, other factors, such as uncompensated ohmic resistance in the cell, electrode geometry, electrode material, activation of electrodes, and different surface phenomena can distort the observed electrochemical response.[27,28]

In spite of the foregoing difficulties, electrochemical techniques play an essential role in various branches of chemistry. Although the following list does not completely exhaust the scope of electrochemical methods, theoretical approaches and instrumentation,[24,25] digital simulation procedures,[26] use of microelectrodes and ultramicroelectrodes, introduction of chemically modified electrodes,[13] etc. have multiplied the number of available techniques. In the next section some extended techniques will be briefly reviewed.

7.3 REVIEW OF SELECTED ELECTROCHEMICAL TECHNIQUES

7.3.1 Chronoamperometry

In this technique, a pulse of potential is applied from a value E_1, at which the electroactive species is stable, to one, E_2, at which it is reduced or oxidized in such a way that its equilibrium concentration at the electrode surface approaches zero (see Fig. 7.3a). Under ordinary conditions, the current is controlled by the diffusion of electroactive species and not by the value of the applied potential. Then the measured current varies with time, as represented in Fig. 7.3b. For a chronoamperometric experiment performed at a planar electrode, the Cottrell equation holds:[6]

$$i = \frac{nFAD^{1/2}c}{(\pi t)^{1/2}} \qquad (14)$$

where n is the number of electrons per molecule oxidized or reduced, A is the electrode area, c is the bulk concentration of the electroactive species, and D its diffusion coefficient.

Interestingly enough, current-time curves are essentially independent of the reversibility of the electron transfer except at short measurement times. Therefore, it can be used to estimate the number of electrons involved in the electrode reaction[29] or, alternatively, to determine the diffusion coefficient of the electroactive species.[30,31]

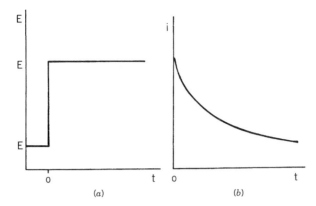

Fig. 7.3. Waveform (a) and resulting current–time response (b) for experiments in which a electroactive species is reduced at a diffusion-limited rate at the potential E_2.

7.3.2 Cyclic Voltammetry

Cyclic voltammetry (or, technically, linear potential sweep chronoamperometry or linear sweep voltammetry)[6] is based on the application of a linearly increasing potential to a working electrode and monitoring of the current flowing as a result of the potential variation. The potential applied to the working electrode is scanned linearly from an initial value E_i to a predetermined limit E_1, where the direction of the scan is reversed. Figure 7.4a shows the time dependence of the applied potential, whereas Fig. 7.4b represents a typical current–potential response for a reversible system.

A few millivolts past the formal potential of the couple, the current reaches a sharp maximum and declines further. Then the reduction of the A^{m-} species is controlled by its diffusion from the bulk of the solution to the electrode surface. Once the switching potential is reached, the potential is sweeping in a positive direction (reverse scan). Then there is a large concentration of $A^{(m+n)-}$ at the vicinity of the electrode. As the potential approaches $E^{O'}$, this anion is reoxidized and an anodic current flows. The following expressions hold, respectively, for the peak current in n-electron reversible and irreversible transfers:

$$i_p = (2.69 \times 10^5) n^{3/2} A c D^{1/2} v^{1/2} \tag{15}$$

$$i_p = (2.99 \times 10^5) n (\alpha n_a)^{1/2} A c D^{1/2} v^{1/2} \tag{16}$$

Beyond the peak the current is controlled by the diffusion of the electroactive species toward the electrode surface. In a potential region sufficiently past the peak, the current verifies the Cottrell equation, being proportional to $t^{-1/2}$.[32,33] In the absence of further electrode processes,[34] the diffusion-controlled region

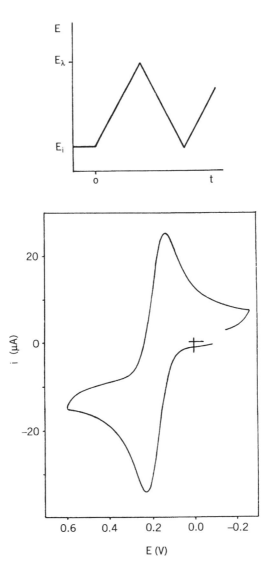

Fig. 7.4. Triangular wave and resulting cyclic voltammogram at the glassy carbon electrode for a 2.0 mM solutions of potassium hexacyanoferrate(II) in neutral aqueous solution (NaClO$_4$ 0.15 M). Scan rate 0.20 V/s.

of cyclic voltammetric curves verify

$$i = \frac{nFAcD^{1/2}v^{1/2}}{\pi^{1/2}(E - E^*)^{1/2}} \tag{17}$$

where E is the actually applied potential and E^* is the potential at which the diffusion control starts. Then, cyclic voltammograms can be used to obtain chronoamperometric data.

The peak potential for reversible systems is related to the formal potential in the forward potential scan as follows[35]:

$$E_p = E^{O'} - \frac{RT}{nF} \ln(D_{ox}/D_{rd})^{1/2} - 1.109\frac{RT}{nF} \qquad (18)$$

Since the ratio of the diffusion coefficients may be considered as unity, it is a common practice to calculate the formal reduction peak of a reversible couple as the average of the cathodic and anodic peak potentials:

$$E^{O'} = (E_{pc} + E_{pa})/2 \qquad (19)$$

This is a satisfactory approximation for completely reversible systems taking into account that the return peak potential depends on the switching potential.[35] Then CV allows one to perform a fast and reasonably accurate determination of $E^{O'}$ values. For irreversible systems, however, the peak potential is:

$$E_p = E^{O'} - \frac{RT}{an_aF}\left(0.78 + \ln\frac{D_{ox}^{1/2}}{k^O} + \ln\frac{\alpha n_a F v^{1/2}}{RT}\right) \qquad (20)$$

and no single correlation with thermodynamic data is available. The morphology of cyclic voltammetric curves is quite sensitive to the mechanism of electrode processes and scan rate variations. Accordingly, diagnostic criteria have been developed for a wide variety of electrochemical processes.[6,35]

7.3.3 Polarography

The invention and development of polarography by Heyrovsky and his successors played a capital role in the historical evolution of electrochemical methods.[1–5] This technique is based on the application of a slow linear potential scan to a dropping mercury electrode. As Fig. 7.5 plots, S-shaped current–potential curves are obtained, the oscillations in current arising from the growth and fall of the individual mercury drops.

Since the electrode surface is continuously renewed by the repeated dropping and stirring action, current plateaus are obtained in the diffusion-limited region of each polarographic wave. For a reversible electron transfer uncomplicated by chemical reactions or undesired surface effects, the following equation holds:

$$E = E_{1/2} + \frac{RT}{nF} \ln \frac{i_L - i}{i} \qquad (21)$$

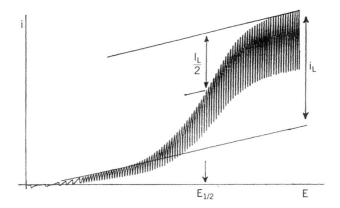

Fig. 7.5. Representative parameters in a typical polarographic wave.

where $E_{1/2}$ is the half-wave potential, directly related to formal potential:

$$E_{1/2} = E^{O'} - \frac{RT}{nF} \ln (D_{ox}/D_{rd})^{1/2} \qquad (22)$$

Polarographic limiting currents (i_L) and half-wave potentials, depicted in Fig. 7.5, are characteristic parameters whose meaning is similar to peak currents and formal cyclic voltammetric potentials. For reversible systems, half-wave potentials are often used as a direct measure of formal electrode potentials. For both reversible and irreversible electron-transfer processes, limiting currents are proportional to the bulk concentration of electroactive species and used, upon calibration, for analytical purposes.

7.4 ELECTROCHEMICAL PATTERN OF ANION COORDINATION

7.4.1 Simple Systems

In general, the electrochemical response of an electroactive species undergoes significant changes upon complexation or adduct formation. The seminal work of Lehn and co-workers on the host–guest interaction of $[Fe(CN)_6]^{4-}$ and $[Ru(CN)_6]^{4-}$ with macrocyclic polyammonium cations (**7.1–7.3**) exemplifies the essential features of anion coordination electrochemistry.[36,37]

The cyclic voltammetric response of $[Fe(CN)_6]^{4-}$ ions in aqueous solution (see Figs. 7.4b and 7.6) consists of a one-electron diffusion-controlled reversible couple that can be described as:

$$[Fe(CN)_6]^{4-} \leftrightarrow [Fe(CN)_6]^{3-} + e^- \qquad (23)$$

Upon addition of increasing amounts of several polyammonium macro-

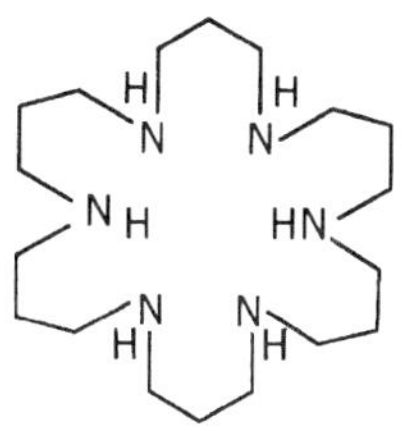

7.2 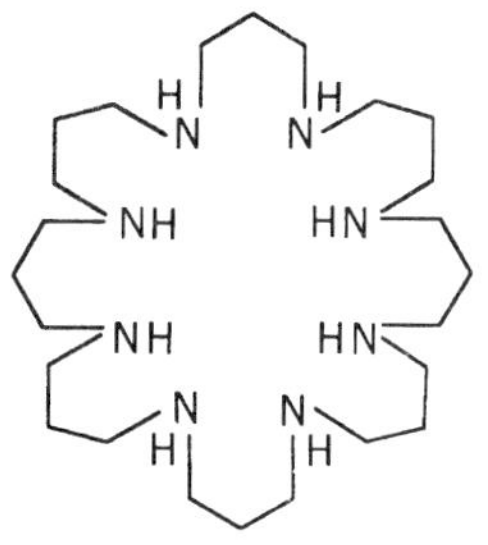   **7.1**

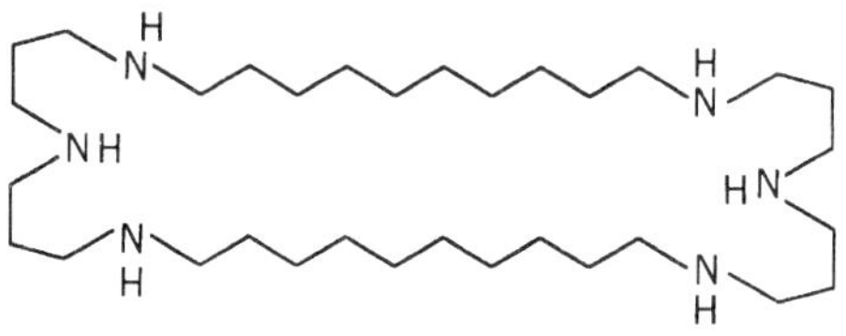

7.3

cycles, CV curves significantly depend on the receptor/hexacyanoferrate(II) molar ratio (R) and the pH of the medium. As can be observed in Fig. 7.6, corresponding to the **7.1**–hexacyanoferrate(II) system at pH 5.5, in the range $0 < R < 1$, the cyclic voltammograms are rather broad and look like the superposition of two waves. Further additions of receptor are accompanied by an enhancement of the more anodic couple at the expense of the less anodic couple. For $R > 1$, only the more positive couple remains, and its electrochemical parameters do not significantly change after subsequent additions of receptor. This behavior can be satisfactorily explained in terms of the formation of stable receptor–$[Fe(CN)_6]^{4-}$ and receptor–$[Fe(CN)_6]^{3-}$ adducts of 1:1 stoichiometry.[36,37] The cyclic voltammetric parameters of the i–E curves for $R > 1$ are as expected for a single one-electron reversible process without chemical complications. This indicates that the coordinated species are electroactive, the electrode process being a monoelectronic Fe(II)/Fe(III) interconversion without dissociation of the complex:

$$[Fe(CN)_6]H_qL^{(4-q)-} \leftrightarrow [Fe(CN)_6]H_qL^{(3-q)-} + e^- \qquad (24)$$

As can be seen in Fig. 7.7, which corresponds to the $[Fe(CN)_6]^{4-}$–**7.3** system, the peak potentials depend on the pH, suggesting the formation of different complex species with different protonation degrees. In the cases in which a small separation exists between the formal potentials of complexed and uncomplexed couples, the cyclic voltammograms look like a single couple whose peak

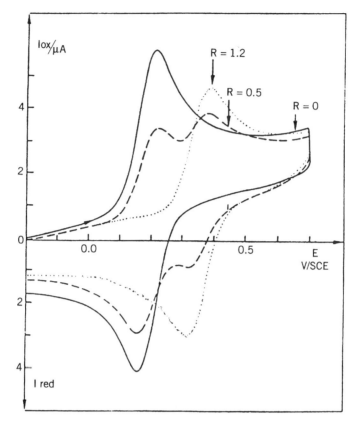

Fig. 7.6. Cyclic voltammograms for the $[Fe(CN)_6]^{4-}$–**7.1** system at pH 5.5 in KCl 0.1 M. Platinum electrode, sweep rate 50 mV/s. $[Fe(CN)_6]^{4-} = 9 \times 10^{-4}$ M; R represents the receptor/$[Fe(CN)_6]^{4-}$ molar ratio. [From F. Peter, M. Gross, M. W. Hosseini, J. M. Lehn, and R. B. Sessions, *J. Chem. Soc. Chem. Commun.*, 1067 (1981), with permission.]

intensity and peak potentials monotonically decrease or increase as the receptor/anion ratio increases.[38,39] Figure 7.8 plots the effect on the formal potential (calculated as the average of the anodic and cathodic peak potentials) and peak current of the receptor/hexacyanoferrate(II) ratio for the $[Fe(CN)_6]^{4-}$–**7.4** system at pH 6.5.

7.4.2 Systems Involving Coupled Chemical Reactions

In contrast with the foregoing systems, there are a number of cases in which the electrochemical pathway is complicated by parallel chemical reactions. For instance, the interaction of polyammonium receptors with dinucleotides is also reflected in significant changes in the electrochemical response. Figure 7.9a presents a cyclic voltammogram at the glassy carbon electrode for an aqueous solution of $NADP^+$ at pH 7.0 (**7.5**).

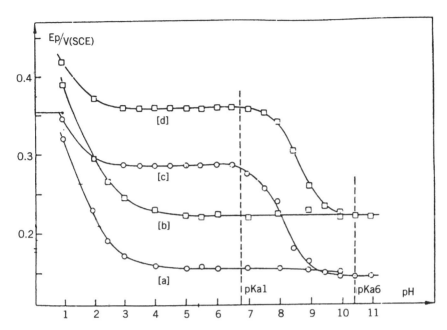

Fig. 7.7. pH dependence of peak potentials ($v = 50\,\text{mV/s}$) for the hexacyanoferrate(II)–**7.3** system in 0.1 M $(CH_3)_4NCl$. (a) Reduction peak of $[Fe(CN)_6]^{3-}$; (b) oxidation peak of $[Fe(CN)_6]^{4-}$, in the absence of macrocyclic ligand; (c) reduction peak of $[Fe(CN)_6]^{3-}$; (d) oxidation peak of $[Fe(CN)_6]^{4-}$, in the presence of macrocyclic ligand. [From F. Peter, M. Gross, M. W. Hosseini, and J. M. Lehn, *J. Electroanal. Chem.* **144**, 279 (1983), with permission.]

The electrochemical redox pattern of NAD^+ and $NADP^+$, summarized in Fig. 7.10, consists of an initial one-electron reduction (cathodic peak A) of the pyridinium ion to a free radical followed by rapid irreversible dimerization[40]:

$$NAD^+ + e^- \leftrightarrow NAD\cdot \quad (25)$$

$$NAD\cdot \rightarrow 1/2\,(NAD)_2 \quad (26)$$

This is an example of an irreversible chemical reaction following a reversible charge transfer. Since the dimerization reaction Eq. 26 is very rapid, a significant amount of $NAD\cdot$ is converted into $(NAD)_2$ during the time scale of the direct voltage scan. On the return scan, therefore, the anodic peak A′, corresponding to the reoxidation of $NAD\cdot$ to NAD^+, is almost entirely absent.

At potentials close to -1.50 V, proton-assisted reduction to dihydropyridine, NADH, takes place (region B in the cyclic voltammogram) as revealed by an ill-defined shoulder in the cyclic voltammograms. This is obscured by hydrogen evolution due to proton reduction at the electrode. The overall electrode

290 ELECTROCHEMICAL ASPECTS OF ANION CHEMISTRY

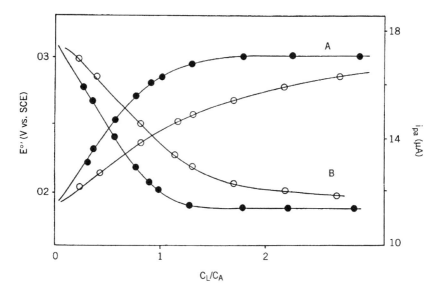

Fig. 7.8. Cyclic voltammetric analysis of the **7.4**–$[Fe(CN)_6]^{4-}$ system in the absence (points) and the presence (circles) of 1,3,5-benzenetricarboxylic acid. Plot of the formal potential (A) and peak current (B) vs. the receptor/substrate molar ratio at pH 6.5. Platinum electrode, $v = 0.10\,\text{V/s}$; $NaClO_4$ 0.15 M. [From A. Bianchi, A. Doménech, E. García-España, and S. V. Luís, *Anal. Chem.* **65**, 3137 (1993), with permission.]

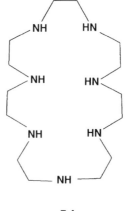

7.4

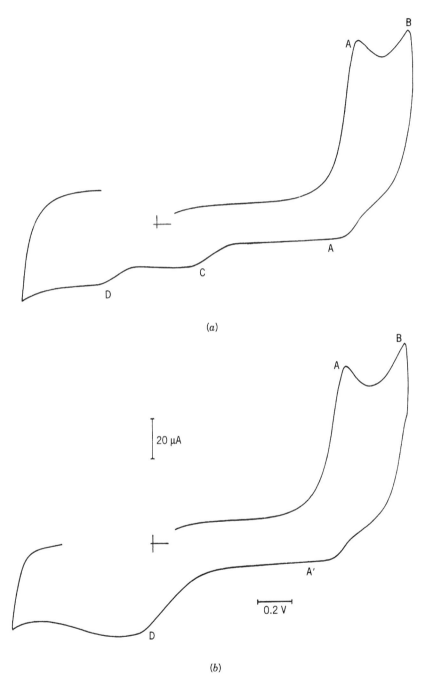

Fig. 7.9. Cyclic voltammograms at the glassy carbon electrode for 0.15 M NaClO$_4$ (pH = 7.0) solutions of (a) NADP$^+$ 1.20 mM; (b) NADP$^+$ 1.20 mM plus **7.4** 1.20 mM. Scan rate 100 mV/s.

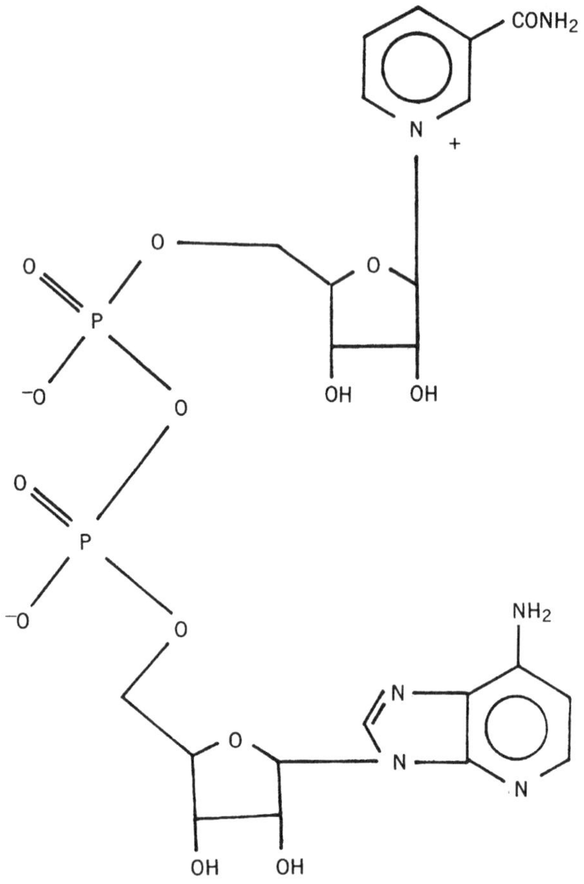

7.5

reduction of NAD$^+$ to NADH can be described as

$$NAD^+ + H^+ + 2e^- \rightarrow NADH \tag{27}$$

After sweep reversal, anodic peaks (C, D) at -0.15 and $+0.32$ V vs. SCE appear. Peak C is attributable to irreversible dimer oxidation to the original nucleotide

$$(NAD)_2 \rightarrow 2\,NAD^+ + 2e^- \tag{28}$$

while peak D corresponds to NADH oxidation, which appears as a result of adsorption at the solution–electrode interface of NAD$^+$. This is confirmed by cyclic voltammograms obtained on NADH solutions. As Fig. 7.11 plots, in a

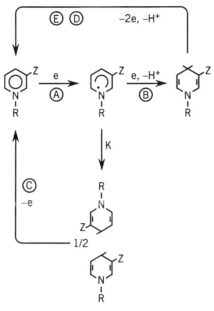

Fig. 7.10. Simplified redox scheme for the electrochemical behavior of nicotinamide adenine dinucleotide and related compounds.

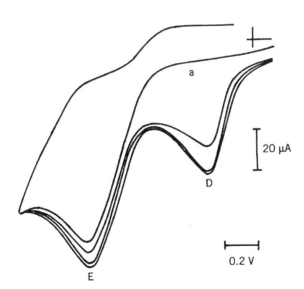

Fig. 7.11. Anodic portion of the cyclic voltammograms for a neutral solution of NADH (1.0 mM) in NaClO$_4$ 0.15 M. First (a) and subsequent scans. Glassy carbon electrode, scan rate 100 mV/s.

first anodic scan, a single peak (E) at +0.72 V appears corresponding to the overall process:

$$\text{NADH} \rightarrow \text{NAD}^+ + \text{H}^+ + 2e^- \qquad (29)$$

In the second and subsequent scans, the peak D precedes oxidation process E, denoting the formation of NAD^+, which remains at least partially adsorbed at the electrode surface. Observation of reduction processes A and B confirms the formation of NAD^+ during electrochemical turnovers. The pathway of NADH oxidation is, however, complicated. Electrochemical studies on NADH/NAD^+ analogues[41] and model compounds[42] in nonaqueous media suggest, as Fig. 7.12 plots, a mechanism consisting essentially of a rate-determining deprotonation step coupled with two electron-transfer processes.[43,44]

Upon addition of an equimolar amount of **7.4** (see Fig. 7.9b), the electrochemical pattern undergoes significant changes.[45] For NAD^+ solutions, the peak A is slightly shifted toward less negative potentials while oxidation peak C disappears. Oxidation peak D is enhanced and displaced toward less positive potentials. For NADH solutions, a shift of peak E toward more positive potentials is observed. These data suggest that substrate–receptor interaction facilitates the reduction of NAD^+ to NADH. This likely results in an increase of the rate of the electrode process B at the expense of the dimerization of the intermediate radical, which appears to be inhibited.

These features are confirmed by using modified carbon-paste electrodes (MCPE) in which powdered NAD^+ or NADH are included in a nujol oil/graphite powder paste. As depicted in Figs. 7.13a and 7.31b, the cyclic

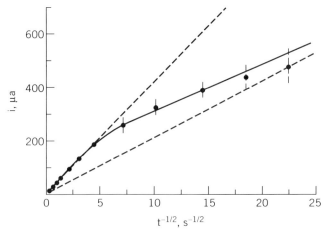

Fig. 7.12. Variation of the current with time on potential-step chronoamperometry at a stationary covered GCE for a 1 mM NADH solution in KCl (0.5 M)-Tris (0.05 M) buffer. [From J. Moiroux and P. J. Elving, *J. Am. Chem. Soc.* **102**, 6533 (1980), with permission.]

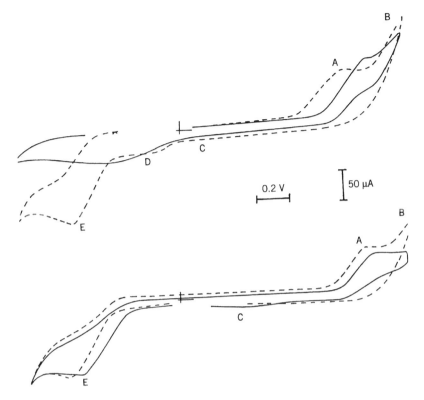

Fig. 7.13. Cyclic voltammograms at modified carbon-paste electrodes immersed into a 0.15 M NaClO$_4$ solution. (a) NAD$^+$ (solid line) and NAD$^+$ plus **7.4** (dotted line). (b) NADH (solid line) and NADH plus **7.4** (dotted line).

voltammograms for MCPE containing NAD$^+$ and NADH (continuous lines) are similar to those recorded at the GCE, the electroactive species being in solution. In the presence of **7.4**, however, the electrochemical pattern changes drastically (dotted lines): for NAD$^+$–MCPE the reduction step occurring at -1.25 V is followed by the oxidation process E, denoting that NADH is the main reduction product.

7.5 ELECTROCHEMICAL ANALYSIS OF ANION COORDINATION EQUILIBRIA

Electrochemical techniques have been extensively used in metal–ligand complexation studies. In particular, polarography has been widely employed for stability constant determination of metal complexes.[1-5,46] All these methods are based on the measurements of shifts in the electrochemical response of the chemical system brought about by complexation, the response being metal or ligand centered.[47] Although potentiometry and UV–vis spectophotometry are, by far, the most commonly employed techniques for studying multiple complex formation equilibria, a wide variety of electrochemical techniques can also be applied to the determination of formation constants. These include cyclic voltammetry,[48] steady-state,[49] and normal[50] pulse voltammetry, as well as others. Extension to systems with adsorption complications[51,52] are also available.

7.5.1 Analysis of Coordination Equilibria from Electrode Potential Data

Classical polarographic methods are available on single[53] and multiple equilibria involving one[54] and mutiple[55,56] ligands. These are based essentially on measuring the shift of polarographic half-wave potentials (a small correction involving diffusion current measurement is required) for the reduction of a metal ion to metallic state or amalgam upon addition of increasing amounts of ligand. Recent works have extended the DeFord and Hume[54] and other methods to cyclic voltammetry[57,58] and general simulation procedures have been devised for the study of pH-dependent complex formation equilibria.[48,59,60]

In aqueous media, formation equilibria of anion receptor adducts occur frequently in stepwise pH-dependent manner:

$$A^{m-} + xL + qH^+ = AH_qL_x^{(m-q)-} \tag{30}$$

$$A^{(m+n)-} + yL + wH^+ = AH_wL_y^{(m+n-w)-} \tag{31}$$

The formation constants are:

$$\beta_{ox} = \frac{[AH_qL_x^{(m-q)-}]}{[A^{m-}][L]^x[H^+]^q} \tag{32}$$

$$\beta_{rd} = \frac{[AH_wL_y^{(m+n-w)-}]}{[A^{(m+n)-}][L]^y[H^+]^w} \tag{33}$$

Then the shift in the formal potential of the $A^{m-}/A^{(m+n)-}$ redox couple (Eq. 11) becomes[46]:

$$\Delta E = \frac{RT}{nF} \ln \frac{\Sigma \beta_{rd}[A^{(m+n)-}][H^+]^w[L]^y}{\Sigma \beta_{ox}[A^{m-}][H^+]^q[L]^x} \tag{34}$$

Taking into account the protonation equilibria of free ligand:

$$L + jH^+ = H_jL \tag{35}$$

for which a series protonation constants, β_j, are known,

$$\beta_j = \frac{[H_j L]}{[L][H^+]} \qquad (36)$$

For a buffered solution containing a sufficiently large excess of ligand, Eq. 26 can be rewritten as:

$$\Delta E = \frac{RT}{nF} \ln \frac{\Sigma \beta_{rd}[H^+]^w}{\Sigma \beta_{ox}[H^+]^q} + (y - x) \frac{RT}{nF} \ln c_L(1 + \Sigma \beta_j [H^+]^j) \qquad (37)$$

where c_L is the total ligand concentration. Assuming that the chemical processes are sufficiently fast so that equilibrium is always established and that the pH of the solution does not change during the potential scan, not even near the electrode surface, cyclic voltammetric peak potentials are usable for a rapid estimate of ΔE. As comparison of Figs. 7.14a and 7.14b, corresponding to the $[Fe(CN)_6]^{4-}$–**7.4** system, reveals, the pH dependence of peak potentials and formal potentials is consistent with the distribution diagrams calculated from potentiometric data. In particular, plateaus in the $E^{O'} = f(pH)$ curve must correspond to pH zones in which equally protonated complexes predominates in both the involved oxidation states; that is $w = q$. Then, Eq. 37 becomes:

$$\Delta E = \frac{RT}{nF} \ln \frac{\beta_{rd}}{\beta_{ox}} \qquad (38)$$

Accordingly, the determined formal potentials can be used to obtain reasonable estimates of the formation constants of the hexacyanoferrate(III) species, providing those of the hexacyanoferrate(II) species are known and that charge transfer does not involve coordination changes.[61]

7.5.2 Analysis of Coordination Equilibria from Current Data

In addition to procedures based on polarographic half-wave potentials and cyclic voltammetric peak potentials, some polarographic methods based on current measurements for diffusion-controlled processes have been devised.[62–69] Extension to cyclic voltammetry is possible in view of the availability of peak current data for measuring concentrations.[70]

Classic stoichiometric methods, such as the molar-ratio[71] and Asmus[72] methods and their generalizations,[73,74] for analyzing single complex formation equilibria involving condensed polymeric species, can be used, at first instance, for electrochemical data. Chronoamperometric data and chronoamperometric analysis of the diffusive portion of cyclic voltammograms[32,33] have been used for applying the molar-ratio method.[75] This procedure can be extended to the

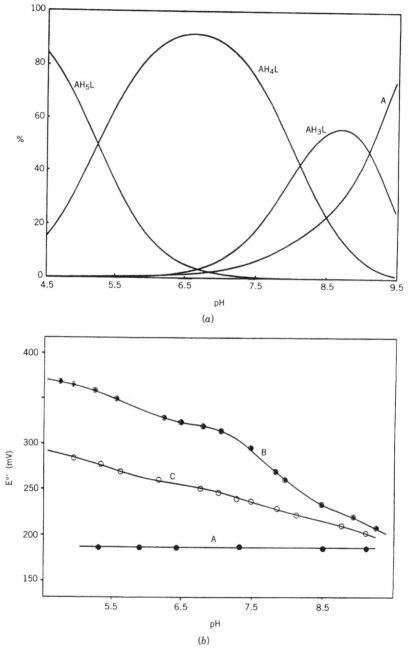

Fig. 7.14. (a) Distribution diagram for the $[Fe(CN)_6]^{4-}$ (2.00 mM)–**7.4** (3.20 mM) system. (b) Formal potentials from cyclic voltammetric data at the Pt working electrode; $v = 0.10$ V/s. A: hexacyanoferrate(II) solution; B: $[Fe(CN)_6]^{4-}$ plus receptor; C: id. plus 1,2,3-benzenetricarboxylic acid (2.20 mM).

Viossat method[76] to estimate the degrees of condensation and protonation of complexes in solution.[77]

The electrochemistry of a number of systems can be described in terms of the oxidation (or reduction) of two interconvertible species, complexed and uncomplexed anion. Both species (represented, respectively, as AL and A, for brevity) are electroactive and exhibit different reduction potentials and diffusion coefficients.[78-81] As previously noted, digital simulation procedures can be applied for estimating the stoichiometry and formation constants of the anion–receptor adducts, providing the electrochemical parameters are known. In the most favorable situation, completely reversible electrode processes are observed with the same diffusion coefficient for all species.[48] In the case where the interconversion reaction is always in equilibrium, chronoamperometric experiments for a two-component system with differing diffusion coefficients can be described as a single system with a mean diffusion coefficient given by[81]:

$$D = \frac{D_A + KD_{AL}}{1 + K} \tag{39}$$

K being the equilibrium constant for the complexation reaction and D_A, D_{AL}, the diffusion coefficients for A and AL. This equation holds for cyclic voltammetric and chronoamperometric experiments when the applied potential is large enough to promote the diffusion-controlled oxidation of A and AL. Under these circumstances, chronoamperometric i t curves or chronoamperometric analysis of the diffusion region in cyclic voltammetric curves provide a mean diffusion coefficient D whose value is representative of the solution composition. As can be seen in curve A of Fig. 7.15, the diffusion coefficient for an aqueous solution of potassium hexacyanoferrate(II) remains constant in neutral and weakly acidic solutions. Below pH 4, the diffusion coefficient increases as the pH decreases, denoting that protonation of $[Fe(CN)_6]^{4-}$ occurs.[37,82] Curve B in Fig. 7.15 plots the pH dependence of the diffusion coefficient measured for an equimolecular solution of $[Fe(CN)_6]^{4-}$ and **7.4**. Below pH 10 the mean diffusion coefficient decreases rapidly, indicating an effective coordination of hexacyanoferrate(II) ions. Below pH 8, $[Fe(CN)_6]^{4-}$ ions are almost quantitatively complexed, and small variations in D appear associated to the formation of different species with different number of protons. D_A and D_{AL} can be calculated at a selected pH from separate experiments in solutions containing the anionic species A before and after addition of a large excess of L. Accordingly, the mean diffusion coefficient obtained for intermediate conditions provides an estimate of the molar fraction of coordinated anion, α_A, when a unique complex exists in solution:

$$\alpha_A = \frac{D - D_A}{D_{AL} - D_A} \tag{40}$$

Equation 40 parallels that of Kacena and Matousek[58] for polarographic

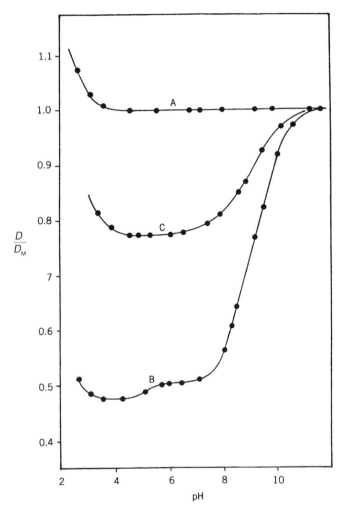

Fig. 7.15. Plots of the pH dependence of the mean diffusion coefficient (D) relative to the diffusion coefficient of uncomplexed $[Fe(CN)_6]^{4-}$ (D_M) at $NaClO_4$ 0.15 M solution for: (A) $[Fe(CN)_6]^{4-}$; (B) id. plus **7.4**; (C) id. plus 1,2-benzenedicarboxylic acid. All reagents in concentration 2.0 mM.

analysis of multiple complex formation equilibria and can be generalized as[46,62,63]:

$$D = \frac{\Sigma D_q \beta_q [L][H^+]^q}{\Sigma \beta_q [L][H^+]^q} \quad (41)$$

Under these conditions, stoichiometric methods can be applied to elucidate the stoichiometry and conditional (pH-dependent) formation constant, as well as the degree of protonation of anion–receptor coordination species. Let us consider the general case in which an anionic species A forms an m-condensed complex with a p-protonated ligand L (charges partially omitted for brevity):

$$mA + nL + qH^+ = A_mL_nH_q \qquad (42)$$

In a molar-ratio experiment, the total concentration of ligand, c_L, is varied, whereas the concentration of A, c_A, is held constant at a given pH. Defining the molar fraction of complexed A as $\alpha_A = m[A_mL_xH_q]/c_A$, the formation constant can be expressed as:

$$\beta = \frac{\alpha_A c_A^{1-m}(1 + \Sigma\beta_j[H^+]^j)^n}{m(1-\alpha_A)^m(c_L - (n/m)\alpha_A c_A)^n[H^+]^q} \qquad (43)$$

Providing the α_A values are determined from experimental data, only the adequate stoichiometric coefficients must satisfy the relationship[69]

$$K = \frac{\alpha_A^{1/n}}{m^{1/n}(1-\alpha_A)^{m/n}(c_L - (n/m)\alpha_A c_A)} = \text{const} \qquad (44)$$

The formation constant can be calculated from the K values, providing the protonation degree of the complex is known:

$$\beta = K^n c_A^{1-m}(1 + \Sigma\beta_j[H^+]^j)^n[H^+]^{-q} \qquad (45)$$

For instance, curves A and B in Fig. 7.8 show the variation of the anodic peak current and the "apparent" formal potential with c_L/c_M ratio for the $[Fe(CN)_6]^{4-}$–**7.4** system at pH 6.5. Application of the molar-ratio method is illustrated in Fig. 7.17. It is observed that the calculated K values remain practically constant only for a 1:1 receptor:hexacyanoferrate(II) stoichiometry, whereas K increases or decreases monotonically for all other stoichiometries.

As previously noted, anion coordination equilibria take place in a pH-dependent stepwise manner. Then the calculated molar fractions of complexed electroactive substrate can be used to determine the stoichiometry and number of protons involved in different formation steps. This requires, in addition to the foregoing commitments, that (1) each of the complex species predominates in a well-defined pH range; (2) the diffusion coefficient of successive complex species are clearly different. This second condition reduces, in practice, the availability of the method to the first protonation step in the majority of cases. In brief, for an experiment in which pH varies whereas the concentrations of anion and ligand remain constant, the molar fraction of complexed A can be calculated from Eq. 40. Then the equation for the equilibrium constant

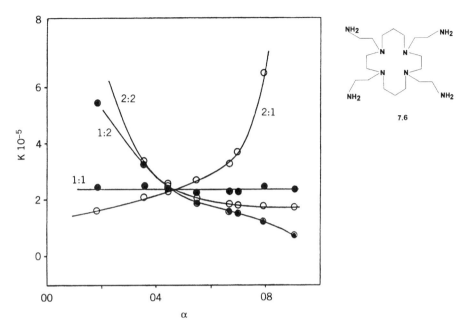

Fig. 7.16. Application of the generalized molar-ratio method for the $[Fe(CN)_6]^{4-}$–**7.6** system at pH 4.5. K versus α plots for different possible stoichiometries. [From A. Bianchi, A. Doménech, E. García-España, and S. V. Luís, *Anal. Chem.* **65**, 3137 (1993), with permission.]

reduces to:

$$f(\alpha_A, H) = \log \frac{\alpha_A^{1/n}(1 + \Sigma\beta_j[H^+]^j)}{m^{1/n}(1-\alpha_A)^{m/n}(c_L - n\alpha_A c_A/m)^n}$$
$$= \frac{1}{n}\log \beta + \frac{m-1}{n}\log C_A + \frac{q}{n}pH \quad (46)$$

This equation predicts a linear dependence between $f(\alpha_A, H)$ and the pH for the adequate values of the stoichiometric coefficients m, n. Figure 7.17 illustrates the application of the method to the $[Fe(CN)_6]^{4-}$–**7.6** system. The slope of such representation yields $q = 4$, in agreement with potentiometric data.

The relevant point to emphasize is that these methods allow a direct procedure to discern between monomeric and dimeric, trimeric, etc., species from accessible electrochemical data. Confirmation of results obtained from potentiometric and spectrophotometric data is thus available. In addition, electrochemical techniques are able to study complex formation equilibria for unstable oxidation states and even for nonelectroactive substrates via competitive complexation with electroactive species. The following sections are devoted to clarifying these applications.

ELECTROCHEMICAL ANALYSIS OF ANION COORDINATION EQUILIBRIA 303

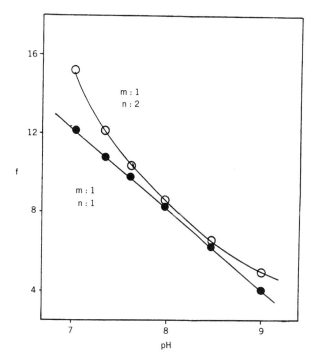

Fig. 7.17. Plots of $f(a_A, H)$ vs. pH for the $[Fe(CN)_6]^{4-}$–**7.6** system for several stoichiometric coefficients.

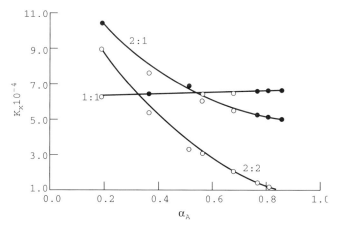

Fig. 7.18. Application of the generalized molar-ratio method to the $[Fe(CN)_6]^{4-}$–Me$_2$ pentae n-ATP system at pH 5.8. [From A. Andrés, J. Aragó, A. Bencini, A. Bianchi, A. Doménech, V. Fusi, E. García España, P. Paoletti, and J. A. Ramírez, *Inorg. Chem.* **32**, 3418 (1993), with permission]

TABLE 7.1. Formal Potentials Calculated from Cyclic Voltammetric Data and Equilibrium Constants for the $[Fe(CN)_6]^{4-}$, $[Fe(CN)_6]^{3-}$, and $[Co(CN)_6]^{3-}$ Adduct Formation with **7.4** in Aqueous ($NaClO_4$, 0.15 M) Solution at 298 K (from Ref. 38)

Reaction	$\log K^a$ $[Fe(CN)_6]^{4-}$	$\Delta E^{O'\,b}$	$\log K^c$ $[Fe(CN)_6]^{3-}$	$\log K^a$ $[Co(CN)_6]^{3-}$
$A + L + 3H$	31.03(4)	35	30.4	30.39(4)
$A + L + 4H$	39.15(4)	105	37.5	37.58(4)
$A + L + 5H$	44.39(4)	165	41.6	41.51(5)

^{a} Calculated from potentiometric data.
^{b} Plateau values in mV vs. SCE in the $E^{O'} = f(pH)$ curve.
^{c} Calculated from Eq. 38. Values in parentheses are standard deviations in the last significant figure.

7.5.3 Analysis of Adduct Formation in Unstable Oxidation States

The use of transient techniques favors the study of the anion-receptor interaction in unstable oxidation states, as an example, the hexacyanoferrate(III)–polyammonium receptor systems. Instability of the hexacyanoferrate(III) aqueous solutions[36,37] makes nonconclusive long-time experiments such as potentiometric titrations. However, $[Fe(CN)_6]^{3-}$ ions and their adducts are sufficiently stable on the time scale of conventional cyclic voltammetry to apply previously described electrochemical methods.

Accordingly, the formation constants of hexacyanoferrate(III)–receptor species can be calculated by inserting into Eq. (38) the cyclic voltammetric potentials and the formation constants for the $[Fe(CN)_6]^{4-}$ adducts determined potentiometrically. Table 7.1 summarizes the results with **7.4** as polyammonium receptor.

Assuming that coordination equilibria are rapidly established near the electrode surface and that hexacyanoferrate(III) species are stable on the time scale of the electrochemical experiment, mean diffusion coefficients determined from double-potential step chronoamperograms or from Cottrell representations of the diffusive cathodic part of the cyclic voltammograms may be used to elucidate the stoichiometry and formation constant of $[Fe(CN)_6]^{3-}$ complexes.[77]

7.6 ELECTROCHEMICAL ANALYSIS OF COORDINATION EQUILIBRIA FOR NONELECTROACTIVE ANIONS: COMPETITIVE METHODS

7.6.1 Effect of Competing Complexation with Metal Ions

Electrochemical methods can also be applied, in some favorable cases, to coordination equilibria involving nonelectroactive species by competitive displacement of an electroactive anion from its complex with a given ligand.

Competitive procedures based on metal ion complexes have been described for polarography[46,83] and used by Kimura and Kodama[84–86] for studying the equilibria of complex formation of linear and cyclic polyamines with mercury(II) and carboxylate ions. In this case, solutions of macrocyclic amines yield reversible two-electron anodic waves corresponding to the overall electrode process

$$Hg + H_jL = HgL^{2+} + jH^+ + 2e^- \quad (47)$$

Measured half-wave potentials are independent of the concentration of ligand and shift in the negative direction on increasing the solution pH. Assuming all the activity coefficients are unity and equal diffusion coefficients for all species in solution, one can write:

$$E_{1/2} = E^{O'}(Hg^{2+}/Hg) + \frac{RT}{2F} \ln [Hg^{2+}] \quad (48)$$

Then inserting the equilibrium constants for ligand protonation and complexation ($\beta_{HgL} = [HgL]/[Hg^{2+}][L]$), one can arrive at:

$$E_{1/2} = E^{O'}(Hg^{2+}/Hg) + \frac{RT}{2F} \ln \frac{1 + \Sigma \beta_j [H^+]^j}{\beta_{HgL}} \quad (49)$$

When a carboxylate ion (A) is added, an 1:1 ion-pair complex is assumed to be formed and Eq. 49 becomes:

$$E_{1/2} = E^{O'}(Hg^{2+}/Hg) + \frac{RT}{2F} \ln \frac{1 + \Sigma \beta_j [H^+]^j + [ALH_j]}{\beta_{HgL}} \quad (50)$$

If the equilibrium constants for ligand protonation and mercury complexation are previously known, the concentration of carboxylate–ligand adducts and the ion-pair association constant can be estimated from the pH dependence of the measured half-wave potentials. Alternatively, molar fractions of uncomplexed (with Hg^{2+}) ligand and conditional formation constant of A-L adducts can be determined from limiting current measurements. As in the case of cyclic voltammetric measurements (Eqs. 40–42), the basic idea is that under pure diffusive control, the mean diffusion coefficient is proportional to the square of polarographic limiting currents. Then, for a given solution of L, the limiting current is $(i_d)_L$; upon addition of an anionic substrate A, the limiting current reaches different (i_d) values until a final value $(i_d)_{AL}$ is attained when the substrate is totally complexed. Then the following equation holds[83]:

$$\frac{[AL]}{[L]} = \frac{(i_d)_L^2 - (i_d)^2}{(i_d)^2 - (i_d)_{LA}^2} \quad (51)$$

Equation 53 is analogous to Eq. 40, taking into account that for diffusion-controlled systems polarographic limiting currents are proportional to the square root of the mean diffusion coefficient of electroactive species.[46]

7.6.2 Effect of Competing Complexation with Anionic Substrates

Application of competitive complexation of different pairs of anions by anionic receptors was proposed by Gross et al. in 1981.[36] Thus it is assumed that an anion–receptor (A–L) system is completely elucidated. Now let us consider a second anionic substrate X, whose protonation and complex formation equilibria with a ligand L are described by:

$$X + tH^+ = H_tX \tag{52}$$

$$rX + zL + sH^+ = X_rL_zH_s \tag{53}$$

If one defines the molar fraction of complexed X as $\alpha_X = r[X_rL_zH_s]/c_X$, and introducing the protonation constants of the ligand, the α_X values can be expressed in terms of the α_A values:

$$\alpha_X = \frac{c_L - (n/m)\alpha_A c_A - \alpha_A^{1/n} K^{-1} m^{1/n} (1 - \alpha_A)^{-m/n}}{(z/r)c_X} \tag{54}$$

By varying the concentration of L in a solution containing constant amounts of A and X, a set of α_A values can be obtained. If the stoichiometry and conditional constant of the A–L complex are known, a set of a_X values can be calculated from Eq. 54 and then only the correct r, z, values must verify[71]:

$$K_X = K_A \frac{m^{1/n} \alpha_X^{1/z} (1 - \alpha_A)^{m/n}}{r^{1/z} (1 - \alpha_X)^{r/z} \alpha_A^{1/n}} = \text{const} \tag{55}$$

Accordingly, the formation constant of the $X_rL_zH_s$ species can be calculated as:

$$\beta_x = K_x^z c_x^{1-r} (1 + \Sigma\, \beta_j [H^+]^j)^r (1 + \Sigma\, \beta_i [H^+]^t)^z \tag{56}$$

These methods can be applied to the study of the interaction between polyammonium receptors and different organic anions X by using hexacyanoferrate(II) ions as a competing substrate. If no interaction exists between $[Fe(CN)_6]^{4-}$ and X, and assuming that only the hexacyanoferrate-containing species are electroactive, it can be expected that addition of X to a solution containing $[Fe(CN)_6]^{4-}$ and L causes an alteration of its electrochemical response. For instance, addition of ATP or dicarboxylic acids to a solution containing equimolecular amounts of $[Fe(CN)_6]^{4-}$ and **7.4**, determines a shift of peak potentials (Fig. 7.14b) and diffusion coefficients (Fig. 7.15) to intermediate

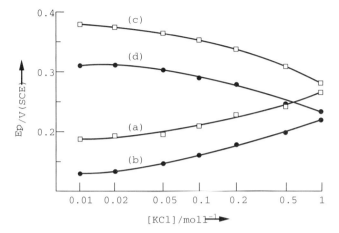

Fig. 7.19. Peak potential shifts induced by changes in the concentration of KCl for the $[Fe(CN)_6]^{3-}/[Fe(CN)_6]^{4-}$ couple in the absence (a) and (b) and in the presence (c) and (d) of 1-6H$^+$. Squares: anodic peak; points: cathodic peak. [Reproduced from F. Peter, M. Gross, M. W. Hosseini, and J.-M. Lehn, *J. Electroanal. Chem.* **144**, 279, (1983), with permission.]

values between those of complexed and uncomplexed hexacyanoferrate(II) ions.[87,88]

This competitive effect can be clearly seen in molar-ratio curves (Fig. 7.8). In the presence of X, formal potentials and peak currents tend to identical limiting values than for the A-L system, but a greater excess of ligand is required to complete the coordination of A. Application of the generalized molar-ratio method is illustrated in Fig. 7.18. Obviously, the protonation equilibria of the *A* and *X* substrates can easily be incorporated, and the method can be extended to the familiar situation in which a series of complexes with identical *X*–*L* stoichiometry but different protonation degree are formed. In favorable cases, the number of protons involved in the first protonation step can be estimated from experiments at a variable pH. Combining the equations for the complex formation equilibria of *A* and *X*, one obtains[73]:

$$g(\alpha_A, \alpha_X, H) = \log \frac{m^{1/n} \alpha_X^{1/z} (1 - \alpha_A)^{m/n}}{r^{1/z} \alpha_A^{1/n} (1 - \alpha_X)^{r/z}}$$

$$= \log \frac{\beta_s^{1/z}}{\beta_q^{1/n}} + \log \frac{c_X^{(1-r)/z}}{c_A^{(1-m)/n}} + (s/z - q/n)\text{pH} \qquad (57)$$

Providing the stoichiometry and number of protons involved in the formation of the *A*–*L* complex, the above equation predicts a linear relationship between $g(\alpha_A, \alpha_X, H)$ and the pH only for the adequate *r*, *z*, and *s* coefficients providing the stoichiometry and formation constants for the *A*–*L* system are known.

7.7 THE SIGNIFICANCE OF ELECTROCHEMICAL DATA

7.7.1 Thermochemical and Structural Considerations

In the mid-1970s thermochemical cycles incorporating electrode potentials were used in order to obtain thermodynamic data: among others, determination of acidity constants[7-9] and bond energies[10,11] for radicals and transient species. In particular, there are various works dealing with the possibility of correlating electrode potentials with HOMO–LUMO energies and optical spectroscopy.[89-94] A series of studies yielded empirical correlations of electrode potentials with ligand substitution effects.[95-97] This includes the parametrization of redox potentials of metal complexes on the basis of separate contributions depending on the structure and coordination sites of ligands.[98-100] This is a promising line of research for designing new species, predicting charge-transfer energies in optical spectroscopy, and checking assignments of observed redox potentials.[89-100]

However, it should be noted that, on solid electrodes, bulky species tend to undergo rapid and irreversible adsorption. The requisite characteristics for macromolecular recognition are not successfully presented on an electrode surface. As a result, the probability for a molecule to approach the electrode in favorable orientations is very low. The redox-active sites are often insulated by redox-inert portions of the molecule. Accordingly, electron transfer to an electrode at appreciable rates is prohibited for all but a few orientations of the molecule, even when it is in "contact" with the electrode.[101] For our purposes, the relevant point to emphasize is that it is possible to obtain significant correlations between electrochemical and structural parameters.[102]

The potential of a given redox couple reflects the relative binding strength of the two oxidation states to the ligand concerned. In the case of hexacyanoferrate(II) coordination by polyammonium receptors, electrode potentials are shifted toward more positive values, denoting that $[Fe(CN)_6]^{4-}$ ions are more strongly binding than $[Fe(CN)_6]^{3-}$ ions. In this context, electrochemical data confirm the idea that electrostatic interactions play a major role in the strength of anion binding. For instance, $[Fe(CN)_6]^{4-}$ and $[Ru(CN)_6]^{4-}$ display an identical shift in the electrode potentials upon adduct formation with fully protonated receptors **7.1** and **7.2** (see Table 7.2).[36,37] Obviously, however, structural commitments are involved.

Structural complementarity between the anionic substrate and the receptor is required; for example, the strong interaction between hexacyanoferrate(II) ions and polyammonium receptors can be interpreted in terms of a favorable structural arrangement. The square-planar fragment in the equatorial plane of the octahedral $[Fe(CN)_6]^{4-}$ anion fit into the macrocycle cavity through hydrogen-bond formation.

The relative stability of the reduced and oxidized complexed forms depends, among other factors, on the degree of protonation of the ligand. As can be seen in Table 7.1, corresponding to the hexacyanoferrate–**7.4** system, the stability of

TABLE 7.2. Formal Potentials for the Anion Complexes Formed by the Polyammonium Receptors 7.1 and 7.2 with the Substrate Anions $[Fe(CN)_6]^{4-}$ and $[Ru(CN)_6]^{4-}$ in Aqueous 0.1 M KCl Solution at pH 5.5; Potentials in mV vs. SCE (from Ref. 34)

Species	$E^{O'}$	$\Delta E^{O'}$
$[Fe(CN)_6]^{4-}$	190	0
7.2·$[Fe(CN)_6]^{4-}$	320	130
7.1·$[Fe(CN)_6]^{4-}$	355	165
$[Ru(CN)_6]^{4-}$	705	0
7.2·$[Ru(CN)_6]^{4-}$	835	130
7.1·$[Ru(CN)_6]^{4-}$	865	165

the $[Fe(CN)_6]^{3-}$–receptor adducts relative to that of $[Fe(CN)_6]^{4-}$ decreases with the degree of protonation of the complex.[38] The similarity between the values for hexacyanoferrate(III) and cobaltocyanate(III) ions denotes the importance of coulombic forces in these systems, though receptor topology and hydrogen bonding are also of significance.

Table 7.3 shows the formal potentials for a series of hexacyanoferrate adducts with polyazacycloalkanes 7.4, 7.7–7.9. The table includes stability constants for the $[Fe(CN)_6]^{4-}$–macrocycle species determined potentiometrically and those for the $[Fe(CN)_6]^{3-}$–receptor adducts calculated from cyclic voltammetric data. On comparing the values for equally protonated polyammonium species of a given receptor with different anionic substrates, one obtains that the anions most strongly complexed are the most highly charged, in agreement with expectations from a single electrostatic model. Comparison of electrode potentials for receptors containing secondary and tertiary amine groups reveals significant differences under similar conditions (pH, degree of protonation). As can be seen in Table 7.3, N-methylation results in a significant decrease of the relative stability of the hexacyanoferrate(III) species with respect

TABLE 7.3. Cyclic Voltammetric Formal Potentials and Logarithms of the Equilibrium Constants for Complex Formation (Reaction $A + H_4L = AH_4L$) of $[Fe(CN)_6]^{3-}$ and $[Fe(CN)_6]^{4-}$ with Different Polazaalkanes: All Data at 298 K in Aqueous Solution (0.15 M $NaClO_4$) (from Refs. 38 and 86)

Ligand	$E^{O'a}$	$\Delta E^{O'}$	$\log \beta^{II\,b}$	$\log \beta^{III\,c}$
7.7 (=[24]aneN$_8$)	330	140	4.1	1.8
7.4 (=[21]aneN$_7$)	305	115	5.1	3.2
7.8 (=[18]aneN$_6$)	275	85	6.3	4.9
7.9 (=Me$_4$[18]aneN$_6$)	250	60	6.4	5.4

[a] From cyclic voltammograms at pH ca. 6.
[b] From potentiometric data.
[c] From cyclic voltammetric data (Eq. 38).

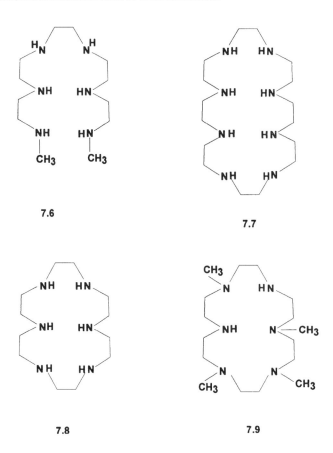

7.6

7.7

7.8

7.9

to the hexacyanoferrate(II) species. Stereochemical hindrance associated with the methyl groups and the poorer sigma-donor character of the tertiary nitrogens in aqueous solutions can be adduced to explain this result. In addition, nitrogen methylation must cause an increase of the hydrophobicity of the molecule. This matter will be treated in connection with solvation in the following.

7.7.2 Solvent and Supporting Electrolyte Effects

As deduced from the thermochemical cycles depicted in Figs. 7.1 and 7.2, solvation plays an important role in electrochemical processes. In the most simple theoretical approach for ion solvation, a single spherical ion is immersed into a homogeneous medium. The Born equation for the solvation free energy (see Chapter 1) of a single ion is[103]:

$$\Delta G_{\text{solv}} = -\frac{B}{r}\left(1 - \frac{1}{\varepsilon}\right) \tag{58}$$

where z represents the charge of the ion, r its radius, ε the dielectric constant of the medium, and the other symbols have their usual significance. It can easily be conjectured that this extremely simple approach is unsatisfactory for a detailed analysis of the ion–solvent interactions. More detailed models are based on incorporation of additive terms of ion–dipole, quadrupole, etc., interactions. For a structural approach, modifications of the inner-sphere interactions by outer-sphere solute–solvent interactions and by entropy changes in solvent structure by the charged species must be accounted for.[104–107]

For anions, generally both forms of the redox couple interact strongly with the solvent, the formal potential being determined by the relative strength of the interactions. From the cycle depicted in Fig. 7.1, one can obtain:

$$\Delta G^0 = \text{const} + (\Delta G_{\text{solv rd}} - \Delta G_{\text{solv ox}}) \qquad (59)$$

For the $A^{m-}/A^{(m+n)-}$ couple, and assuming a similar ionic radius for both reduced and oxidized anions, application of the single Born electrostatic model reduces the solvation term approximately to: const $\times (2mn + n^2)$. Since both the solvation free energies are negative, an increase of the overall ionic charge must make the solvation term more negative. Then the electrode potential becomes more positive, denoting the reduced species is stabilized in relation to the oxidized one. A comparison of the ionic contribution to solvation energy (as a function of the net charge of oxidized and reduced ionic species) and the shift in the formal potential of the $A^{m-}/A^{(m+n)-}$ couple upon complexation is presented in Table 7.4. As expected, the shift in the formal potential increases as the ionic factor increases.

Electrode potentials in different solvents are determined to a great extent by the interaction of the solvent with both the oxidized and reduced form of a redox couple. A simple example is provided by the reversible reduction of dioxygen to superoxide ions in aprotic solvents.[108] As can be seen in Table 7.5, the redox potential for the O_2/O_2^- couple presents relatively large variations depending on the solvent. Assuming that the solvation energy for O_2 is small

TABLE 7.4. Comparison of Cyclic Voltammetric Formal Potentials, Net Ionic Charge of Reduced and Oxidized Forms, and Expected Free Enthalpy Variations for Complex Formation of $[Fe(CN)_2]^{4-}$ 1,20-*bis*(Methylamino)-3,6,9,12,15,18-Hexaazanicosane; Data at 298 K in Aqueous Solution (0.15 M NaClO$_4$) (from Ref. 38)

Number of protons	Net charge		Theoretical ΔG_{solv}	Observed $\Delta E^{0'}$ (mV)
	reduced	oxidized		
4	0	+1	C^a	30
5	+1	+2	3C	90
6	+2	+3	5C	110
7	+3	+4	5C	150

^{a} C = const.

TABLE 7.5. Comparison of the Formal Potentials for O_2/O_2^- (*A*) and the $[Fe(CN)_6]^{3-}$/ $[Fe(CN)_6]^{4-}$ (*B*) Couples in Various Solvents with their Dielectric Constant at 298 K and Acceptor Number; Potentials in Volts vs. SCE (Data from Refs. 108, 109)

Solvent	AN	ε	$E^{O'}(A)^a$	$E^{O'}(B)^b$
water	54.8	78.5	−0.41	+0.19
MeOH	41.3	32.6	−0.77	+0.47
DMSO	19.3	46.7	−0.78	−0.27
MeCN	19.3	36.0	−0.87	−0.27
DMF	16.0	36.7	−0.86	−0.31

^{a} In 0.10 M $(Et_4N)ClO_4$.
^{b} In 0.10 M $(Me_4N)ClO_4$.

and about the same for the different solvents,[108] the formal potential must depend principally on the degree of solvation of the superoxide ion. Then a positive shift of $E^{O'}$ parallel to the increase of the anion solvation ability of the solvent can be expected.

It is convenient to note that ion solvation in different solvents may be empirically characterized by different solvent-dependent parameters. In particular the donor number (DN) or the acceptor number (AN) can be correlated with the electrode potentials for selected redox couples.[109] With regard to anion coordination chemistry, formation of anion–receptor adducts may be described as an outer-sphere interaction. This situation can be described in terms of a competition between the receptor and the solvent molecules for coordination sites on the redox active species.

Coordination by the solvent results in the shortening of the ligand–receptor bonds. Solvent–ion interaction depends on the net charge of the ion and its structure and size; that is, in addition to purely electrostatic interactions, structural factors could be present.[106,107] For the $([Fe(CN)_6]^{3-}/[Fe(CN)_6]^{4-})$ system, the solvent interaction with the reduced form is much stronger than with the oxidized form. This can be rationalized taking into account that the reduced form carries a higher negative charge and by the enhancement of the donor ability of nitrogen atoms on each of the cyano groups. Then, increasing acceptor properties of the solvent eases the reduction of hexacyanoferrate(III) to hexacyanoferrate(II) ions. As can be seen in Table 7.5, the formal potential for the $([Fe(CN)_6]^{3-}/[Fe(CN)_6]^{4-})$ couple is shifted to more positive values by an increase of the acceptor number of the solvent. For the $([Fe(CN)_6]^{3-}/[Fe(CN)_6]^{4-})$ system, the solvent interaction with the reduced form is much stronger than with the oxidized form.

On the other hand, electrochemical measurements are carried out in the presence of a large excess of a supporting electrolyte. Therefore, a significant competitive receptor binding may occur between the anionic substrate and the anion of the supporting electrolyte. This can be seen in Fig. 7.19, which reflects the peak potential shifts for the $[Fe(CN)_6]^{3-}/[Fe(CN)_6]^{4-}$ system in the absence and in the presence of an anionic receptor, associated with changes in the concentration of supporting electrolyte.[37] For equally charged anions, this competitive effect, however, is about the same at a given ionic strength regardless of their chemical identity. Nevertheless, the chemical identity of the cation of the supporting electrolyte (at a given ionic strength and a given counterion) can affect the observed redox potentials. Such an effect suggests that formation of ion pairs between the anionic substrate and the cation of the background electrolyte ion pairs occurs.[37]

Similar considerations can be applied to the observed changes in formal potentials upon variation of the cation of the supporting electrolyte.[109]

7.8 OTHER ISSUES

As noted elsewhere, a variety of applications dealing with the field of electrochemistry derive from the host–guest interactions. Anion coordination electrochemistry is relevant in preparation of molecular sensory devices, redox catalysts, and modeling electron-transfer reactions in biological systems.[15]

In brief, electrochemical sensors rely on the relationship between a measurable electrical property—typically current (amperometric) or emf (potentiometric)—of an electrochemical cell and the concentration of the problem species in the sample. Redox-responsive macrocycles containing electrochemically reducible substrates for binding metal cations,[110–112] anions,[113] and redox-active cavitand host molecules for inclusion of organic guest substrates[114,115] have been developed. Significant effort has recently been placed on the development of ion-selective electrodes (ISEs) from the preparation of modified electrodes. In particular, the tactic of incorporating electroactive ions into polyionic films on electrodes has been extensively employed to prepare electroactive films.[116–118]

Ionophore-impregnated polymers are employed in a variety of analyses. As is well known, electrode selectivity is usually governed by simple anion lipophilicity, according to Hofmeister series. To develop electrodes whose selectivity differs from the Hofmeister series, a specific interaction must occur between the ion and the ionophore. Thus a number of anion-selective electrodes have been developed on the basis of reversible coordination of the anionic guests to the vacant coordination sites of the central metal ions in alkyltin compounds,[119,120] vitamin B_{12} and metalloporphyrin derivatives.[121] Alternatively, electrostatic interaction between the anionic guests and the cationic hosts, as is the case of diphosphonium[122] and diammonium cations,[123] and macrocyclic polyamines,[124] provides the basis for chemical sensing.

Ion-selective membranes can also be prepared by electropolymerization. This is based on the electrodeposition of a polymeric film on the surface of an electrode that is immersed in a solution of an appropriate monomer.[125] The monomer needs to contain groups that can be electrochemically oxidized or reduced to form a polymer. For example, membranes prepared by electropolymerization of cobalt(II) tetrakis(o-aminophenyl) porphyrin on a glassy-carbon surface exhibit a near-Nernstian response and good selectivity properties for anions.[126]

There are a number of electrochemical processes that can be accelerated by addition of an adequate catalyst to the solution or inserting it into a modified electrode. For instance, water-soluble Ni complexes with azacyclam-like ligands, which have been shown catalytic activity in DNA modification under oxidative conditions,[127] are able to mediate electrochemical reduction of CO_2[128–131] and alkyl halides.[132,133] A surface complex formed between the adsorbed metal–ligand complex and the substrate is generally recognized as the initial step of substrate reduction,[134] involving, eventually, intermediate oxidation states.[135] In any case, the ligand structure appears to be strictly critical; thus modifications of the ring size, introduction of alkyl substituents, etc. drastically modify the catalytic efficiency in the reported processes.[135]

Alternatively, catalysts can be coated on electrode surfaces. For instance, the catalytic reduction of NO_2^- and NO by iron, ruthenium complexes with EDTA, and porphyrins,[136–138] and nickel and cobalt with cyclam[139] have been reported. It appears that adducts formed between substrate and adsorbed catalyst are formed in the initial steps on the reduction.[140]

Although the above examples do not exhaust the projection of electrochemistry into the framework of anion coordination chemistry, they are illustrative of the existence of a wide variety of applications that make electrochemistry an active field of research in various chemistry areas and, in particular, in anion supramolecular chemistry.

7.9 REFERENCES

1. I. M. Kolthoff and J. J. Lingane, *Polarography*, 2nd ed., Wiley-Interscience, New York, 1952.
2. P. Delahay, *New Instrumental Methods in Electrochemistry*, Wiley-Interscience, New York, 1954.
3. J. J. Lingane, *Electroanalytical Chemistry*, 2nd ed., Wiley-Interscience, New York, 1958.
4. L. Meites, *Polarographic Techniques*, 2nd ed., Wiley-Interscience, New York, 1958.
5. J. Heyrovsky and J. Kuta, *Principles of Polarography*, Academic Press, New York, 1966.
6. A. J. Bard and L. R. Faulkner, *Electrochemical Methods*, John Wiley & Sons, New York, 1980.
7. R. Breslow and K. Balasubramanian, *J. Am. Chem. Soc.* **91**, 5182 (1969).
8. R. Breslow and W. Chu, *J. Am. Chem. Soc.* **92**, 2165 (1970).
9. R. Breslow and J. Grant, *J. Am. Chem. Soc.* **99**, 7745 (1977).

10. F. G. Bordwell, X. Zhang, and J. P. Cheng, *J. Org. Chem.* **56**, 3216 (1991).
11. D. D. Wayner and V. D. Parker, *Acc. Chem. Res.* **26**, 287 (1993).
12. J. Janata, M. Josowicz, and D. M. DeVaney, *Anal. Chem.* **66**, 207R (1994).
13. M. D. Ryan, E. F. Bowden, and J. Q. Chambers, *Anal. Chem.* **66**, 360R (1994).
14. J.-M. Lehn, *Pure Appl. Chem.* **50**, 871 (1978).
15. P. D. Beer, *Chem. Soc. Rev.* **18**, 409 (1989).
16. S. F. Nelsen, C. R. Kessel, D. J. Brien, and F. Weinhold, *J. Org. Chem.* **45**, 2116 (1980).
17. J. M. Bobbitt and J. P. Wills, *J. Org. Chem.* **45**, 1978 (1980).
18. C. A. Rice and J. T. Spence, *Inorg. Chem.* **19**, 2845 (1980).
19. M. J. Powers and T. J. Meyer, *J. Am. Chem. Soc.* **102**, 1289 (1980).
20. A. M. Nicholas and D. R. Arnold, *Can. J. Chem.* **60**, 2165 (1982).
21. W. C. Barrette, H. W. Johnson, and D. T. Sawyer, *Anal. Chem.* **56**, 1890 (1984).
22. V. D. Parker, K. L. Handoo, F. Roness, and M. Tilset, *J. Am. Chem. Soc.* **113**, 7493 (1991).
23. H. Matsuda and Y. Ayabe, *Z. Elektrochem.* **59**, 494 (1955).
24. D. D. Macdonald. *Transient Techniques in Electrochemistry*, Plenum Press, New York, 1977.
25. A. Bond, *Modern Polarographic Methods in Analytical Chemistry*, Marcel Dekker, New York, 1980.
26. M. Rudolph, D. P. Redy, and S. W. Feldberg, *Anal. Chem.* **66**, 589A (1994).
27. R. S. Nicholson, *Anal. Chem.* **37**, 667 (1965).
28. W. E. Geiger, *Progr. Inor. Chem.* **33**, 275 (1986).
29. C. Amatore, M. Azzabi, P. Colas, A. Jutand, C. Lefrou, and Y. Rollin, *J. Electroanal. Chem.* **288**, 45 (1990).
30. W. T. Yap and L. M. Doane, *Anal. Chem.* **54**, 1437 (1982).
31. D. E. Weisshaar and D. E. Tallman, *Anal. Chem.* **55**, 383 (1983).
32. G. Ginzburg. *Anal. Chem.* **50**, 375 (1978).
33. B. R. Eggins and N. H. Smith, *Anal. Chem.* **51**, 2282 (1979).
34. D. S. Polcyn and I. Shain, *Anal. Chem.* **38**, 370 (1966).
35. R. S. Nicholson and I. Shain, *Anal. Chem.* **36**, 706 (1964).
36. F. Peter, M. Gross, M. W. Hosseini, J. M. Lehn, and R. B. Sessions, *J. Chem. Soc. Chem. Commun.*, 1067 (1981).
37. F. Peter, M. Gross, M. W. Hosseini, and J. M. Lehn, *J. Electroanal. Chem.* **144**, 279 (1983).
38. J. Aragó, A. Bencini, A. Bianchi, A. Doménech, and E. García-España, *J. Chem. Soc. Dalton Trans.*, 319 (1991).
39. A. Bencini, A. Bianchi, M. I. Burguete, A. Doménech, E. García-España, S. V. Luís, M. A. Niño, and J. A. Ramírez, *J. Chem. Soc. Perkin Trans.*, 1445 (1991).
40. C. O. Schmakel, K. S. V. Santhanam, and P. J. Elving, *J. Am. Chem. Soc.* **97**, 5083 (1975).
41. K. S. V. Santhanam and P. J. Elving, *J. Am. Chem. Soc.* **95**, 5482 (1973).
42. P. Hapiot, J. Moiroux, and J. M. Savéant, *J. Am. Chem. Soc.* **112**, 1337 (1990).

43. J. Moiroux and P. J. Elving, *J. Am. Chem. Soc.* **102**, 6533 (1980).
44. G. S. Alberts and I. Shain, *Anal. Chem.* **35**, 1859 (1963).
45. Unpublished results.
46. D. R. Crow, *Polarography of Metal Complexes*, Academic Press, London, 1970.
47. E. Casassas and M. Esteban, *J. Electroanal. Chem.* **194**, 11 (1985).
48. H. Gampp, *Anal. Chem.* **59**, 2456 (1987).
49. M. M. C. DosSantos and M. L. S. Goncalves, *Electrochim. Acta* **37**, 1413 (1992).
50. K. Wikiel, M. M. DosSantos, and J. Osteryoung, *Electrochim. Acta* **38**, 1555 (1993).
51. E. Casassas and C. Ariño, *J. Electroanal. Chem.* **213**, 235 (1986).
52. M. Esteban, E. Casassas, and L. Fernández, *Talanta* **33**, 847 (1986).
53. J. J. Lingane, *Chem. Rev.* **29**, 1 (1941).
54. D. D. DeFord and D. N. Hume, *J. Am. Chem. Soc.* **73**, 5321 (1951).
55. J. I. Watters and R. J. DeWitt, *J. Am. Chem. Soc.* **82**, 1333 (1960).
56. W. B. Schaap and D. L. McMasters, *J. Am. Chem. Soc.* **83**, 4699 (1961).
57. H. M. Killa, E. E. Mercer, and R. H. Philp, *Anal. Chem.* **56**, 2401 (1984).
58. H. M. Killa, *J. Chem. Soc. Faraday Trans.* **81**, 2659 (1985).
59. H. M. Killa, *Polyhedr.* **8**, 2299 (1989).
60. H. M. Killa and R. H. Philp, *J. Electroanal. Chem.* **175**, 223 (1984).
61. C. G. Butler and R. C. Kaye, *J. Electroanal. Chem.* **8**, 463 (1964).
62. V. Kacena and L. Matousek, *Collect. Czech. Chem. Commun.* **18**, 294 (1953).
63. D. R. Crow, *J. Electroanal. Chem.* **16**, 137 (1968).
64. D. R. Crow, *Talanta* **29**, 733 (1982).
65. D. R. Crow, *Talanta* **29**, 739 (1982).
66. D. R. Crow, *Talanta* **30**, 659 (1983).
67. D. R. Crow, *Talanta* **31**, 421 (1984).
68. D. R. Crow, *Talanta* **33**, 553 (1986).
69. D. R. Crow, *Electrochim. Actal* **28**, 1799 (1983).
70. K. Ogura, Y. Fukusima, and C. Aomizu, *J. Electroanal. Chem.* **107**, 271 (1980).
71. J. H. Yoe and A. L. Jones, *Ind. Eng. Chem. Anal.* **16**, 11 (1944).
72. E. Asmus, *Z. Anal. Chem.* **178**, 104 (1960).
73. A. Beltrán, J. Beltrán, A. Cervilla, and J. A. Ramírez, *Talanta* **30**, 124 (1983).
74. A. Beltrán and J. A. Ramírez, *Can. J. Chem.* **61**, 1100 (1983).
75. A. Bianchi, A. Doménech, E. García-España, and S. V. Luís, *Anal. Chem.* **65**, 3137 (1993).
76. B. Viossat, *Rev. Chim. Miner.* **9**, 737 (1972).
77. A. Doménech, E. García-España, and J. A. Ramirez, *Talanta* **42**,1663 (1995).
78. R. Guidelli, *J. Electroanal. Chem.* **33**, 291 (1971).
79. R. Guidelli, *J. Electroanal. Chem.* **33**, 303 (1971).
80. K. B. Oldham, *J. Electroanal. Chem.* **313**, 3 (1991).
81. D. N. Blauch and F. C. Anson, *J. Electroanal. Chem.* **309**, 313 (1991).

82. E. García-España, M. Micheloni, P. Paoletti, and A. Bianchi, *Inorg. Chim. Acta* **102**, L9 (1985).
83. G. Schwarzenbach, R. Gut, and G. Anderegg, *Helv. Chim. Acta* **37**, 937 (1954).
84. M. Kodama and E. Kimura, *J. Chem. Soc. Dalton Tran.*, 2335 (1976).
85. M. Kodama and A. Kimura, *Inorg. Chem.* **17**, 2446 (1978).
86. E. Kimura, A. Sakonaka, T. Yatsunami, and M. Kodama, *J. Am. Chem. Soc.* **103**, 3041 (1981).
87. A. Andrés, J. Aragó, A. Bencini, A. Bianchi, A. Doménech, V. Fusi, E. García-España, P. Paoletti, and J. A. Ramírez, *Inorg. Chem.* **32**, 3418 (1993).
88. A. Bencini, A. Bianchi, M. I. Burguete, P. Dapporto, A. Doménech, E. García-España, S. V. Luís, P. Paoli, and J. A. Ramírez, *J. Chem. Soc. Perkin Trans.*, 569 (1994).
89. D. C. Olson, V. P. Mayweg, and G. N. Schrauzer, *J. Am. Chem. Soc.* **88**, 4876 (1966).
90. C. J. Pickett and D. Pletcher, *J. Organomet. Chem.* **102**, 327 (1975).
91. B. E. Burster, *J. Am. Chem. Soc.* **104**, 1299 (1982).
92. A. A. Vlcek, *Coord. Chem. Rev.* **43**, 39 (1982).
93. G. S. Wilson, *J. Am. Chem. Soc.* **111**, 4036 (1989).
94. C. S. J. Chang, A. Rai-Chaudhuri, D. L. Lichtenberger, and J. H. Enemark, *Polyhedr.* **9**, 1965 (1990).
95. P. M. Treichel, G. E. Durren, and H. J. Mueh, *J. Organomet. Chem.* **44**, 339 (1972).
96. A. Sarapu and R. F. Fenske, *Inorg. Chem.* **14**, 247 (1975).
97. A. B. P. Lever, *Inorg. Chem.* **29**, 1271 (1990).
98. A. W. Addison, *Inorg. Chim. Acta* **162**, 217 (1989).
99. M. M. Bernardo, M. J. Hccg, R. R. Schroeder, L. A. Ochrymowcycz, and D. B. Rorabacher, *Inorg. Chem.* **31**, 191 (1992).
100. E. S. Doodsworth, A. A. Vlcek, and A. B. P. Lever, *Inorg. Chem.* **33**, 1045 (1994).
101. J. F. Endicott, K. Kumar, T. Ramasami, and F. P. Rotzinger, *Progr. Inorg Chem.* **30**, 141 (1983).
102. W. L. Reynolds and R. W. Lumry, *Mechanics of Electron Transfer*, Ronald, New York, 1966.
103. J. O'M. Bockris and A. K. N. Reddy, *Modern Electrochemistry*, Plenum Press, New York, 1970.
104. J. T. Hupp and M. J. Weaver, *Inorg. Chem.* **23**, 3639 (1984).
105. E. L. Yee, R. J. Cave, K. L. Guyer, P. D. Tyma, and M. J. Weaver, *J. Am. Chem. Soc.* **101**, 1131 (1979).
106. R. H. Stokes and R. Mills, *Viscosity of Electrolytes and Related Properties*, Pergamon, London, 1965.
107. M. M. Graziani, F. Sánchez, A. Rodríguez, and M. L. Moyá, *Inorg. Chim. Acta* **208**, 213 (1993).
108. D. T. Sawyer, G. Chiericato, C. T. Angelis, E. J. Nanni, and T. Tsuchiya, *Anal. Chem.* **54**, 1720 (1982).

109. V. Gutmann, *The Donor–Acceptor Approach to Molecular Interactions*, Plenum Press, New York, 1978.
110. H. Bock, B. Hierholzer, F. Votgle, and G. Hollman, *Angew. Chem. Int. Ed. Engl.* **23**, 57 (1984).
111. R. E. Wolf and S. R. Cooper, *J. Am. Chem. Soc.* **106**, 4646 (1984).
112. D. A. Custowski, M. Delgado, V. J. Gatto, L. Echegoyen, and G. W. Gokel, *J. Am. Chem Soc.* **108**, 7553 (1986).
113. P. D. Beer and A. D. Keefe, *J. Organomet. Chem.* **375**, C40 (1989).
114. P. D. Beer, M. G. B. Drew, and A. D. Keefe, *J. Organomet. Chem.* **353**, C10 (1988).
115. P. D. Beer, M. G. B. Drew, A. Ibbotson, and E. L. Tite, *J. Chem. Soc. Chem. Commun.*, 1498 (1988).
116. N. Oyama, T. Shimomura, K. Sigehara, and F. C. Anson, *J. Electroanal. Chem.* **112**, 271 (1980).
117. D. A. Buttry, F. C. Anson, *J. Am. Chem. Soc.* **105**, 685 (1983).
118. F. C. Anson, T. Ohsaka, and J. M. Saveant, *J. Am. Chem. Soc.* **105**, 4883 (1983).
119. U. Wuthier, H. V. Pham, R. Zünd, R. Welti, R. J. J. Funck, A. Bezegh, D. Ammann, E. Pretsch, and W. Simon, *Anal. Chem.* **56**, 535 (1984).
120. S. C. Glazier, M. A. Arnold, *Anal. Chem.* **63**, 754 (1991).
121. P. Schulthess, D. Ammann, B. Kraütler, C. Caderas, R. Stepánek, and W. Simon, *Anal. Chem.* **57**, 1397 (1985).
122. A. Ohki, M. Yamura, S. Kamamoto, S. Maeda, T. Takeshita, and M. Tagaki, *Chem. Lett.* 95 (1989).
123. V. J. Wotring, D. M. Johnson, and L. G. Bachas, *Anal. Chem.* **62**, 1506 (1990).
124. Y. Umezawa, M. Kataoka, W. Takami, E. Kimura, T. Koike, and H. Nada, *Anal. Chem.* **60**, 2392 (1988).
125. J. Heinze, *Top. Curr. Chem.* **152**, 1 (1980).
126. S. Daumert, S. Wallace, A. Florido, and L.G. Bachas, *Anal. Chem.* **63**, 1676 (1991).
127. X. Chen, S. E. Rovita, and C. J. Burrows, *J. Am. Chem. Soc.* **113**, 5884 1991.
128. B. Fisher, R. Eisenberg, *J. Am. Chem. Soc.* **102**, 7361 (1980).
129. M. Beley, J. P. Collin, R. Ruppert, and J. P. Sauvage, *J. Chem. Soc. Chem. Commun.*, 1315 (1984).
130. M. Beley, J. P. Collin, R. Ruppert, and J. P. Sauvage, *J. Am. Chem. Soc.* **108**, 7461 (1986).
131. H. Fujihara, Y. Hirata, and K. Suga, *J. Electroanal. Chem.* **292**, 199 (1990).
132. C. Gosden, K. P. Healy, D. Pletcher, and R. Rosas, *J. Chem. Soc. Dalton Trans.*, 972 (1978).
133. C. Gosden, J. B. Kerr, D. Pletcher, and R. Rosas, *J. Electroanal. Chem.* **117**, 101 (1978).
134. J. E. Toth, and F. C. Anson, *J. Am. Chem. Soc.* **111**, 2444 (1989).
135. F. Abbà, G. De Santis, L. Fabbrizzi, M. Licchelli, A. M. Manotti, P. Pallavicini, A. Poggi, and F. Ugozzoli, *Inorg. Chem.* **33**, 1366 (1994).
136. K. Ogura and H. Ishikawa, *J. Chem. Soc. Faraday Trans.* **80**, 2243 (1980).

137. S. Ochiyama and G. Muto, *J. Electroanal. Chem.* **127**, 275 (1981).
138. I. Taniguchi, N. Nakashima, K. Matsushita, and K. Yasukouchi, *J. Electroanal. Chem.* **224**, 19 (1987).
139. H. L. Li, J. Q. Chamber, and D. T. Hobbs, *J. Electroanal. Chem.* **256**, 447 (1988).
140. J. Zhang, A. B. P. Lever, and W. J. Pietro, *Inorg. Chem.* **33**, 1392 (1994).

CHAPTER 8

Photochemistry and Photophysics of Supramolecular Species Containing Anions

L. MOGGI and M. F. MANFRIN

8.1 Introduction
8.2 Supramolecular Systems Involving Anionic Components
8.3 Ion Pairs
8.4 Adducts of Polyammonium Macrocycles
8.5 Other Studies
8.6 References

8.1 INTRODUCTION

The photochemical and photophysical properties of a chemical species concern the events that follow the production of an electronic excited state of the species. Essentially, an excited state may be considered as an unstable species that tends to lose its excess energy through a sequence of monomolecular or bimolecular steps. The former steps may be classified as:

1. Radiationless deactivation that converts the excited state to a lower-energy excited state or to the ground electronic state of the same molecule;

$$^{**}A \rightarrow {}^*A \quad \text{or} \quad A + \text{heat}$$

2. Radiative deactivation (luminescence) that generally leads to a ground-state molecule;

$$^*A \rightarrow A + h\nu$$

Supramolecular Chemistry of Anions, Edited by Antonio Bianchi, Kristin Bowman-James, and Enrique García-España.
ISBN 0-471-18622-8. © 1997 Wiley-VCH, Inc.

3. Intramolecular photoreaction (mainly, photodissociation or photoisomerization).

$$*A \rightarrow \text{products}$$

On the other hand, the most common types of bimolecular processes are the electronic energy transfer

$$*A + B \rightarrow A + *B$$

and the photochemical electron transfer.

$$*A + B \rightarrow A^+ + B^- \quad \text{or} \quad *A + B \rightarrow A^- + B^+$$

Since an excited state may simultaneously undergo more than one of the above-mentioned steps, the ultimate fate of the original excitation is the result of a delicate kinetic balance among the various steps that may originate from each one of the excited states involved in the overall process.

In order to discuss the photochemical and photophysical properties of supramolecular systems, it is convenient to consider a supermolecule as formed by molecular components that are capable, with minor modifications, of separate existence and that experience only a small electronic interaction with each other in the supermolecule. Under such conditions, each component is expected to maintain its own sequence of electronic excited states, slightly modified in energy and equilibrium configuration by the presence of the other components, and the photochemistry of a supermolecule may therefore be considered as a modification of the photochemistry of one (or more) of its components induced by the other components and usefully compared to the photochemical properties of the photoactive isolated component.

While the light absorption properties of a component, and thus the optical population of its excited states, are often a little affected by the other components of the system (the absorption spectrum of a supermolecule is often similar to the sum of the spectra of the components), drastic changes may easily occur in the sequence of deactivation steps that follow the electronic excitation, and thus in the photochemical and photophysical properties. There are three main ways by which a second component affects the properties of a photoactive component. The first is simply due to the electronic interaction between the components, which causes a perturbation of the electronic states of the photoactive one, with a change in the efficiency of the various deactivation modes; large changes in the quantum yields of photoreaction and/or luminescence may ultimately result. The second way is a consequence of the presence of electronic excited states of the second component in the energy range of the excited states involved in the photochemistry (or photophysics) of the photoactive component. An intramolecular energy transfer between the two components may thus take place,

$$*A\text{-}B \rightarrow A\text{-}*B \tag{1}$$

which competes with the "normal" deactivation modes of *A and reduces their efficiencies. In other words, the component B acts as a "quencher" of the excited state *A. The final result is the reduction (as a limit, the disappearance) of the photoreaction and/or emission of the component A and, possibly, the appearance of some photochemical or photophysical property of B. The latter way occurs when two components of a supramolecular system are arranged in a geometry that imposes a constraint to large-amplitude nuclear motions of the photoactive component; such a constraint, in fact, may reduce the possibility that photochemical reactions involving large geometrical modifications, such as, for example, dissociation and isomerization, of the photoactive component occur in the supramolecular system. In terms of potential energy curves of the excited states of the reactive component, this effect does not cause perturbations near their equilibrium geometries, so that the photophysical properties of the component are often not affected at all. Host–guest supramolecular systems appear to be the most appropriate candidates for this type of effect.

In some cases electronic excited states are present in supramolecular systems, which are a consequence of intercomponent interactions and so are not present in the isolated components. They are generally excited states that, with respect to the ground state, involve the transfer of one electron from one component to another one,

$$A-B \rightarrow A^+-B^- \qquad (2)$$

and therefore are related to the redox properties of the components. These intercomponent excited states may be populated by light absorption,

$$A-B + h\nu \rightarrow A^+-B^- \qquad (3)$$

and in such a case a new absorption band appears in the spectrum of the supramolecule that is not present in the spectra of the isolated components. They can also be populated by nonradiative conversion of a higher-energy excited state localized on one component:

$$*A-B \rightarrow A^+-B^- \qquad (4)$$

This process may be considered as a light-induced charge separation in the supramolecular system, a fundamental step for the conversion of light energy into chemical energy.

From a practical point of view, two simple applications of the photochemical properties of supramolecular systems may be considered. The first is the tuning of the photochemical and/or photophysical properties of a chemical species by means of its interaction with a second appropriate component in order to improve these properties; the second is the use of the change in the emission of a luminofore induced by supramolecular complexation as a tool for molecular recognition. Photochemical applications in more complex supramolecular systems have been widely discussed in a recent monograph.[1] They include (1) photoinduced electron-transfer processes (reaction 4) for light energy conversion, electron collection, and switching of electric signals, (2) intramolecular energy transfer (reaction 1) as a way to induce antenna effects, light energy upconversion and remote photosensitization, and (3) photoinduced structural changes in a component, which may cause changes in the receptor ability of a supramolecular system or in the size of a cavity as well as the switching of electric signals.

8.2 SUPRAMOLECULAR SYSTEMS INVOLVING ANIONIC COMPONENTS

Photochemical and photophysical studies on supramolecular systems have received a great deal of attention recently,[2] for both theoretical purposes related to the modification of electronic excited states in supramolecular environments and practical applications. Moreover, luminescence has been widely used in molecular recognition by means of the formation of host–guest adducts.

In spite of this promising presentation, only a very small fraction of the work done in this field concerns supramolecular systems involving anions as fundamental components. The largest part of the publications regards the photochemical properties of ion pairs, that is, the simplest form of supramolecular systems involving ions; they will be briefly reviewed in the next section. In addition, only one line of research has systematically been investigated, namely, the photochemical properties of the adducts between anionic cyano complexes and ammonium macrocycles, which will be dealt with in Section 8.4. Finally, some significant scattered studies found in the literature will be collected in the last section.

8.3 ION PAIRS

The photochemical properties of ion pairs have been widely examined in Refs. 1 and 3, and so we will discuss here only their main general features. Ion pairing is a common property in the field of inorganic chemistry, but the species of interest from our point of view are, however, limited to those that involve at least one photoactive coordination compound, mainly because of spectroscopic reasons. The photochemistry of ion pairs is obviously less developed in the field of organic chemistry, which mainly concerns neutral species; most of the studies deal with pyridinium-type cations, and particularly the 1,1'-dimethyl-4,4'-dipyridinium dication (methylviologen, MV^{2+}) **8.1**, which is commonly used as an electron relay in systems for the photochemical conversion and storage of light energy.

One of the most common and important features of ion pairs is the presence of an intercomponent electron-transfer excited state, called an ion-pair

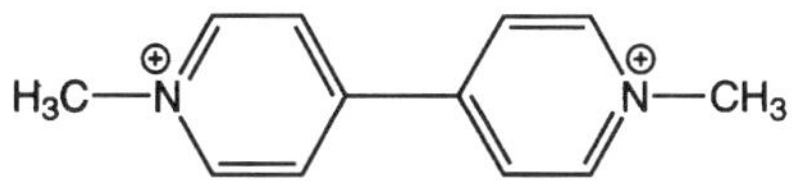

methylviologen
(paraquat)
8.1

charge-transfer (IPCT) state, at low energy; the corresponding transition from the ground state

$$A^{n+}\text{-}B^{m-} + h\nu \rightarrow *[A^{(n-1)+}\text{-}B^{(m-1)-}]$$

often appears as a new band in the absorption spectrum of the system, in the range of or even below the bands corresponding to transitions within each parent component. The major part of the studies on the photochemical properties of ion pairs concerns systems characterized by IPCT excited states. When such a state is the lowest-energy excited state (or it is close to the lowest one), it may determine the final fate of the excited-state deactivation even if the original excitation leads to an excited state localized on a single component, because of efficient radiationless transitions to the IPCT state (reaction 4); quenching of the photoreaction and/or emission of the originally excited components are thus a common consequence of ion pairing. On the other hand, an IPCT state often gives rise to the so-called "back-electron-transfer" thermal reaction to the ground state

$$*[A^{(n-1)+}\text{-}B^{(m-1)-}] \rightarrow A^{n+}\text{-}B^{m-}$$

in a time scale shorter than the time required by the components to dissociate and escape the solvent cage; no permanent chemical change nor emission can be observed in this case upon stationary excitation, and only flash spectroscopy is able to indicate the presence of the IPCT state. An overall photoreaction may, however, be observed whenever the $A^{(n-1)+}$ or the $B^{(m-1)-}$ species is unstable and can rapidly be converted to a different species, in competition with the back electron transfer. The one-electron reduction of coordination compounds leads sometimes to labile complexes that undergo a rapid ligand release. Typical in this sense is the different behavior of the I^- ion pairs of $Co(NH_3)_6^{3+}$ and its cage analogous $Co(sep)^{3+}$ (sep = sepulchrate: 1,3,6,8,10,13,16,19-octaazabicyclo-[6,6,6]-icosane), **8.2**. The IPCT state of the former pair undergoes a rapid dissociation of the Co(II) complex leading to $Co^{2+}(aq)$ and I_2 as final photochemical products.[4]

$$Co(NH_3)_6^{3+}\text{-}I^- + h\nu \rightarrow Co(NH_3)_6^{2+}\text{-}I$$
$$Co(NH_3)_6^{2+}\text{-}I \rightarrow Co^{2+} + 6NH_3 + \tfrac{1}{2}I_2$$

in the case of $Co(sep)^{2+}$, the cage nature of the ligand prevents the analogous

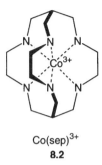

Co(sep)³⁺
8.2

dissociation, and back electron transfer occurs with practically unitary efficiency[5]:

$$\text{Co(sep)}^{3+}\text{-I}^- + h\nu \rightarrow \text{Co(sep)}^{2+}\text{-I}$$

$$\text{Co(sep)}^{2+}\text{-I} \rightarrow \text{Co(sep)}^{3+}\text{-I}^-$$

A similar situation occurs in the case of some anions (e.g., $C_2O_4^{2-}$) that become labile species upon one-electron oxidation. The use of "sacrificial" electron donors in systems used for investigating the photochemical conversion and storage of solar energy is based on similar properties.

Another interesting feature is the possibility of tuning the excited-state properties of an ionic species by coupling it to an appropriate counterion. Three examples merit reporting. The first is given by the emission intensity and lifetime of Os(phen)_3^{2+}, which, in solvents of low dielectric constant, depend on the nature of the anion.[6] The effect has been considered to be electrostatic in nature, related to the localization of the excited electron into a single phenanthroline ligand.

The second example is given by the changes in the absorption spectra, emission spectra, and lifetimes from the $[\text{Eu} = \mathbf{8.3}]^{3+}$ and $[\text{Tb} = \mathbf{8.3}]^{3+}$ cryptates induced by the addition of F^- ions.[7] The observed effects are attributed to the formation of 1:1 and 1:2 adducts between the complex and the anion, in which F^- is partially coordinated to the metal ion. These adducts seem therefore to represent a boundary situation between ion pair and a new complex.

Finally, the third example concerns the association between the lasalocid A anion (LAS^-) and some cationic complexes, Cr(bpy)_3^{3+}, Cr(phen)_3^{3+},

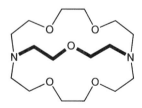

2.2.1 cryptand
8.3

Ru(bpy)$_3^{2+}$, and Pt(bpy)(NH$_3$)$_2^{2+}$, investigated by means of absorption and emission spectrophotometric measurements.[8] The most interesting results are those obtained with Cr(bpy)$_3^{3+}$. With respect to the behavior of an isolated complex or anion, the solution containing both components shows a quenching of LAS$^-$ luminescence, an increase of the complex emission and a clear nonadditivity of the absorption spectrum. All these results were interpreted by assuming the formation of a 1:1 adduct, where electronic interaction causes a strong perturbation in the ligand-centered transition of the complex. A similar perturbation was not observed in the 1:2 adduct between LAS$^-$ and Pt(bpy)(NH$_3$)$_2^{2+}$; the other two complexes do not form an adduct with the anion.

8.4 ADDUCTS OF POLYAMMONIUM MACROCYCLES

Polyazamacrocycles, in their protonated forms, are known to form supramolecular 1:1 adducts with several anionic compounds. The adducts, formed by anionic cyano complexes, such as, for example, Co(CN)$_6^{3-}$, are of particular interest, because they may involve multiple hydrogen bonds between the CN$^-$ ligands and the ammonium groups of the macrocycle, besides the obvious coulombic anion–cation interaction. Photochemical studies were supposed to provide useful information about the nature of these bonds, as well as on the geometry of the adducts, and the supramolecular derivatives of Co(CN)$_6^{3-}$ were selected at first because the free complex, in aqueous acidic solutions, undergoes a clean photoaquation reaction,

$$Co(CN)_6^{3-} + H_2O + h\nu \rightarrow Co(CN)_5(H_2O)^{2-} + CN^-$$

which can be easily followed by means of spectrophotometric measurements and shows a relatively high quantum yield, practically independent on the pH and the excitation wavelength. The same reaction was observed upon irradiation of the adducts between Co(CN)$_6^{3-}$ and several polyazamacrocycles (Fig. 8.1),[9] but the quantum yields Φ of the photoaquation, collected in Table 8.1, are always lower than that of the free complex ($\Phi^\circ = 0.30$).

The inspection of this table clearly shows that the quantum yields depend on the size of the macrocycle, but have no relation with its ionic charge; moreover, the ratio of the quantum yield of the free complex to the quantum yield of the adduct, Φ°/Φ, is practically equal to 2 for the small receptors and equal to 3 for the large ones. Such a discrete quenching effect is an indication that the adducts possess rather well-defined supramolecular structures and strongly suggests that the polyammonium macrocycles are responsible for the quantum yield decrease by restraining the escape out of the adduct of the hydrogen-bound CN$^-$ ligands. In this light, the ratio Φ°/Φ is an indirect measure of the number

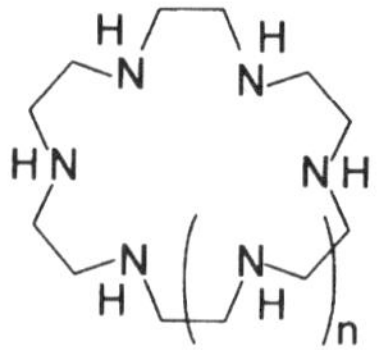

n = 3 [24]aneN$_8$
n = 5 [30]aneN$_{10}$

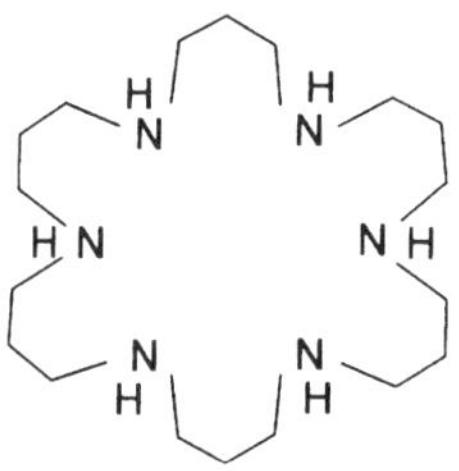

[24]aneN$_6$

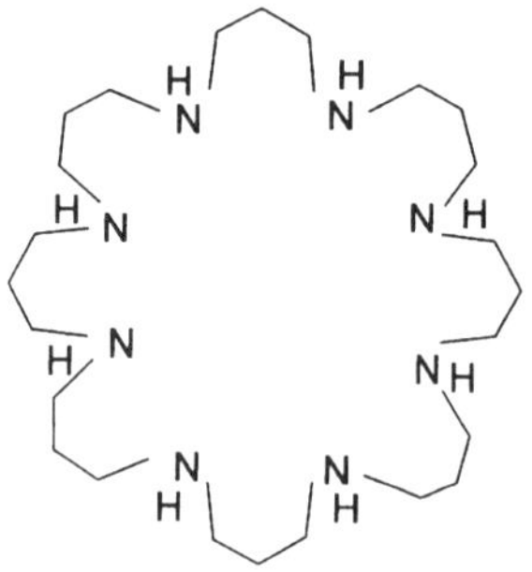

[32]aneN$_8$

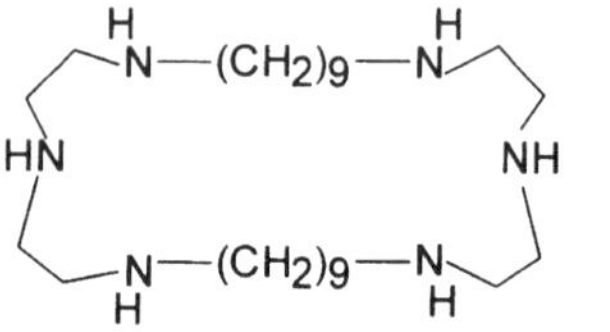

Fig. 8.1. Polyaza macrocycles.

n of ligands bound to the macrocycle,

$$\Phi^\circ/\Phi = 6/(6-n)$$

which should therefore be $n = 3$ for the first four macrocycles of Table 8.1 and $n = 4$ for the other two.

Indirect support for this hypothesis was given by studies on the Br$^-$ photoaquation of Co(CN)$_5$Br^{3-}.[10] The quantum yield for this reaction is 0.23 for the free complex, and it is reduced to 0.05, 0.01, and 0.02 in the adducts with H$_6$[24]aneH$_6^{6+}$, H$_8$[24]aneN$_8^{8+}$, and H$_8$[32]aneH$_8^{8+}$, respectively; not only is

TABLE 8.1. Quantum Yields of the $Co(CN)_6^{3-}$ Photo-aquation Reactions in the Presence of Polyazamacrocycles

Macrocycle	Φ	Φ°/Φ
$H_6[24]aneN_6^{6+}$	0.15	2
$H_7[24]aneN_8^{7+}$	0.15	2
$H_8[24]aneN_8^{8+}$	0.16	2
$H_{10}[30]aneN_{10}^{10+}$	0.14	2
$H_6[32]aneN_6-C_9^{6+}$	0.11	3
$H_8[32]aneN_8^{8+}$	0.10	3

the Φ°/Φ ratio relatively large, but also it parallels the ionic charge of the receptor rather than its size. Since the Br^- ligand is not involved in the hydrogen bonds with the macrocycle, its photoinduced release cannot be affected by the steric limitations responsible for the quenching of $Co(CN)_6^{3-}$ photoaquation; the quenching of the $Co(CN)_5Br^{3-}$ photoreaction is therefore ascribed to the positive charge of the adduct, which increases the rate of the back recombination between Br^- and $Co(CN)_5^{2-}$ formed in the primary photodissociation step.

Steric hindrance to the release of the hydrogen-bound CN^- ligands is also responsible for the quenching of $Cr(CN)_6^{3-}$ photoaquation by $H_8[32]aneN_8^{8+}$ [$\Phi^\circ/\Phi = 3$, equal to the value for the adduct of the same receptor with $Co(CN)_6^{3-}$],[11] and for the quenching of $Co(CN)_6^{3-}$ photoaquation by some polyazacyclophanes ($\Phi^\circ/\Phi = 2$).[12] In these last adducts, a complete static quenching of the luminescence of the receptor was also observed, so that emission intensity measurements could be used to evaluate the association constants of the adducts.

The photochemical results discussed above provide information on the nature and number of bonds in the adducts between cyano complexes and polyazamacrocycles, but give no indication of the geometry of the adducts. In the case of the adducts of $H_8[32]aneN_8^{8+}$ with $M(CN)_6^{3-}$ complexes ($M = Co$, Cr), for example, which have four hydrogen bonds between CN^- ligands and ammonium groups, two structures can be suggested (Fig. 8.2): In the first four adjacent CN^- groups are bound to the receptor ("boat" structure), while four equatorial CN^- are bound in the second structure ("belt" structure), leaving two CN^- ligands *trans* to each other unbound. Studies on the CN^- photoaquation of the mixed-ligand complexes $Co(CN)_5(H_2O)^{2-}$ (Ref. 13) and $Cr(CN)_5(H_2O)^{2-}$ (Ref. 11) in the presence of $H_8[32]aneN_8^{8+}$ provided information on this topic. In the case of the chromium complex, no change of the quantum yield was observed in the presence of the macrocycle; this result is in good agreement with the theoretical expectation for the release of the CN^- ligand *trans* to H_2O, only if the adduct has the "belt" structure. For the cobalt complex, a Φ°/Φ value of 2.8 was obtained, which is more consistent with the "belt" structure than with the "boat" one. Even if definite evidence was not obtained, it seems likely that the adducts between hexacoordinated cyano complexes and large polyazamacrocycles have a "belt" structure.

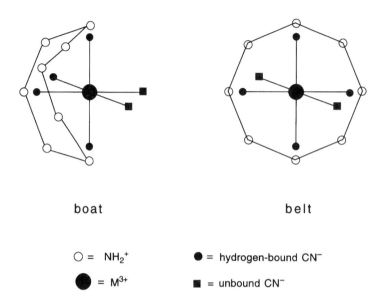

Fig. 8.2. "Boat" and "belt" structures for the adducts between $H_8[32]aneN_8^{8+}$ and $M(CN)_6^{3-}$ complex anions.

This type of investigation has also been extended[14] from molecular cyclic polyazamacrocycles to their branched polymeric analogous (PEIs), which show open cavities quite similar to the rings of the above-mentioned macrocycles. The photoaquation of $Co(CN)_6^{3-}$ was used even in this case as a tool for evaluating the complex–polymer interaction, and its quantum yield was measured as a function of the molecular weight of the polymer, the polymer concentration, and the pH. The results obtained indicate that, under the most appropriate experimental conditions, almost all the amino groups of the polymer are involved in hydrogen bonds with the complex, which, however, cannot contribute to more than four bonds per molecule. PEIs therefore offer the same protection against $Co(CN)_6^{3-}$ photoaquation as the cyclic receptors discussed above, with the advantage that they can be used even at relatively high pHs.

Finally, it is important to note that a comparison of the effects of polyazamacrocycles on the photochemical and thermal behavior of cyano complexes may be useful for the determination of the reaction mechanisms. The thermal aquation of $Cr(CN)_6^{3-}$ in its adduct with $H_8[32]aneN_8^{8+}$, for example, occurs with a rate constant about 40 times smaller than that of the free complex,[10] while the $\Phi°/\Phi$ ratio for the photoaquation of the complex is only 3. Since the thermal reaction occurs as a H^+-assisted mechanism, the large effect of the macrocycle can be safely attributed to its positive charge, which obviously causes a decrease of the basicity of the CN^- ligands in the adduct. A similar effect cannot be observed in the photochemical reaction, which occurs by a dissociative, pH-independent mechanism. The drastic change in the ionic charge in going from the free complex to its adduct with $H_8[32]aneN_8^{8+}$ is probably also responsible for the fact that the thermal aquation of $Cr(CN)_5(H_2O)^{2-}$ is practically unaffected by the receptor.[10] In this case, in fact, the obvious negative effect on the H^+-assisted path of the reaction [analogous to that observed in the thermal aquation of $Cr(CN)_6^{3-}$] is accompanied by a positive effect on a parallel reaction path that is pH independent and involves the coordinated water molecule. Note that the lack of influence by the macrocycle on the photoreaction of the complex is due to a completely different reason.

8.5 OTHER STUDIES

Molecular recognition of anions through the formation of supramolecular adducts with appropriate receptors has been the subject of several studies, and in few cases emission measurements were used to evaluate the formation of the adducts. Changes in the fluorescence intensity and lifetime of the diprotonated form of sapphyrin were used, in parallel with other techniques, to evaluate the association of this receptor to fluoride, chloride, and bromide anions.[15] Steady-state fluorescence spectroscopy was also used to evaluate the interaction of sapphyrin to DNA phosphates.[16] An interesting example of a fluorescent probe for anions is given[17] by protonated anthrylpolyamines like **8.4**. The free probes show a low fluorescence from the aromatic chromophore because of an efficient intramolecular electron-transfer quenching by the close amine group. The complexation of the probe to phosphate, citrate, sulfate, acetate, or dimethyl phosphate anions causes a strong increase of the emission intensity, because the amine group is involved in hydrogen bonds with the anion, and thus it is no longer able to quench the emission.

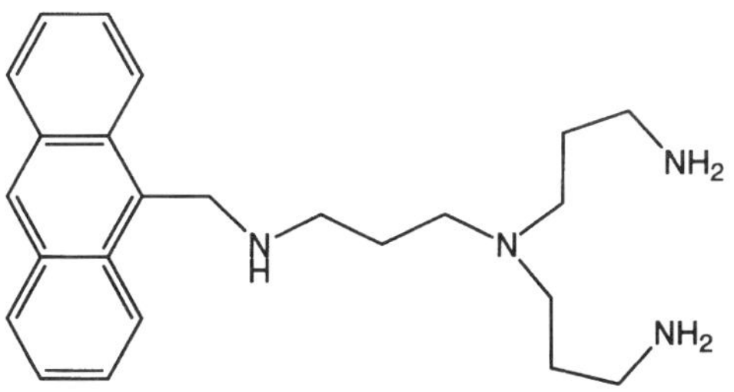

8.4

Time-resolved fluorescence measurements were used[18] to study the transfer of the excitation energy in a system comprising a protonated sapphyrin and the anion of a carboxylated porphyrin. Excitation of the porphyrin moiety in this system was in fact expected to cause energy transfer to the sapphyrin, analogous to that already observed in covalently linked sapphyrin–porphyrin systems. Actually, a biexponential fluorescence decay was observed, the short component of which ($\tau = 500$ ps) increased in fractional amplitude with increasing sapphyrin concentration, and so it was ascribed to the expected singlet–singlet energy transfer. The rate constant (1.8×10^9 s^{-1}) and efficiency ($\Phi = 0.96$) of this process were also measured.

The photostimulated transport of several anions across a liquid–liquid membrane was achieved[19] by using the copolymer of styrene and 4-vinyl Malachite green as carrier. The system comprises two aqueous phases separated by a CCl_4 solution containing the carrier: irradiation of the organic phase causes the conversion of the copolymer to its cationic form, which can thus act as carrier by ion pairing of the anion.

The influence of the protonated forms of the polyazamacrocycles [24]aneN$_6$, [32]aneN$_8$, and [30]aneN$_{10}$ on the spectral and photochemical properties of the Co(EDTA)$^-$–I$^-$ system were also investigated.[20] This system may be considered as an ion pair that exhibits an IPCT excited state; light excitation in the corresponding absorption band causes an intramolecular photooxidation reduction:

$$Co^{III}(EDTA)^- \text{-} I^- + h\nu \rightarrow Co^{II}(EDTA)^{2-}\text{-}I^-$$

followed by secondary processes.[21] The addition of the macrocycles causes an increase of the absorbance of the ion pair and of the observed quantum yield of its photoreaction. Analysis of these data as a function of macrocycle or I$^-$ concentration shows that the macrocycles cause an increase of the association constant between Co(EDTA)$^-$ and I$^-$, while they have no effect on the photochemical efficiency of the Co(EDTA)$^-$–I$^-$ ion pair. The observed effect can be easily explained by the formation of an adduct between macrocycle and complex, which results in a high positive charge. In other words, the effect of the polyazamacrocycles on this system is due to their high positive ionic charge, and thus is analogous to the effects of the same macrocycles on the photochemistry of Co(CN)$_5$Br^{3-} and on the thermal reactivity of the chromium-cyano complexes discussed above.

8.6 REFERENCES

1. V. Balzani and F. Scandola, *Supramolecular Photochemistry*, Ellis Horwood, New York, 1991
2. A large review of these studies has been reported in Ref. 1.
3. A. Vogler and H. Kunkely, *Topics Curr. Chem.* **158**, 1 (1990).
4. A. W. Adamson and A. H. Sporer, *J. Am. Chem. Soc.* **80**, 3865 (1958).
5. F. Pina, M. Ciano, L. Moggi, and V. Balzani, *Inorg. Chem.* **24**, 844 (1985).
6. W. J. Vining, J. V. Caspar, and T. J. Meyer, *J. Phys. Chem.* **89**, 1095 (1985).
7. N. Sabbatini, S. Perathoner, G. Lattanzi, S. Dellonte, and V. Balzani, *J. Phys. Chem.* **91**, 6136 (1987).

8. R. Ballardini, M. T. Gandolfi, M. L. Moya, L. Prodi, and V. Balzani, *Israel J. Chem.* **32**, 47 (1992).
9. (a) M. F. Manfrin, N. Sabbatini, L. Moggi, V. Balzani, M. W. Hosseini, and J.-M. Lehn, *J. Chem. Soc., Chem. Commun.*, 555 (1984). (b) N. Manfrin, L. Moggi, V. Castelvetro, V. Balzani, M. W. Hosseini, and J.-M. Lehn, *J. Am. Chem. Soc.* **107**, 6888 (1985); (c) F. Pina, L. Moggi, M. F. Manfrin, V. Balzani, M. W. Hosseini, and J.-M. Lehn, *Gazz. Chim. Ital.* **119**, 65 (1989).
10. F. Pina, L. Moggi, M. F. Manfrin, V. Balzani, M. W. Hosseini, and J.-M. Lehn, *XII JUPAC Symposium on Photochemistry*, July 1988, Bologna, Italy, Abstracts of papers, 248.
11. J. Sotomajo, A. J. Parola, F. Pina, E. Zinato, P. Riccieri, M. F. Manfrin, and L. Moggi, *Inorg. Chem.* **34**, 6532 (1985).
12. M. A. Bernardo, A. J. Parola, F. Pina, E. García-España, V. Marcelino, S. V. Luis, and J. F. Miravet, *J. Chem. Soc., Dalton Trans.*, 993 (1995).
13. A. J. Parola and F. Pina, *J. Photochem. Photobiol. A: Chem.* **66**, 337 (1992).
14. M. F. Manfrin, L. Setti, and L. Moggi, *Inorg. Chem.* **31**, 2768 (1992).
15. M. Shionoya, H. Furuta, V. Lynch, A. Harriman, and J. L. Sessler, *J. Am. Chem. Soc.* **114**, 5714 (1992).
16. B. L. Iverson, K. Shreder, V. Král, and J. L. Sessler, *J. Am. Chem. Soc.* **115**, 11022 (1993).
17. M. E. Huston, E. U. Akkaya, and A. W. Czarnik, *J. Am. Chem. Soc.* **111**, 8735 (1989).
18. V. Král, S. L. Springs, and J. L. Sessler, *J. Am. Chem. Soc.* **117**, 8881 (1995).
19. I. Willner, S. Sussan, and S. Rubin, *J. Chem. Soc., Chem. Commun.*, 100 (1992).
20. F. Pina, A. J. Parola, A. Bencini, M. Micheloni, M. F. Manfrin, and L. Moggi, *Inorg. Chim. Acta* **195**, 139 (1992).
21. (a) F. Pina and M. Maestri, *Inorg. Chim. Acta* **142**, 223 (1988); (b) F. Pina and J. Costa, *J. Photochem. Photobiol. A: Chem.* **48**, 233 (1989).

CHAPTER 9

Anion Binding Receptors: Theoretical Studies

JOANNA WIÓRKIEWICZ-KUCZERA and KRISTIN BOWMAN-JAMES

9.1 Introduction
9.2 Polyammonium Receptors
   9.2.1 Macromonocycles
   9.2.2 Macrobicycles
   9.2.3 Macrotricycles
9.3 Calixarenes
9.4 Organoboron Macrocyclic Receptors
9.5 Organotin Macrocyclic Receptors
9.6 Electrides
9.7 Conclusion
9.8 References

9.1 INTRODUCTION

Theoretical studies of supramolecular systems have provided much-needed insight to the conformational preferences of the hosts complexed with different guests, as well as ligand selectivity.[1] In addition to molecular-mechanics modeling of host–guest systems, molecular-dynamics simulations have become increasingly popular in recent years, providing even greater understanding of the solution structure, as well as the preorganization and complexation mechanisms of host–guest systems.[2] However, theoretical studies have mainly focused on macrocyclic receptors for cations, since anion chemistry has developed more slowly. Hence, most of the investigations have been carried out for crown ethers,[3,4] thia crown ethers,[5] cryptands,[3] and cavitands[6,7] as receptors for cations and neutral molecules.

Supramolecular Chemistry of Anions, Edited by Antonio Bianchi, Kristin Bowman-James, and Enrique García-España.
ISBN 0-471-18622-8.   © 1997 Wiley-VCH, Inc.

The focus of this chapter is on the theoretical studies that have been reported for anion receptors. Included are positively charged polyammonium macrocycles, which interact via both electrostatic and hydrogen bonding, calixarenes, as dual cation–anion receptors, neutral Lewis acids such as boron and tin macrocycles, as well as the intriguing case of electrides, isolated electrons stabilized by the macrocyclic crown ethers.

9.2 POLYAMMONIUM RECEPTORS

A prominent class of macrocyclic receptors for anions consists of polyammonium macrocycles, positively charged compounds containing NH_2^+ groups separated by hydrocarbon chains or heteroatomic subunits. These cations are capable of strong electrostatic in addition to hydrogen-bonding interactions with anions.

9.2.1 Macromonocycles

A study of complexes involving a molecular-mechanics approach on a polyammonium macrocycle used this method to find energy minima for unliganded neutral 1,4,7,10,13,16-hexaazacyclooctadecane ([18]aneN$_6$) (**9.1**) and its complexes with carbonate anion and boric and silicic acids.[8] The MM2 package[9] was used in these calculations, with additional parameters for the macrocycle derived by the authors. Several possible conformations of the neutral macrocycle were considered, the most favorable having D_{3d} symmetry with an almost circular structure. This conformation as well as one boatlike (C_2 symmetry) and two chairlike (C_i symmetry) low-energy structures were further used to study the binding of ligands by the triprotonated [18]aneN$_6$ in vacuo. Protonation was assumed to be on alternating nitrogen atoms for the study, and the conformation of the protonated receptor was assumed to be identical to that of the neutral species. After a rigid docking of the guest ligands was undertaken, all

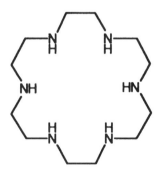

9.1

TABLE 9.1. Results of Calculations on 9.1·3H⁺ Uncomplexed and Complexed with Carbonate and Boric and Silicic Acids[a]

Compound	9.1·3H⁺	[9.1·3H⁺][CO₃²⁻]	[9.1·3H⁺][BO₃H₃]	[9.1·3H⁺][SiO₄H₄]
E_{tot}	13.55	−0.12	0.27	−20.27
$E_{complex}$	—	−13.67	−12.19	−26.38
NH···O (Å)		2.09, 2.09, 2.09	2.26, 2.23, 2.16	2.16, 2.23, 2.13

[a] The starting conformation is the D_{3d} conformation resulting in the most circular structure for the macrocycle. Energies are in kcal mol⁻¹.

three substrates, the carbonate anion and silicic and boric acids, were found to form hydrogen bonds with **9.1·3H⁺**. Complexation in all three cases resulted in lowering the energy of the combined species compared to the sums of the energies of separate macrocyclic cation and either anionic or neutral species (Table 9.1).

While for carbonate and boric acid the difference was ca. −13 kcal mol⁻¹, a considerably larger stabilization was observed for silicic acid, −26.38 kcal mol⁻¹. The latter effect was attributed to the possibility of forming three rather than two strong hydrogen bonds between anion and cation (Fig. 9.1).

Since charged ammonium groups in protonated receptors exert electrostatic repulsion upon one another, theoretical studies of polyammonium macrocycles have also focused on the role of water in stabilizing conformations propitious to cavity formation and ligand binding.[10,11] Based on initial crystal structure coordinates, a short 10 ps trajectory was calculated for the hexaprotonated form of 1,4,10,13,16,22-hexaaza-7,19-dioxacyclotetracosane ([24]N₆O₂), **9.2** in water without counterions.[10]

The AMBER-OPLS package was used for this study, and electrostatic water–macrocycle interactions were found to stabilize the nonplanar shape of the receptor (see Chapter 5, Fig. 5.3A). The ammonium groups were oriented inside the macrocycle in a conformation suitable for anion complexation, and a water molecule was found to form multiple hydrogen bonds with two opposite ammonium nitrogens and one ring oxygen. By contrast, during a longer 50 ps

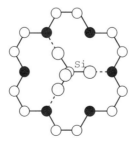

Fig. 9.1. Proposed hydrogen bonding interactions of the structure of SiO₄H₄ with **9.1·3H⁺**.

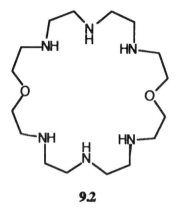

9.2

trajectory in vacuo, starting from the same crystal structure data, the conformation of the compound evolved toward one in which the ammonium groups were oriented outside the macrocycle.[10]

A subsequent report compared expanded studies of **9.2**, together with a related compound, the 1,4,7,12,15,18-hexaazacyclodocosane ([22]aneN$_6$), **9.3**,[11] using the crystal structure data obtained for the hexahydrochloride salts. The smaller macrocycle **9.3** has an almost planar structure and specifically binds two Cl$^-$ anions above and below the macrocycle plane (see Chapter 5, Fig. 5.3B). The larger macrocycle **9.2** adopts a boatlike conformation in the solid state, with one Cl$^-$ anion located in the center of the macrocycle, and two others in proximity (Chapter 5, Fig. 5.3A). Molecular-dynamics studies were undertaken using the AMBER package, with additional potential energy parameters derived by the authors.[10] In 50 ps trajectories two sets of calculations were performed for each macrocycle: simulation in a continuous dielectric medium and in water without counterions, and additionally simulations for **9.2** in water with counterions. While **9.3** underwent no significant conformational changes in either simulation, **9.2** showed considerable changes during the simulations. In both studies a "breathing" (opening and closing) of the macrocycle pocket was

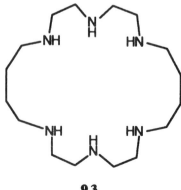

9.3

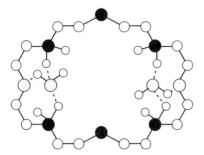

Fig. 9.2. Water structure in simulations of the hexaprotonated form of **9.2** showing the linkage between the ether oxygens and the ammonium groups.

observed in the continuous medium trajectory for **9.2**.[10,11] In simulations with 600 TIP3P water molecules and a rectangular box of $30 \times 27.5 \times 23.5$ Å, the pocket was found to open in water to allow for hydration, and water molecules were found to form a bridge linking adjacent ammonium groups with the ether oxygen atoms (Fig. 9.2).[10,11] In the simulation with counterions, three out of the six Cl$^-$ ions drifted away freely from the macrocycle, while the other three ions stayed in its proximity and were found to interact symmetrically with two NH$^+$ groups.[11] The conclusions drawn were that ring size and thus conformational flexibility are important factors for the dynamics of polyammonium receptors in solution and their interactions with anionic guests.

A recent extensive molecular-dynamics study on two polyammonium macrocycles with nitrate anions in water[12] sheds further light on the role of solvent and ring size in the interactions of the receptors with anions. Crystal structure data for the nitrate complexes of tetraprotonated **9.2**·4HNO$_3$ and 1,4,10,13-tetraaza-7,16-dioxacyclooctadecane ([18]aneN$_4$O$_4$·4HNO$_3$) **9.4**·4HNO$_3$ were used for molecular-dynamics simulations out to 400 ps for **9.2** and 200 ps for **9.4**. The CHARMM program[13] was used with additional potential energy functions

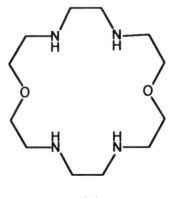

9.4

derived by the authors. Crystal structure data indicated that the smaller macrocycle, **9.4**, is relatively flat in the solid state, with two nitrates above and two below the plane of the macrocycle (see Chapter 5, Fig. 5.4A and B). As observed in the earlier study by Wipff and co-workers for the smaller **9.3**, hydration had very little effect on the conformation of **9.4**, which fluctuated around the original C_{2h} conformation throughout the simulation.

Again, however, the larger macrocycle **9.2** exhibited more interesting features during the simulation, attributed to the larger and thus more flexible ring as well as the presence of the ether oxygen atoms. The nitrate salt of **9.2** adopts a boatlike conformation in the solid state, with one of the four nitrates situated inside the macrocyclic cavity, with O–N–O bonds aligned along the C–N–C bonds of the central amine, almost fitting like a glove (Fig. 5.4A). No significant conformational changes were observed for the macrocycle during the first 120 ps of the trajectory, except that the nitrate rotated 90° around the vertical axis to be aligned along the C–O–C bonds (Fig. 9.3A). Hydration of this macrocycle was found to be a slow process. Only after 200 ps did the hydrogens of the ammonium groups turn outside toward the bulk solvent and the ring begin to open and flatten (Fig. 9.3B). Between 300 and 400 ps, the ring opening reached its maximum, although the nitrate originally in the cavity remained (Fig. 9.3C). As in the earlier simulation with the hexachloride salt, water molecules were found to enter the cavity as the simulation progressed (Fig. 9.3D). One of the most striking features observed in this study was the observed formation of contact, close and far ion pairs with water molecules bridging the electrostatic interaction between the cation and anion. This is the first instance in which long-distance electrostatic relay interactions in water have been reported from molecular-dynamics simulations. The formation of these ion pairs may help in the understanding of reaction mechanisms in multicharged systems in aqueous solution, in which the initial step involves the formation of an outer-sphere complex.

9.2.2 Macrobicycles

The importance of size selectivity has been exceptionally well illustrated in a study of the binding of halides by the octaazacryptand **9.5**.[14] Proton, fluoride, and chloride binding were examined using both potentiometric and NMR techniques. Findings indicated an exceptionally high affinity of the macrocycle for fluoride ion, with $\log K_S = 3.6$ and 11.2 for the tri- and hexaprotonated forms of the macrocycle, respectively, a factor of 10^6 times stronger than observed for chloride ion. MM2 calculations were used to estimate the strain energy of **9.5** with respect to cavity size in a range from 1.9 to 3.6 Å for anion–nitrogen distances, using the MM2 program. The findings indicated that the minimum strain energy for the ligand occurs at an anion–nitrogen distance of 2.91 Å (Fig. 9.4). This is strikingly close to the 2.81 Å observed for the N–F distance in the crystal structure of the encapsulated fluoride complex of **9.5**.[15] The estimated increase in strain energy for substitution of Cl⁻ in the cavity was

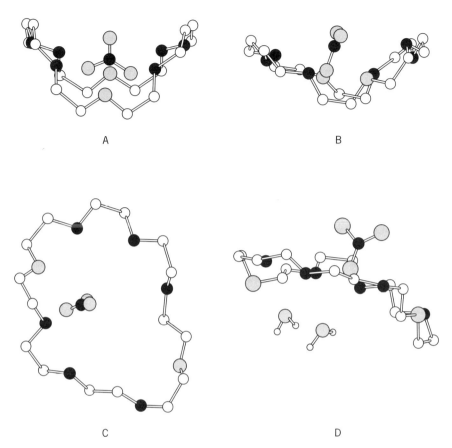

Fig. 9.3. Average structures from snapshots of the nitrate complex of **9.2** at (A) 0–120, (B) 120-200, and (C) 300–400 ps. (D) Final snapshot showing two water molecules, responsible for the unfolding of the macrocycle. (Only one nitrate has been included in the drawings for simplicity.)

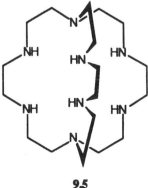

9.5

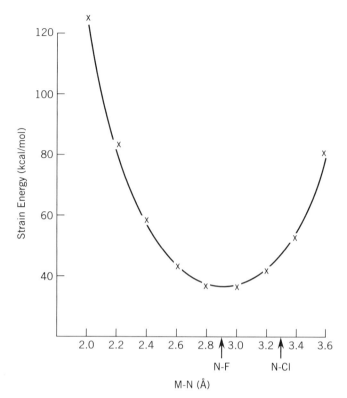

Fig. 9.4. Plot of the approximate strain energy of **9.5** as a function of substrate–nitrogen bond length.

found to be 10.2 kcal mol^{-1}, corresponding to the increased Cl–N distance of 3.23 Å.

9.2.3 Macrotricycles

The macrotricycle SC24 ($C_{24}H_{48}N_4O_6$), **9.6**, can be viewed as a tetrahedron with four nitrogen atoms at its vertices with the nitrogen atoms connected by six $-CH_2CH_2OCH_2CH_2-$ bridges. The resulting tricycle has a rigid polycyclic architecture, which ensures high stability of the ion–cryptand complex. Ion binding affinity of **9.6** is a function of pH. In basic or neutral environments **9.6** selectively binds NH_4^+, while at low pH tri- or tetraprotonated **9.6** preferentially binds Cl$^-$. Theoretical studies of anion binding have focused on the mechanism of ion capture from water by the cryptand[16] and on the binding affinity of different anions to the receptor.[17] In these studies **9.6** served as a model of an ionophore participating in ion transfer from an aqueous environment through a nonpolar membrane interior.

The binding and mechanism of dehydration of a chloride ion by **9.6·4H$^+$**

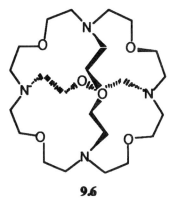

9.6

have been studied in water.[16] Due to steric limitations, an ion must undergo complete dehydration to form the inclusion complex. To analyze the ion–receptor interaction, a reaction coordinate r_{rc} was defined as the distance of the chloride ion from the center of one of the four equivalent faces of the tetraprotonated **9.6**·4H$^+$ tetrahedron.[16a] Two types of calculations were performed. In one set of calculations the chloride ion was allowed to move freely, and the system was allowed to explore fully the potential energy surface, while in the other the motion of the chloride ion was restricted to a spherical shell centered at the tetrahedron center. Trajectories ranging in length from 60 to 125 ps were calculated for each shell. The trajectories in which no restrictions were imposed on the system were used to characterize the cryptand binding sites. Two binding sites were found: one inside the tetrahedron ($r_{rc} = 0.6$ Å) and the other outside the cryptand ($r_{rc} = 4.0$ Å). The energy minimum inside the cryptand is mostly associated with chloride–**9.6** interactions, while the outer binding site is stabilized by solute–water interactions. Energy calculations along the reaction coordinate r_{rc} yielded a broad barrier of about 20 kcal mol^{-1} separating the two binding sites. This barrier arises from the steric hindrance experienced by the ion as it enters the cavity and from the dehydration of the anion, which was found to provide the major contribution. The unfavorable dehydration contribution is compensated by the electrostatic interaction between the ion and cryptand and energetically favorable rearrangements of the water structure. The ion dehydration was found to occur in several steps, with the number of waters decreasing as the anion approached **9.6**. Conformational changes in **9.6** were also found to play a role in the capture process. From experimental data, it is known that there is an equilibrium between *exo* and *endo* forms of the macrocyclic amines,[18] and it has been proposed that the *exo* conformation may assist the initial dehydration.

The structure of the cryptand is significantly affected by the complexation process as noted above, and as described more precisely in a related article.[16b] In this study a reaction coordinate r_c was defined as the distance between the chloride ion and the macrocyclic center. When the cryptand and chloride ion are separated by large distances, the ligand is essentially tetrahedral based on the

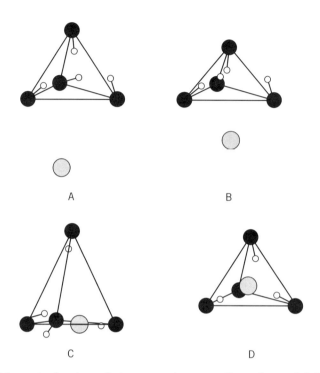

Fig. 9.5. Schematic drawing of the approximate conformations of **9.6** as chloride approaches from distances of (A) $r_c = 4.4$ Å, (B) 2.9 Å, (C) 2.0 Å, and (D) 0.2 Å.

placement of the NH atoms, with N···N distances on the order of 5.5 Å (Fig. 9.5A). Although the N–H bonds are all *endo*, the hydrogen atoms do not point toward the cavity center, allowing for an energetically favorable reduction in the electrostatic repulsion between the positively charged ammonium ions. As chloride approaches the macrocycle from a distance $r_c = 40$ to 2.6 Å, the cryptand contracts, decreasing the distance between the nitrogens by about 0.8 Å and increasing the space for the chloride ion in the direction of its approach (Fig. 9.5B). As the chloride distance lessens to $r_c = 2.0$ Å, the three N–H bonds along the face of the chloride approach shift toward the anion, and the macrocycle becomes elongated rather than contracted (Fig. 9.5C). As the chloride moves toward the center of the macrocycle, the conformation undergoes another change, relaxing to a structure more similar to the initial conformation, and the conformation approaches that of the crystal structure (Fig. 9.5D).

These two closely related studies provide considerable insight into the mechanism of ion transport, in that they indicate the importance of desolvation in the recognition step, as well as a stepwise mechanism for this process, in addition to shedding light on the conformational fluctuations of the macrocycle upon complexation.

A related study by Wipff and co-workers used the thermodynamic cycle-perturbation approach to study the influence of the anion on binding, by examining the relative affinity of chloride and bromide ions.[17] This approach involved mutating Cl⁻ into Br⁻ and vice versa in solution with the uncomplexed and complexed receptor to calculate the relative free energy of binding for the two anions to the macrocycle (Scheme 9.1, where A is the Helmholtz free energy). Molecular-dynamics calculations were carried out using the GROMOS program,[19] with potential energy functions adjusted for **9.6**. Production dynamics were carried out for 30 ps after heating and equilibration. A examination of the relative binding affinity of Cl⁻ and Br⁻ anions to the macrocycle **9.6** in water suggests that ion size is an important factor for the selective binding of ions by the cryptand. The calculated relative free energy of hydration for Cl⁻ and Br⁻ (ΔA_3) was found to be 3.35 kcal mol⁻¹, in good agreement with the experimental value of 3.3 kcal mol⁻¹ and indicating that desolvation of Br⁻ is accompanied by a smaller change in free energy than in the case of Cl⁻. The relative free energy of binding to **9.6** for Cl⁻ and Br⁻ was estimated at

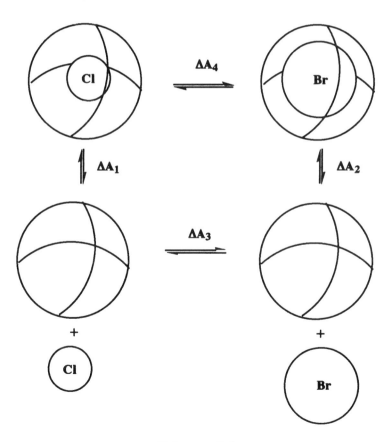

Scheme 9.1

4.15 kcal mol^{-1}, also in accord with the experimental result of 4.3 kcal mol^{-1}, making intermolecular interactions between **9.6** and Cl$^-$ more favorable than those for Br$^-$. It was concluded that selective binding of Cl$^-$ over Br$^-$ by **9.6** is thus due to the favorable interaction of Cl$^-$ with the cryptand compared to Br$^-$, with the anion–receptor interaction more than compensating for the relatively unfavorable dehydration process.

9.3 CALIXARENES

Another class of macrocyclic receptors that have received growing interest are calixarenes, phenol–formaldehyde cyclic oligomers built of several phenolic units connected by methylene bridges, **9.7**. Functional groups may be present in the *para* position with respect to the hydroxyl group of the phenol ring, and the hydroxyl group may also be substituted. The phenolic OH groups and the *para* positions of the phenols are called the lower and upper rim, respectively. Since calixarenes contain both polar and apolar molecular regions, they are an interesting class of synthetic host systems.

Calix[4]arenes with four phenolic subunits have received the most attention to date. Calixarene derivatives with amides, ketones, and esters at the lower rim have been shown to have significant cation affinities,[20,21] and uranyl-containing calix salen crown ethers have been found to form complexes with neutral molecules.[20,22] Recently calix[4]arene sulfonamides have been synthesized that form complexes with anions (Cl$^-$, NO$_3^-$, HSO$_4^-$).[21,23] For the larger calix[6]-arenes and calix[8]arenes no cavity is observed in most cases, due to the increased flexibility of those systems related to the size of the ring. Selective functionalization of the macrocycles has led in several cases to a more rigid conformation with guest binding potential, however.[20,21,24]

NMR and crystallographic measurements suggest that conformations adopted by calix[4]arenes may be divided in four categories: a cone (all phenyl rings oriented in the same direction, forming a cuplike structure), a partial cone (two phenyl rings pointing up, one phenyl pointing down, the fourth phenyl adopting an intermediate orientation), a 1,2-alternate structure (two neighboring rings pointing up, two down), and a 1,3-alternate conformation (with opposite phenyl rings pointing up or down) (Fig. 9.6). While not involving the anion complexing ability of these macrocycles, the conformational aspects in the absence of anions are in themselves of interest, and were found to depend somewhat on the substituents at both the upper and lower rims. The study of unsubstituted and substituted calixarenes (methyl ethers of *p-tert*-butylcalix[4]arene) undertaken by Reinhoudt and co-workers[25] aimed at analyzing the relative stability of different conformations of various calixarenes using Allinger's MM2 program. The structures and energies of different conformers were then compared for one particular calixarene obtained from several other force fields (AMBER, MM2P, and CHARMM).[26] The results indicated that the preferred conformations were dependent on the substituents

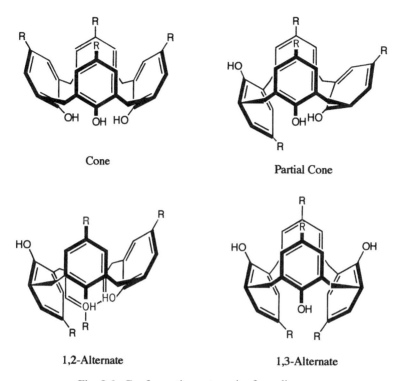

Fig. 9.6. Conformation categories for calixarenes.

on the phenolic oxygen atoms (X), and that *para* substituents had little influence on conformation. Previous studies have shown that substituents larger than methoxy have been found to rigidify the conformation of the tetrasubstituted calix[4]arene.[26]

More recent molecular-dynamics simulations have also been carried out for a *p-tert*-butylcalix[4]arene-tetraamide **9.7**, $X = CH_2C(O)NEt_2$, with one neutral

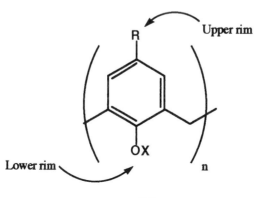

9.7

(H_2O, MeOH, and MeCN) or anionic (SCN^-) guest inside the cone and cations (Li^+, Na^+, K^+, Cs^+, or Eu^{3+}) at the lower rim.[27] The AMBER package was used in this study, and calculations were performed for 50–300 ps in vacuo and in water. Trajectories in vacuo and in water for the free calixarene show a stable cone conformation with C_{2v} symmetry as the most favored state, with the C_{4v} cone being a saddle point, as previously suggested.[25] No water was found inside the cone during a 50 ps simulation, probably due to the hydrophobic influence of the four *tert*-butyl groups. The structure and dynamics of the complexed receptor, however, was found to depend on the size and charge of the guest molecules. Vacuum simulations of the complex with acetonitrile and methanol showed that the guest molecules fit easily inside the cone with average C_{4v} cone conformations. Simulations with neutral guests in the cone and Na^+ in the polar lower rim pseudocavity showed that the orientation of the guest molecule inside the cone was determined by its interactions with the cone rather than the cation. Replacement of a neutral guest by SCN^- in the cone was found to lead to a strong electrostatic attraction between SCN^- and Na^+, which resulted in the cation being pulled inside the cone to form a close ion pair. A similar result was obtained when Na^+ was replaced by Li^+. Only Cs^+, being too large to move into the cone, remained in the polar pseudocavity during the simulation. Hence the structure and dynamics of the complexed receptor were thus found to depend on the charge and size of the guest ions.

9.4 ORGANOBORON MACROCYCLIC RECEPTORS

The macrocycles investigated in the previously cited studies contained nitrogen and/or oxygen donor atoms, and in the case of polyamines, anion coordination involves protonation of nitrogen atoms. Direct anion coordination by neutral multidentate Lewis acids has been gaining interest recently, with a number of experimental studies of systems with two to six Lewis acid sites, and tin,[28] mercury,[29] and boron[30] as Lewis acid centers. Because of the covalent or "real" bond formation, these compounds do not strictly fall into the "supramolecular" category.

A semiempirical study of interactions between organoboron macrobi- and tricycles and halide or oxide anions was carried out to investigate anion binding and selectivity in these systems.[30] The structure and interaction energy in vacuo of boron-containing macrocycles $B[(CH_2)_n]_3B$ ($n = 3$–10) **9.8** and $B_4[(CH_2)_n]_6$ ($n = 2$–4) **9.9** with H^-, F^-, Cl^-, and O^{2-} anions were calculated using the AM1 molecular orbital method.[31]

All **9.8** macrobicycles with $n = 3$–10 form complexes with H^-. A cavity with $n = 4$ is the smallest one to accommodate F^-, with the most stable complexes calculated for $n = 5$–6 (Table 9.2). For Cl^- the smallest cavity for inclusion was found for $n = 6$, with the most stable structure obtained for $n = 8$. In the case of H^- and F^- complexes with **9.8** the most stable complexes were found to involve the formation of an almost linear B–X–B bridge (μ–X complexes), with a considerable decrease in distance between the two boron atoms compared to the unliganded state (2.74 Å for H^- and 2.8 Å for F^- compared to 4.16 Å in the unliganded state). For larger macrocycles the distance between the two borons is too large (5 Å), and H^- binds to only one boron. In O^{2-} complexes the anion–host interaction leads extremely strong B–O–B bonds, with even the C10 linkage forming a symmetrical B–O–B bridge with a B–B distance of 2.74 Å compared to 6.81 Å in the unliganded case. Compared to simple adducts with $(CH_3)_3B$, the binding energies for the macrocyclic counterparts and oxygen are significantly larger. Charge distribution analysis of the complexes indicated that complexation in these systems is better described in terms of formation of covalent B–X bonds rather than purely electrostatic ionic interactions dependent on cage size. This effectively allows anions to fit into cavities that appear too small to accommodate them.

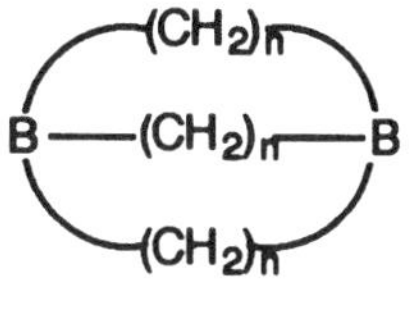

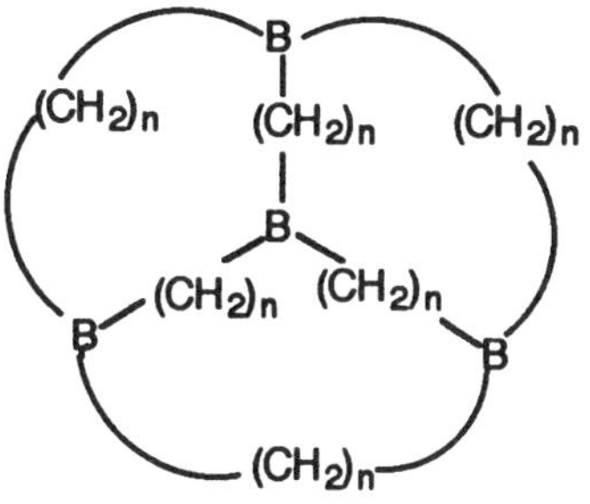

Covalent bond formation was also found to occur in the complexation of anions with **9.9** ($n = 2-4$) macrotricycles, with μ–X bridges calculated in many cases. Significantly increased anion specificity was observed for the tricyclic systems, as can be seen from Table 9.2, as well as increases in the binding energy in cases where the anion binds to more than three or more boron atoms. In **9.9** uncomplexed structures the four boron atoms form a tetrahedral unit with a B–B separation averaging 3.05(6) Å. The number of borons involved in covalent bonds with an anion was found to be a function of cage size: two for $n = 2$ in complexes with H^- and F^-, for $n = 3$ with F^-, and $n = 4$ with O^{2-};

TABLE 9.2. Binding Energies in kcal mol^{-1} for Anion Complexes with 9.8 and 9.9

Compound	n value	H^-	F^-	Cl^-	O^{2-}
9.8	3	−42.6	—	—	−136.8
	4	−75.5	−63.0	—	−217.1
	5	−76.0	−78.9	—	−245.0
	6	−78.6	−88.3	−10.8	−256.7
	8	−63.9	−51.0	−13.2	−237.0
	10	−61.8	−51.0	−12.1	−233.8
9.9	2	−76.9	−60.6	—	−332.0
	3	−62.1	−77.1	−29.8	−289.7
	4	−61.8	−53.3	−16.7	−267.3

three in a complex with Cl^- for $n = 3$; and four with O^{2-} for $n = 2$ and 3. When no bridge was formed linking two boron atoms, the anion formed a covalent bond to only one boron atom. The results of these studies suggest that different Lewis acid macrocyclic hosts may in fact be optimal for specific anions.

9.5 ORGANOTIN MACROCYCLIC RECEPTORS

Bicyclic Lewis acidic tin complexes, **9.10**, the analogs of the catapinands, comprise another class of anion receptors.[29] These compounds have also been examined by molecular mechanics (MM2) methods to assess whether the

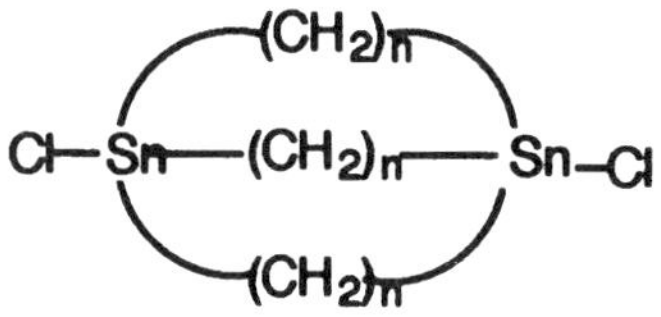

$$Cl-Sn \underset{(CH_2)_n}{\overset{(CH_2)_n}{\rightleftarrows}} Sn-Cl$$

9.10

modeling efforts could reproduce the experimentally observed structures, and to ascertain if minimum energy structures had certain predictive elements. These compounds are again somewhat outside the scope of the "supramolecular" concept of this chapter, since the bonds are really covalent in nature as opposed to the diammonium catapinand counterparts. The force field parameters used for tetrahedral tin were obtained from a study in which simple stannanes were modeled.[32] The joining methylene chains, as anticipated, played a critical role in the overall energies, and the authors divided the study into even and odd numbers of carbon atoms in the linkers. For odd chains the low-energy conformation generally was found to be C_3 symmetry, with the major axis passing through the two tin atoms, and MM2 calculations agreed quite well with the crystal structure data (Fig. 9.7). For example, for the C_5 case, a Sn–Sn distance of 5.892(2) Å was observed experimentally and MM2 calculations yielded a distance of 5.926 Å. The even chains generally gave a C_2 structure of lower energy with two chains in antiparallel conformations. A good correlation between the experimentally and theoretically determined structures was obtained, with the largest differences occurring for structures in which the crystallographic data indicated large thermal motions in the bridges.

9.6 ELECTRIDES

One of the most interesting series of anionic complexes comprises the electrides, crown ether complexes of alkali metals, in which the positive charge of the alkali cation is balanced by an electron. A major question regarding these compounds

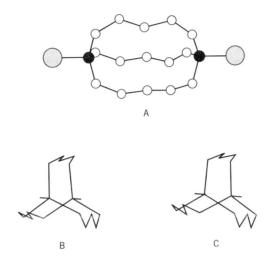

Fig. 9.7. (A) Model of the chloride adduct of the simple five-carbon bridge stannane. (B) Stick drawing based on the crystal structure of (A). (C) The C_3 symmetry structure resulting from MM2 calculations performed using the crystal structure coordinates from (B).

is whether the electron is trapped interstitially or lies near the alkali cation. A study using self-consistent local-density-functional approximation calculations for the complex $Cs^+(15\text{-crown }5)_2 \cdot e^-$ was performed.[33] Results indicated that the electron, which derives from the $6s$ electron of metallic cesium, is located interstitially, which makes these salts analogous to ionic insulators with "*F*-center" electronic defects. The crystal structure contains Cs^+ ions (at 0,0,0 fractional coordinate) sandwiched between two crown ethers. The unit cell contains a large cavity at $\frac{1}{2},\frac{1}{2},\frac{1}{2}$, which is the presumed site of the electride. The position of the electride is in a cavity where the potential is flat but actually repulsive, but the authors conclude that the electride is forced into this region because it can lower its kinetic energy by "spreading out." The resulting band-structure diagram is shown in Fig. 9.8, where the electride state is the half-filled narrow band at $\sim 4.5\,\text{eV}$.

9.7 CONCLUSION

As noted earlier, anion chemistry and the supramolecular chemistry of anions are rapidly expanding fields. Theoretical treatments of these systems are now beginning to shed considerable light on the effects of preorganization, solvation, and desolvation, as well as the electrostatic and hydrogen-bonding aspects of complexation. Development of new computational algorithms and computer hardware is certain to make molecular-mechanics and molecular-dynamics

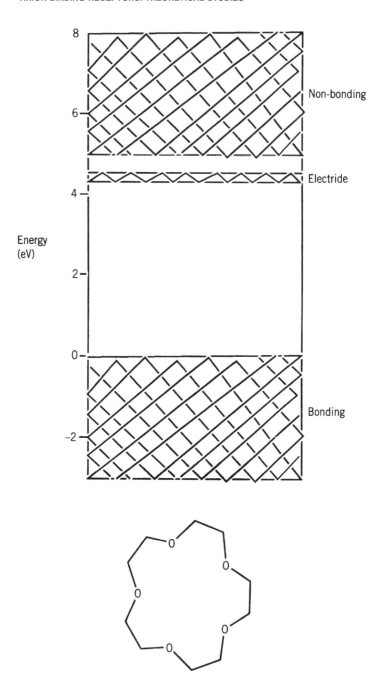

Fig. 9.8. Schematic drawing of the electronic energy bands for the 15-crown-5 electride.

methods complement experimental approaches even more effectively in the future.

9.8 REFERENCES

1. R. D. Hancock, *Acc. Chem. Res.* **23**, 253 (1990).
2. W. F. van Gunsteren and H. J. C. Berendsen, *Angew. Chem. Int. Ed. Engl.* **29**, 992 (1990).
3. (a) G. Rangino, M. S. Romano, J.-M. Lehn, and G. Wipff, *J. Am. Chem. Soc.* **107**, 7873 (1985); (b) T. P. Straatsma and J. A. McCammon, *J. Chem. Phys.* **91**, 3631 (1989); (c) P. Cieplak and P. Kollman, *J. Chem. Phys.* **99**, 55 (1995); (e) Y. L. Ha and A. K. Chakraborty, *J. Chem. Phys.* **95**, 10781 (1991).
4. L. Troxler and G. Wipff, *J. Am. Chem. Soc.* **116**, 1468 (1994).
5. G. A. Forsyth and J. C. Lockhart, *J. Chem. Soc., Dalton Trans.*, 697 (1994).
6. (a) C. I. Bayly and P. Kollman, *J. Am. Chem. Soc.* **116**, 697 (1994); (b) B. E. Thomas IV and P. Kollman, *J. Am. Chem. Soc.* **113**, 3449 (1994).
7. R. M. Izatt, J. S. Bradshaw, K. Pawlak, R. L. Breuning, and B. J. Tarbet, *Chem. Rev.* **92**, 1261 (1992)..
8. M. A. Santos and M. G. B. Drew, *J. Chem. Soc., Faraday Trans.* **87**, 1321 (1991).
9. (a) J. T. Sprague, J. C. Tai, Y. Yuh, and N. L. Allinger, *J. Comput. Chem.* **8**, 581 (1987); (b) U. C. Singh, P. K. Weiner, J. Caldwell, and P. A. Kollman, AMBER 3.0, University of California, San Francisco,1987; (c) QUANTA 2.1A, Polygen Corp., Waltham, MA, 1989.
10. S. Boudon and G. Wipff, *J. Chim. Phys.* **88**, 2443 (1991).
11. S. Boudon, A. Decian, J. Fischer, M. W. Hosseini, J.-M. Lehn, and G. Wipff, *J. Coord. Chem.* **23**, 113 (1991).
12. G. Papoyan, K. Gu, J. Wiórkiewicz-Kuczera, K. Kuczera, and K. Bowman-James, *J. Am. Chem. Soc.* **118**, 1354 (1996).
13. M. Karplus, private communication.
14. S. D. Reilly, G. R. K. Khalsa, D. K. Ford, J. R. Brainard, B. P. Hay, and P. H. Smith, *Inorg. Chem.* **34**, 569 (1995).
15. B. Dietrich, J.-M. Lehn, J. Guilhem, and C. Pascard, *Tetrahedron Lett.* **30**, 4125 (1989).
16. (a) B. Owenson, R. D. MacElroy, and A. Pohorille, *J. Mol. Struct. (THEOCHEM)* **179**, 467 (1988); (b) B. Owenson, R. D. MacElroy, and A. Pohorille, *J. Am. Chem. Soc.* **110**, 6992 (1988).
17. T. P. Lybrand, J. A. McCammon, and G. Wipff, *Proc. Natl. Acad. Sci. USA* **83**, 833 (1986).
18. J. P. Kintzinger, J.-M. Lehn, E. Kauffmann, J. L. Dye, and A. I. Popov, *J. Am. Chem. Soc.* **105**, 7549 (1983).
19. H. J. C. Berendsen, J. P. M. Potsma, W. F. van Gunsteren, and J. Hermans, *Intermolecular Forces*, B. Pullman, Ed., Reidel, Dordrecht, Holland, 331.
20. W. Verboom, R. H. Vreekamp, D. M. Rudkevich, D. N. Rudkevich, and D. N. Reinhoudt, *NATO ASI Ser., Ser. C* **456**, 119 (1995).

21. R. Ungaro A. Arduini, A. Casnati O. Ori, A. Pochini, and F. Ugozzoli, *NATO ASI Ser., Ser. C* **426**, 277 (1994).
22. D. M. Rudkevich, W. T. S. Huck, F. C. J. M. van Veggel, and D. N. Reinhoudt, *NATO ASI Ser., Ser. C* **448**, 329 (1994).
23. J. Scheerder, M. Fochi, J. F. J. Engbersen, and D. N. Reinhoudt, *J. Org. Chem.* **59**, 7815 (1994).
24. J. Scheerder, J. F. J. Engbersen, A. Casnati, R. Ungaro, and D. N. Reinhoudt, *J. Org. Chem.* **60**, 6448 (1995).
25. P. D. J. Grootenhuis, P. A. Kollman, L. C. Groenen, D. A. Reinhoudt, G. J. van Hummel, F. Ugozzolli, and G. D. Andreetti, *J. Am. Chem. Soc.* **112**, 4165 (1990).
26. (a) C. D. Gutsche, B. Dhawan, J. A. Levine, K. H. No, and L. J. Bauer, *Tetrahedron* **39**, 409 (1983); (b) V. Bocchi, D. Foina, A. Pochini, T. Ungara, and G. E. Andreetti, *Tetrahedron* **38**, 373 (1982).
27. P. Guilbaud, A. Varnek, and G. Wipff, *J. Am. Chem. Soc.* **115**, 8298 (1993).
28. (a) M. Simard, J. Vaugeois, and J. D. Wuest, *J. Am. Chem. Soc.* **115**, 370 (1993); (b) W. Alsina, W. Clegg, K. A. Fraser, and J. Sola, *J. Chem. Soc., Chem. Commun.*, 1010 (1992); (b) X. Yang, S. E. Johnson, S. I. Khan, and M. F. Hawthorne, *Angew. Chem., Int. Ed. Engl.* **31**, 893 (1992).
29. J. H. Horner, P. J. Squattrito, N. McGuire, J. P. Rickenspies, and M. Newcomb, *Organometallics* **10**, 1741 (1991).
30. S. Jacobson and R. Pizer, *J. Am. Chem. Soc.* **115**, 11216 (1993).
31. (a) M. J. S. Dewar, E. G. Zoebisch, E. F. Healy, and J. J. P. Stewart, *J. Am. Chem. Soc.* **107**, 3902 (1985); (b) M. J. S. Dewar, C. Jie, and E. G. Zoebisch, *Organometallics* **7**, 513 (1988).
32. J. H. Horner and M. Newcomb, *Organometallics*, **10**, 1732 (1991).
33. D. J. Singh, H. K. Krakauer, C. Haas, and W. E. Pickett, *Nature* **365**, 39 (1993).

CHAPTER 10

Application Aspects Involving the Supramolecular Chemistry of Anions

J. L. SESSLER, P. I. SANSOM, A. ANDRIEVSKY, and V. KRAL

10.1 Introduction
10.2 General Classes of Synthetic Anion Receptors
10.3 Binding and Transport of Physiologically Important Anions
    10.3.1 Phosphates and Nucleotide Analogues
    10.3.2 Oligonucleotides
    10.3.3 Chloride
    10.3.4 Carboxylates and Amino Acids
    10.3.5 Other anions
10.4 Anion Receptors in Catalysis
10.5 Anion Receptors in Analytical Chemistry
    10.5.1 Chromatographic Methods
    10.5.2 Selective Electrodes
    10.5.3 Modified Surfaces
10.6 Waste Management
10.7 Anion-Templated Reactions
10.8 References

10.1 INTRODUCTION

Anions play an important role in various areas including nature, industry, and medicine. In the past decade, there has been a boom in the study of synthetic receptors directed towards the molecular recognition of anions. With these advances have come the development of applications of anion binding receptors. Such applications are the focus of this chapter. Accordingly, we will first describe the major known classes of synthetic receptors (Section 10.2). This will

---

Supramolecular Chemistry of Anions, Edited by Antonio Bianchi, Kristin Bowman-James, and Enrique García-España.
ISBN 0-471-18622-8. © 1997 Wiley-VCH, Inc.

be followed by a discussion of recognition, binding, and transport processes involving physiologically important anions, specifically phosphates and nucleotide analogues, chlorides, carboxylates and amino acids, and finally other anions such as fluoride, bromide, and sulfates (Section 10.3). Further topics covered include anion receptors for use in catalysis (Section 10.4) and their applications in analytical chemistry (Section 10.5). Finally, applications of anion binding in waste management (Section 10.6) and for templated syntheses (Section 10.7) will be covered. In all cases, an effort has been made to review the literature thoroughly however, much of the exemplary discussion will necessarily focus on work produced in the laboratories of the authors.

10.2 GENERAL CLASSES OF SYNTHETIC ANION RECEPTORS

Anion binding ligands have been designed to incorporate a vast array of different types of binding sites. These include ammonium, guanidinium, and Lewis acid moieties, among others. These classes will be described briefly with more specific examples given in conjunction with their respective applications. The ammonium receptors include linear polyammoniums, macrocyclic, macrobicyclic, and macrotricyclic derivatives. Also in this class are the pyridine- and pyrimidinium-containing receptors and bisintercalators. The guanidinium compounds include acyclic and macrocyclic receptors containing one to four guanidinium units. Among the Lewis acid receptors are those containing tin, mercury, boron, silicon, and germanium. Additionally, receptors with amide functionalities, porphyrin trimers, flourinated macrocyclic ethers, metal centers, and several expanded porphyrin anion receptors have been studied.

10.3 BINDING AND TRANSPORT OF PHYSIOLOGICALLY IMPORTANT ANIONS

10.3.1 Phosphates and nucleotide analogues

Anionic phosphorylated entities are of ubiquitous importance in biology. They play critical roles in a variety of fundamental processes ranging from gene replication to energy transduction.[1] Additionally, certain phosphate-bearing nucleotide analogues are known to display antiviral activity against a wide range of disorders, including in certain instances herpes simplex and AIDS.[2] Many of these analogues, however, in spite of being active in vitro, are inactive in vivo due to an inability to cross hydrophobic cell membranes.[3] Therefore, generally, the corresponding nucleosides are administered in place of the phosphorylated form of the drug. For instance, two members of the family of dideoxynucleosides, namely, $3'$-azido-$2'$-deoxythymidine (AZT) and $2',3'$-dideoxyinosine (DDI), are approved for use in AIDS therapy. For these nucleosides to be active, they must be phosphorylated through the action of

BINDING AND TRANSPORT OF PHYSIOLOGICALLY IMPORTANT ANIONS **357**

cellular nucleoside kinases, which at least in the case of AZT may not be very efficient.[2] It is thus considered desirable to develop methods allowing the phosphorylated nucleotide analogue to be administered directly. Indeed, in recent years, increasing effort has been devoted to this problem and to the more general one of phosphate recognition and transport.

In early work, Tabushi used the lipophilic, diazabicyclooctane-derived quartenary amine system **10.1** to effect adenine nucleotide transport.[4] This system, however, failed to transport guanosine-derived nucleotides. Since this time, several systems have been designed that include various acyclic, macrocyclic, and macrobicyclic polyaza systems,[5–11] such as **10.2–10.6** introduced, respectively, by Lehn and Mertes,[5] Kimura,[6] Schmidtchen,[7] Burrows,[8] and Diederich,[9] and the *bis*-intercalands of Lehn (e.g., **10.7**).[11]

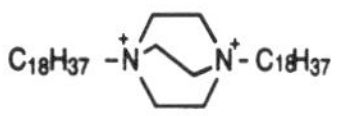

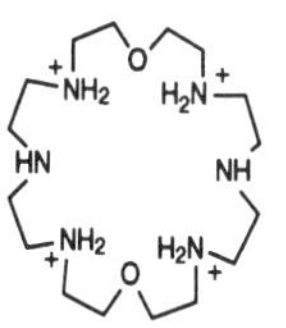

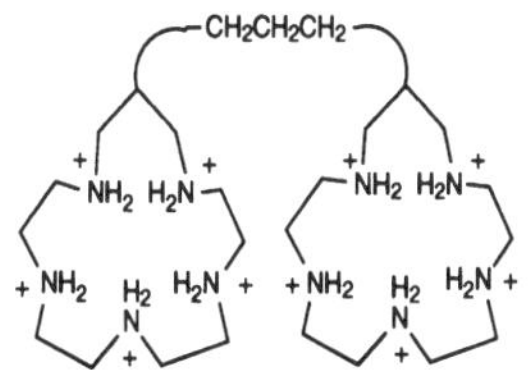

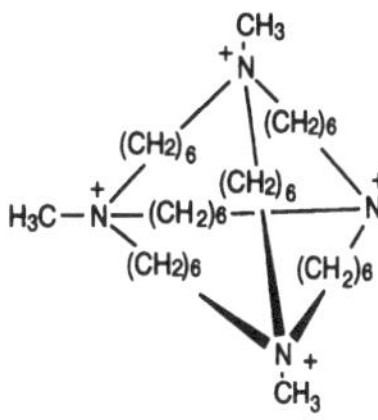

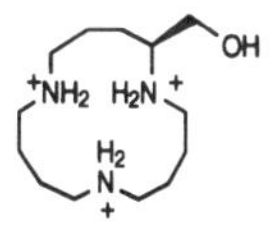

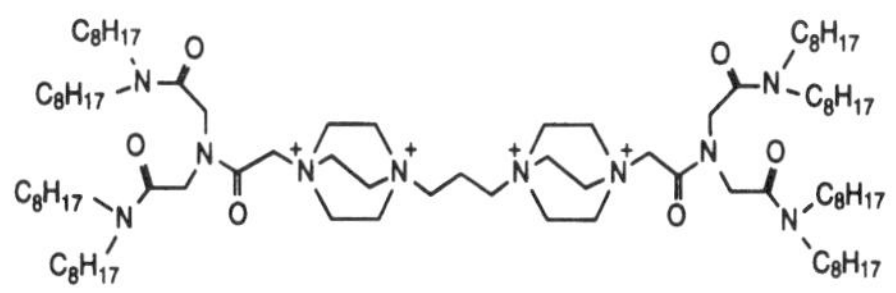

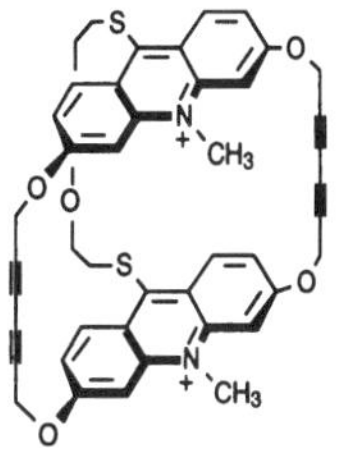

Extensions of this idea include anthrylpolyamines such as **10.8** from Czarnik,[12] cyclophanes from Schneider (e.g., **10.9**)[13] and García-España and Luis,[14] as well as the porphyrins with ammonium side chains introduced by Ogoshi **10.10**.[15] Other systems include the guanidinium-based receptors (e.g., **10.11–10.13**). This is a class of anion binding agents first developed by Lehn,[16] Schmidtchen,[17] and de Mendoza.[18] More recently, it has been elaborated on by Rebek and collaborators, to produce an adenine–adenine dinucleotide binding system **10.14**[19] and a cAMP binding and adenine mono- and dinucleoside monophosphate transporting system **10.15**,[20] as well as independently by Anslyn[21] and Hamilton[22] to produce the phosphodiester receptors **10.16** and **10.17**. Additionally, Kunitake has recently observed binding of ATP and AMP to guanidinium-functionalized monolayers (e.g., building block **10.18**).[23]

Other systems include the UO_2–salenes reported by Reinhoudt (e.g., **10.19**),[24] the amido binding systems of Reinhoudt (e.g., **10.20**),[25] the amido-

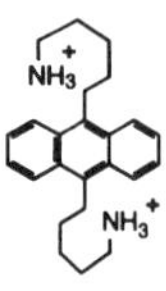

358 APPLICATION ASPECTS INVOLVING THE SUPRAMOLECULAR CHEMISTRY OF ANIONS

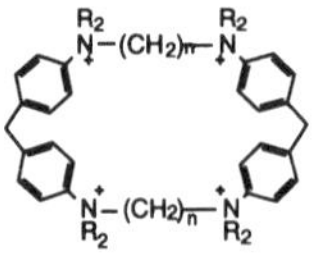

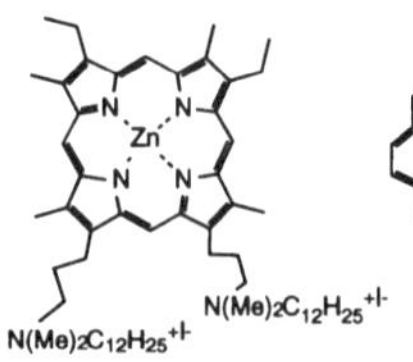

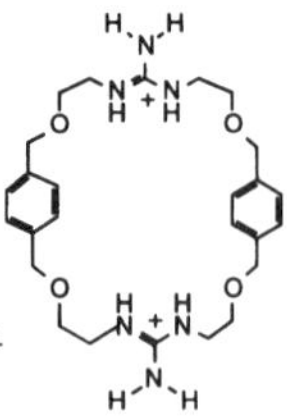

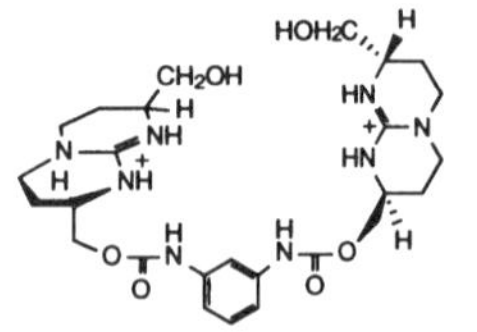

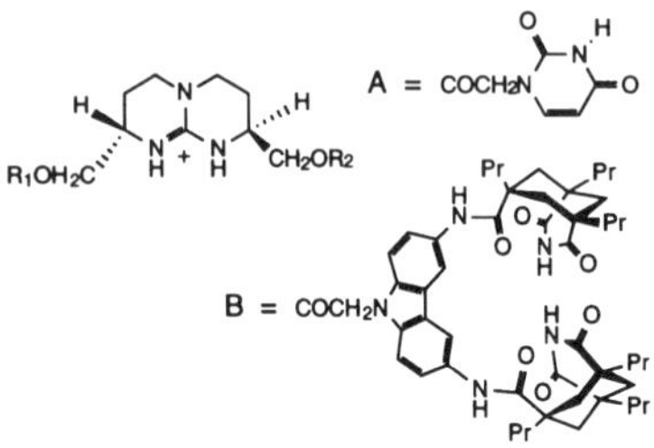

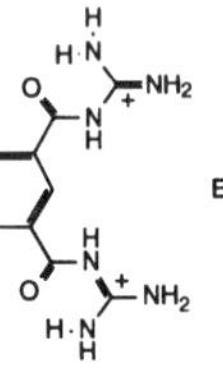

binding *bis*-calix[4]arenes of Beer,[26] the aminocyclodextrins of Schneider (e.g., **10.21**),[27] the ferrocene receptors of Beer (e.g., **10.22**),[28] and Koga's three-hydrogen-bonding system **10.23**.[29]

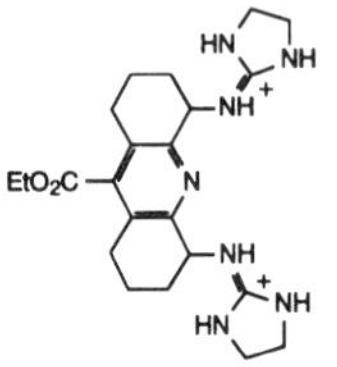

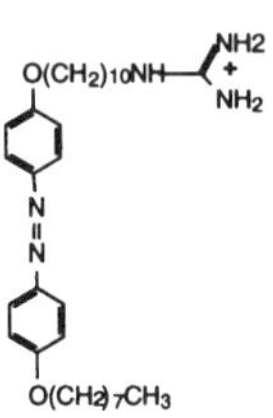

Despite an impressive amount of work in the phosphate anion binding field, we are currently aware of only relatively few systems that have been investigated thoroughly with regard to their ability to transport nucleotide phosphates. Carriers that actually work for nucleotide transport through model liquid membrane systems include Diederich's DABCO-based carriers **10.6**, which transport nucleotide 5′-triphosphates,[9] Ogoshi's dicationic porphyrin Zn complexes **10.10**, which transport dianionic monophosphate nucleotides,[15] and Rebek's carrier **10.15**, which transports adenine mono- and dinucleotide monophosphates.[20]

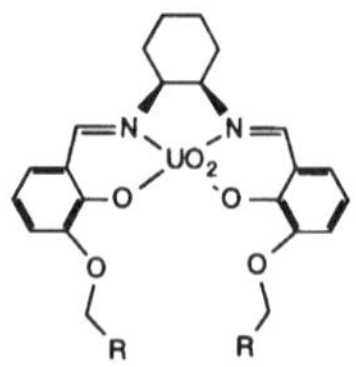

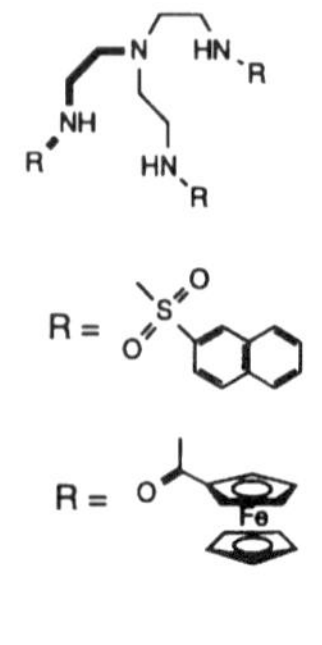

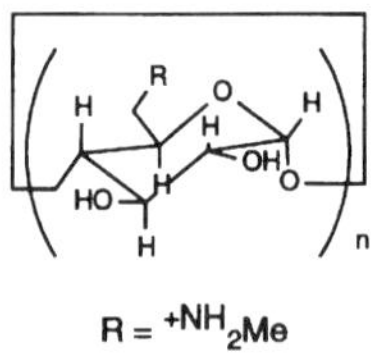

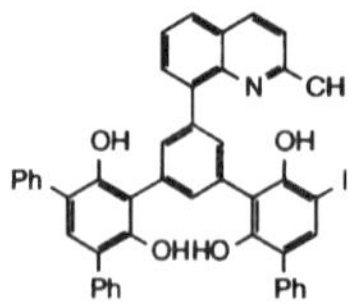

An alternative approach to nucleotide transport is being developed in the authors' laboratories. It is based on the use of so-called "expanded porphyrins," large pyrrole-containing macrocycles. These systems chelate phosphate in a way that is without precedent in simpler porphyrin-type systems, as manifest by the prototypic systems, sapphyrin **10.24**[30] and rubyrin **10.25**.[31] In the solid state, sapphyrin has been shown to complex monobasic phosphate (cf. Fig. 10.1) and phenylphosphate (cf. Fig. 10.2).[30] Evidence for sapphyrin-based phosphate anion binding in solution has come from a variety of transport experiments using simple aqueous I–organic–aqueous II U-tube-type model membrane systems.[32] This work has shown that sapphyrin is capable of transporting GMP and other nucleotides under conditions where it remains doubly protonated, that is, at pH ≤ 3.5. Other evidence for solution-phase binding has come from ^{31}P-NMR measurements and from studies involving various optical spectroscopies (e.g., UV–vis, fluorescence).[33] In addition, this system has also been found to be effective as a mediator for enhancing the through-membrane transport of cyclic-AMP at neutral pH.[30a] Indeed, sapphyrin has also been shown to bind cAMP in the solid state (cf. Fig. 10.3).[33] Rubyrin, on the other hand, in the presence of added solubilized Watson–Crick molecular recognition co-carriers, acts as an efficient carrier for the through-membrane transport of GMP and other mononucleotide species.[31]

Using an extension of this approach, it has also become possible, by functionalizing sapphyrins with nucleic acid bases (forming carriers such as **10.26** and **10.27**), to achieve the through-membrane transport of mononucleotides at neutral pH.[34] As expected, these systems display selectivity for substrates containing the appropriate Watson–Crick complement. Thus the cytosine-bearing sapphyrin nucleobase conjugate **10.26a** effects the selective transport of GMP over CMP and AMP at neutral pH. Interestingly, this same monosubstituted cytosine-derivatized system displays an intrinsic preference for the 2'-, as opposed to 3'- or 5'-, isomers of GMP.[35] While the reasons for this latter preference remain recondite, it is nonetheless important to appreciate that these carriers are the first to be reported that are able to effect the selective through-membrane transport of a chosen nucleotide at neutral pH.

A further extension of this approach involves the synthesis of oligosapphyrins, in particular two trimers (linear **10.28** and branched **10.29**) and a tetramer

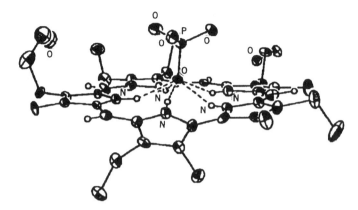

Fig. 10.1. Single crystal X-ray structure of the 1:1 chelated complex formed between monobasic phosphoric acid and sapphyrin **10.24**. For each diprotonated **10.24** there is also a nonligated monobasic phosphate (not shown). (Taken with permission from Ref. 30a.)

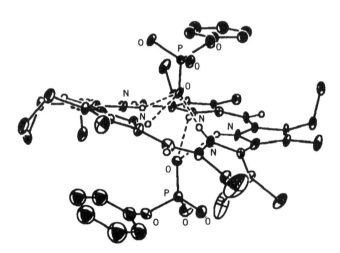

Fig. 10.2. Single crystal X-ray structure of the bisphenylphosphate salt of sapphyrin **10.24**. (Taken with permission from Ref. 30a.)

10.30.[36] These new systems are effective receptors for the binding and through-model-membrane transport of mono-, di-, and triphosphates at neutral pH. Here, ^{31}P-NMR spectroscopic studies served to support the suggestion that the key recognition motif associated with both binding and transport involves nucleotide phosphate oxyanion chelation by the sapphyrin subunits present in the receptors.

A further medical application of phosphate receptors is their use as calculi solubilizers in the treatment of urinary lithiasis, a disease in which calculi form in the urinary tract. Roughly 5% of the human population suffers to some

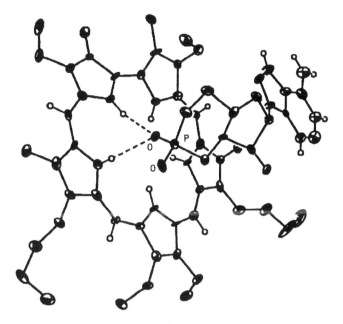

Fig. 10.3. Single crystal X-ray structure of cAMP bound to diprotonated sapphyrin **10.24**.

10.28

10.29

10.30

degree from urinary lithiasis. At this time, treatment is generally palliative and surgical. Human calculi can grow to a size of a few centimeters in diameter and are composed to the 90% level of insoluble inorganic salts such as calcium phosphate, calcium oxalate, mixed oxalate phosphate, and calcium phosphate–magnesium phosphate. Many attempts have been made to find a nonsurgical method for removal of calculi. One involves the chemical dissolution of human calculi via irrigation of the renal pelvis. Anion chelators such as the polyammonium system **10.31**, functioning via the presumed removal of phosphate or oxalate anions from insoluble calculi, have been investigated in this context as possible litholytic agents.[6b,37] The success obtained in the course of these studies support the contention that anion chelators have a possible role to play in the design of yet-improved renal irrigation fluids.

10.31

10.3.2 Oligonucleotides

Molecular recognition involving the phosphate groups of oligonucleotides is an important area in the field. It is relevant to the study of chemotherapeutic agents,[38] the development of synthetic transport carriers for oligonucleotide antisense agents,[39] and for the design of specific affinity cleavage agents. In this context, it is worth noting that the strength of the electrostatic interactions between acyclic and cyclic peralkylammonium compounds and DNA has been determined quantitatively by Schneider[40] and Stewart.[41] Rebek has also completed a set of preliminary extraction studies involving oligonucleotides of varying lengths, using his carrier **10.15**.[20] As yet, however, neither these compounds nor other synthetic phosphate receptors have been shown to be effective for transport of oligonucleotides, even though the facilitated transport of these substrates using polycations is well established.[42]

The fact that the protonated forms of sapphyrin were found to bind simple phosphates, such as diphenylphosphate (cf. Fig. 10.4),[33] led to the consideration that sapphyrins, by virtue of remaining monoprotonated at neutral pH, could bind to polyphosphate esters such as DNA. Indeed, it was found that the monoprotonated form of the water-solubilized sapphyrin does in fact bind the anionic phosphate backbone of DNA in a specific, rigid fashion.[43] Further, by using a variety of spectroscopic and biochemical techniques, it was established that sapphyrin does not bind DNA by the traditional modes of intercalation or groove binding. Rather, it chelates the anionic phosphate backbone in a manner

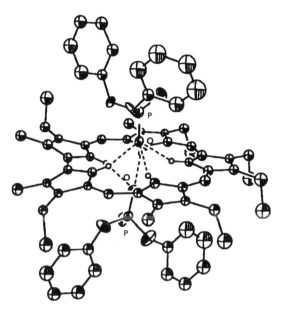

Fig. 10.4. Single crystal X-ray structure of diphenylphosphate with diprotonated sapphyrin **10.24**.

analogous to the sapphrin–phosphate structure shown in Fig. 10.4.[33] At present, therefore, efforts are being made to exploit this new mode of recognition in a range of applications including specific affinity cleavage and antisense oligonucleotide transport.

10.3.3 Chloride

Chloride anion is both ubiquitous in the biosphere and critical for a large number of biological processes.[44,45] Among these, there is none perhaps so important as that associated with cystic fibrosis.[46–50] Cystic fibrosis (CF) is the most common lethal genetic disease in Caucasians, striking about one in 2,500 U.S. infants.[49] This disease, which is becoming increasingly well understood as the result of the recent identification of the responsible gene mutation,[50,51] is characterized by an inability to produce proper, working chloride anion channel proteins (the so-called cystic fibrosis transmembrane regulator or CFTR protein),[48,52–54] which is believed to effect insufficient chloride and fluid excretion from, among others, pulmonary epithelia cells.[47,48] This, in turn, leads to a thick buildup of mucous deposits in the lungs and to a higher than normal susceptibility towards fatal pulmonary infections.[46–49] It is these infections, often of the *Pseudomonas aeruginosa* type,[47] that are generally the causative agents of cystic fibrosis–related death.[46–48]

At present, the established treatment protocols for cystic fibrosis involve treating these secondary infections with appropriate antibiotics, as well as adjusting diet and removing by physical means the deleterious buildup of mucocilliary secretions.[47] Unfortunately, these methods have not succeeded in prolonging the median life expectancy of cystic fibrosis patients past the age of 25.[47] Thus considerable current effort is being devoted to developing treatments that operate by attacking the underlying cause of disease. Here, a variety of approaches have been explored.[55-58] These range from attempts at gene therapy (incorporating the normal, wild-type cystic fibrosis gene into epithelia cells)[55] to the administration of agents that restore electrolyte balance either by opening up other non-CFTR-dependent chloride anion channels[56] or by inhibiting cellular uptake of sodium cations.[57] Unfortunately, however, at present the viability of this latter electrolyte balance restoration approach remains limited.[56] This is in large measure a result of the fact that, in the words of Knowles, a leading practitioner in the area: "There are no available therapeutic agents to improve chloride anion secretion."[56] Thus, as is true for the administration of antiviral agents, the existence of a chloride-selective carrier could be of prime clinical utility in terms of treating a major public health problem.

The other important fact to consider is that throughout the biosphere phosphate and halide transport are often interlinked[59] and that such a biomimetic linking (wherein transport of one anionic species is achieved at the expense of the other) could confer considerable clinical advantage in the case of purely synthetic nucleotide carriers: It would allow one to achieve into-cell delivery of an antiviral agent without creating a deleterious osmotic imbalance. Such considerations of osmotic balance, of course, are not relevant to CF treatment; here it would be best to effect concurrent out-of-cell chloride and sodium ion transport and use this as a trigger for subsequent fluid excretion. However, simultaneous into-cell diffusion of, for example, phosphorylated entities could be used to overcome problems associated with out-of-cell chloride anion transport in the case of unfavorable chloride anion concentration gradients.

Although, as mentioned above, no clinically usable chloride anion transport agent currently exists, it is important to appreciate that considerable effort has, nonetheless, been devoted to the problem of preparing halide-binding receptors, and a number of such systems are now known.[60-87] For instance, in what is now recognized as being the first description of a synthetic anion receptor, Park and Simmons reported in 1968 that 3° ammonium bicycles, such as **10.32**, could be used to bind chloride anion with a certain degree of specificity (at least relative to bromide and iodide).[60] Unfortunately, however, only very low affinity constants were recorded for these systems, and even then media of low pH were required. In later work, Lehn and co-workers found that the structurally related but intellectually independent cryptand cations (e.g., **10.33**[61] and **10.34**[62]) could be used to bind halide anions with both high affinity and high selectivity.[61-63] In fact, by "tuning" the size and geometry of his systems, Lehn was able to generate receptors with near-exclusive halide specificities (e.g., **10.33**

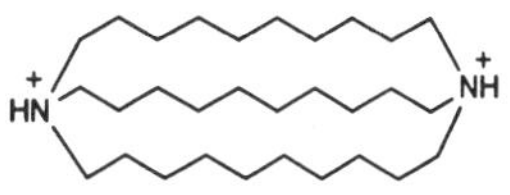

10.32 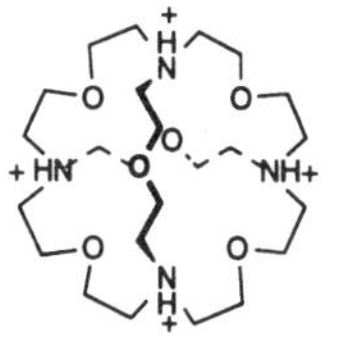 **10.33** 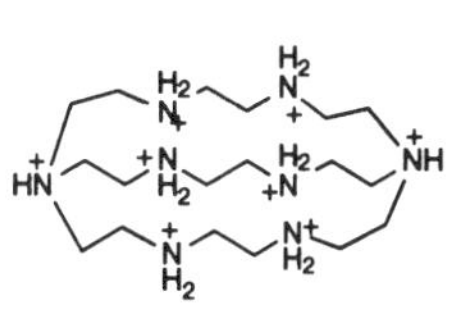 **10.34**

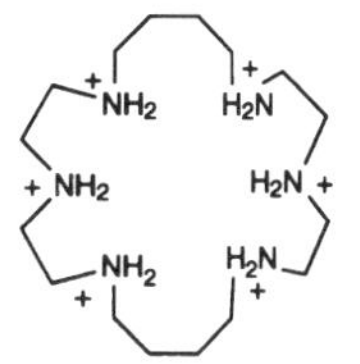

10.35 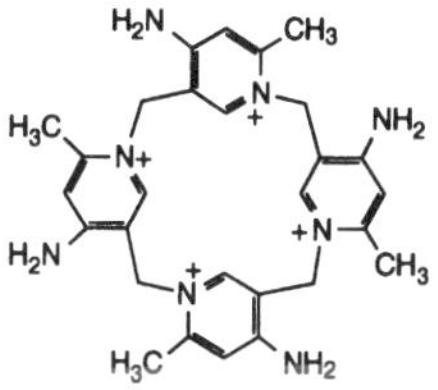 **10.36** 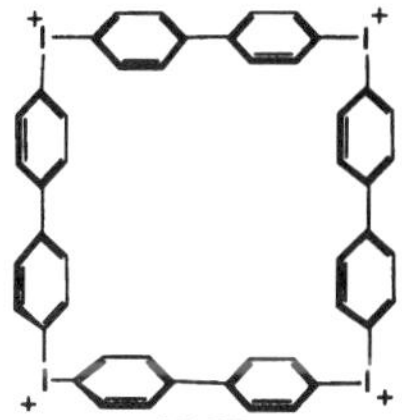 **10.37**

for chloride[61] and **10.34** for fluoride[62]). Unfortunately, these and other more simple "monocyclic" polyammonium macrocycle systems (e.g., **10.31**[64] and **10.35**[63d]) suffer from a serious drawback as far as anion transport is concerned: They rely on a high net positive charge to effect the coordination of a single anion. As a result, the supramolecular anion-to-receptor complex formed upon anion binding is still highly charged and relatively insoluble in organic media. Thus through-membrane transport is only achieved if a large, organic soluble "helper anion" is added to the halide-bearing transport medium.[65–67] It also plagues receptors of the pyridinium-based cyclophane class (i.e., **10.36**) reported by Cramer and co-workers,[68] the macrocyclic tetrabiphenyldiyltetraiodum cationic molecular boxes (e.g., **10.37**) reported recently by Stang and Zhdankin[69] and the amide-linked tridentate cobalticinium-based receptor **10.38** synthesized in the group of Beer.[70] In the latter case, an interesting participation of the amide NH groups in the anion-binding process was suggested.

Hydrogen bonding of anions, in particular by the NH groups of (thio)urea, was further explored by Reinhoudt and co-workers; they achieved selective

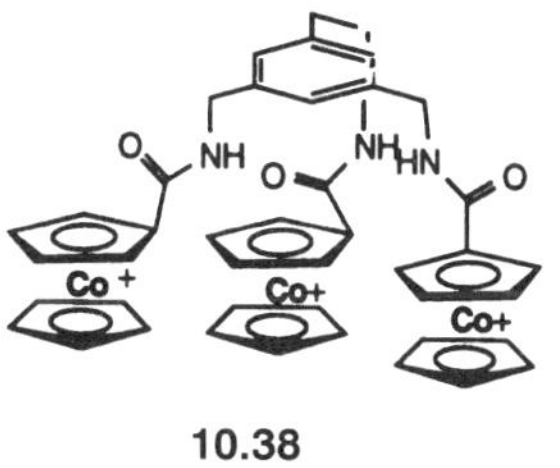

10.38 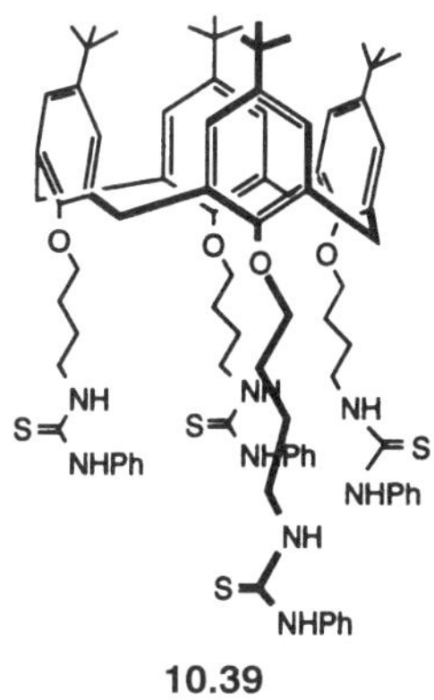 **10.39**

366 APPLICATION ASPECTS INVOLVING THE SUPRAMOLECULAR CHEMISTRY OF ANIONS

binding of chloride over bromide and iodide using calix[4]rene derivative **10.39**.[71]

Considerations of charge are of less concern in the case of the newer metal- and metalloid-derived anion receptors,[72–79] examples of which include **10.40**–**10.47** developed, respectively, by the groups of Wuest,[72] Newcomb,[73] Kuivila,[74] Jung,[75] Takeuchi,[76] Pizer,[77] Katz,[78] and Hawthorne.[79] In this instance, however, questions of heavy metal toxicity cloud considerations of possible clinical utility.

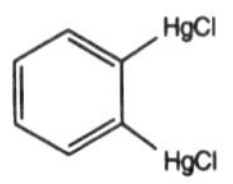

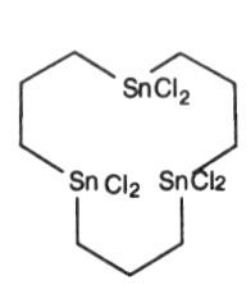

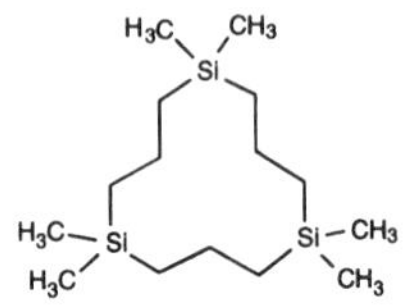

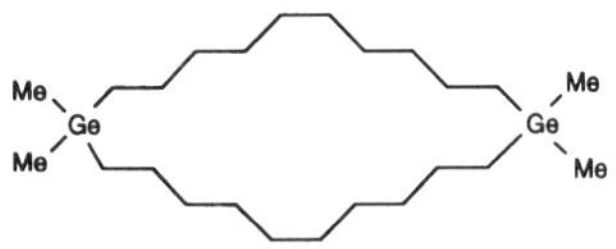

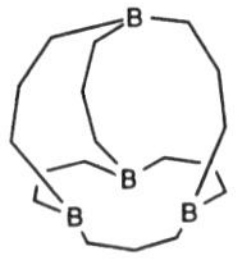

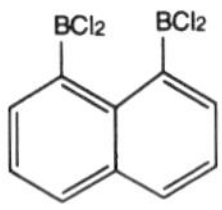

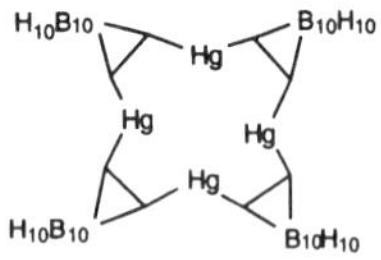

A new approach to anion chelation, based on ion–dipole interactions, was developed recently in the Schmidtchen group. Here, neutral zwitterionic polyazamacrocycles (e.g., **10.48**[80a] and **10.49**[80b]) that incapsulate anions in their cages were used. While elegant, this non–Lewis acid approach necessarily requires a near-perfect preorganization of the host. Depending on the targeted applications, this can be either good or bad.

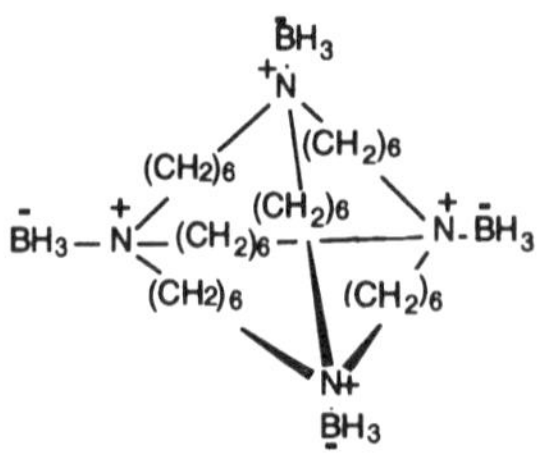

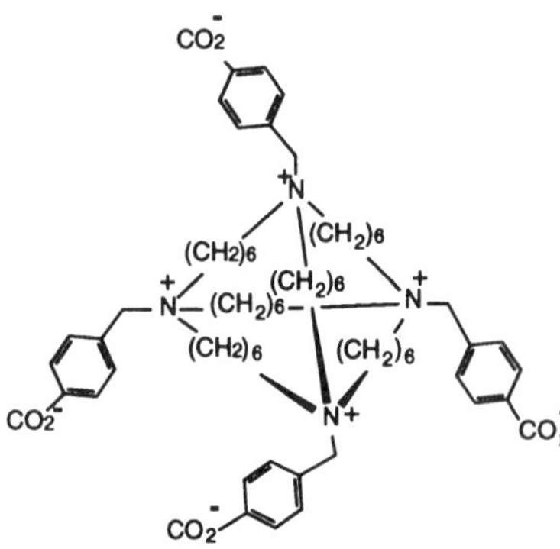

Expanded porphyrins developed in the Sessler group have also proved effective for chloride anion recognition, a point that has been established by X-ray structural studies and solution-phase binding experiments. Specifically, sapphyrin **10.24**, rubyrin **10.25**, and anthraphyrin **10.50** were found to bind chloride anion both in solution and in the solid state (c.f. Figs. 10.5–10.7).[30a,82–84] In the case of the sapphyrin–chloride complexes, two X-ray structures, corresponding to mono- and diprotonated sapphyrin species were obtained (cf. Figs. 10.5a and 10.5b).[30a]

As might be expected from the structural differences, the sapphyrin, rubyrin, and anthraphyrin macrocycles display inherently different affinities for chloride and other anions. This is something that has been established not only by direct affinity measurements, but also by transport experiments carried out using a standard aqueous I–organic–aqueous II model U-tube membrane system.[82,85] Interestingly, in these studies it was found that the anion that is less well bound is transported at the greatest rate (i.e., chloride by sapphyrin and fluoride by anthraphyrin). Moreover, it was found that through-transport could be effected

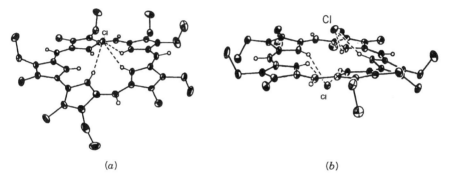

Fig. 10.5. (a) Single crystal X-ray structure of the chloride anion complex of monoprotonated sapphyrin **10.24**. (b) Single crystal X-ray structure of the bishydrochloride salt of sapphyrin **10.24**. (Taken with permission from Ref. 30a.)

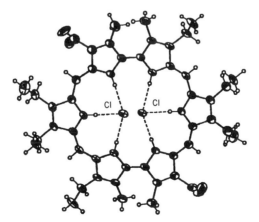

Fig. 10.6. Single X-ray structure of the bishydrochloride salt of rubyrin **10.25**. (Taken with permission from Ref. 84.)

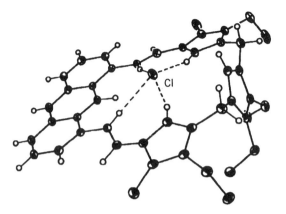

Fig. 10.7. Single crystal X-ray structure of the chloride anion complex of diprotonated anthraphyrin **10.50**. (Taken with permission from Ref. 30a.)

not only via simple synport (wherein the target anion is cotransported with some other counter cation), but also via antiport, in which a particular target anion is transported through a membrane at the expense of backtransmission of some other negatively charged chemical entity. Although not surprising, this ability to effect anion transport by two different mechanisms has important implication as far as the possibility to couple chloride and phosphate transport is concerned (see the discussion above). Finally, higher-order (i.e., three- and fourfold) chloride anion binding has also been recently reported for the triply and quadruply protonated forms of rosarin **10.51** and turcasarin **10.52** (c.f. Figs. 10.8 and 10.9).[86,87] Thus these various studies not only support the suggestion that it should be possible to effect anion-specific transport under physiological conditions, they also support the contention that, by proper synthetic "selection" or "fine tuning," it should be possible to design expanded porphyrin anion receptors that are selective for either phosphate-derived antivirals, or chloride anion, or both.

10.50   10.51   10.52

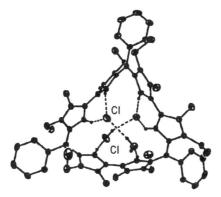

Fig. 10.8. Single crystal X-ray structure of rosarin **10.51**. In addition to the two chloride anions shown, for each unit of **10.51**, a third nonchelated chloride anion is found within the crystal lattice. (Taken with permission from Ref. 86.)

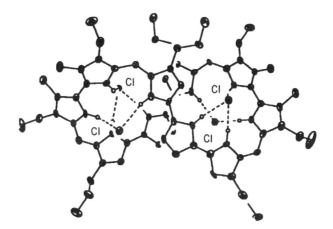

Fig. 10.9. Single crystal X-ray structure of turcasarin **10.52** bound to four chloride anions. (Taken with permission from Ref. 87.)

10.3.4 Carboxylates and amino acids

The carboxylate group is another anionic entity of prime importance in nature. Enzymes, antibodies, amino acids, and metabolic intermediates, as well as the other natural products, contain a range of carboxylate functionalities that, in some instances, account for the characteristic biochemical behavior.[88] Depending on the structure of the compound and the microenvironment, the carboxyl groups in question can either be in their neutral (protonated) or anionic (deprotonated) forms.

In terms of molecular recognition, carboxylates are important since they chelate various positively charged species, such as metal cations and organic, mostly nitrogen-based, cations. Also, hydrogen-bonding arrays, involving one or both oxygens of carboxylic acid (or carboxylate anion) take an important part in noncovalent organization of secondary and tertiary structures of complex biological molecules as well as in chelation of "small" substrates. Not surprisingly, therefore, the design and synthesis of biomimetic receptors for carboxylates have received much attention over the past two decades. Here, an "extra" impetus for study has come from a realization that carboxylate-binding processes could be of prime import in terms of the design of chromatographic materials for analysis and separation of important carboxylate-containing substrates, such as, for instance, amino acids.[89]

There are several groups of synthetic receptors for carboxylates currently known.[6a,17a,63a,b,17a,90–99] These may be classified according to their charge, nature of chelation, and/or stereochemical dependence of the interactions. The first group consists of positively charged polyammonium (e.g., **10.2**, **10.3**, **10.53**)[6a,63a,b,90] and guanidinium (e.g., **10.54**, **10.55**)[17a,91] anchor groups of different complexity that interact with the carboxylate group via a combination of electrostatic and dipole–dipole attractions as well as through hydrogen bonding. Historically, these two functional groups were initially examined by the groups of Lehn,[63a,b,90a–c,e,91d,g] Kimura,[6a,90i–k] and Schmidtchen.[17a,91c,e,f,i,j]

Anion chelation by Lewis acids, together with additional amide hydrogen-bonding interaction, has allowed a degree of dicarboxylate binding selectivity to be achieved in the case of the redox-responsive ditopic bis(cobalticenium) calix[4]rene receptor **10.56** prepared by Beer's group.[92] Additionally, neutral dinuclear metallomacrocycles (c.f. **10.57**) prepared in the Reinhoudt group chelate carboxylates by their Lewis acid uranyl anchor groups.[93] A similar approach has allowed for the enantioselective recognition of N-protected amino acids in the case of the chiral, capped metalloporphyrin **10.58** synthesized in Inoue's laboratory.[94] Finally, a two-point electrostatic bonding approach, based on the binding of dibasic dicarboxylates by the two lithiated cationic crown moieties of the double-looped crownophane **10.59**, has been proposed by Nishimura and co-workers; it appears to allow for the transport of dicarboxylate anions through lipophilic bulk membranes.[95]

Complementing the above approaches are those involving *neutral* receptors. These, which have been extensively studied by Hamilton,[96a–d,i,97a]

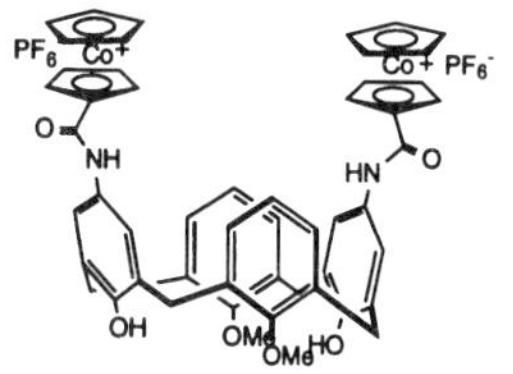

10.56 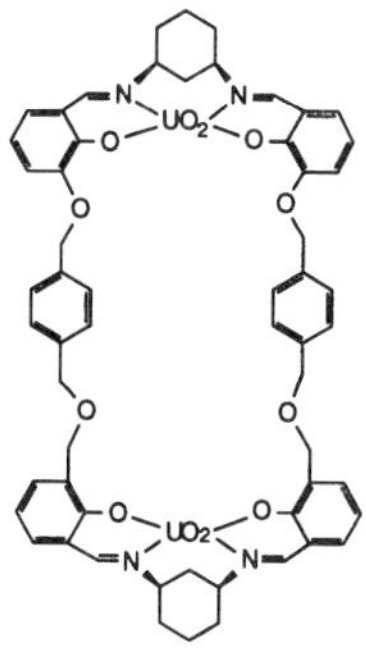 **10.57** 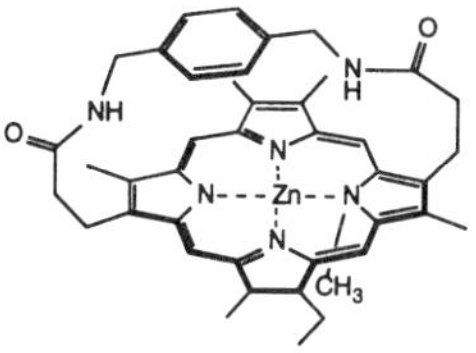 **10.58** 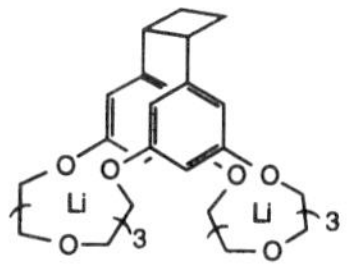 **10.59**

Diederich,[97b–e] and others,[96e–h] are represented by a plethora of receptor systems containing multiple urea, thiourea, amidopyridine, and simple amide binding sites within well-defined structural arrays.[93,96–98] Here, in addition to covalent multidentate receptors (e.g., **10.60–10.66**),[96] systems have also been reported that were designed to possess chirality (e.g., **10.67–10.69**).[97,98] Chelation of anions by these receptors is based on the formation of several cooperative hydrogen bonds. Also, in some selected cases, additional interactions, such as π–π stacking, provide all-important selectivity. Related to these types of receptors are the charged metal complexes, which assemble noncovalently (e.g., **10.70**[98a] and **10.71**[98b]) and possess chirality. These bind carboxylate anions via their neutral appendages.

An alternative approach to dicarboxylate binding has been pursued recently in the Sessler group. It is based on the use of sapphyrins as the binding site(s). Here, for instance, dicarboxylate binding and transport were demonstrated using the covalently linked sapphyrin dimer **10.72**.[99] Interestingly, this system displays a preference for linear over bent substrates and aromatic over aliphatic substrates. Such high selectivity was not necessarily expected, given the inherently floppy nature of this receptor, but does at least serve to highlight that there are subtleties associated with this binding motif that remain to be explored.

Binding and transport of amino acids in their cationic,[100] zwitterionic[101–109] or anionic[94,110–113] forms constitutes an important problem in supramolecular chemistry from both a biomimetic and analytical point of view. At present, some information is available about the dynamics and regulation of amino acid transport in nature, but a detailed mechanistic or structural understanding remains lacking.[114] Hence, the design of synthetic receptors has been guided as much by intuition as by an analysis of natural amino acid binding systems. Making matters further challenging is the fact that the strong influence of the nearby protonated amino group makes it hard to bind the zwitterionic form of amino acids (the natural state of these species) via the use of "normal" carboxylate receptors. Also, as Rebek has pointed out, in water, the best solvent for these doubly ionic species, the formation of strong hydrogen bonds between the receptor and the substrate is energetically unfavorable.[101] On the other

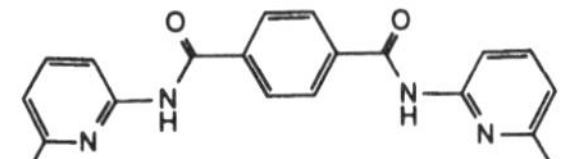

372 APPLICATION ASPECTS INVOLVING THE SUPRAMOLECULAR CHEMISTRY OF ANIONS

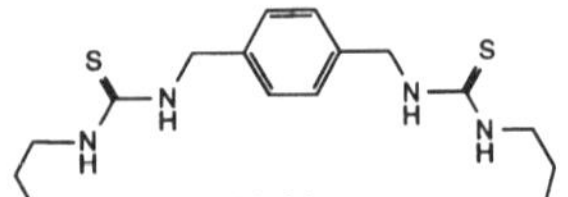

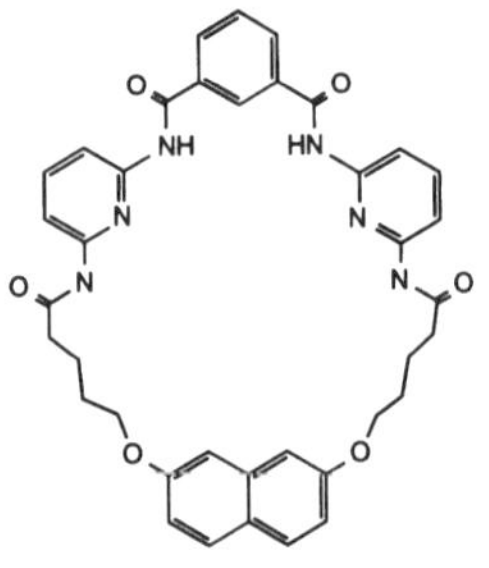

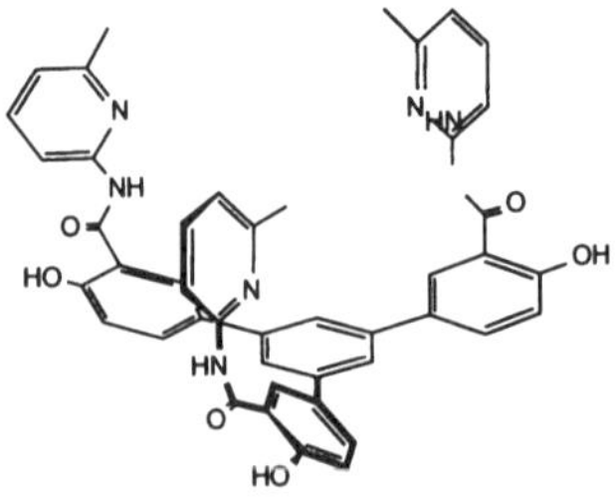

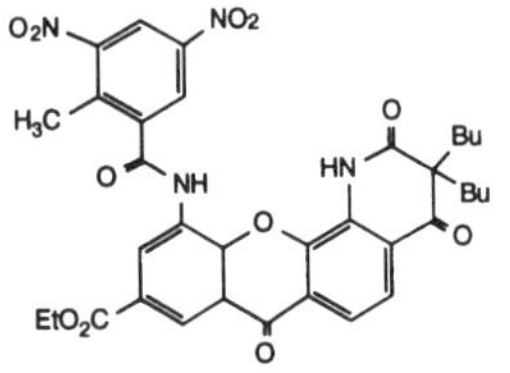

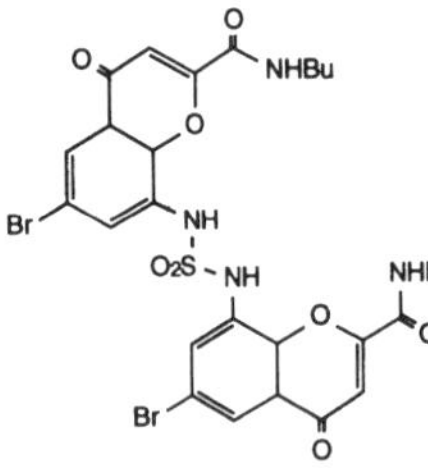

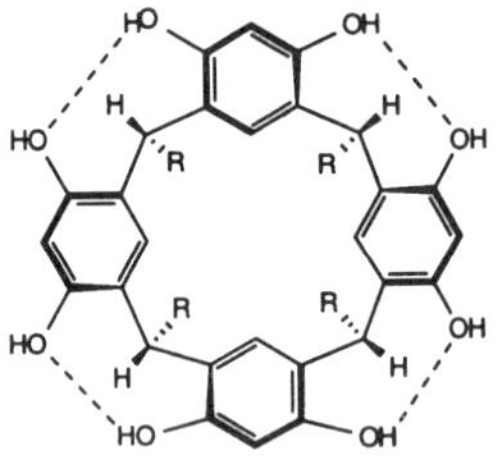

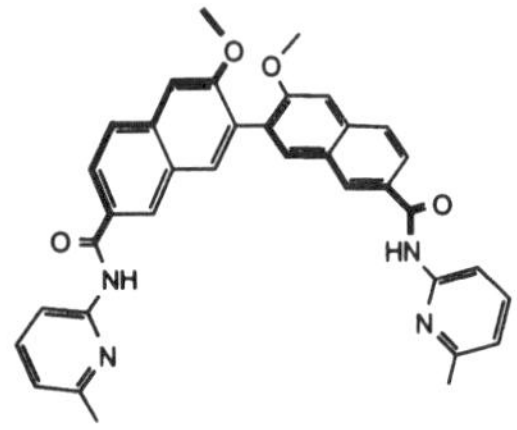

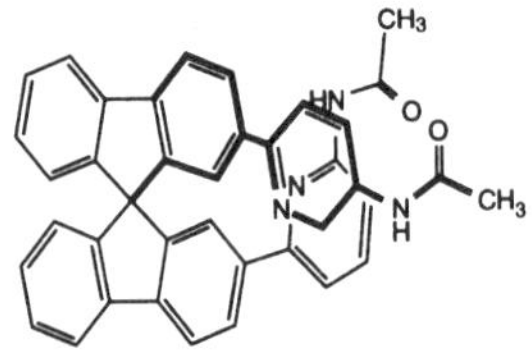

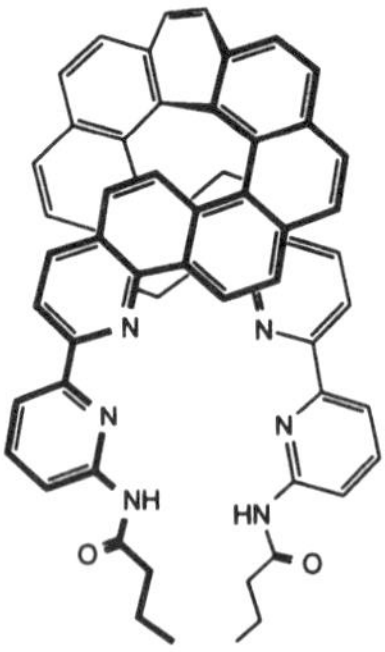

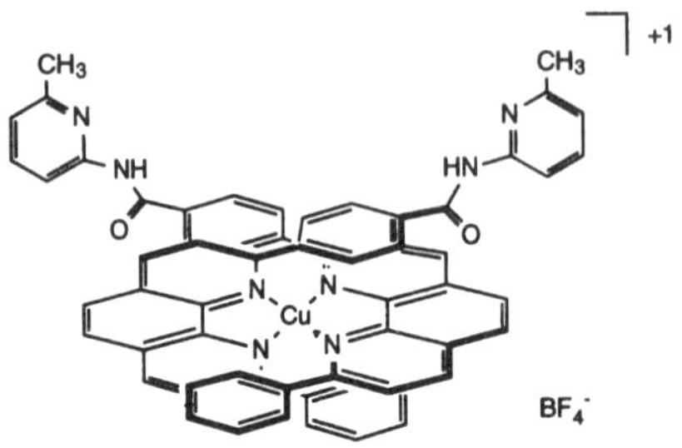

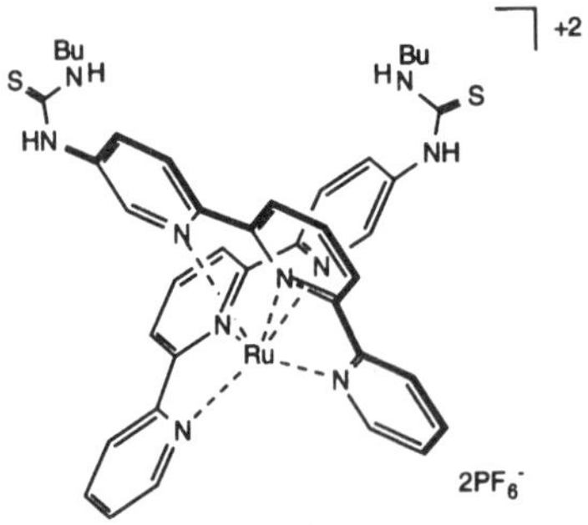

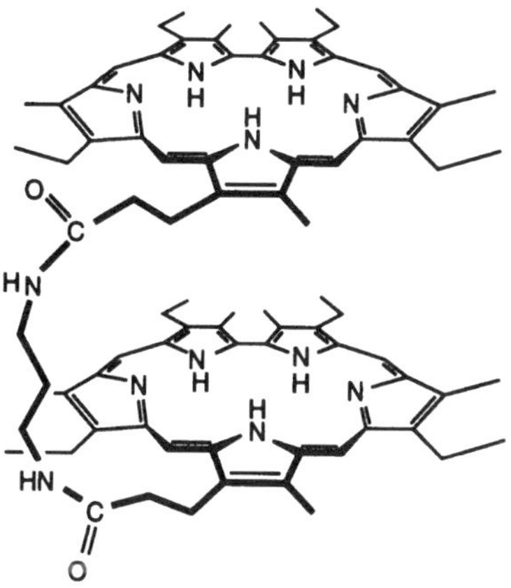

10.72

hand, the low solubility of amino acids in organic solvents imposes a costly energetic burden on their recognition in nonaqueous solutions. Thus, in order to chelate an amino acid in its zwitterionic form, *bidentate* receptors, containing recognition sites for both carboxylate anion and protonated amino group, have to be employed. These, as in the case of simple carboxylate binding discussed above, could be neutral, charged, or zwitterionic.

Several elegant receptors for amino acids have been reported. For instance, in what is recognized as being the first well-documented example of work along these lines, Baczuk reported in 1971 that *zwitterionic* L-arginine, when bound covalently to Sephadex resin, was capable of effecting chromatographically the separation of racemic dihydroxyphenylalanine (DOPA).[102] Subsequently, de Mendoza and co-workers reported an enantiomerically pure receptor **10.73** that incorporates a guanidinium group, a crown ether, and a naphthalene substituent so as to bind the carboxylate anion, protonated amine, and aromatic side chain portions, respectively, of phenylalanine and tryptophane.[103] With this system, chiral recognition with excellent enantioselectivity was achieved. A highly effective carrier system **10.74**, based on the zwitterionic acridine derivative of Kemp's diacid, has also been described by Rebek.[101] In this system, a combination of electrostatic attractions, multiple hydrogen bonding,

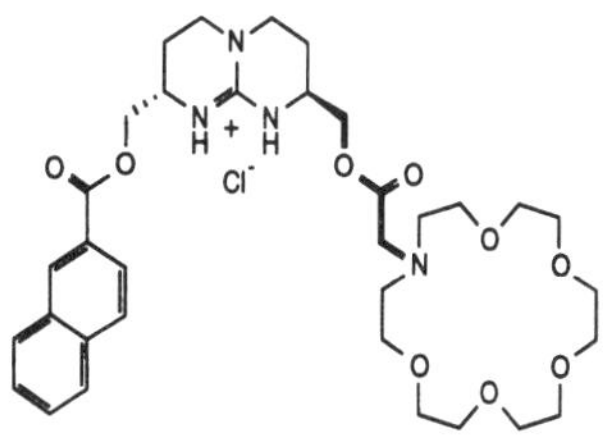

10.73 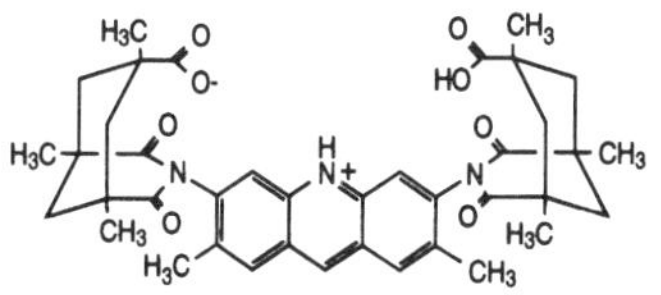   **10.74** 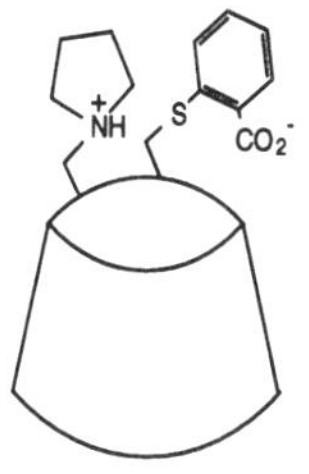   **10.75**

and aryl–aryl stacking interactions provides the key recognition elements required to achieve the binding and through-membrane transport of phenylalanine, tryptophane, and tyrosine methyl ether under neutral conditions. Functionalized zwitterionic (i.e., **10.75**) and plain cyclodextrins have also been used by Tabushi[104] and others[115] to bind amino acids in aqueous solution. Here, amino acid side chain inclusion within the cyclodextrin cavity provides an all important complement to the recognition process.

In a different approach, the photoresponsive spiropyrane system **10.76**, which readily undergoes reversible isomerization to the zwitterionic merocyanine **10.77** upon irradiation with UV light, was employed by Sunamoto and co-workers to effect the transport of phenylalanine across a liposomal bilayer (Fig. 10.10).[105] Later, this same molecule was used to prepare a tryptophane-imprinted merocyanine copolymer membrane with photoregulated permeability.[106]

Another approach to amino acid binding is based on the use of Lewis acids, such as copper complexes[107] or arylboronic acid species, either free[108] or in combination with crown ethers.[109] Here it is to be noted that the copper-based complexes, when derived from appropriate chiral ligands, have found wide use in cromatography-based resolution of racemic amino acids.[116]

While not as challenging as zwitterionic binding, it is nonetheless worth noting that N-protected amino acids and amino acid anions can be chelated readily by "classic" carboxylate receptors, such as thioureas,[110] metalloporphyrins,[94] cationic macrocycles,[111] quaternary ammonium salts,[112] or guanidinium anchor groups.[113] N-CBz-protected dicarboxylic (aspartic, glutamic) and aromatic (phenylalanine, tryptophane, and tyrosine) amino acids can also be bound by cyclic and acyclic sapphyrin dimers, including ones such as **10.78–10.80** that feature chiral linking bridging units.[117]

The above findings lead the authors to predict that carboxylate receptors or their analogues could find application in the analytical detection and/or chromatographic analysis and separation protocols. Such carboxylate receptors could also prove useful in medicine, since they could function as artificial carriers allowing the treatment of diseases caused by an improper regulation of processes involving carboxylate-containing substrates.[118] However, a great deal of experimental studies have to be done before these or other receptors are applicable to these kinds of "real life" problems.

Less long range than the above applications is the use of carboxylate

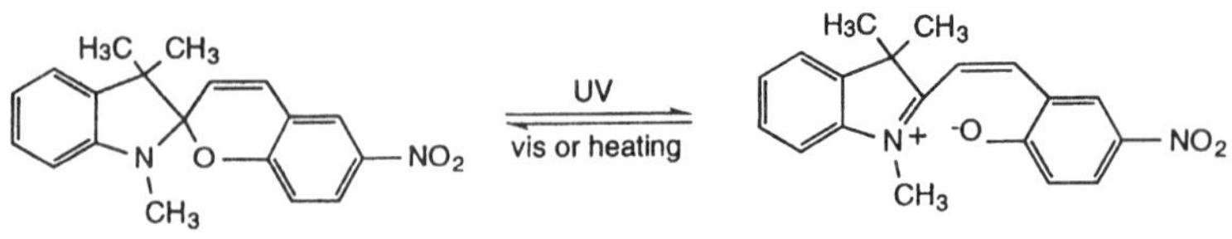

Fig. 10.10. Photoisomerization of spiropyrane **10.76** to merocyanine **10.77**.

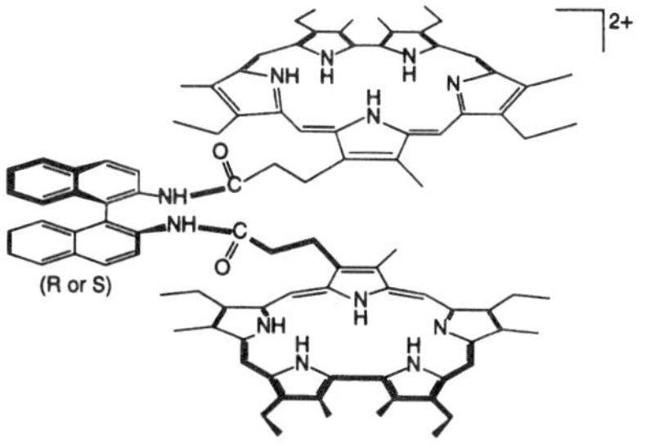

BINDING AND TRANSPORT OF PHYSIOLOGICALLY IMPORTANT ANIONS **375**

10.78

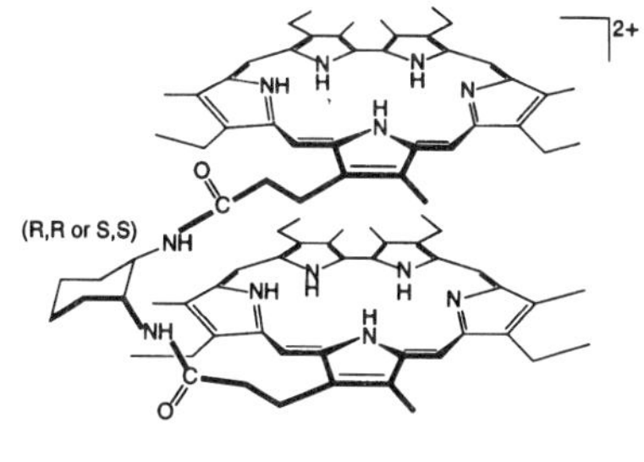

10.79

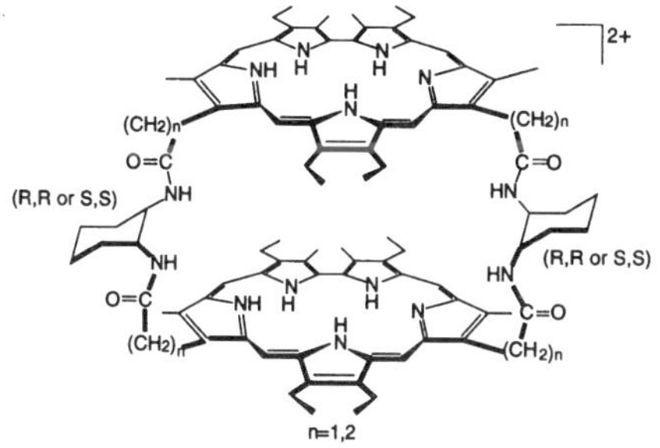

10.80

receptors to develop new electron-transfer/photosynthetic model systems. This is an aspect that has been explored in the authors' laboratory.[119] Specifically, a new noncovalent ensemble for energy transfer has been prepared that is based on the binding of a carboxyl-containing, free-base porphyrin photodonor by a monoprotonated sapphyrin acceptor (Fig. 10.11). Upon irradiation at 573 nm, singlet–singlet energy transfer from the porphyrin to the sapphyrin subunits takes place readily with energy-transfer dynamics that are consistent with a Föster-type mechanism. This system thus appears to be prototypic of a new approach to energy and electron-transfer modeling that is predicated on the use of anion chelation. As such, it provides an important hint of other things yet to come. Indeed, one can easily imagine the use of such recognition processes to prepare complex, multiple, chromophore arrays wherein the key redox active groups are all oriented with respect to each other via noncovalent interactions.

10.3.5 Other anions

Anions other than phosphates, carboxylates, and chloride are also present in most biological systems, albeit in far lower concentrations. They nonetheless play critical roles in a large number of biological processes. For instance, fluoride anion activation is important for a variety of enzymatic systems, including phosphatases, adenylate cyclase, and the GTP-binding protein (G-protein).[120] Iodide anion is also required for a proper functioning of the thyroid gland.[121]

Examples such as these, which are representative, serve to highlight the

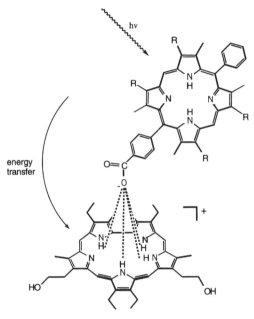

Fig. 10.11. Proposed noncovalent ensemble for energy transfer.

need to make receptors for other "unusual" anions. Further motivativation for the construction and study of such putative receptors comes from an appreciation that waste removal, especially in heavy industry, but also in electronic, food, and other industries, is a problem of prime import (see Section 10.6).

As in the case of phosphate, chloride, and carboxylate binding (discussed above), various macrocyclic polyaza systems (e.g., **10.2**, **10.3**, **10.31**, **10.33**, **10.34**) were found useful for complexation of other anions.[122] These include anions with one, two, or more negative charges, as well as even more complex anions, including ones containing metal centers.[90d] Unfortunately, most of these receptors, in their "normal" highly protonated states, are selective for the *charge* of the targeted complexing anion, rather than for the substrate per se. Further, because the major driving force for binding in these kinds of systems is electrostatic in nature, it is usually the anions with larger charge that are bound more strongly. However, a few polyaza systems are known that were designed for the specific binding of certain select anions. For example, cryptands **10.33** and **10.34**, prepared in the Lehn group, show some size-based selectivity for chloride[61] and fluoride[62] anions, respectively. The polyammonium macrotricycles of type **10.4**, **10.48**, and **10.49** prepared in the Schmidtchen group contain cavities of different sizes and also display differing selectivities for various halides.[7,80] Neutral macrocycles with Lewis acidic moieties, such as **10.41**, **10.43**, and **10.45**, also possess a certain degree of selectivity towards different halide anions.[73,75,77] Finally, calixarene-derived receptors of Reinhoudt[71] and Beer[26] show selectivities for Cl^- and F^- anions, respectively.

Sapphyrin, the most extensively studied anion-binding expanded porphyrin, forms complexes with various structurally distinct anions (see discussion in the previous paragraphs). Stability constants of the complexes formed between sapphyrin and halides are strong ($K_S \geq 10^7$ M^{-1} for F^- in CH_2Cl_2), and the selectivity is as follows: $F^- > Cl^- > Br^-$.[83] Also, the mono- and diprotonated forms of sapphyrin are known to chelate $CF_3CO_2^-$, NO_3^-, ClO_4^-, and p-$CH_3(C_6H_4)SO_3^-$, in addition to binding phosphate-type species. This chelation was proved via spectroscopic binding experiments[123] and, in the case of the fluoride,[83] p-toluenesulfonate,[123] and azide complexes, via single crystal X-ray diffraction analysis (Figs. 10.12–10.14). Interestingly, rubyrin, a hexapyrrolic

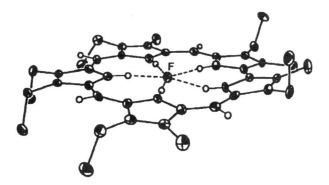

Fig. 10.12. Single crystal X-ray structure of the fluoride anion encapsulated mixed hydrofluoride–hydrohexafluorophosphate salt of diprotonated sapphyrin **10.24**. (Taken with permission from Ref. 30a.)

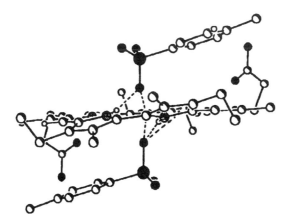

Fig. 10.13. Single crystal X-ray structure of the *bis*-tosylate complex of diprotonated sapphyrin **10.24**.

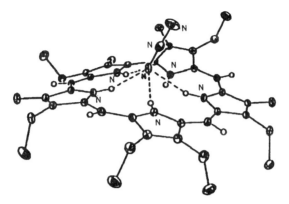

Fig. 10.14. Single crystal X-ray structure of the azide anion complex of monoprotonated sapphyrin **10.24**. (Taken with permission from Ref. 30a.)

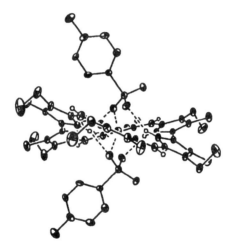

Fig. 10.15. Single crystal X-ray structure of *bis*-tosylate complex of diprotonated rubyrin **10.25**.

expanded porphyrin, was also found to bind *p*-toluenesulfonic acid in the solid state (Fig. 10.15),[124] a finding that supports the contention that anion binding by expanded porphyrins is not limited to the "special case" of sapphyrin.

10.4 ANION RECEPTORS IN CATALYSIS

Catalysis represents a major goal in supramolecular chemistry.[125] This is no less true in the area of anion recognition. In both cases, the predicative problem is to design a molecular receptor with specificity for a given substrate. This design, once realized in terms of synthesis, should then lead to (1) binding of the substrate, and (2) transformation of the bound species into products that are released along with the regenerated catalyst. In the context of anion binding, supramolecular catalysts have been largely limited to systems that promote phosphoryl transfer reactions as well as to those that promote hydrolytic cleavage transformations of phosphodiesters and RNA.

Polyammonium macrocycles, such as those described previously in the phosphate binding and transport section (e.g., **10.2**), have been investigated in great detail by Lehn and Mertes as enzyme mimics for phosphoryl transfer processes.[5a] These compounds have several attributes that combine to make them effective catalysts. These include a high positive charge density and the presence of numerous potential hydrogen-bonding sites that, in concert, can act to promote complex formation with biologically relevant anionic substrates. Not surprisingly, therefore, these systems have been found to be efficient catalysts for the dephosphorylation of ATP (forming inorganic phosphate and ADP).[126,5c] Additionally, kinase activity, as manifest by phosphorylation of inorganic phosphate to form pyrophosphate, is observed in the presence of magnesium(II) and calcium(II).[5a] Likewise, ATP was found to be generated catalytically from acetyl phosphate and ADP in the presence of magnesium(II).[5b] Finally, this type of basic catalysis has been expanded in its scope so as to generate a designed multistep pathway wherein dephosphorylation of ATP is coupled to a formyl transfer reaction.[127]

The catalysis of phosphodiester hydrolysis has been extensively studied due to its relevance to DNA and RNA hydrolysis. Although some work involving simple amines and imidazole units[128] on non-metal-mediated catalysis of phosphodiesters has appeared, most of what has been considered to date has involved the use of guanidinium-based receptors.[91a] Hamilton has reported acceleration of both intramolecular and intermolecular phosphodiester cleavage using the receptor **10.16**.[129] *bis*-Guanidinium receptors are also useful for biological substrates. Anslyn has found that receptor **10.17** enhances the hydrolytic reaction of poly-RNA in the presence of imidazole.[130] Finally, Göbel found that compound **10.81** complexes strained cyclic phosphates via its amidinium group, a moiety that plays a key role in effecting the intracomplex transesterification processes observed in the presence of guanidinium ions.[131]

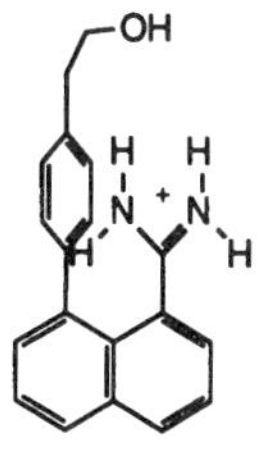

10.81

10.5 ANION RECEPTORS IN ANALYTICAL CHEMISTRY

10.5.1 Chromatographic methods

The separation of biologically important anions represents an ongoing challenge for analytical chemists. To date, the primary focus has been on the resolution of amino acids[132] and nucleotides[133–136] with a recent emphasis on chiral separation.[137,138] In this context, a variety of chromatographic methods have been utilized that include ion-exchange chromatography,[139–141] dynamic anion-exchange chromatography,[142] ion-pair chromatography, reversed-phase ion-pair and ion-interaction chromatography,[143,144] affinity chromatography,[145] and boronate affinity chromatography.[146]

Anion binding is widely used in common chromatographic techniques such as those mentioned above.[147] For instance, ion-exchange chromatography uses cationic or anionic stationary phases with aqueous solutions for the mobile phase, and ion-pair reversed-phase chromatography uses a reversed-phase stationary phase with an aqueous–organic mixture for the mobile phase, to which ion-pairing agent is added. This latter technique is referred to in different modifications as a soap chromatography,[148] solvent-generated ion-exchange chromatography,[149] hetaeric chromatography,[150] hydrophobic chromatography with dynamically coated stationary phase,[151] ion-pair chromatography,[152] ion-pair adsorption,[153] dynamically coated ion-interaction chromatography,[154] or dynamic-complex-exchange chromatography.[155] They are often considered as alternative separation techniques to ion chromatography.[156] In the majority of these methods, the ion-pairing agents are added to an aqueous organic mobile phase to form an ion pair with the analyte. The ion pair has increased lipophilic properties and is thus retained on the reverse-phase column. This mechanism is probably based on a dynamic modification induced on the surface of the reversed-phase packing material. This is the case when alkylammonium salts are added to the mobile phase.[157] The hydrophobic alkylammonium ion is adsorbed onto the alkyl-bonded apolar surface of the stationary phase (e.g., C_{18}), forming a positively charged ion layer. In a second step, the counteranion is also bound through electrostatic forces to form an oppositely charged diffuse layer. The alkylammonium ion and its counteranion give rise to an ion pair that is adsorbed onto the stationary phase to form a new stationary phase. Separation of anionic species may be obtained by this method.

In anion exchange and ion-pair chromatography, ion-pair formation between the eluted molecules and oppositely charged counteranions is assumed to be responsible for the modulation of retention. Because anionic, cationic, and zwitterionic molecules can all undergo ion pairing, the scope of these kinds of chromatography has expanded considerably. This method is thus very well suited for the separation of anionic species like nucleotides[158,159] and amino acids.[160,161]

Another separation method of prime import is affinity chromatography. Currently, it is used to purify both biological macromolecules and small molecules. The development of this technique has benefitted from the knowledge that surface recognition of folded macromolecules is both a recurrent and unifying theme in biology, as well as, in general, a type of interaction of high specificity and strength. The basic approach involves attaching known molecular recognition units to an insoluble matrix and then using the resulting derivatized supports to absorb selectively sought-after molecular species, multimolecular assemblies, or even cells from mixtures eluted through the affinity matrix. It has also been used to effect the purification of antibodies, genes, and enzymes.

Anion recognition mediated by phenylboronate matrices is another strategy that has been elaborated in terms of effecting the separation of nucleotides, oligonucleotides,[162,163] modified nucleotides,[164] and their corresponding species. This approach has also been used to purify nucleotidyl peptides[165] and specific RNAs.[166] In terms of underlying chemistry, this approach to anion purification is based on the interaction of boronates with cis-diols. Borate–diol complexes exhibit an anionic charge at pH > 9 and therefore readily participate in anion-exchange reactions. In terms of borate-derived anion-exchange columns, therefore, the processes involve adding the mixtures in question followed by washing. Only solutes having affinity for the borate anion bind to the borate column, while others elute freely from the column. In order to maintain the boric acid in its anionic form, the operation must be carried out under alkaline conditions (pH 9.0–9.5). The molecules bound to the borate matrix are eluted by lowering pH or by increasing the salt concentration of eluent. In this way not just ribonucleotides[167] but also large biomolecules such as aminoacyl-tRNA,[168] oligoribonucleotides,[169] and transfer RNAs have been separated/purified.[164,166]

Recently, several macrocycle-derivatized sorbents have been used for selective anion separation. The binding modes include anion-exchange reversed-phase chromatography, axial ligand (anion) exchange, and specific anion binding by positively charged macrocycles. Although the focus will be on types of supports incorporating metalloporphyrins and the expanded porphyrins, it should be mentioned that covalently polymer-bound azacrowns have been utilized for anion separation in the column chromatographic or isotachophoretic modes.[170,171]

In the case of the metalloporphyrin solid supports, two general methods of preparation have been employed. The first involves using electrostatic interactions between ionic metalloporphyrins and counterionic groups to effect the key surface modification process; the second, the use of covalent bonds between the porphyrinic species and the solid support. The main advantages of an electrostatic approach is a generally strong binding interaction and a significant ease in surface preparation.[172–174] On the other hand, metalloporphyrins covalently bound to surfaces offer supports that are for the most part even more stable than those made via electrostatic binding. Amine[175,176] and amide[177–179] linkages are the most common method for covalent attachment. Other methods, however, that have been employed include the formation of esters,[180] ethers,[181] and quarternary ammonium ions.[182] In all cases, the nature of the linker unit, the choice of metal, and kind of surface employed provide a means for modulating the properties of the solid support. This in turn allows the properties of these kinds of supports to be tailored to specific needs.[175]

Anion ligand exchange reactions involving metalloporphyrins are well known and provide the basis for the separations effected by metalloporphyrin-modified surfaces.[183,184] In the case of a polymer-bound ferri-protoporphyrin IX chloride complex, for instance, efficacy as a cyanide and phosphate ion-exchanger system has been reported. Likewise, in the case of an oxochloromolybdenum(V) tetraphenylporphyrin complex [MoO(Cl)TPP] entrapped in polystyrene (PSt), an inherent selectivity for dihydrogenphosphate could be established both by adsorption and column experiments.[185] Here, the relevant chemistry was found to follow a straightforward reversible ligand (anion) exchange process:

$$\text{MoO(Cl)TPP} + \text{H}_2\text{PO}_4^- = \text{MoO(H}_2\text{PO}_4)\text{TPP} + \text{Cl}^-$$

Interestingly, the specifics of this exchange process were found to vary when the complex was dissolved in organic solvents as opposed to being entrapped in polymers. In both instances, however, studies of pH dependence and experiments in the presence of other anions like NO_2^-, NO_3^-, HCO_3^-, and SO_4^{2-} revealed the established selectivity for dihydrogenphosphate anion. Polymer membranes doped with Sn(IV) and In(III) tetraphenylporphyin have also been reported.[186,187] These show selectivity for salicylate and other aromatic carboxylates over more lipophilic anions.

Recently, a tetraphenylporphyrin-based stationary phase for high-performance liquid chromatography was prepared by attaching a p-carboxyphenyl derivative of the tetraphenylporphyrin to aminopropyl silica gel through an amide bond. The resulting functionalized support was found useful in achieving the reverse-phase separation of several aromatic hydrocarbons,[188] including metallo-fullerenes.[189] Immobilized silica gel metallotetraphenylporphyrins in which the metal is In(III) and Sn(IV) are also known, and they too have been found useful for anion separations. In this case, the acetylated aminopropyl silica blank and unmetallated TPP stationary phases were tested; both showed little retention for any of the anions tested. It is thus apparent that the presence of the metal center plays a role in mediating the critical anion retention chemistry.

Both of the above metalloporphyrin-silica stationary phases show higher affinities for aromatic carboxylates than for aromatic sulfonates or inorganic anions.[186,187] Presumably, in the case of the aromatic substrates, the stabilizing interactions involve both anion binding to the metalloporphyrin and $\pi-\pi$ overlap between the aromatic porphyrin macrocycle and the phenyl ring(s) of the substrates. In any event the empirically observed selectivities make these supports, when used in a reverse-phase mode, convenient for the separation of aromatic anions. While such species can be purified/analyzed using traditional separation by ion-pairing chromatography, the advantage of this approach is higher efficiency. Additionally, the separation of aromatic anions is achieved without the use of mobile-phase modifiers. As a result, the chromatographic reproducibility is enhanced, the mobile-phase reagent cost is lowered, and column lifetime is lengthened.

Related to the covalently bound metalloporphyrin supports are interesting studies of polymers bearing noncovalently bound porphyrins on their surface. Groups that have been used as metal-binding ligands to fix the porphyrins to the surface include silanol (silica gel), pyridine N-oxide [from poly(4-vinyl)pyridine N-oxide], trialkylamine N-oxide, thiol, and carboxylate.[174,190] For the most part, it has been iron protoporphyrin IX that has been bound by these ligands.[191–193] Recently, however, Meunier[194] has devised a new approach to polymer modification that is based on a combination of metal-centered coordination and electrostatic interactions involving anionic groups on the porphyrin and countercationic groups on the support (Fig. 10.16).

As described in previous sections, we have discovered a new and selective interaction between the monoprotonated form of sapphyrin,[195,196] a pentapyrrolic expanded porphyrin, and certain anionic species including phosphate-type anions.[30a,83,197] To exploit this interaction, we have synthesized sapphyrin-functionalized silica gels as shown in Fig. 10.17.[198] One such sapphyrin-functionalized silica gel was found to be capable of separating polyphosphorylated oligonucleotides under standard, isochratic HPLC conditions at neutral pH. As expected, the retention time correlates quantitatively with the total phosphate number. Thus the sapphyrin-modified silica gel columns provide a new nonelectrophoretic, HPLC method for oligonucleotide separations that complements and extends the types of separations that may be achieved by using diethylaminoethyl (DEAE)-type systems.[199]

The sapphyrin-modified silica gels also provide a method of separating by HPLC various anions and, as such, could also serve to supplement the metalated tetraphenylporphyrin-silica gel based approaches.[200,201] In particular, it was found that, although neutral species such as benzophenone were not significantly retained on these sapphyrin-functionalized silica gel columns, various monoanionic entities such as diphenyl phosphate, benzene sulfonate,

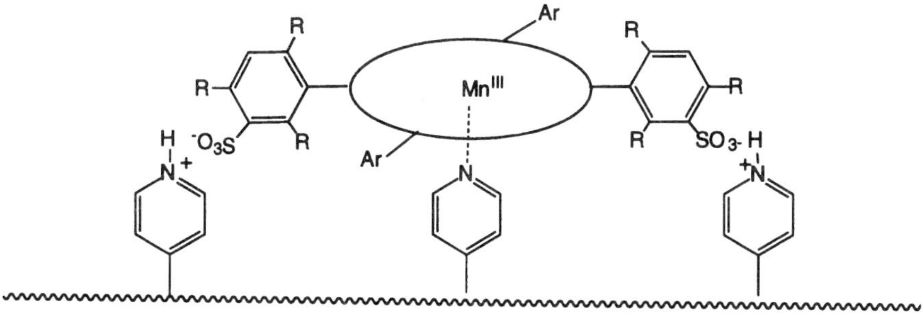

Fig. 10.16. Schematic representation of how anionic metalloporphyrins may be anchored to a partially protonated poly(vinylpyridine) via a combination of coordinative and electrostatic bonding; this figure was modified from Ref. 194.

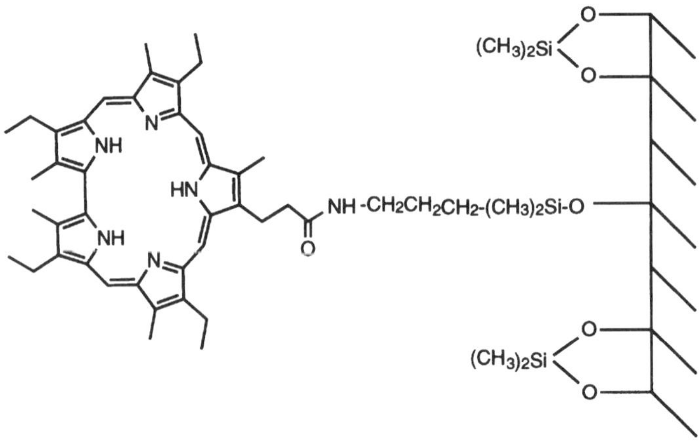

Fig. 10.17. Sapphyrin-functionalized silica gel.

benzene phosphonate, and benzoate could all be separated from each other. Various phosphorylated mononucleotides, including AMP, ADP, and ATP, were also retained on these columns and found to be readily separable from one another under isochratic HPLC conditions. In both cases, presumably, the key anion-retaining molecular recognition interactions involve a specific interaction between the relevant oxyanion portion of the substrate and the positively charged sapphyrin core just as is observed in the solid state.[30a,202] In any event, these new modified silica gels offer an interesting alternative to supports containing metalated tetraphenylporphyrins.[200,201]

Recently, Gd-texaphyrin and Eu-texaphyrin bonded solid supports have been prepared in the authors' laboratory. In this case, the synthetic strategy is based on the coupling of the appropriate metallotexaphyrin monocarboxylic acid with aminopropylsilica gel followed by silylation of the remaining free silanol groups with trimethylsilyl-imidazole. These novel chromatographic phases, which represent immobilized metal ions, are currently undergoing testing for use in various anion-separation interactions, including reverse-phase HPLC. In addition, other modified silica gels bearing other metal and/or anion chelating expanded porphyrins are also being considered for synthetic construction by the authors. These new supports could extend the scope of this approach to anion binding and purification.

10.5.2 Selective electrodes

Ion-selective electrodes (ISEs) represent an important tool in analytical chemistry. They have also allowed for significant advances in the field of molecular recognition. This is because in supramolecular chemistry, the analogy between abiotic synthetic molecules and biological receptors has often been limited to the selective recognition only of a guest molecule. This, however, represents but the first step in the cascade of events produced by biological receptors. ISEs provide a more active means of studying molecular recognition phenomena. They, therefore, represent an important new tool for shaping our thinking about biomimetic receptors. In this review, we will concentrate on molecular recognition phenomena involving anion-selective electrodes, specifically those based on polymer matrix liquid membranes, electropolymerized films, and Langmuir–Blodgett (LB) molecular assemblies deposited directly on glassy carbon electrodes.

There are two generalized modes of signal transduction that may be induced by host–guest complexation at a membrane surface: namely, the membrane potential change and the permeability change.[203] Interestingly, in the case of ion-selective electrodes, it is mostly induced membrane potential changes that have been used as the basis of the principle of signal transduction. For such membrane potential changes there are several established principles for discrimination of organic substrates by host–guest complexation at the membrane surfaces.[204] Two such principles involve (1) potentiometric discrimination based on hydrogen bonding and (2) charged group interactions with specific functional groups present on the substrates. A third principle involves potentiometric discrimination based on shape recognition arising from steric interactions between the targeted substrates and the "walls" of the membrane-bound receptors sites (e.g., calixarenes). Additional sources of discrimination include ones based on lipophilicity,[205–207] chirality,[208–210] and axial ligand binding.[211]

A number of ISEs based on polymeric matrix liquid membranes have been investigated. Many of these display high selectivity for particular target substances and are now commercially available.[212–217] In most of these systems, the lipophilic organic liquid membrane contains a hydrophobic host molecule that is generally supported by a PVC (polyvinyl chloride) polymer matrix. This combination produces a liquid membrane aqueous interface, the charge separation across which is modulated by guest binding.

The majority of the polymer matrix liquid membranes developed thus far have focused on the recognition of alkali and alkaline earth metal cations by the use of natural and synthetic cyclic and acyclic ionophores as sensory elements.[218–220] By contrast only a few solvent polymeric membranes have been prepared that are based on compounds that reversibly bind anions.[221]

It is well known that solvent-polymeric membrane electrodes prepared with conventional dissociated anion exchangers (e.g., tetraalkylammonium or similar compounds) exhibit potentiometric anion selectivity patterns in accord with those predicted by the Hofmeister series.[222] The classical Hofmeister sequence (i.e., emf response to $ClO_4^- > IO_4^- > SCN^- > NO_3^- > I^- > Br^- > Cl^- > HCO_3^- = H_2PO_4^-$) shows the maximum response to lipophilic anions. Since such quarternary ammonium species function as purely dissociated ion exchangers, equilibrium anion extraction into membranes is solely based on the difference between the free energy of solvation for an anion in the aqueous and organic membrane phases (single ion partition coefficient). Non-Hofmeister behavior, however, can occur when neutral lipophilic species are incorporated within the membrane. Various metal–ligand complexes and organometallic host systems have been used to construct just these kinds of polymeric membrane electrodes. Not surprisingly, they display the expected non-Hofmeister anion selectivity.[222–225]

Another employed class of molecules that has been employed as active membrane components in liquid/polymeric anion-selective electrodes[223,226–229] is the metalated porphyrins. Metalloporphyrin-based devices display anion selectivities that differ significantly from those of conventional anion-selective, ion-exchange electrodes derived from quarternary ammonium salts (vide infra).[230] At present the exact mechanism of anion response for the porphyrin-derived electrodes is still subject to debate and, indeed, could vary depending on the nature of the porphyrin species doped into membrane.[231] Nonetheless, several reports indicate that anion selectivity patterns are dictated by the relative strength of interaction of anions as axial ligands with the metal center of the metalloporphyrin,[229,230] as well as by structural influences from the surrounding porphyrin macrocycle.[227]

Previous study of polymeric membranes doped with various Sn(IV), Mn(III), In(III), and Co(III) metalloporphyrins[223,226–229,231] have demonstrated that such membranes display potentiometric responses to anions that deviate significantly from the classical Hofmeister sequence.[223,226–229] The unique potentiometric anion selectivity of these metalloporphyrin-based membranes[223,226–229] has been attributed to the fact that the anions in question can act as effective axial ligands. Consistent with this supposition is the finding that observed potentiometric response to anions is dependent on the nature of metal center, its oxidation state, and the exact structure of the porphyrin framework surrounding the metal ion. In the specific case of membranes containing In(III) porphyrins (e.g., **10.82**), a non-Hofmeister selectivity was noted with a remarkable enhancement in potentiometric response being observed in the case of chloride.

To date, several chloride-selective electrodes have been proposed that are based on liquid and polymer membrane-based anion-binding receptors.[232–235] In most cases, the active membrane components are dissociated quarternary

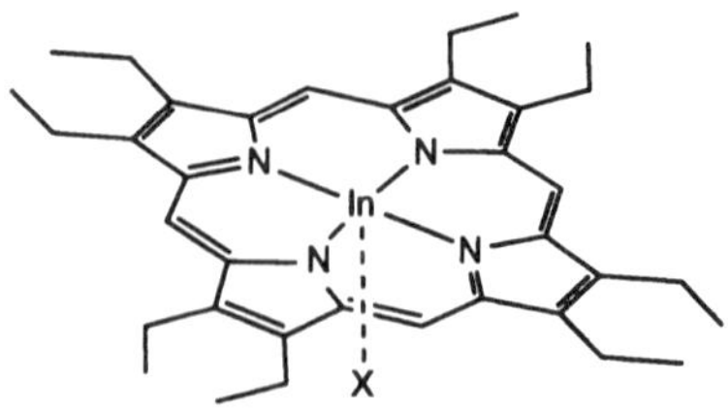

10.82a X = Cl⁻
10.82b X = CN⁻
10.82c X = N₃⁻

ammonium centers. Unfortunately, these kinds of systems are often rendered inaccurate by virtue of interference from other anions.[236] This, however, is not the case in systems based on In(III) porphyrins. Here, the resulting sensors can be used to determine chloride anion concentrations accurately even in human serum samples.[231]

One further example of the use of response modulation in bioanalytical applications is based on the use of sterically hindered Mn(III) porphyrins such as Mn(III) tetrakis(2,4,6-triphenyl-phenyl)porphyrins (e.g., **10.83**). These, when incorporated within a solvent polymer membrane, form highly selective thiocyanate sensors. The same is true, but to a lesser extent, for the corresponding Co(III) tetraphenyl porphyrin species.[223]

10.83

The above selectivity is both noteworthy and important; thiocyanate is a principal metabolite of cyanide[237] and may be used as an indicator of cyanide exposure.[238] Furthermore, the amount of thiocyanate present in biological fluids can be used to estimate exposure to tobacco smoke. This is because it is known that such smoke contains significant levels of hydrogen cyanide.[239] The half-life of thiocyanate in physiological fluids is about two weeks[240] and is much longer than that of other tobacco products (i.e., carbon monoxide[241] and nicotine or its metabolites[242]). Thus its measurement in blood, urine, or saliva provides a convenient means of assessing chronic exposure to tobacco smoke. In this context, it is to be noted that these electrode-based detection approaches have many advantages over other more classic methods for measuring thiocyanate concentration including colorimetric complexation, GC, GC/MS, and HPLC.[243,244] These latter methods are at times nonspecific and often suffer from interference due to cyanide or halide anions. Further, they require the removal of proteins. This, in turn, necessitates extensive sample preparation and/or derivatization.

Also interesting are polymeric membranes doped with organotin compounds.[245] These have shown useful selectivities for hydrogen phosphate and chloride.[246] Another type of selective electrode has been reported by Simon.[247] He reported that mercury-containing organic compounds such as **10.84** and **10.85**, when incorporated into solvent polymeric membranes, act as selective electrodes for chloride and thiocyanate, while showing little response to nitrate and perchlorate anions. In the case of chloride, the response time was less than ten seconds. This combination of speed and selectivity makes this kind of electrode useful as a chloride sensing tool in biological applications.

388 APPLICATION ASPECTS INVOLVING THE SUPRAMOLECULAR CHEMISTRY OF ANIONS

[Structures 10.84 and 10.85]

Plasticized polyvinyl chloride membranes doped with 5,10,15,20-tetraphenyl(porphyrinato)tin(IV) dichloride [Sn(Cl$_2$)TPP] represent another type of ISE. They exhibit selective potentiometric response toward salicylate over a wide range of other anions,[248] including thiocyanate and perchlorate. A detailed mechanistic study of the relevant anion binding was carried out for these systems using ^{119}Sn-NMR and ^{3}H$_2$O uptake. This led to the conclusion that complex mechanisms were operative, especially in water, where the observed response reflects the outer-sphere coordination of the salicylate anion to the diaquo-ligated form of the metalloporphyrin. On the other hand, simple axial ligand-exchange mechanisms were found to be operative in pure organic solvents.[248] (This is summarized schematically in Fig. 10.18).

Macrocyclic polyamines have also been observed to function as anion

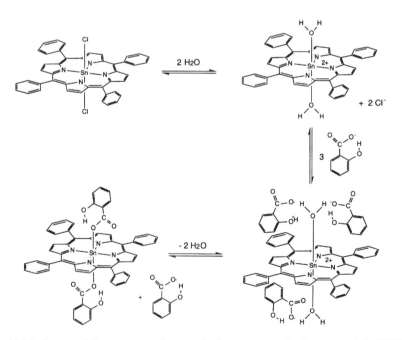

Fig. 10.18. Proposed three-step mechanism for interactions of salicylate with Sn(TPP) in the membrane phase of porphyrin-based ISE.

receptors. Along with the alkyltin compounds, metalloporphyrin derivatives, and bis(quarternary) ammonium salts highlighted above,[247] these compounds have been used to generate anion-specific sensors. Macrocyclic polyamines are generally multiply protonated in acidic to neutral aqueous media. As discussed earlier, the resulting polycations often act as electrostatic-based anion binding agents.[6b,63c,249–251] Not surprisingly, therefore, lipophilic derivatives of these species may be used as sensory elements for anion-selective electrodes. For instance, it was found that pentaamine-based electrodes can be used for the analysis of phosphates, including various biologically relevant phosphates such as ATP.[90k,252] Here, one of the remarkable findings was the discovery that these and other macrocyclic polyamine-based electrodes respond to the adenosine nucleotides in proportion to the number of negative charges present in the system being examined. By far the strongest potentiometric response is observed for ATP^{4-}, as opposed to ADP^{3-} or AMP^{2-}. Consistent with this is the finding that protonated macrocyclic polyamines appear to bind mainly to the phosphate groups of nucleotides.

One would not expect this type of interaction to permit discrimination among similarly charged nucleotides bearing different kinds of nucleobases. Umezawa et al.,[90k,204,252–254] has, however, recently reported several polyammonium systems that, when made up into anion-selective ISEs in conjunction with appropriate nucleobase recognizing components, will allow for a potentiometric discrimination between complementary and noncomplementary nucleotides.

In aqueous solution, protonated polyamines are also known to discriminate between potential substrates on the basis of shape.[6b] The same is true for polyamine-derived ISEs. For instance, for the positional phthalic acid isomers, the magnitude of the observed ISE-mediated potentiometric response decreased in the order of *ortho* > *meta* > *para*.[204] Interestingly, the exact opposite selectivity was observed in solution-phase binding studies utilizing sapphyrin dimers.[99]

Other sensory elements for which potentiometric responses have been reported include vitamin B_{12}, bis(quarternary ammonium), and diphosphonium salts.[255,256] In general, however, the macrocyclic polyamine-derived electrodes displayed much better selectivities for organic anions than electrodes derived from these other compounds.[90k,204,252] For instance, liquid membrane electrodes containing Capriaquat, a conventional anion exchanger, showed some potentiometric discrimination between dicarboxylate anions.[204] Still analogous polyamine-derived systems displayed superior selectivity, sensitivity, dynamic range, and response time.[90k,204,252]

In order to obtain a better discrimination for and between nucleotides, Umezawa, in conjunction with the authors, prepared cytosine-pendant triamines and used them as putative ditopic (electrostatic and base-pairing) recognition liquid membrane sensors for guanidine nucleotides.[254,257] As expected, the resulting liquid membrane electrode showed a selective potentiometric response for 5′-GMP and 5′-GTP, whereas the response to 5′-AMP and 5′-ATP was almost negligible. Likewise, double-receptor-type liquid membranes composed of separate cytosine neutral derivatives and dioxotetra-amines gave a strong response to 5′-GMP, whereas membranes containing only one of the components gave only a very weak response. These findings highlight the utility of the ditopic base-pairing mode and electrostatic recognition motif when trying to design nucleotide-selective electrodes.

In collaboration with the Umezawa group, the authors have also been considering the use of expanded porphyrins as the basis for preparing anion-specific sensors. Here, on the one hand, it was appreciated that the metallo-texaphyrins, in analogy to the metalloporphyrins, could recognize anions, perhaps selectively via axial ligand anion exchange. However, since the texaphyrins form stable complexes with a number of "unusual" metal cations, such as the lanthanides, it is possible to conceive that these systems could display unexpected selectivities for anions. On the other hand, certain nonmetalated expanded porphyrins, such as sapphyrin and rubyrin, display inherent anion-binding capabilities at physiological pH and could thus provide a means of preparing electrodes capable of detecting biologically relevant species such as chloride, carbonate, bicarbonate, and dibasic phosphate (Cl^-, HCO_3^-, CO_3^{2-}, HPO_4^{2-}) as well as, perhaps, nucleotides.

At present, the texaphyrins have yet to be tested as electrode-making materials. However, systems based on sapphyrin have been made and, at pH 6.6, found to allow the detection of 5′-GMP and 5′-AMP with some slight selectivity in favor of the latter.[254] Also, liquid membrane electrodes containing cytosine-pendant sapphyrins have been prepared. These show promise for the selective recognition of guanosine nucleotides.[258]

Electrochemical sensors based on the use of electropolymerized organic films have received considerable attention over the past decade.[259–261] While much of this effort has focused on the use of these kinds of chemically modified electrodes (CMEs) in amperometric and voltametric configurations,[263,263] there have been several reports describing the potentiometric ion responses of these systems. A portion of this latter effort has been devoted to anion detection. For instance, Dong et al. showed that the potentiometric anion selectivity pattern of glassy electrodes modified with poly(pyrrole) film appears to vary, depending on the specific doping anion used during the electropolymerization process.[264,265] Their data suggest that anion channels with specific pore sizes for the doped anion are formed within the electropolymerized film; here the anion serves as a template during the film formation.

In another example relevant to anion recognition, Daunert et al. studied the potentiometric anion responses of electrodes modified with electropolymerized Co(II)tetrakis(o-aminophenyl)porphyrin films.[266] He found that such modified electrodes, when poised at a reduced state after anodic polymerization of metalloporphyrin, exhibit selective potentiometric responses toward thiocyanate. Meyerhoff et al. has also examined the anion response properties of unmetallated poly(porphyrins) in their fully conducting (oxidized) and non-conducting (reduced) states.[248] Here, the electrodes consisted of polymeric films of tetrakis(p-aminophenyl)porphyrin formed on the surface of glassy carbon electrodes.[248] They displayed rather selective potentiometric responses favoring iodide over a wide range of other anions, when the film was studied in its oxidized form, but relatively little potentiometric response to anions when the film was used in its reduced state. Selectivity for iodide was also recently found for conducting poly(3-methylthiophene) films on graphite electrodes.[267]

The selective and sensitive processes of molecular recognition and signal transduction and/or amplification in living systems, especially that displayed by membrane-bound ion-channel proteins, continues to provide the inspiration for yet-improved chemical sensors. In this context, the recently described biomimetic "ion-channel sensors" of Umezawa are particularly noteworthy.[268] These systems were constructed by incorporating several kinds of receptor molecules into Langmuir–Blodgett molecular assemblies deposited directly on glassy carbon electrodes[268] and are believed to work via "stimulant"-controlled model channel switching.[253,269] In particular it is thought that, for a charged membrane system composed of lipid and receptor deposited directly on a glassy carbon electrode, the binding of a charged "stimulant" receptor site will engender a distortion in the membrane assembly and that this, in turn, will be reflected in an increase (or decrease) in the ion permeability of the membrane. This corresponds to model "channel opening" (or "channel closing"). By using appropriate redox-active ions as markers for membrane permeability, the stimulant concentration in a solution can be quantitatively determined by means of cyclic voltametry carried out using the underlying electrode. Further, as the result of the cascade effect of membrane distortions as reflected in rapid ion flux, this approach can offer an amplified measure of stimulant concentration.

As a specific illustration of the above concepts, anion-responsive sensors were constructed from a composite membranes consisting of either dipalmitoylphosphatidylcholine (DPPC) and lipophilic macrocyclic polyamines or cyclodextrin derivatives.[90k,252] The response of these sensors to anionic guests was observed as a decrease or increase in the peak current in cyclic voltametry when $[Fe(CN)_6]^{4-}$ or $[Ru(bpy)_3]^{2+}$ was used as a marker ion. Specifically, it was found that model channel closing (opening) could be triggered in this case by the partial neutralization of the membrane charge due to complexation between the cationic polyammonium receptor and the anionic guest. Good discrimination among adenosine nucleotides was observed. Distinctions between geometrical and positional isomers of dicarboxylate anions could also be effected. In both cases, the principal mode of interaction was thought to be electrostatic. Thus, in order to enhance the degree of steric-based recognition, an anion-responsive sensor was constructed from an LB membrane containing a lipophilic cyclodextrin polyamine, $(6-C_{12}H_{35}NH)_7-\beta-CD$.[270] For it, too, channel-closing behavior could be induced by the addition of an anionic guest (as indicated, again, by cyclic voltametry). Interestingly, the observed response order for phthalate ions, namely, *meta* > *para* > *ortho*, is quite different from that found for simpler systems based only on macrocyclic polyamines. This could be a reflection of the fact that these modified cyclodextrin receptors effect recognition not only via electrostatic effects, but also through steric-sensitive interactions involving the cyclodextrin cavity.

10.5.3 Modified surfaces

One of the challenges of supramolecular chemistry is the development of highly ordered molecular assemblies on surfaces.[271–273] This field has attracted great interest recently since it can lead to materials that have a wide range of potential analytical applications such as biosensing, cell guidance, molecular electronics, nonlinear optics, electrochemistry, and monolayer catalysis.[274–279]

An elegant strategy for the creation of such systems is the self-assembling of molecules in mono-[280] and multilayer[281] formations by incorporating built-in functionalities such as multidentate ligands to give higher-order superstructures. Specifically, such systems may be built up from alkylsilanes on glass or alkanethiols on gold; this allows one to fabricate highly ordered multifunctional thin films. The central advantage of using such self-assemblies is that the covalent bond between the interfaces is stronger than the van der Waals or hydrogen bonding interactions utilized in other systems such as Langmuir–Blodgett films. Moreover, uniform film thickness and dense coverage can be achieved with these systems.

Due to their easy preparation and relative stability, thiolate monolayers on gold are well suited for such use and thus are particularly interesting (Fig. 10.19). These surface modifications are based on a two-step synthetic principle. First, long spacers such as ω-amino or carboxy-functionalized thiols are introduced to the surface. (It is known that organosulfur compounds bind strongly to gold.[282,283]) In the next step, the compound of interest is then attached on the gold surface. Monolayers of this type, which incorporate calixarenes, resorcinarenes, ferrocenyl groups, (2,6-diaminopyridine)amides of isophthalic acid, cyclodextrins, porphyrins, and expanded porphyrins, will be highlighted in this section. These kinds of systems are of interest in view of their documented or proposed anion binding ability.

Calixarenes attached to gold surfaces are among the most extensively studied of the substrate-binding chemically modified monolayers known.[284,285] They have been used as synthetic receptors for binding cations, anions, and neutral

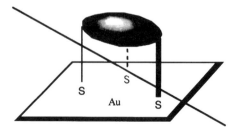

Fig. 10.19. General representation of self-assembled monolayers of receptor adsorbates on gold surfaces.

molecules.[24c,286] Indeed, calixarenes seem to be very convenient building blocks for the construction of monolayers bearing specific recognition sites. In the exemplary case of a calix[4]arene derivative, monolayers were formed by immersing a gold substrate into a solution of the calixarene in ethanol:chloroform (1:1). The layers were characterized by FT-IR spectroscopy, contact angle measurements, and surface plasmon resonance (to determine its thickness).[285]

Reinhoudt et al. have reported that resorcin[4]arene-based receptor molecules containing four alkyl sulfide chains will self-assemble into stable, well-packed monolayers on gold surfaces (**10.86**).[287,288] The influence of temperature affecting the organization of monolayers has been investigated. It was determined that thermal rearrangements lead to a higher organization on the gold surface. Related systems, including the tetrathio derivative **10.87**, have also been prepared and deposited on gold surfaces under standard conditions.

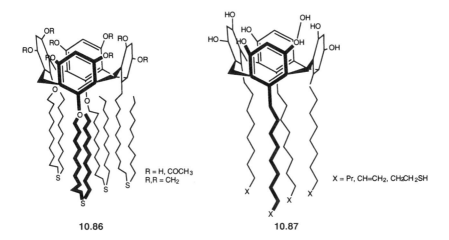

The binding of different substrates was studied with these systems. To do this the gold-thiol monolayers were treated with aqueous solutions of a series of potential substrates. While simple anions like azides were not bound, organic anions and dianions such as carboxylates and dicarboxylates (e.g., glutarate or gluconate) were bound. The main binding mode for polyhydroxy compounds is most likely multiple hydrogen-bonding interactions involving the rim of the calix[4]resorcinarene bowl. Finally, with some substrates, including vitamin C, binding was irreversible due to covalent bond ester formation to the substrate. However, as in the case of perchloroethylene binding by systems based on **10.86**, binding was also presumably achieved via more hydrophobic-type interactions.[289]

Comparative binding studies carried out with calix[4]resorcinarene receptors in solution and with monolayers derived therefrom revealed that the binding effects observed on the surfaces are much stronger. On monolayer surfaces, the receptor is believed to be better preorganized for binding. Further, in solution (particularly in hydrophobic solvents), the calix[4]resorcinarenes are strongly aggregated[290] in a bowl-to-bowl fashion. This makes substrate binding entropically disfavored.[290] Thus, in more general terms, these results tend to highlight that other receptor systems that are strongly aggregated in solution and bind substrates with low affinity may show stronger binding characteristics when oriented as monolayers on surfaces.

Another set of compounds that offers possibilities for molecular recognition on modified surfaces consists of ferrocene derivatives. When placed on gold surfaces via appended alkanethiol groups, these cationic agents interact with amphiphilic receptors in aqueous solutions.[291] By varying the length of the alkanethiol chains, one can easily change the degree of accessibility of the ferrocenyl groups to both the contacting solution and the receptors contained therein.[292]

Turning to specifics, it is known that sulfonated calix[6]arenes are excellent hosts for several ferrocene derivatives in aqueous media.[293] However, these same anionic species failed to bind to cationic monolayers bearing thiol-anchored ferrocenyl groups. The reason seems to be that this ionic host is well solvated by water molecules. This reduces the degree of interaction between these species and the cationic layer/solution interface. By contrast, amphilic analogues, prepared by O-alkylation of the parent calixarene, showed a strong interaction with the ferrocenyl surfaces, as judged by the strong voltametric response of monolayer (Fig. 10.20). The change in the oxidation potential observed upon addition of calixarene are due to variations in the ferrocene microenvironment. While in the solution the hydrophilic calixarene ($R = H$) acts as an anionic host for ferrocene in its oxidized form, in this case, the amphilic calixarene ($R = C_{12}H_{25}$) probably interacts with the ferrocenyl group primarily by inclusion in its "octopuslike" aliphatic tentacles. Still, this binding process serves to modulate the potentiometric response.

A system for probing molecular recognition on interfaces using fluorescence receptors incorporated in mixed self-assembled monolayers on gold has also been reported.[294] The system studied involved the interaction between barbituates and the (2,6-diaminopyridine)amide of isophthalic acid. In solution,

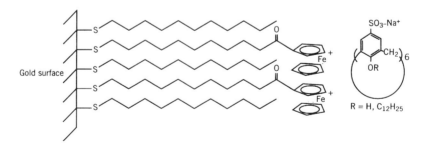

Fig. 10.20. An example of modified surfaces showing amphilic calixarenes with cationic monolayers bearing thiol-anchored ferrocenyl groups.

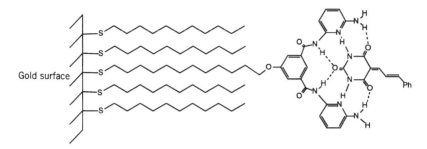

Fig. 10.21. An example of modified surfaces for probing molecular recognition using fluorescence receptors.

these latter receptors bind barbituric acid derivatives readily, as was demonstrated early on by Hamilton.[295] In terms of the monolayer work, Myles et al.[294] reported that these same receptors may be modified so as to be assembled on gold surfaces. Here, the basic diaminopyridine framework was appended with an alkyl thiol tail. The resulting species were then diluted with simple octanethiol to form monolayers in which interactions between receptor molecules are minimized (Fig. 10.21). Once the monolayers were prepared, complex formation was followed by monitoring the change in receptor fluorescence occurring upon barbiturate binding. The resulting fluorescence spectra in solution and on the monolayer are very similar, which showed evidence for the reversible binding of the ligand to the monolayer-containing receptor. Such a fluorescence monitoring method, which has often been used in the study of biological systems, proved very useful in this instance where the complexity of the assembly precluded direct examination of the basic recognition events.[296] However, to the best of the authors' knowledge this approach has not been applied to the binding of anions per se.

Kaifer et al. has reported the formation of defined mixed monolayers of β-cyclodextrin and pentanethiol and showed that ferrocene was complexed into the resulting surface-supported cavities.[297] Here, per-6-thio-β-cyclodextrin was prepared in two steps from β-cyclodextrin and chemisorbed onto gold surfaces via the formation of six sulfur–gold bonds per receptor molecule **(10.88)**.

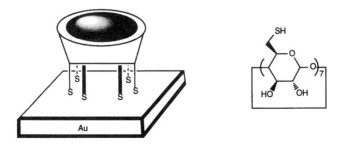

10.88

Although this derivatization process leads to imperfect monolayers in which a substantial portion of the gold surface remains uncovered, the monolayer defects could be covered by treating the gold surface with a solution of pentanethiol and ferrocene. Ferrocene, an excellent substrate for β-cyclodextrin, protects the monolayer binding sites and directs the pentanethiol molecules to seal selectively the defective sites.

Electrodes derivatized by this procedure showed effective response to low concentrations of ferrocene molecules. Presumably, this is the result of an into-cyclodextrin binding process. Consistent with this proposal, the process of interfacial ferrocene complexation was found to have the expected dynamic character, as demonstrated by competition experiments involving m-toluic acid. In terms of the present chapter, it is interesting to speculate as to whether these monolayers will act as anion binding systems as well. Schneider has already shown[27] that positively charged cyclodextrin derivatives may be used for selective anion binding in solution. Thus this logical extension may well be one that is soon realized.

Finally, a number of porphyrin derivatives have been attached to gold surfaces[298] (Fig. 10.22) as well as to silica gels (c.f. Fig. 10.23 and Section

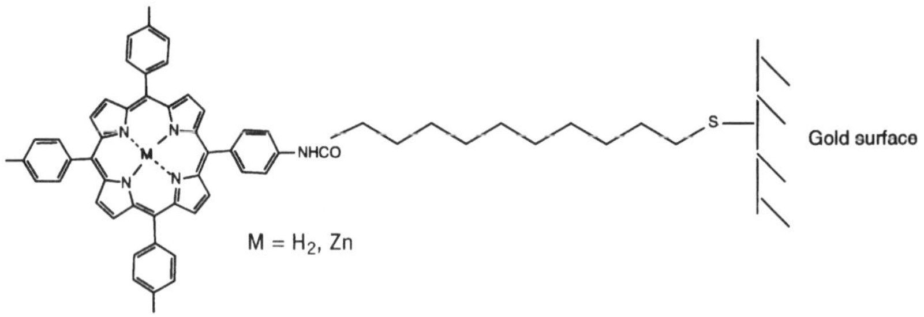

Fig. 10.22. The structure of porphyrin molecular assembly on gold (Ref. 298).

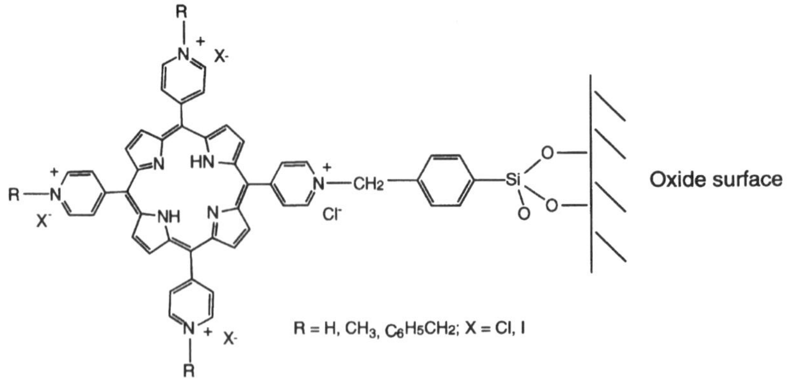

Fig. 10.23. Porphyrin–silica gel (Ref. 273).

10.5.1).[174,273,298] Although not yet studied in this context, these systems are potentially interesting as anion recognition units. This should also be true for systems based on expanded porphyrins, as these latter are known to be anion binding agents *par excellence*. As yet, however, no experimental work along these lines is known to the authors.

10.6 WASTE MANAGEMENT

The cleanup of anthropogenic contaminants that are present in the environment is one of the most important problems facing mankind today. These contaminants can cause a wide range of pollution problems, including global climate change, ozone depletion, ecological deterioration, and groundwater contamination. Since World War II, waste management problems have been exacerbated by two factors: quantity and toxicity.[299] Waste quantities have steadily increased due to population growth, while increasing waste toxicities are relatively new phenomena. Radioactive wastes, for example, were virtually nonexistent prior to World War II. The remediation of existing waste, along with the concerns over the fate of waste, is recognized by the public as being the leading environmental issue confronting the civilized world at present.[300] Cleanup costs are estimated to be as high as a trillion dollars in the United States alone.[301] The use of anion binding chemistry in the environmental restoration problem area is thus an application that should be well received. This is likely to be especially true in the case of the most environmentally dangerous anions.

There are several waste anions that, when present in soil, food, and water, present a long-term health risk to the general population and which engender a negative effect on the environment. Some examples include nitrates and phosphates from agricultural products and detergents that are commonly found in wastewater streams and as soil contaminants. Additionally, nitrates are known to form nitrosamines through in vivo reduction. The latter are very potent carcinogens and thus are extremely dangerous. Sulfates and heavy metal–based anions (like arsenates) are also species that are deleterious when present in water at significant concentration. Additionally, sulfides, thiosulfides, and cyanides are common industrial inorganic pollutants. Removal of these anions from agricultural and industrial effluent thus defines an important challenge for chemists. A related challenge, which also involves waste, consists of finding ways to capture and recycle commercially important anions such as $AuCl_4^-$.

At present, these critical problems are generally being approached either via diffusive means or by attaching one or more putative anion binding receptors to a solid support such as modified silica gels, resins, and zeolites and then using the resulting materials for the actual waste removal. Diffusive methods have largely relied on supported liquid membranes (SLM). The procedures developed employ various techniques, including precipitation, complexation, adsorption, and oxidation. Also, supported liquid membrane processes have been considered, along with other possible options, for the removal of contaminants from groundwater. While not discussed at length, these methods do offer some advantages over other competing techniques (e.g., solvent extraction, ion exchange, polymer membrane process, etc.), including: (1) high concentration factors (achieved through the use of a high feed-to-strip-volume ratio), (2) low carrier inventory, (3) minimal phase-separation problems, (4) negligible into-organic phase entrainment, (5) and simplicity of operation. On the other hand, a lack of long-term stability is a general typical drawback of SLM processes.

The principle behind the use of SLM approaches to remove nitrates, chromates, pertechnetates (in their acidic form), and other anions (as acids) from low-pH groundwater is based on formation of membrane-soluble salts. After diffusing through a liquid membrane, these salts are released at the strip side of the membrane, where an alkaline stripping solution (e.g., NaOH solution) ensures that the free carrier is regenerated. Three commercially available long-chain aliphatic amines (primary, secondary, and tertiary) were tested as membrane carriers for nitrate, pertechnetate, and chromate anions.[302] The primary amine-based membrane carrier was found to be very efficient for nitrate removal, whereas secondary and tertiary amines were effective for removing other anions.[302]

As far as solid-support-based approaches are concerned, well-studied guanidinium derivatives, protonated polyazacrowns, metalloporphyrins, expanded porphyrins (e.g., texaphyrins and sapphyrins), and other anion binding compounds may be considered as potential key anion binding portions of these modified sorbent surfaces. In terms of categorization, two general modes of anion binding can be considered as being applicable to this problem. These are axial ligand exchange and anion chelation, with the former approach relying on the use of metalloporphyrins and metalated expanded porphyrins, and the latter on the use of protonated guanidinium derivatives, polyazacrowns, and expanded porphyrins. In both cases, however, the fundamental premise is that by changing the pH of the eluent, it should prove possible to release the adsorbed anions and concentrate them into a substantially smaller volume. This is shown schematically in Fig. 10.24. Here it is important to appreciate that waste material in question is usually dissipated in a large volume of liquid or solid; therefore, concentrating represents the major challenge. What follows now, therefore, is a discussion of how this is being done in certain water-producing industrial and agricultural situations. This will include several specific examples relating to potential applications of known anion binders in this field.

Fluoride anion is considered an essential nutrient in human metabolism.[303] In addition, it is added to many drinking waters in small quantities to prevent dental caries. However, if the concentration of fluoride anion in the water becomes too high, dental fluorosis (mottling of teeth) occurs. Intake of excessively large amount of fluoride over prolonged periods of time may also

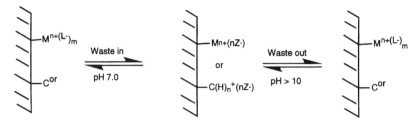

Fig. 10.24. Schematic representation of methods used to effect anionic waste control. Z, M, and C are meant to denote waste, metalated macrocycles, and unmetalated macrocycles, respectively.

produce other health problems, including bone changes, crippling fluorosis, and death from a single dose of 2250 to 4500 mg. As a result, fluoride anion concentration in drinking water is typically held at between 1.4 and 2.4 mg/L, whereas in most natural waters, it ranges from 1 µg/L to 1–2 mg/L. There are some natural waters with unusually high fluoride content (as high as 32 mg/L in Arizona, for instance, and 67 mg/L in South Africa).[304] But the major source of "excess" fluoride anion concentration in water is now due to waste contamination arising from industrial facilities that use or process fluoride-rich mineral resources. These include coal-fired power plants, aluminum smelteries, and phosphorous and fluorine manufacturing plants.[305] In addition, the use of hydrogen fluoride in the manufacturing of fluoroplastics, the cleaning of surfaces, and metal and glass production is leading to an increased concern about the possible adverse ecological effects of water-borne fluoride anion.

Wastewater containing fluoride anion can be decontaminated in several different ways: (1) by coagulating precipitation, in which calcium hydroxide or calcium chloride are added so that calcium fluoride is obtained as a precipitate[305]; (2) adsorption processes, using activated alumina, activated carbon, etc.[303]; (3) use of chelating resins for formation of metal–fluoride complexes[305]; and (4) application of inorganic anion exchangers of various types[305] (e.g., rare earth oxides). While all these approaches show promise, they also suffer from limitations. It is the opinion of the authors, therefore, that the problem of fluoride anion remediation is one that could benefit from solid-supported molecular recognition approaches. To the extent this is true, it is likely that the sapphyrins will have a critical role to play. They bind fluoride anion well[83] and are easily appended to silica gel and polystyrene supports.

The same prognosis could also hold true in the case of arsenic detoxifications. Most forms of arsenic are toxic, and acute symptoms following ingestion relate to irritation of the gastrointestinal tract, which can progress to shock and death.[306] Arsenic compounds occur as pollutants in water as the anionic forms of weak arsenious [As(III)] and arsenic [As(V)] acids. Since most arsenic compounds occur in nature in their anionic forms, removal by activated carbon is not effective (only about 3–8% can generally be removed in this way).[307] Thus chemical coagulation and adsorption on alumina or bauxite are commonly used for decontamination. However, these approaches are "nonrecyclable" in that agents used to detoxify the solutions in question are not reusable. Thus alternative methods for arsenate removal from aqueous solutions need to be found. In this context, the discovery that arsenates are bound to sapphyrins with high affinity with K_S values even higher than those for phosphates may be relevant. Indeed, it is tempting to propose that sapphyrins, when attached to solid supports, could function as arsenate-removal agents. In this case, the arsenate-containing supports could be subsequently regenerated by means of a low-pH wash.

Chromium content in natural waters is limited by the low solubility of Cr(III) oxides. Therefore, the only essential chromium aqueous chemistry is its occurrence in +6 oxidation state. Removal of chromates from industrial wastewaters has been attempted via several adsorption techniques. However, the results were not very promising.[308] Clearly, there is need for developing alternative methods to conventional coagulation techniques. These could be based on specific chromate recognition by receptors attached to solid supports. In this particular case, many protonated polyazacrowns would have the features needed to function as potential receptors for chromate anions.[90d] Also, recent reports from the Sanders group indicate that cyclic porphyrin oligomers bind a variety of large metallic anions such as $Fe(CN)_6^{4-}$, $Co(CN)_6^{3-}$, $Pt(CN)_4^{2-}$, $PtCl_6^{2-}$.[315] These receptors thus show promise for applications involving chromate anion removal.

Hydrogen sulfide is an infamous contributor to odorous water. It is not a normal constituent of natural waters, but is instead associated with waters with high sulfate and organic matter contents.[303] H_2S also comes into water as the result of microbial-mediated oxidation of organic matter by organisms that use sulfates as their requisite electron acceptors. Currently, H_2S in water is removed via oxidation or adsorption onto carbon. However, it is also conceivable that anion receptors could be developed that would bind/trap monobasic hydrogen sulfide (HS^-) and other related species such as organic thiolates (e.g., RS^-). So far, though, little work along these lines has been carried out.

Phenolic compounds, especially chlorinated derivatives, constitute an important ecological problem. Most of these compound come from the chlorination treatment of drinking water. Powdered activated carbon (PAC) and granulated activated carbon (GAC) have been successfully used in the water treatment industry for many years for the phenolic compounds removal.[309] However, as is true for sulfur-containing species, it is possible that molecular recognition approaches could be used to help with remediation in this area. In this context, recent findings that sapphyrins, rubyrins, and certain metalloporphyrins are capable of forming complexes with phenol-type substrates is of interest. Such species, if attached to solid supports, could be used to remove phenolic pollutants from waste streams.

10.7 ANION-TEMPLATED REACTIONS

Many biological "supramolecules," like DNA, are produced in nature by means of templated syntheses. Likewise, a number of complex abiotic materials, including crown ethers,[310] coronands,[311] catenanes,[312] zeolites,[313] functionalized polymers,[314] and cyclic porphyrin oligomers[315] as well as biomimetic models for enzymatic catalysis[316] and self-replication,[317] are prepared using substrate molecules as templates for directing the assembly of complementary host structures.[318] In most cases, the relevant templates are either cationic or netural in character. Recently, however, a few examples of anion-templated syntheses have appeared in the literature.

Currently, there are three well-defined examples of anion-templated macrocyclization. First, Hawthorne and co-workers found that halide anions act as templates facilitating the exclusive formation of the "[12]-crown-4-like" macrocycle from 1,2-dilithio-carboranes with mercury halides (Fig. 10.25).[79a,319] Second, Müller and co-workers utilized anions such as Cl^-, CO_3^{2-}, ClO_4^-, N_3^-, and NO_3^- in the construction of polyoxovanadate cages.[320] The organization of oxovanadium–organophosphate clusters around chloride anions has also been

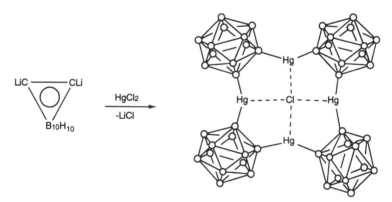

Fig. 10.25. Formation of a [12]-crown-4-like macrocycle from 1,2-dilithio-carboaranes and $HgCl_2$.

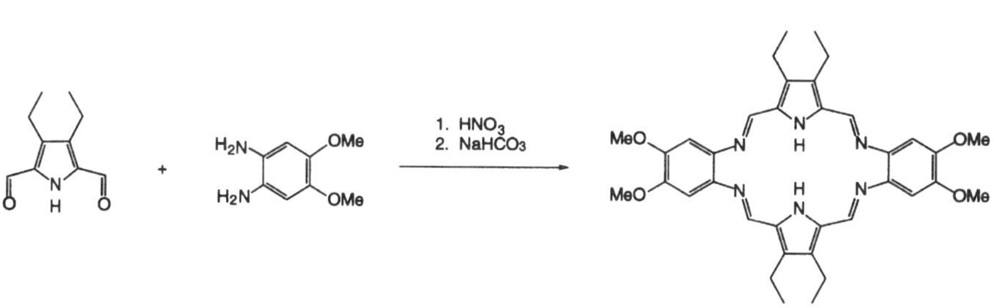

Fig. 10.26. Nitrate template-assisted formation of an expanded porphyrin.

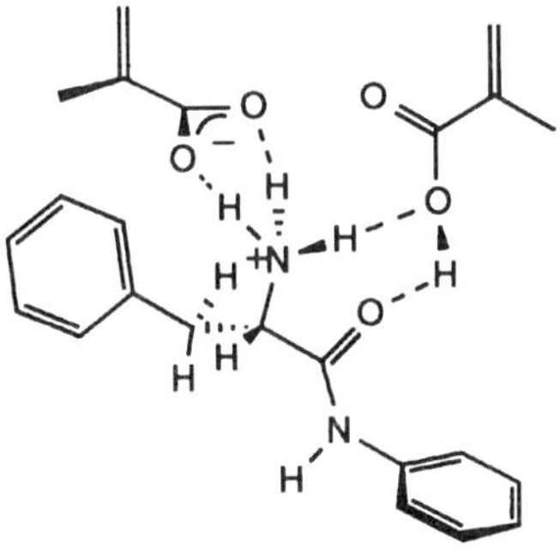

Fig. 10.27. Anion applications in imprinted polymers: the use of L-phenylalanine anilide print molecules to achieve highly enantioselective and substrate selective polymers.

reported.[321] These species would not form in the absence of anions.[322] Finally, Sessler and Mody determined that the condensation reaction of Fig. 10.26 could be ameliorated in terms of product yield and purity by the use of Brønsted acids that had nitrate anion as their conjugate bases.[323] Given the excitement attendant to these results, it is safe to assume that in the near future other examples of anionic-template syntheses will appear. This seems likely given the fact that the supramolecular chemistry of anions is much younger than that of cations.

Another application of anion-binding effects in synthesis is in the area of imprinted polymers.[324] Here, ion pairing between a carboxylate anion and a protonated amino group has been used to prepare imprinted polymers. More specifically, highly enantioselective and substrate-selective polymers were obtained by Sellergren and Mosbach when methacrylic acid monomer was allowed to polymerize in the presence of L-phenylalanine anilide print molecules (i.e., Fig. 10.27).[324c] Although carboxylate-to-protonated amine chelation is not the only possible interaction between the monomer and the print molecules, it was nevertheless considered critical in terms of mediating a presumably templated polymerization.

A final example of anion-related transformational chemistry comes from the Breslow group. They reported that selective remote functionalization of flexible dicarboxylates could be achieved in yields up to 93% if a "pseudotemplate" *bis*-quaternary amine was used.[325] The effect is attributed to double ion pairing occurring prior to the reaction as shown in Fig. 10.28.

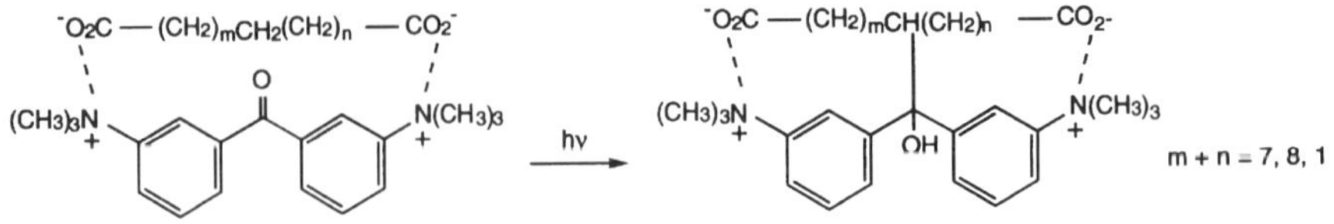

Fig. 10.28. Use of a "pseudotemplate" *bis*-quarternary amine to achieve selective remote functionalization of flexible dicarboxylates.

10.8 REFERENCES

1. *The Biochemistry of the Nucleic Acids*, 10th ed., R. L. P. Adams, J. T. Knowler, and D. P. Leader, Eds., Chapman and Hall, New York, 1986.

2. (a) *Nucleotide Analogues as Antiviral Agents, ACS Symposium Series 401*, J. C. Martin, Ed., American Chemical Society, Washington, DC, 1989; (b) P. A. Furman, J. A. Fyfe, M. H. St. Clair, K. Weinhold, J. L. Rideout, G. A. Freeman, S. N. Lehrman, D. P. Bolognesi, S. Broder, H. Mitsuya, and D. W. Barry, *Proc. Natl. Acad. Sci. U.S.A.* **83**, 8333 (1986).

3. S. N. Farrow, A. S. Jones, A. Kumar, R. T. Walker, J. Balzarini, and E. de Clerq, *J. Med. Chem.* **33**, 1400 (1990).

4. (a) I. Tabushi, Y. Kobuke, and J. Imuta, *J. Amer. Chem. Soc.* **103**, 6152 (1981); (b) I. Tabushi, Y. Kobuke, and J. Imuta, *J. Am. Chem. Soc.* **102**, 1744 (1980).

5. (a) M. P. Mertes and K. B. Mertes, *Acc. Chem. Res.* **23**, 413 (1990); (b) M. W. Hosseini and J.-M. Lehn, *J. Chem. Soc., Chem. Commun.*, 451 (1991); (c) M. W. Hosseini, J.-M. Lehn, K. C. Jones, K. E. Plute, K. B. Mertes, and M. P. Mertes, *J. Am. Chem. Soc.* **111**, 6330 (1989).

6. (a) E. Kimura, Y. Kuramoto, T. Koike, H. Fujioka, and M. Kodama, *J. Org. Chem.* **55**, 42 (1990); (b) E. Kimura, *Top. Curr. Chem.* **128**, 113 (1985) and refs. therein.

7. F. P. Schmidtchen, *Top. Curr. Chem.* **132**, 101 (1986) and refs. therein.

8. J. F. Marecek, P. A. Fischer, and C. J. Burrows, *Tetrahedron Letters* **29**, 6231 (1988).

9. (a) T. Li, S. J. Krasne, B. Persson, H. R. Kaback, and F. Diederich, *J. Org. Chem.* **58**, 380 (1993); (b) T. Li and F. Diederich, *J. Org. Chem.* **57**, 3449 (1992).

10. For examples of nucleotide/phosphate binding with simple acyclic polyamines, see, e.g.: (a) C. Nakai and W. Glinsmann, *Biochemistry* **25**, 5636 (1977); (b) N. H. Woo, N. C. Seeman, and A. Rich, *Biopolymers* **18**, 539 (1979); (c) H. R. Wilson and R. J. P. Williams, *J. Chem. Soc., Farada Trans. 1* **83**, 1885 (1987).

11. S. Claude, J.-M. Lehn, F. Schmidt, and J.-P. Vigneron, *J. Chem. Soc., Chem. Commun.*, 1182 (1991).

12. (a) S. A. van Arman and A. W. Czarnik, *Supramol. Chem.* **1**, 99 (1993); (b) S. A. van Arman and A. W. Czarnik, *J. Am. Chem. Soc.* **116**, 9397 (1994).

13. H.-J. Schneider, T. Blatter, B. Palm, U. Pfingstag, V. Rüdiger, and I. Theis, *J. Am. Chem. Soc.* **114**, 7704 (1992).

14. A. Andrés, M. I. Burguete, E. García-España, S. V. Luis, J. F. Miravet, and C. Soriano, *J. Chem. Soc. Perkin Trans. 2*, 749 (1993).

15. (a) Y. Aoyama, S. Nonaka, T. Motomura, H. Toi, and H. Ogoshi, *Chemistry Letters*, 1241 (1991); (b) Y. Kuroda, H. Hatakeyama, H. Seshimo, and H. Ogoshi, *Supramol. Chem.* **3**, 267 (1994).

16. B. Dietrich, D. L. Fyles, T. M. Fyles, and J.-M. Lehn, *Helv. Chim. Acta* **62**, 2763 (1979).

17. (a) F. P. Schmidtchen, *Tetrahedron Lett.* **30**, 4493 (1989); (b) P. Schiessel and F. P. Schmidtchen, *J. Org. Chem.* **59**, 509 (1994).

18. A. Galán, E. Pueyo, A. Salmerón, and J. de Mendoza, *Tetrahedron Letters* **32**, 1827 (1991).
19. A. Galán, J. de Mendoza, C. Toiron, M. Bruix, G. Deslongchamps, and J. Rebek, Jr., *J. Amer. Chem. Soc.* **113**, 9424 (1991).
20. (a) G. Deslongchamps, A. Galán, J. de Mendoza, and J. Rebek, Jr., *Angew. Chem. Int. Ed. Engl.* **31**, 61 (1992); (b) C. Andreu, A. Galán, K. Kobiro, J. de Mendoza, T. K. Park, J. Rebek, Jr., A. Salmerón, and N. Usman, *J. Amer. Chem. Soc.* **116**, 5501 (1994).
21. (a) K. Ariga and E. V. Anslyn, *J. Org. Chem.* **57**, 417 (1992); (b) L. S. Flatt, V. Lynch, and E. V. Anslyn, *Tetrahedron Letters* **33**, 2785 (1992); (c) D. M. Kneeland, K. Ariga, V. M. Lynch, C.-Y. Huang, and E. V. Anslyn, *J. Amer. Chem. Soc.* **115**, 10042 (1993); (d) F. Chu, L. S. Flatt, and E. V. Anslyn, *J. Amer. Chem. Soc.* **116**, 4194 (1994).
22. (a) R. P. Dixon, S. J. Geib, and A. D. Hamilton, *J. Amer. Chem. Soc.* **114**, 365 (1992); (b) S. C. Hirst, P. Tecilla, S. J. Geib, E. Fan, and A. D. Hamilton, *Israel J. Chem.* **32**, 105 (1992).
23. D. Y. Sasaki, K. Kurihara, and T. Kunitake, *J. Amer. Chem. Soc.* **113**, 9685 (1991).
24. (a) H. C. Visser, D. M. Rudkevich, W. Verboom, F. de Jong, and D. N. Reinhoudt, *J. Amer. Chem. Soc.* **116**, 11554 (1994); (b) D. M. Rudkevich, W. Verboom, Z. Brzozka, M. J. Palys, W. P. R. V. Stauthamer, G. J. van Hummel, S. M. Franken, S. Harkema, J. F. J. Engbersen, and D. N. Reinhoudt, *J. Amer. Chem. Soc.* **116**, 4341 (1994); (c) D. M. Rudkevich, W. Verboom, and D. N. Reinhoudt, *J. Org. Chem.* **59**, 3683 (1994); (d) D. M. Rudkevich, Z. Brzozka, M. Palys, H. C. Visser, W. Verboom, and D. N. Reinhoudt, *Angew. Chem. Int. Ed. Engl.* **33**, 467 (1994).
25. S. Valiyaveettil, J. F. J. Engbersen, W. Verboom, and D. N. Reinhoudt, *Angew. Chem. Int. Ed. Engl.* **32**, 900 (1993).
26. P. D. Beer, P. A. Gale, and D. Hesek, *Tetrahedron Letters* **36**, 767 (1995).
27. A. V. Eliseev and H.-J. Schneider, *Angew. Chem. Int. Ed. Engl.* **32**, 1331 (1993).
28. P. D. Beer, Z. Chen, A. J. Goulden, A. Graydon, S. E. Stokes, and T. Wear, *J. Chem. Soc., Chem. Commun.*, 1834 (1993).
29. K. Manabe, K. Okamura, T. Date, and K. Koga, *J. Amer. Chem. Soc.* **114**, 6940 (1992).
30. (a) J. L. Sessler, M. Cyr, H. Furuta, V. Král, T. Mody, T. Morishima, M. Shionoya, and S. Weghorn, *Pure & Appl. Chem.* **65**, 393 (1993); (b) J. L. Sessler, A. K. Burrell, H. Furuta, G. W. Hemmi, B. L. Iverson, V. Král, D. J. Magda, T. D. Mody, K. Shreder, D. Smith, and S. J. Weghorn, accepted for publication, *Metal Ions in Supramolecular Chemistry*, NATO ASI Series, Kluwer, Amsterdam, **448**, 391 (1994).
31. H. Furuta, T. Morishima, V. Král, and J. L. Sessler, *Supramolecular Chem.* **3**, 5 (1993).
32. H. Furuta, M. J. Cyr, and J. L. Sessler, *J. Amer. Chem. Soc.* **113**, 6677 (1991).
33. H. Furuta, V. Král, and J. L. Sessler, unpublished results.
34. (a) V. Král, J. L. Sessler, and H. Furuta, *J. Amer. Chem. Soc.* **114**, 8704 (1992); (b) J. L. Sessler, H. Furuta, and V. Král, *Supramolecular Chem.* **1**, 209 (1993).

35. V. Král and J. L. Sessler, *Tetrahedron* **51**, 539 (1995).
36. V. Král, A. Andrievsky, and J. L. Sessler, *J. Chem. Soc., Chem. Commun.*, 2349 (1995).
37. E. Kimura, A. Watanabe, and H. Nihira, *Chem. Pharm. Bull.* **31**, 3264 (1983).
38. *Chemistry and Physics of DNA–Ligand Interaction*, N. R. Kallenbach, Ed., Adenine Press, Guilderland, NY, 1989.
39. (a) E. Uhlmann and A. Peyman, *Chem. Reviews* **90**, 543 (1990); (b) J. Goodchild, *Bioconjugate Chem.* **1**, 165 (1990).
40. (a) H.-J. Schneider and T. Blatter, *Angew. Chem. Int. Ed. Engl.* **31**, 1207 (1992); (b) H.-J. Schneider, T. Blatter, A. Eliseev, V. Rüdiger, and O. A. Raevsky, *Pure & Appl. Chem.* **65**, 2329 (1993).
41. K. D. Stewart, "Designed DNA Interactions," in *Inclusion Phenomena and Molecular Recognition*, J. L. Atwood, Ed., Plenum Press, New York, 1990.
42. A. V. Kabanov and V. A. Kabanov, *Bioconjugate Chem.* **6**, 7 (1995).
43. (a) B. L. Iverson, K. Shreder, V. Král, and J. L. Sessler, *J. Amer. Chem. Soc.* **115**, 11022 (1993); (b) B. L. Iverson, K. Shreder, V. Král, D. A. Smith, J. Smith, and J. L. Sessler, *Pure & Appl. Chem.* **66**, 845 (1994).
44. For a particularly useful and general introduction, see: K. L. Kirk, *Biochemistry of the Elemental Halogens and Inorganic Halides*, Plenum Press, New York, 1991. For a book specifically devoted to the problem of chloride anion transport, see Ref. 45.
45. G. Gerenscser, Ed., *Chloride Transport Coupling in Biological Membranes and Epithelia*, Elsevier, Amsterdam, 1984.
46. For an introduction to cystic fibrosis, see: T. F. Boat, M. J. Welsh, and A. L. Beaudet, in *The Metabolic Basis of Inherited Disease*, 6th ed., C. R. Scriver, A. L. Beaudet, W. S. Sly, and D. Valle, Eds., McGraw-Hill, New York, 1989, p. 2649. See also Ref. 44, pp. 88–90 and Refs. 47–50.
47. P. B. Davis, *N. Engl. J. Med.* **325**, 575 (1991).
48. P. M. Quinton, *FASEB* **4**, 2709 (1990).
49. R. Weiss, *Science News* **139**, 132 (1991).
50. For a conference proceedings book describing the discovery of the CFTR gene and its implications, see: L.-C. Tsui, G. Romeo, R. Greger, and S. Gorini, Eds., *The Identification of The CF (Cystic Fibrosis) Gene, Recent Progress and New Research Strategies*, Plenum, New York, 1991. See also Ref. 51.
51. (a) J. M. Rommens, M. C. Iannuzzi, B. Kerem, M. L. Drumm, G. Melmer, M. Dean, R. Rozmahel, J. L. Cole, D. Kennedy, N. Hidaka, M. Zsiga, M. Buchwald, J. R. Riordan, L.-C. Tsui, and F. S. Collins, *Science* **245**, 1059 (1989); (b) J. R. Riordan, J. M. Rommens, B. Kerem, N. Alon, R. Rozmahel, Z. Grzelczak, J. Zielenski, S. Lok, N. Plavsic, J.-L. Chou, M. L. Drumm, M. C. Iannuzzi, F. S. Collins, and L.-C. Tsui, *Science* **245**, 1066 (1989); (c) B. Kerem, J. M. Rommens, J. A. Buchanan, D. Markiewicz, T. K. Cox, A. Chakravarti, M. Buchwald, and L.-C. Tsui, *Science* **245**, 1073 (1989).
52. (a) M. J. Welsh, A. E. Smith, R. J. Gregory, D. P. Rich, and M. P. Anderson, *Science* **251**, 679 (1991); (b) J. R. Riordan, C. E. Bear, J. M. Rommens, L.-C. Tsui, E. F. Reyes, C. A. Ackerley, S. Sun, A. L. Naismith, T. T. Jensen, J. W. Hanrahan,

and N. Kartner, *Cell* **64**, 681 (1991). For a brief account, see: R. Nowak, *J. NIH Res.* **3**, 30 (1991).

53. (a) M. J. Welsh, A. E. Smith, P. Manavalan, M. P. Anderson, R. J. Gregory, and D. P. Rich, *Science* **253**, 205 (1991); (b) M. J. Welsh, A. E. Smith, R. C. Mulligan, D. W. Souza, S. Paul, S. Thompson, R. J. Gregory, and M. P. Anderson, *Science* **253**, 202 (1991).

54. (a) S. H. Cheng, D. P. Rich, J. Marshall, R. J. Gregory, M. J. Welsh, and A. E. Smith, *Cell* **66**, 1027 (1991); (b) M. P. Anderson, H. A. Barger, D. P. Rich, R. J. Gregory, A. E. Smith, and M. J. Welsh, *Cell* **67**, 775 (1991).

55. (a) M. L. Drumm, H. A. Pope, W. H. Cliff, J. M. Rommens, S. A. Marvin, L.-C. Tsui, F. S. Collins, R. A. Frizzell, and J. M. Wilson, *Cell* **62**, 1227 (1990); (b) R. J. Gregory, S. H. Cheng, D. P. Rich, J. Marshall, S. Paul, K. Hehir, L. Ostedgaard, K. W. Klinger, M. J. Welsh, and A. E. Smith, *Nature* **347**, 382 (1990); (c) D. P. Rich, M. P. Anderson, R. J. Gregory, S. H. Cheng, S. Paul, D. M. Jefferson, J. D. McCann, K. W. Klinger, A. E. Smith, and M. J. Welsh, *Nature* **347**, 358 (1990).

56. M. R. Knowles, L. L. Clarke, and R. C. Boucher, *N. Engl. J. Med.* **325**, 533 (1991).

57. M. R. Knowles, N. L. Church, W. E. Waltner, J. R. Yankaskas, P. Gilligan, M. King, L. J. Edwards, R. W. Helms, and R. C. Boucher, *N. Engl. J. Med.* **322**, 1189 (1990).

58. For other innovative preclinical and clinical approaches to cystic fibrosis treatment, see: (a) S. Shak, D. J. Capon, R. Hellmiss, S. A. Marsters, and C. L. Baker, *Proc. Natl. Acad. Sci. USA* **87**, 9188 (1990); (b) N. G. Mcelvaney, R. C. Hubbard, P. Birrer, M. S. Chernick, D. B. Caplan, M. M. Frank, and A. G. Crystal, *Lancet* **337**, 392 (1991). See also Ref. 47.

59. Cellular uptake of phosphorylated entities, for instance, is often achieved at the expense of chloride anion counter transport. Additionally, many channels for chloride, including that of the CFTR protein,[54] are activated by cyclic nucleotides and/or require the concomitant hydrolysis of nucleotide triphosphates for effective function.[44,45]

60. C. H. Park and H. E. Simmons, *J. Am. Chem. Soc.* **90**, 2431 (1968).

61. (a) E. Graf and J.-M. Lehn, *J. Am. Chem. Soc.* **98**, 6403 (1976); (b) J.-P. Kintzinger, J.-M. Lehn, E. Kauffmann, J. L. Dye, and A. I. Popov, *J. Am. Chem. Soc.* **105**, 7549 (1983); (c) M. W. Hosseini, J.-P. Kintzinger, J.-M. Lehn, and A. Zahidi, *Helv. Chim. Acta* **72**, 1078 (1989). For the cryptand **10.3(2)** chloride anion crystal structure, see: B. Metz, M. J. Rosalky, and R. Weiss, *J. Chem. Soc., Chem. Commun.*, 533 (1976).

62. B. Dietrich, J.-M. Lehn, J. Guilhem, and C. Pascard, *Tetrahedron Lett.* **30**, 4125 (1989).

63. For other relevant halide anion chelation work from the Lehn group, see: (a) M. W. Hosseini, and J.-M. Lehn, *Helv. Chim. Acta* **71**, 749 (1988); (b) D. Heyer and J. M. Lehn, *Tetrahedron Lett.* **27**, 5869 (1986); (c) J.-M. Lehn, *Science* **227**, 849 (1985), and references therein; (d) S. Boudon, A. Decian, J. Fischer, M. W. Hosseini, J.-M. Lehn, and G. Wipff, *J. Coord. Chem.* **23**, 113 (1991).

64. (a) J. Cullinane, R. I. Gelb, T. N. Margulis, and L. J. Zompa, *J. Am. Chem. Soc.* **104**, 3048 (1982); (b) R. I. Gelb, B. T. Lee, and L. J. Zompa, *J. Am. Chem. Soc.* **107**, 909 (1985).

65. B. Dietrich, T. M. Fyles, M. W. Hosseini, J.-M. Lehn, and K. C. Kaye, *J. Chem. Soc., Chem. Commun.*, 691 (1988).
66. Charge neutralization as a requirement for efficient carrier mediated transport is a corollary of Fick's First Law. See Ref. 67, p. 78.
67. T. Araki and H. Tsukube, *Liquid Membranes: Chemical Applications*, CRC Press, Boca Raton, 1990.
68. R. E. Cramer, V. Fermin, E. Kuwabara, R. Kirkup, M. Selman, K. Aoki, A. Adeyemo, and H. Yamazaki, *J. Am. Chem. Soc.* **113**, 7033 (1991).
69. (a) P. J. Stang and V. V. Zhdankin, *J. Am. Chem. Soc.* **115**, 9808 (1993); (b) P. J. Stang, D. H. Cao, S. Saito, and A. M. Arif, *J. Am. Chem. Soc.* **117**, 6273 (1995).
70. P. D. Beer, D. Hesek, J. Hodacova, and S. E. Stokes, *J. Chem. Soc., Chem. Commun.*, 270 (1992).
71. J. Scheerder, M. Fochi, J. F. J. Engbersen, and D. N. Reinhoudt, *J. Org. Chem.* **59**, 7815 (1994).
72. A. L. Beauchamp, M. J. Olievier, J. D. Wuest, and B. Zacharie, *J. Am. Chem. Soc.* **108**, 73 (1986). For a later related macrocyclic (but non-halide-binding) receptor approach, see: J. D. Wuest and B. Zacharie, *J. Am. Chem. Soc.* **109**, 4714 (1987).
73. (a) M. T. Blanda and M. Newcomb, *Tetrahedron Lett.* **27**, 3501 (1989). For related work, see: (b) M. Newcomb, J. H. Horner, M. T. Blanda, and P. J. Squattrito, *J. Am. Chem. Soc.* **111**, 6294 (1989), and references therein.
74. K. Jurkschat, H. G. Kuivila, S. Liu, and J. A. Zubieta, *Organometallics* **8**, 2755 (1989).
75. M. E. Jung and H. Xia, *Tetrahedron Lett.* **29**, 297 (1988).
76. S. Aoyagi, K. Ogawa, K. Tanaka, and Y. Takeuchi, *J. Chem. Soc. Perkin Trans. 2*, 355 (1995).
77. S. Jacobson and R. Pizer, *J. Am. Chem. Soc.* **115**, 11216 (1993).
78. (a) H. E. Katz, *J. Am. Chem. Soc.* **108**, 7640 (1986); (b) H. E. Katz, *Organometallics* **6**, 1134 (1987).
79. (a) X. Yang, C. B. Knobler, and M. F. Hawthorne, *Angew. Chem. Intl. Ed. Engl.* **30**, 1507 (1991); (b) X. Yang, C. B. Knobler, and M. F. Hawthorne, *J. Am. Chem. Soc.* **114**, 380 (1992).
80. (a) K. Worm, F. P. Schmidtchen, A. Schier, A. Schafer, and M. Hesse, *Angew. Chem. Int. Ed. Engl.* **33**, 327 (1994); (b) K. Worm and F. P. Schmidtchen, *Angew. Chem. Int. Ed. Engl.* **34**, 65 (1995).
81. For examples of anion transport or recognition achieved via chelation to a metal center, see: (a) M. Huser, W. E. Morf, K. Fluri, K. Seiler, P. Schulthess, and W. Simon, *Helv. Chim. Acta* **73**, 1481 (1990); (b) P. Schulthess, D. Ammann, W. Simon, C. Caderas, R. Stepánek, and B. Kräutler, *Helv. Chim. Acta* **67**, 1026 (1984); (c) K. M. Kadish and R. K. Rhodes, *Inorg. Chem.* **22**, 1090 (1983); (b) L. A. Bottomley and K. M. Kadish, *Inorg. Chem.* **20**, 1348 (1981).
82. J. L. Sessler, T. D. Mody, D. A. Ford, and V. Lynch, *Angew. Chem. Int. Ed. Engl.* **31**, 452 (1992).
83. M. Shionoya, H. Furuta, V. Lynch, A. Harriman, and J. L. Sessler, *J. Am. Chem. Soc.* **114**, 5714 (1992).

84. J. L. Sessler, T. Morishima, and V. Lynch, *Angew. Chem. Int. Ed. Engl.* **30**, 977 (1991).
85. J. L. Sessler, D. A. Ford, M. J. Cyr, and H. Furuta, *J. Chem. Soc., Chem. Commun.*, 1733 (1991).
86. J. L. Sessler, S. J. Weghorn, T. Morishima, M. Rosingana, V. Lynch, and V. Lee, *J. Am. Chem. Soc.* **114**, 8306 (1992).
87. J. L. Sessler, S. J. Weghorn, V. Lynch, and M. R. Johnson, *Angew. Chem. Int. Ed. Engl.* **33**, 1509 (1994).
88. D. Voet and J. G. Voet, *Biochemistry*, John Wiley & Sons, New York, 1990.
89. See the paragraph about chromatographic applications of anion binding appearing later in this chapter. See also Ref. 116.
90. (a) B. Dietrich, M. W. Hosseini, and J.-M. Lehn, *J. Am. Chem. Soc.* **103**, 1282 (1981); (b) B. Dietrich, M. W. Hosseini, and J.-M. Lehn, *Helv. Chim. Acta* **66**, 1262 (1983); (c) M. W. Hosseini, J.-M. Lehn, *Helv. Chim. Acta* **69**, 587 (1986). For other synthetic polyammonium receptor systems for mono-, di-, and tricarboxylates, see: anion binding review: (d) B. Dietrich, *Pure & Appl. Chem.* **65**, 1464 (1993) and original publications: (e) B. Dietrich, J. Guilhem, J.-M. Lehn, C. Pascard, and E. Sonveaux, *Helv. Chim. Acta* **67**, 91 (1984); (f) J. Jazwinski, J.-M. Lehn, D. Lilienbaum, R. Ziessel, J. Guilhem, and C. Pascard, *J. Chem. Soc. Chem. Commun.*, 1691 (1987); (g) J.-M. Lehn, R. Méric, J.-P. Vigneron, I. Bkouche-Waksman, and C. Pascard, *J. Chem. Soc., Chem. Commun.*, 62 (1991); (h) M. W. Hosseini, and J.-M. Lehn, *J. Am. Chem. Soc.* **104**, 3525 (1982); (i) E. Kimura, A. Sakonaka, T. Yatsunami, and M. Kodama, *J. Am. Chem. Soc.* **103**, 3041 (1981); (j) E. Kimura, *Pure Appl. Chem.* **61**, 823 (1989), and references therein; (k) M. Kataoka, R. Naganawa, K. Odashima, Y. Umezawa, E. Kimura, and T. Koike, *Anal. Lett.* **22**, 1089 (1989); (l) A. Bencini, A. Bianchi, M. I. Burguete, P. Dapporto, A. Domenech, E. García-España, S. V. Luis, P. Paoli, and J. A. Ramirez, *J. Chem. Soc. Perkin Trans. 2*, 569 (1994).
91. For guanidinium receptor reviews: (a) C. L. Hannon and E. V. Anslyn "The Guanidinium Group: Its Biological Role and Synthetic Analogues," in *Bioorganic Chemistry Frontiers*, Springer-Verlag, Berlin, 1993; (b) B. Dietrich, D. L. Fyles, and J.-M. Lehn, *Helv. Chim. Acta* **62**, 2763 (1979). See also: (c) G. Müller, J. Riede, and F. P. Schmidtchen, *Angew. Chem. Int. Ed. Engl.* **27**, 1516 (1988); (d) A. Echavarren, A. Galan, J. de Mendoza, A. Salmeron, and J.-M. Lehn, *Helv. Chim. Acta* **71**, 685 (1988); (e) H. Kurzmeier and F. P. Schmidtchen, *J. Org. Chem.* **55**, 3749 (1990); (f) A. Gleich and F. P. Schmidtchen, *Chem. Ber.* **123**, 907 (1990); (g) A. Echavarren, A. Galan, J.-M. Lehn, and J. de Mendoza, *J. Am. Chem. Soc.* **111**, 4994 (1989); (h) A. Galan, E. Pueyo, A. Salmeron, and J. de Mendoza, *Tetrahedron Lett.* **32**, 1827 (1991); (i) A. Gleich, F. P. Schmidtchen, P. Mikulcik, and G. Müller, *J. Chem. Soc., Chem. Commun.*, 55 (1990); (j) P. Schiessel and F. P. Schmidtchen, *Tetrahedron Lett.* **34**, 2449 (1993). For synthesis of more rigid dibenzoguanidine receptor for oxoanion recognition, see: (k) J.-L. Chicharro, P. Prados, and J. de Mendoza, *J. Chem. Soc., Chem. Commun.*, 1193 (1994).
92. P. D. Beer, M. G. B. Drew, C. Hazlewood, D. Hesek, J. Hodacova, and S. E. Strokes, *J. Chem. Soc., Chem. Commun.*, 229 (1993).
93. S. M. Lacy, D. M. Rudkevich, W. Verboom, and D. N. Reinhoudt, *J. Chem. Soc. Perkin Trans. 2*, 135 (1995).

94. K. Konishi, K. Yahara, H. Toshishige, T. Aida, and S. Inoue, *J. Am. Chem. Soc.* **116**, 1337 (1994).

95. S. Inokuma, S. Sakai, T. Yamamoto, and J. Nishimura, *J. Membrane Sci.* **97**, 175 (1994).

96. (a) S. J. Geib, C. Vicent, E. Fan, and A. D. Hamilton, *Angew. Chem. Int. Ed. Engl.* **32**, 119 (1993); (b) E. Fan, S. A. Van Arman, S. Kincaid, and A. D. Hamilton, *J. Am. Chem. Soc.* **115**, 369 (1993); (c) F. Garcia-Tellado, S. Goswami, S.-K. Chang, S. J. Geib, and A. D. Hamilton, *J. Am. Chem. Soc.* **112**, 7393 (1990); (d) A. D. Hamilton, E. Fan, S. A. Van Arman, C. Vicent, F. Garcia-Tellado, and S. J. Geib, *Supramol. Chem.* **1**, 247 (1993); (e) P. Ballester, A. Costa, P. M. Deyà, J. F. Gonzáles, M. C. Rotger, *Tetrahedron Lett.* **35**, 3813 (1994); (f) M. Crego, C. Raposo, M. C. Caballero, E. Garcia, J. G. Saez, and J. R. Morán, *Tetrahedron Lett.* **33**, 7437 (1992); (g) C. Raposo, M. Crego, M. L. Mussons, M. C. Caballero, and J. R. Morán, *Tetrahedron Lett.* **35**, 3409 (1994); (h) Y. Tanaka, Y. Kato, and Y. Aoyama, *J. Am. Chem. Soc.* **112**, 2807 (1990). For a new family of receptors for carboxylates based on the multidentate recognition of ristocetin, see: (i) J. S. Albert and A. D. Hamilton, *Tetrahedron Lett.* **34**, 7363 (1993).

97. (a) F. Garcia-Tellado, J. Albert, and A. D. Hamilton, *J. Chem. Soc. Chem. Commun.*, 1761 (1991) and Ref. 96d; (b) V. Alcazar, L. Tomlinson, K. N. Houk, and F. Diederich, *Tetrahedron Lett.* **32**, 5309 (1991); (c) V. Alcazar and F. Diederich, *Angew. Chem. Int. Ed. Engl.* **31**, 1521 (1992); (d) V. Alcazar, J. R. Morán, and F. Diederich, *Isr. J. Chem.* **32**, 69 (1992); (e) L. Owens, C. Thilgen, F. Diederich, and C. B. Knobler, *Helv. Chim. Acta* **76**, 2757 (1993).

98. (a) M. S. Goodman, J. Weiss, and A. D. Hamilton, *Tetrahedron Lett.* **35**, 8943 (1994); (b) M. S. Goodman, V. Jubian, and A. D. Hamilton, *Tetrahedron Lett.* **36**, 2551 (1995).

99. V. Král, A. Andrievsky, and J. L. Sessler, *J. Am. Chem. Soc.* **117**, 2953 (1995).

100. Binding and transport of cationic amino acids and amino acid esters was pioneered independently by Lehn and Cram. Lehn used a negatively charged carrier, dinonylnaphthalenesulfonate, to transport amino acid salts in a U-tube experimental setup: (a) J.-P. Behr and J.-M. Lehn, *J. Am. Chem. Soc.* **95**, 6108 (1973). Cram and co-workers have employed chiral crown ethers for chiral recognition of amino acid cations and amino acid esters: (b) M. Newcomb, J. L. Toner, R. C. Helgeson, and D. J. Cram, *J. Am. Chem. Soc.* **101**, 4941 (1979), and references therein; (c) G. D. Y. Sogah and D. J. Cram, *J. Am. Chem. Soc.* **101**, 3035 (1979), and references therein. Other examples of crown ether use are: (d) J.-P. Behr, J.-M. Lehn, and P. Vierling, *Helv. Chim. Acta* **65**, 1853 (1982); (e) R. B. Davidson, J. S. Bradshaw, B. A. Jones, N. K. Dalley, J. J. Christensen, and R. M. Izatt, *J. Org. Chem.* **49**, 353 (1984); (f) K. Naemura, R. Fukunaga, and M. Yamanaka, *J. Chem. Soc., Chem. Commun.*, 1560 (1985); (g) M. Sawada, Y. Takai, H. Yamada, T. Kaneda, K. Kamada, T. Mizooku, K. Hirose, Y. Tobe, and K. Naemura, *J. Chem. Soc., Chem. Commun.*, 2497 (1994). Macrocycles other than crown ethers were used for amino ester transport through the lipophylic liquid membranes: (h) H. Kataoka and T. Katagi, *Tetrahedron* **43**, 4519 (1987); (i) S.-K. Chang, H.-S. Hwang, H. Son, J. Youk, and Y. S. Kang, *J. Chem. Soc. Chem. Commun.*, 217 (1991). Chemically modified monensins bearing neutral terminal groups were employed for effective chiral complex formation with several amino ester salts in a liquid membrane–type

electrode system: (j) K. Maruyama, H. Sohmiya, and H. Tsukube, *J. Chem. Soc., Chem. Commun.*, 864 (1989).

101. J. Rebel, Jr., B. Askew, D. Nemeth, and K. Parris, *J. Am. Chem. Soc.* **109**, 2432 (1987).

102. R. J. Baczuk, G. K. Landram, R. J. Dubois, and H. C. Dehm, *J. Chromatog.* **60**, 351 (1971).

103. A. Galán, D. Andreu, A. M. Echavarren, P. Prados, and J. de Mendoza, *J. Am. Chem. Soc.* **114**, 1511 (1992).

104. I. Tabushi, Y. Kuroda, and T. Mizutani, *J. Am. Chem. Soc.* **108**, 4514 (1986).

105. J. Sunamoto, K. Iwamoto, Y. Mohri, and T. Kominato, *J. Am. Chem. Soc.* **104**, 5502 (1982).

106. S. Marx-Tibbon and I. Willner, *J. Chem. Soc., Chem. Commun.*, 1261 (1994).

107. (a) K. Maruyama, H. Tsukube, and T. Araki, *J. Am. Chem. Soc.* **104**, 5197 (1982), and references therein; (b) R. P. Bonomo, V. Cucinotta, F. D'Allessandro, G. Impellizzeri, G. Maccarrone, E. Rizzarelli, and G. Vecchio, *J. Inclusion Phenomena* **15**, 167 (1993); (c) Y. N. Belokon, L. K. Pritula, V. I. Tararov, V. I. Bakhmutov, D. G. Gusev, M. B. Saporovskaya, and V. M. Belikov, *J. Chem. Soc. Dalton Trans.*, 1873 (1990); (d) R. Corradini, A. Dossena, G. Impellizzeri, G. Maccarrone, R. Marchelli, E. Rizzarelli, G. Sartor, and G. Vecchio, *J. Am. Chem. Soc.* **116**, 10267 (1994); (e) P. Scrimin, P. Tecilla, and U. Tonellato, *Tetrahedron* **51**, 217 (1995); (f) C. L. Gatlin, F. Turecek, and T. Vaisar, *J. Am. Chem. Soc.* **117**, 3637 (1995).

108. L. K. Mohler and A. W. Czarnik, *J. Am. Chem. Soc.* **115**, 7037 (1993).

109. M. T. Reetz, J. Huff, J. Rudolph, K. Töllner, A. Deege, and R. Goddard, *J. Am. Chem. Soc.* **116**, 11588 (1994).

110. G. J. Pernía, J. D. Kilburn, and M. Rowley, *J. Chem. Soc., Chem. Commun.*, 305 (1995).

111. (a) H. Tsukube, *Tetrahedron Lett.* **22**, 3981 (1981); (b) H. Tsukube, *J. Chem. Soc. Perkin Trans. 1*, 2359 (1982); (c) E. Kimura, *J. Inclusion Phenomena* **7**, 183 (1989); (d) P. Chaudhuri, M. Winter, P. Fleischhauer, W. Haase, U. Flörke, and H.-J. Haupt, *J. Chem. Soc., Chem. Commun.*, 1728 (1990). See also Ref. 95.

112. Quaternary ammonium salt, Aliquat 336, was used to effect amino acid carboxylate transport through a lipophilic toluene membrane; see Ref. 100(a).

113. F. P. Schmidtchen, A. Gleich, and A. Schummer, *Pure Appl. Chem.* **61**, 1535 (1989).

114. For reviews of biochemical amino acid transport systems, see: (a) M. S. Kilberg and D. Häussinger, Eds., *Mammalian Amino Acid Transport. Mechanisms and Control*, Plenum Press, New York,1992; (b) M. S. Kilberg, B. S. Stevens, and D. A. Novak, *Annu. Rev. Nutr.* **13**, 137 (1993); (c) K. Ring, *Angew. Chem. Int. Ed. Engl.* **9**, 345 (1970).

115. K. B. Lipkowits, S. Raghothama, and J. Yang, *J. Am. Chem. Soc.* **114**, 1554 (1992).

116. For excellent general reviews on the topic, see: (a) W. H. Pirkle and T. C. Pochapsky, *Chem. Rev.* **89**, 347 (1989); (b) W. H. Pirkle and T. C. Pochapsky, in *Advances in Chromatography*, Vol. 27, J. C. Giddings, E. Grushka, and P. R. Brown, Eds., Marcel Dekker, New York, 1987, p. 73. See also Refs. 107(b) and 107(d).

117. A. Andrievsky and J. L. Sessler, unpublished results.
118. For instance, hyperornithinaemia, hyperammonemia, and homocitrullinuria syndromes (HHH disorder) are believed to be caused by a defective mitochondrial ornithine transport system [Ref. 114(a), p. 109]. Also, chronic metabolic acidosis [Ref. 114(a), p. 249], Hartnup disease [Ref. 114(a), p. 254], phenylketonurea and maple-syrup urine syndrome are caused by imbalances in amino acids [Ref. 114(a), p. 165]. Further, uncontrolled release of exitatory amino acids has been linked to the neuronal degeneration and death that occurs in ischemia, hypoxia, and hypoglycemia [Ref. 114(a), p. 165].
119. V. Král, S. L. Springs, and J. L. Sessler, *J. Am. Chem. Soc.*, **117**, 8881 (1995).
120. J. Bigay, P. Deterre, C. Pfister, and M. Chabre, *EMBO J.* **6**, 2907 (1987).
121. (a) R. Hall and J. Köbberling, Eds., *Thyroid Disorders Associated with Iodine Deficiency and Excess*, Raven Press, New York, 1985; (b) B. S. Hetzel, *The Story of Iodine Deficiency. An International Challenge in Nutrition*, Oxford University Press, New York, 1989.
122. See, for example, the review: Refs. 90d, 6, 61–64, 90.
123. V. Král and J. L. Sessler, unpublished results.
124. V. Král, D. Schweitzer, and J. L. Sessler, unpublished results.
125. J.-M. Lehn "Supramolecular Reactivity and Catalysis of Phosphoryl Transfer," in *Bioorganic Chemistry in Healthcare and Technology*, U. K. Pandit and F. C. Alderweireldt, Eds., Plenum Press, New York, 1991.
126. (a) M. W. Hosseini, J.-M. Lehn, L. Maggiora, K. B. Mertes, and M. P. Mertes, *J. Amer. Chem. Soc.* **109**, 537 (1987); (b) M. W. Hosseini, A. J. Blacker, and J.-M. Lehn, *J. Amer. Chem. Soc.* **112**, 3896 (1990).
127. H. Jahansouz, Z. Jiang, R. H. Himes, M. P. Mertes, and K. B. Mertes, *J. Amer. Chem. Soc.* **111**, 1409 (1989).
128. (a) K. Shinozuka, K. Shimizu, Y. Nakashima, and H. Sawai, *Bioorganic & Medicinal Chemistry Letters* **4**, 1979 (1994); (b) K. Yoshinari, K. Yamazaki, M. Komiyama, *J. Amer. Chem. Soc.* **113**, 5899 (1991); (c) M. Komiyama and K. Yoshinari, *J. Chem. Soc. Chem. Commun.*, 1880 (1989).
129. V. Jubian, R. P. Dixon, and A. D. Hamilton, *J. Amer. Chem. Soc.* **114**, 1120 (1992).
130. J. Smith, K. Ariga, and E. V. Anslyn, *J. Amer. Chem. Soc.* **115**, 362 (1993).
131. M. W. Gobel, J. W. Bats, and G. Dürner, *Angew. Chem. Int. Ed. Engl.* **31**, 207 (1992).
132. M. C. J. Wilce, M. I. Aguilar, and M. T. W. Heran, *J. Chromatography* **632**, 11 (1993) and references therein.
133. J. D. Pearson and F. E. Regnier, *J. Chromatography* **255**, 137 (1983).
134. D. R. Ramos and A. M. Schoffstall, *J. Chromatography* **261**, 83 (1983).
135. B. J. Bergot and W. Egan, *J. Chromatography* **599**, 35 (1992).
136. M. Polverelli, M. Berger, F. Odin, and J. Cadet, *J. Chromatography* **613**, 257 (1993).
137. S. Allenmark, *Chromatographic Enantioseparation, Methods and Application*, Ellis Horwood, New York, 1991.
138. A. Krstulovic, *Chiral Separations by HPLC: Application to Pharmaceutical Compounds*, John Wiley & Sons, New York, 1989.

139. *Advances in Chromatogaphy, Vol. 1*, J. C. Giddings and R. A. Keller, Eds., Marcel Dekker, New York, 1965, pp. 3–60.
140. C. S. Knight, *Advances in Chromatography, Vol. 4*, Marcel Dekker, New York, 1967, pp. 61–108.
141. N. E. Hoffman, in *Advances in Chromatography, Vol. 34*, P. R. Brown and E. Grushka, Eds., Marcel Dekker, New York, 1994, pp. 310–46.
142. R. H. A. Sorel and A. Hulshoff, in *Advances in Chromatography, Vol. 21*, J. C. Giddings, E. Griska, J. Cazes, and P. R. Brown, Eds., Marcel Dekker, New York, 1983, pp. 87–124.
143. M. T. W. Hearn, in *Advances in Chromatography, Vol. 18*, J. C. Giddings, E. Griska, J. Cazes, and P. R. Brown, Eds., Marcel Dekker, New York, 1980, pp. 60–100.
144. M. C. Gennaro, in *Advances in Chromatography, Vol. 35*, P. R. Brown and E. Grushka, Eds., Marcel Dekker, New York, 1995, pp. 344–81.
145. G. Fassina and I. M. Chaiken, in *Advances in Chromatography, Vol. 27*, J. C. Giddings, E. Griska, and P. R. Brown, Eds., Marcel Dekker, New York, 1987, pp. 248–93.
146. R. P. Singhal and S. S. M. DeSilva, in *Advances in Chromatography, Vol. 31*, J. C. Giddings, E. Griska, and P. R. Brown, Eds., Marcel Dekker, New York, 1992, pp. 294–330.
147. L. R. Snyder, J. L. Glajch, and J. J. Kirkland, *Practical HPLC Method Development*, John Wiley & Sons, New York, 1988.
148. J. H. Knox and G. R. Laird, *J. Chromatogr.* **122**, 17 (1976).
149. J. C. Kraak, K. M. Jonker, and J. F. K. Huber, *J. Chromatogr.* **142**, 671 (1977).
150. C. Horvath, W. Melander, I. Molanr, and P. Molnar, *Anal. Chem.* **49**, 2295 (1977).
151. R. A. Wall, *J. Chromatogr.* **194**, 353 (1980).
152. R. Gloor and E. L. Johnson, *J. Chromatogr. Sci.* **15**, 413 (1977).
153. T. Fornstedt, *J. Chromatogr.* **612**, 137 (1993).
154. M. Mulholland, P. R. Haddad, and D. B. Hibbert, *J. Chromatogr.* **602**, 9 (1992).
155. W. R. Melander and C. Horvath, *J. Chromatogr.* **201**, 211 (1980).
156. P. R. Haddad and C. Kalambaheti, *Anal. Chim. Acta* **250**, 31 (1991).
157. J. Bidlingmeyer, *J. Chromatogr. Sci.* **18**, 525 (1980).
158. A. P. Deleenheer, M. C. Cosyns-Duyck, and P. M. Van Vaerenberg, *J. Pharm. Sci.* **66**, 1190 (1977).
159. I. Lurie, *J. Assoc. Anal. Chem.* **60**, 1035 (1977).
160. M. T. W. Hearn and W. S. Hancock, *Trends in Biochemica Science* **4**, 58 (1978).
161. M. T. W. Hearn and W. S. Hancock, in *Biological–Biomedical Applications of Liquid Chromatography*, G. L. Hawk, Ed., Marcel Dekker, New York, 1979, p. 243.
162. B. Pace and N. R. Pace, *Anal. Biochem.* **107**, 128 (1980) and references therein.
163. N. W. Y. Ho, R. E. Duncan, and P. T. Gilham, *Biochemistry* **20**, 64 (1981).
164. R. P. Singhal, R. K. Bajaj, C. M. Buess, D. B. Smoll, and V. N. Vakharia, *Anal. Biochem.* **109**, 1 (1980).
165. A. E. Annamalai, P. K. Pal, and R. F. Colman, *Anal. Biochem.* **99**, 85 (1979).
166. H.-E. Wilk, N. Kecskemethy, and K. P. Schafer, *Nucl. Acid. Res.* **10**, 7621 (1982).

167. R. Alvarez-Gonzalez, H. Juarez-Salinas, E. L. Jacobson, and M. K. Jacobson, *Anal. Biochem.* **135**, 69 (1983).
168. B. A. Roe, A. F. Stankewitz, and C. Y. Chen, *Nucl. Acid. Res.* **4**, 2191 (1977).
169. S. Ackerman, B. Cool, and J. J. Furth, *Anal. Biochem.* **100**, 174 (1970).
170. R. M. Izatt, R. L. Bruening, B. J. Tarbet, D. Griffin, M. L. Bruening, K. E. Krakowiak, and J. S. Bradshaw, *Pure Appl. Chem.* **62**, 1115 (1990).
171. J. S. Bradshaw, K. Krakowiak, B. J. Tarbet, R. L. Bruening, J. F. Biernat, M. Bochenska, R. M. Izatt, and J. J. Christensen, *Pure Appl. Chem.* **61**, 1619 (1989).
172. Y. Saito, T. Mifune, T. Kawaguchi, J. Odo, Y. Tanaka, M. Chikuma, and H. Tanaka, *Chem. Pharm. Bull.* **34**, 2885 (1986).
173. G. Labat and G. B. Meunier, *J. Chem. Soc. Chem. Commun.*, 1414 (1990).
174. L. Smith in *Metalloporphyrins in Catalytic Oxidations*, R. A. Sheldon, Ed., Marcel Dekker, New York, 1995, p. 325.
175. L. D. Rollman, *J. Am. Chem. Soc.*, 2132 (1975).
176. P. Battioni, J. F. Bartoli, D. Mansuy, Y. S. Buyn, and T. G. Traylor, *J. Chem. Soc. Commun.*, 1051 (1991).
177. K. Maruyama, H. Tamiaki, and S. Kawabata, *J. Chem. Soc. Perkin Trans.* **2**, 543 (1986).
178. T. Mori, T. Santa, and M. Hirobe, *Tetrahedron Lett.* **26**, 5555 (1985).
179. T. Fujimoto, H. Umekawa, and N. Nishino, *Chem. Lett.*, 37 (1992).
180. K. Takahashi, A. Matsushima, Y. Saito, and Y. Inada, *Biochem. Biophys. Res. Commun.* **138**, 283 (1986).
181. A. W. van der Made, J. W. H. Smeets, and R. J. M. Nolte, *J. Mol. Cat.* **31**, 271 (1985).
182. H. S. Hilal, M. L. Sito, and A. F. Schreiner, *Inorg. Chim. Acta* **189**, 141 (1991).
183. E. Kokufuta, N. Watanabe, and I. J. Nakamura, *Appl. Polym. Sci.* **26**, 2601 (1981).
184. E. Kokufuta, H. Watanabe, K. Saito, and I. Nakamura, *J. Appl. Polym. Sci.* **30**, 3557 (1985).
185. E. Kokufuta, T. Sodeyama, and S. Takeda, *Polym. Bull.* **15**, 479 (1986).
186. N. Chatoniotakis, S. Park, and M. E. Meyerhoff, *Anal. Chem.* **61**, 566 (1989).
187. S. Park, W. Matuszewski, M. E. Meyerhoff, Y. H. Lui, and K. M. Kadish, *Electroanalysis* **3**, 909 (1991).
188. C. E. Kibbey and M. E. Meyerhoff, *Anal. Chem.* **65**, 2188 (1993).
189. J. Xiao, R. Savina, G. B. Martin, A. H. Francis, and M. E. Meyerhoff, *J. Am. Chem. Soc.* **116**, 9341 (1994).
190. O. Leal, D. L. Anderson, R. C. Bowman, F. Basolo, and R. L. Burwel, *J. Am. Chem. Soc.* **97**, 5125 (1975).
191. H.-H. Furhop, S. Besecke, W. Voght, J. Ernst, and J. Subramanian, *Macromol. Chem.* **178**, 1621 (1997).
192. J. P. Collman and C. A. Reed, *J. Am. Chem. Soc.* **95**, 2048 (1973).
193. E. Tsuchida, E. Hasegawa, and K. Honda, *Biochim. Biophys. Acta* **393** 483 (1975).
194. G. Labat and B. Meunier, *C.R. Acad. Sci. Paris* **311**, 625 (1990).

195. V. J. Bauer, D. L. J. Clive, D. Dolphin, J. B. Paine III, F. L. Harris, M. M. King, J. Loder, S. W. C. Wang, and R. B. Woodward, *J. Amer. Chem. Soc.* **105**, 6429 (1983).

196. M. J. Broadhurst, R. Grigg, and A. W. J. Johnson, *J. Chem. Soc. Perkin Trans.* **1**, 2111 (1972).

197. J. L. Sessler, M. Cyr, V. Lynch, E. McGhee, and J. A. Ibers, *J. Am. Chem. Soc.* **112**, 2810 (1990).

198. B. L. Iverson, R. E. Thomas, V. Král, and J. L. Sessler, *J. Am. Chem. Soc.* **116**, 2664 (1994).

199. J. D. Pearson and F. E. Regnier, *J. Chrom.* **255**, 137 (1983).

200. C. E. Kibbey and M. E. Meyerhoff, *Anal. Chem.* **65**, 2189 (1993).

201. E. Kokufuta, T. Sodeyama, and S. Takeda, *Polym. Bull.* **15**, 479 (1986).

202. The ability of phosphates to bind sapphyrin-substituted silica gels was confirmed by solid-state ^{31}P-NMR spectroscopy. Briefly, samples of both the sapphyrin-functionalized and simple silyl-capped silica gels were prepared by incubating with 5′-AMP. Relative to the control, the sapphyrin-containing sample showed a 4 ppm shift in the ^{31}P phosphate resonance. Such a shift is consistent with the proposed "phosphate chelation" interaction.

203. (a) K. Odashima and Y. Umezawa, in *Biosensor Technology, Fundamentals and Applications*, R. P. Buck, W. E. Hatfield, M. Umana, and E. F. Bowden, Eds., Marcel Dekker, New York, 1990, pp. 71–93; (b) K. Odashima, M. Sugawara, and Y. Umezawa, *Trends Anal. Chem.* **10**, 207 (1991).

204. K. Odashima, R. Naganawa, H. Radecka, M. Kataoka, E. Kimura, T. Koike, K. Tohda, M. Tange, H. Furuta, J. L. Sessler, K. Yagi, and Y. Umezawa, *Supramolecular Chem.* **4**, 101 (1994).

205. T. Maeda, M. Ikeda, M. Shibahara, T. Haruta, and I. Satake, *Bull. Chem. Soc. Jpn.* **54**, 94 (1981).

206. M. Bochenska and J. F. Biernat, *Anal. Chim. Acta* **162**, 369 (1984).

207. S. S. M. Hassan and E. M. Elnemma, *Anal. Chem.* **61**, 2189 (1989).

208. W. Bussmann, J.-M. Lehn, U. Oesch, P. Plummeré, and W. Simon, *Helv. Chim. Acta* **64**, 657 (1981).

209. H. Tsukube and H. Sohmiya, *J. Org. Chem.* **56**, 875 (1991).

210. K. Maruyama, H. Sohmia, and H. Tsukube, *Tetrahedron* **48**, 805 (1992).

211. F. Bedioui, *Coord. Chem. Rev.*, **144**, 39 (1995).

212. J. Koryta, *Anal. Chim. Acta* **223**, 1 (1990).

213. R. L. Solsky, *Anal. Chem.* **62**, 21R (1990).

214. J. Janata, *Anal. Chem.* **62**, 33R (1990).

215. M. E. Collison and M. E. Meyerhoff, *Anal. Chem.* **62**, 425A (1990).

216. E. Pungor, E. Lindner, and K. Tóth, *Fresenius J. Anal. Chem.* **337**, 503 (1990).

217. J. Janata, *Chem. Rev.* **90**, 691 (1990).

218. D. Ammann, W. E. Morf, P. Anker, P. C. Meier, E. Pretsch, and W. Simon, *Ion-Selective Electrode Rev.* **5**, 3 (1983).

219. T. Shono, *Bunseki Kagaku* **33**, E449 (1984).

220. The selective recognition of heavy metal ions by calixarene-based sensors has been

accomplished: P. L. H. M. Cobben, R. J. M. Egbering, J. G. Bomer, P. Bergveld, W. Verboom, and D. N. Reinhoudt, *J. Am. Chem. Soc.* **114**, 10573 (1992).
221. F. P. Schmidtchen, *Nachr. Chem. Tech. Lab.* **36**, 8 (1988).
222. (a) F. Hofmeister, *Arch. Ex. Pathol. Pharmakol.* **24**, 247 (1988); (b) U. Wuthier, H. Pham, E. Pretsch, D. Ammann, A. Beck, D. Seebach, and W. Simon, *Helv. Chim. Acta* **68**, 1822 (1985).
223. A. Hodinár and A. Jyo, *Chem. Lett.*, 993 (1988).
224. S. Glazier and M. Arnold, *Anal. Chem.* **60**, 2540 (1988).
225. S. Daunert and L. Bachas, *Anal. Chem.* **61**, 499 (1989).
226. D. Ammann, M. Huser, B. Krautler, B. Rusterholtz, P. Schulthes, P. Lindemann, E. Halder, and W. Simon, *Helv. Chim. Acta* **69**, 849 (1986).
227. N. Chaniotakis, S. B. Park, and M. Meyerhoff, *Anal. Chem.* **61**, 566 (1989).
228. N. Chaniotakis, A. Chasser, M. Meyerhoff, and J. Groves, *Anal. Chem.* **60**, 185 (1988).
229. D. V. Brown, N. A. Chaniotakis, I. H. Lee, S. C. Ma, S. B. Park, M. E. Meyerhoff, R J. Nick, and J. T. Groves, *Electroanalysis* **1**, 477 (1989).
230. W. E. Morf, *The Principles of Ion-Selective Electrodes and Membrane Transport*, Elsevier, Amsterdam, 1981.
231. S. B. Park, W. Matuszewski, M. E. Meyerhoff, Y. H. Liu, and K. M. Kadish, *Electroanalysis* **3**, 909 (1991).
232. S. Oka, Y. Sibazaki, and S. Tahara, *Anal. Chem.* **53**, 588 (1981).
233. K. Hartman, S. Luterotti, H. Osswald, M. Oehme, P. Meier, D. Ammann, and W. Simon, *Microchimica Acta II*, 235 (1978).
234. J. Willis, C. Young, R. Martin, P. Stearns, M. Pelosi, and D. Magnanti, *Clin. Chem.* **29**, 1193 (1983).
235. C. V. Coetzee and H. Freiser, *Anal. Chem.* **41**, 1128 (1969).
236. R. Lewandowski, T. Sokalski, and A. Hulanicki, *Clin. Chem.* **35**, 2146 (1989).
237. M. Bodansky and M. Levy, *Arch. Int. Med.* **31**, 373 (1973).
238. P. Pranitis and A. Stolman, *J. Forensic Sci.* **17**, 148 (1970).
239. N. Haley, C. Axelrad, and K. Tilton, *Am. J. Public Health* **73**, 1204 (2983).
240. A. Pettigrew and G. Fell, *Clin. Chem.* **18**, 996 (1972).
241. J. Rea, P. Tyre, H. Kosap, and S. Beresford, *Br. J. Prev. Soc. Med.* **27**, 114 (1973).
242. J. Langone, H. Gjika, and H. Van Vanakis, *Biochemistry* **12**, 5025 (1973).
243. R. Bowler, *Biochem. J.* **38**, 385 (1944).
244. E. Dalferes, L. Webber, B. Rhadhakrishnamurthy, and G. Berenson, *Clin. Chem.* **26**, 493 (1980).
245. U. Wuthier, H. Pham, R. Zünd, D. Welti, R. Funk, A. Bezegh, D. Ammann, E. Pretsch, and W. Simon, *Anal. Chem.* **56**, 535 (1984).
246. (a) S. A. Glazier and M. A. Arnold, *Anal. Chem.* **63**, 754 (1991); (b) H. V. Pham, E. Pretsch, K. Fluri, A. Bezegh, and W. Simon, *Helv. Chim. Acta* **73**, 1894 (1990).
247. M. Rothmaier and W. Simon, *Analytica Chimica Acta* **271**, 135 (1993).
248. D. M. Kliza and M. E. Meyerhoff, *Electroanalysis* **4**, 841 (1992).

249. E. Kimura, in *Crown Ethers and Analogous Compounds: Studies in Organic Chemistry, Vol. 45*, M. Hiraoka, Ed., Elsevier, Tokyo, 1992, pp. 381–478.

250. J.-M. Lehn, *Angew. Chem. Int. Ed. Engl.* **27**, 89 (1988).

251. (a) M. W. Hosseini and J.-M. Lehn, *Helv. Chim. Acta* **70**, 1312 (1987); (b) M. W. Hosseini and J.-M. Lehn, *Helv. Chim. Acta.* **71**, 749 (1988).

252. Y. Umezawa, M. Kataoka, W. Takami, E. Kimura, T. Koike, and H. Nada, *Anal. Chem.* **60**, 2392 (1988).

253. Y. Umezawa, M. Sugawara, M. Kataoka, and K. Odashima, *Ion-Selective Electrodes., Vol. 5*, Pundor, Budapest, 1989, pp. 211–34.

254. K. Tohda, R. Naganawa, X. M. Lin, M. Tange, K. Umezawa, K. Odashima, Y. Umezawa, H. Furuta, and J. L. Sessler, *Sensors and Actuators B* **13–14**, 669 (1993).

255. (a) A. Ohki, M. Yamura, S. Kumamoto, and S. Maeda, *Chem. Lett.*, 95 (1989).

256. V. J. Wotring, D. M. Johnson, and L. G. Bachas, *Anal. Chem.* **62**, 1506 (1990).

257. K. Tohda, M. Tange, K. Odashima, Y. Umezawa, H. Furuta, and J. L. Sessler, *Anal. Chem.* **64**, 960 (1992).

258. U. Umezawa and J. L. Sessler, unpublished results.

259. *Handbook of Conducting Polymers*, T. A. Skotheim, Ed., Marcel Dekker, New York, 1986.

260. A. Ivaska, *Electroanalysis* **3**, 247 (1991).

261. M. Imisides, R. John, P. Riley, and G. G. Wallace, *Electroanalysis* **3**, 879 (1991).

262. M. Umana and J. Waller, *Anal. Chem.* **58**, 2979 (1986).

263. M. Meaney, M. R. Smith, J. G. Vos, and G. G. Wallace, *Electroanalysis* **1**, 357 (1989).

264. S. Dong, Z. Sun, and Z. Lu, *J. Chem. Soc. Chem. Commun.*, 993 (1988).

265. Z. Lu, Z. Sun, and S. Dong, *Electroanalysis* **1**, 271 (1989).

266. S. Daunert, S. Wallace, A. Florido, and L. Bachas, *Anal. Chem.* **63**, 1676 (1991).

267. A. Karagozler, O. Ataman, A. Gala, Z. Xue, H. Zimmer, and H. B. Mark, *Anal. Chim. Acta* **248**, 163 (1991).

268. M. Sugawara, M. Kataoka, Y. Odashima, and Y. Umezawa, *Thin Solid Films* **180**, 129 (1989).

269. M. Sugawara, K. Kojima, H. Sazawa, and U. Umezawa, *Anal. Chem.* **59**, 2842 (1987).

270. (a) Y. Kawabata, M. Matsumoto, H. Tanaka, H. Takahashi, Y. Iranatsu, S. Tamura, W. Tagaki, H. Nakahara, and H. Fukuda, *Chem. Lett.*, 1933 (1986); (b) H. Niino, A. Yabe, A. Ouchi, M. Tanaka, Y. Kawabata, S. Tamura, T. Miyasaka, W. Tagaki, H. Nakahara, and K. Fukuda, *Chem. Lett.*, 1227 (1988); (c) S. Taneva, K. Ariga, Y. Okahata, and W. Tagaki, *Langmuir* **5**, 111 (1989).

271. H. Lee, L. J. Kepley, H. G. Hong, and T. E. Mallouk, *J. Am. Chem. Soc.* **110**, 618 (1988).

272. D. L. Allara, S. V. Atre, C. A. Elliger, and R. G. Snyder, *J. Am. Chem. Soc.* **113**, 1852 (1991).

273. D. Li, B. I. Swanson, J. M. Robinson, and M. A. Hoffbauer, *J. Am. Chem. Soc.* **115**, 6975 (1993).
274. I. Willner, R. Blonder, and A. Dagan, *J. Am. Chem. Soc.* **116**, 9365 (1994).
275. J. K. Whitesell and H. K. Chang, *Science* **261**, 73 (1993).
276. J. K. Whitesell, H. K. Chang, and C. S. Whitesell, *Angew. Chem. Int. Ed. Engl.* **33**, 871 (1994).
277. D. J. Pritchard, H. Morgan, and J. M. Cooper, *Angew. Chem. Int. Ed. Engl.* **34**, 91 (1995).
278. J. Kim, R. M. Crooks, M. Tsen, and L. Sun, *J. Am. Chem. Soc.* **117**, 3963 (1995).
279. L. H. Dubois and R. G. Nuzzo, *Annu. Rev. Phys. Chem.* **43**, 437 (1992).
280. R. Maoz and J. Sagiv, *J. Colloid, Interface Sci.* **100**, 465 (1989).
281. D. Li, M. A. Ratner, T. J. Marks, C. Zhang, J. Yang, and C. G. Wang, *J. Am. Chem. Soc.* **112**, 7389 (1990).
282. A. Ulman, *Ultrathin Organic Films*, Academic Press, Boston, 1991, and references therein.
283. *Supramolecular Architecture; Synthetic Control in Thin Films and Solids*, T. Bein, Ed. ACS Symposium Series 499, American Chemical Society, Washington DC, 1992.
284. C. D. Gutsche, *Calixarenes, Monographs in Supramolecular Chemistry*, J. F. Stoddart, Ed., The Royal Society of Chemistry, London, 1989.
285. B.-H. Huisman, E. U. Thoden van Velzen, F. C. J. M. van Veggel, J. F. J. Engbersen, and D. N. Reinhoudt, *Tetrahedron Lett.* **36**, 3273 (1995).
286. P. Timmerman, W. Verboom, F. C. J. M. van Veggel, J. P. M. van Duynhoven, and D. N. Reinhoudt, *Angew. Chem. Int. Ed. Engl.* **33**, 2345 (1994).
287. E. U. Thoden van Velzen, J. F. J. Engbersen, and D. N. Reinhoudt, *J. Am. Chem. Soc.* **116**, 3597 (1994).
288. E. Van Dienst, W. I. Iwema Bakker, J. F. Engbersen, W. Verboom, and D. N. Reinhoudt, *Pure Appl. Chem.* **65**, 387 (1993).
289. K.-D. Schierbaum, T. Weib, E. U. Thoden van Welzen, J. F. J. Engbersen, D. N. Reinhoudt, and W. Göpel, *Science* **265**, 1413 (1994).
290. H. Adams, F. Davis, and C. J. M. Stirling, *J. Chem. Soc. Chem. Commun.*, 2527 (1994).
291. L. Zhang, L. A. Godinez, T. Lu, G. W. Gokel, and A. E. Kaifer, *Angew. Chem. Int. Ed. Engl.* **34**, 253 (1995).
292. G. K. Rowe and S. E. Creager, *Langmuir* **7**, 2307 (1991).
293. L. Zhang, A. Macias, T. Lu, J. I. Gordon, G. W. Gokel, and A. E. Kaifer, *J. Chem. Soc. Chem. Commun.*, 1017 (1993).
294. K. Motesharei and D. C. Myles, *J. Am. Chem. Soc.* **116**, 7413 (1994).
295. S.-K. Chang and A. D. Hamilton, *J. Am. Chem. Soc.* **110**, 1318 (1988).
296. T. R. Tosteson, in *Biotechnology of Marine Polysaccharides*, R. R. Colwell, E. R. Parisen, and A. J. Sinskey, Eds., Hemisphere, New York, 1985, p. 101 and references therein.
297. M. T. Rojas, R. Koniger, J. F. Stoddart, and A. E. Kaifer, *J. Am. Chem. Soc.* **117**, 336 (1995).

298. T. Akiyama, H. Imahori, and Y. Sakata, *Chem. Lett.*, 1447 (1994).
299. *Emerging Technologies in Hazardous Waste Management*, D. W. Tedder and F. G. Pohland, Eds., ACS Symposium Series 422, Washington, DC, 1990, and references therein.
300. P. H. Abelson, *Science* **255**, 901 (1992).
301. *Environmental Remediation. Removing Organic and Metal Ion Pollutants*, G. F. Vandegrift, D. T. Reed, and I. R. Tasker, Eds., ACS Symposium Series 509, Washington, DC, 1992.
302. R. Chiarizia, E. P. Horwitz, and K. M. Hodgson, *Environmental Remediation. Removing Organic and Metal Ion Pollutants*, G. F. Vandegrift, D. T. Reed, and I. R. Tasker, Eds., ACS Symposium Series 509, Washington, DC, 1992, pp. 27–33.
303. S. D. Faust and O. M. Aly, *Chemisry of Water Treatment*, Butterworths, Stoneham, MA, 1983.
304. S. D. Faust and O. M. Aly, *Chemistry of Natural Waters*, Butterworths, Stoneham, MA, 1981.
305. J. Nomura, H. Imai, and T. Miyake, in *Emerging Technologies in Hazardous Waste Management*, D. W. Tedder and F. G. Pohland, Eds., ACS Symposium Series 422, Washington, DC, 1990.
306. *The Merck Index*, S. Budavari, Ed., Merck & Co., Inc., Rahway, New Jersey, 1989, p. 126.
307. S. D. Faust and O. M. Aly, *Adsorption Processes for Water Treatment*, Butterworths, Stoneham, MA, 1987, p. 288.
308. S. D. Alexandros, D. W. Crick, D. R. Quilen, and C. E. Grady, in *Environmental Remediation. Removing Organic and Metal Ion Pollutants*, G. F. Vandegrift, D. T. Reed, and I. R. Tasker, Eds., ACS Symposium Series 509, Washington, DC, 1992, p. 295.
309. S. D. Alexandros, D. W. Crick, D. R. Quilen, and C. E. Grady, in *Environmental Remediation Removing Organic and Metal Ion Pollutants*, G. F. Vandegrift, D. T. Reed, and I. R. Tasker, Eds., ACS Symposium Series 509, Washington, DC, 1992, p. 244.
310. For a review concerning crown ether synthesis, see: L. F. Lindoy, *The Chemistry of Macrocyclic Ligand Complexes*, Cambridge University Press, Cambridge, 1989.
311. D. H. Busch and N. A. Stephenson, *Coord. Chem. Rev.* **100**, 119 (1990). See also Ref. 310.
312. See, for example: (a) C. O. Dietrich-Büchecker, and J.-P. Sauvage, *Tetrahedron* **46**, 503 (1990); (b) P. R. Ashton, T. T. Goodnow, A. E. Kaifer, M. V. Reddington, A. M. Z. Slawin, N. Spencer, J. F. Stoddart, C. Vincent, and D. J. Williams, *Angew. Chem. Int. Ed. Engl.* **28**, 1396 (1989).
313. S. L. Lawton and W. J. Rohrbaugh, *Science* **247**, 1319 (1990).
314. K. J. Shea and D. Y. Sasaki, *J. Am. Chem. Soc.* **111**, 3442 (1989).
315. H. L. Anderson and J. K. M. Sanders, *Angew. Chem. Int. Ed. Engl.* **29**, 1400 (1990), and references therein.
316. (a) T. R. Kelly, C. Zhao, and G. J. Bridger, *J. Am. Chem. Soc.* **111**, 3744 (1989); (b) M. Hattori, H. Nakagawa, and M. Kinoshita, *Macromol. Chem.* **181**, 2325 (1980).

317. (a) T. Tijivikua, P. Ballester, and J. Rebek, Jr., *J. Am. Chem. Soc.* **112**, 1249 (1990); (b) J. S. Nowick, Q. Feng, T. Tijivikua, P. Ballester, and J. Rebek, Jr., *J. Am. Chem. Soc.* **113**, 8831 (1991). For more discussion on this topic, see: (c) F. M. Menger, A. V. Eliseev, N. A. Khanjin, *J. Am. Chem. Soc.* **116**, 3613 (1994); (d) M. M. Conn, E. A. Winter, and J. Rebek, Jr., *J. Am. Chem. Soc.* **116**, 8823 (1994); (e) F. M. Menger, A. V. Eliseev, N. A. Khanjin, and M. J. Sherrod, *J. Org. Chem.* **60**, 2870 (1995).

318. For a review, see: S. Anderson, H. L. Anderson and J. K. M. Sanders, *Acc. Chem. Res.* **26**, 469 (1993).

319. (a) Z. Zheng, X. Yang, C. B. Knobler, M. F. Hawthorne, *J. Am. Chem. Soc.* **115**, 5320 (1993); (b) Z. Zheng, C. B. Knobler, and M. F. Hawthorne, *J. Am. Chem. Soc.* **117**, 5105 (1995).

320. (a) A. Müller, R. Rohlfing, E. Krickemeyer, M. Penk, and H. Bogge, *Angew. Chem., Int. Ed. Engl.* **32**, 909 (1993); (b) A. Müller, K. Hovemeier, and R. Rohlfing, *Angew. Chem., Int. Ed. Engl.* **31**, 1192 (1992).

321. J. Salta, Q. Chen, Y. Chang, and J. Zubieta, *Angew. Chem., Int. Ed. Engl.* **33**, 757 (1994).

322. A. Müller, *Nature* **352**, 115 (1991).

323. J. L. Sessler, T. D. Mody, and V. Lynch, *Inorg. Chem.* **31**, 529 (1992).

324. (a) R. Arshady and K. Mosbach, *Macromol. Chem.* **182**, 687 (1981); (b) T. Rosatzin, L. I. Andersson, W. Simon, and K. Mosbach, *J. Chem. Soc. Perkin Trans. 2*, 1261 (1991); (c) B. Sellergen, M. Lepistö, and K. Mosbach, *J. Am. Chem. Soc.* **110**, 5853 (1988); (d) O. Nörrlow, M. Glad, and K. Mosbach, *J. Chromatog.* **299**, 29 (1984); (e) L. Fischer, R. Müller, B. Ekberg, and K. Mosbach, *J. Am. Chem. Soc.* **113**, 9358 (1991); (f) D. K. Robinson and K. Mosbach, *J. Chem. Soc., Chem. Commun.*, 969 (1989). See also Ref. 106.

325. R. Breslow, R. Rajagopalan, and J. Schwarts, *J. Am. Chem. Soc.* **103**, 2905 (1981).

CHAPTER 11

Supramolecular Catalysis of Phosphoryl Anion Transfer Processes

MIR WAIS HOSSEINI

11.1 Introduction
11.2 Molecular Recognition of Nucleotides
    11.2.1 Binding Stoichiometry
    11.2.2 Binding Stability
    11.2.3 Structure of the Complex
    11.2.4 Selectivity Enhancement
    11.2.5 Polymer-Supported Nucleotides Receptor Molecules
    11.2.6 Receptor Molecules for Extraction of Nucleotides
11.3 From Recognition to Transformation
    11.3.1 Dephosphorylation Processes (P–O–P Bond Breaking)
    11.3.2 Phosphorylation Processes (P–O–P Bond Formation)
11.4 Conclusion
11.5 References

11.1 INTRODUCTION

Molecular recognition, supramolecular catalysis, and reactivity, as well as transport processes, represent the basic functional features of supramolecular chemistry.[1–17] Molecular recognition of a substrate results from the readout of specific information concerning the substrate to be bound. This information must be stored at the molecular level within the structure of the receptor molecule.

Supramolecular catalysis, the chemical transformation of the bound substrate, involves first a binding step for which molecular recognition is a prerequisite, followed by the transformation of the complexed species, and

---

Supramolecular Chemistry of Anions, Edited by Antonio Bianchi, Kristin Bowman-James, and Enrique García-España.
ISBN 0-471-18622-8. © 1997 Wiley-VCH, Inc.

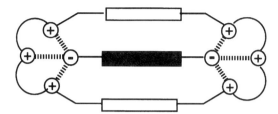

Fig. 11.1. Electrostatic charge–charge interactions between positively charged ammonium binding sites of the receptor and the negatively charged centers of the substrate.

finally the release of the product with regeneration of the catalytic unit.[3–12] The design of selective molecular catalysts may reveal important factors contributing to enzymatic catalysis and may provide useful synthetic tools for specific chemical transformations.

Supramolecular reactivity, the modification of chemical and/or physical properties of the bound substrate, results from the formation of a receptor–substrate complex, and may be tuned by changing the structure of the receptor molecule. The design of specific complexing agents may generate specific chemical and physical reactivities leading to new and/or controlled properties.[1,4,18]

Anions play important roles in both chemical and biochemical processes.[19] Anion coordination chemistry, the binding of anions by organic receptor molecules, has recently been recognized and developed as a new area of coordination chemistry.[20–22] Complexation of anions by synthetic macro-(poly)cyclicpolyammonium molecules has been explored over the past decade.[17,22,50] These preorganized organic polycations form stable and selective complexes by electrostatic attractive charge–charge interactions and H bonding (Fig. 11.1) with a variety of inorganic as well as organic anions.[17,22–50]

11.2 MOLECULAR RECOGNITION OF NUCLEOTIDES

Among the biologically important anions are nucleotide polyphosphates; in particular adenosine mono-, di-, and triphosphate (Fig. 11.2) are basic components in the bioenergetics of all living organisms, the center for chemical energy storage and transfer being their polyphosphate chains.

Macrocyclic polyamines, when protonated, bind strongly and selectively to nucleotides via electrostatic interactions and hydrogen bonding between the cationic binding sites (ammonium groups) of the receptor and the negatively charged polyphosphate groups of the substrate.[22,27,31,45,48,49] Among various polyamines investigated, the ditopic macrocyclic hexaamine [24]N_6O_2 (**11.1**) (Fig. 11.3) was found to bind nucleotide polyphosphates strongly.[22,45]

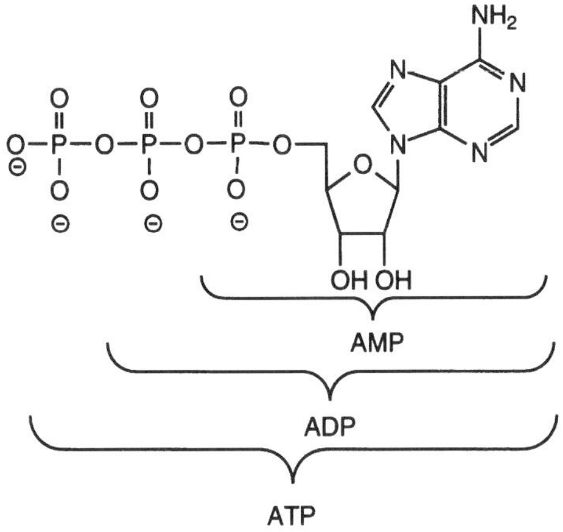

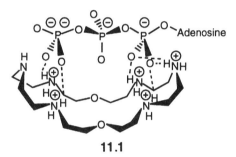

Fig. 11.2. Structures of nucleotide polyphosphate.

Fig. 11.3. Hypothetical structure of (**11.1**-5H$^+$, ATP) complex.

11.2.1 Binding Stoichiometry

Knowing the number of receptors and substrates involved in the complexes is vital for understanding both the complexation processes and the supramolecular catalysis when it takes place. For binding studies involving nucleotides and polyphosphates, ^{31}P-NMR spectroscopy may be a powerful observation technique, since complexation of phosphate-containing molecules by protonated polyamines induces shifts in the ^{13}P-NMR signals. In the presence of polyammonium receptors, the $P(\alpha)$ signal of ATP is usually almost unaffected, whereas the $P(\beta)$ and particularly the terminal phosphate $[P(\gamma)]$ signals are considerably shifted (Fig. 11.4).[22,45]

The plot of [receptor]/[substrate] ratio (R) versus the difference in the chemical shifts (Δ in ppm) between the complexed and free substrate leads to titration curves, which allow the examination of the stoichiometry of the complex. The stoichiometry of a variety of binary nucleotide–polyammonium

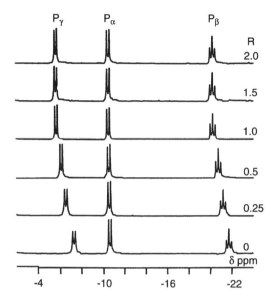

Fig. 11.4. Titration of ATP by protonated **11.1** in aqueous solution at pH 7 followed by ^{31}P-NMR spectroscopy.

complexes has been studied using ^{31}P-NMR spectroscopy. The result obtained for spermine, a natural tetraamine, was in agreement with the results published using other methods.[51,52]

In the presence of Ca^{2+} ions, at pH 7.6 compound **11.1** forms the ternary complexes (**11.1**, ATP, Ca^{2+}) with 1:1:1 **11.1**:ATP:Ca^{2+} stoichiometry.[69,70] The stoichiometry of ATP binding by the analogue of **11.1** bearing an acridine moiety, **11.2** (Fig. 11.5) was established to be 1:1 using fluorescence emission spectroscopy.[53]

11.2.2 Binding Stability

Polyamines form many types of complexes with nucleotides in solution, mainly differing by the number (n) of H^+ involved in the complexation processes.

The stability constants for ADP and ATP complexation by protonated **11.1**-nH^+ were determined using pH-metric measurements (Fig. 11.6).[22,45] These complexes are strong in aqueous solution; a value up to 10^{11} is obtained for (**11.1**-$6H^+$, ATP^{4-}) complex. Since the interactions involved in the binding processes are mainly electrostatic, usually the most stable complexes are formed between the most highly charged species. Thus increasing the number of ammonium sites on the receptor by decreasing the pH of the solution, or increasing the total negative charges on the substrate, leads to the formation of the most stable complexes: for example, see in Fig. 11.6 (**11.1**-$6H^+$, ATP^{4-}) vs. (**11.1**-$5H^+$, ATP^{4-}) or (**11.1**-$6H^+$, $HATP^{3-}$).

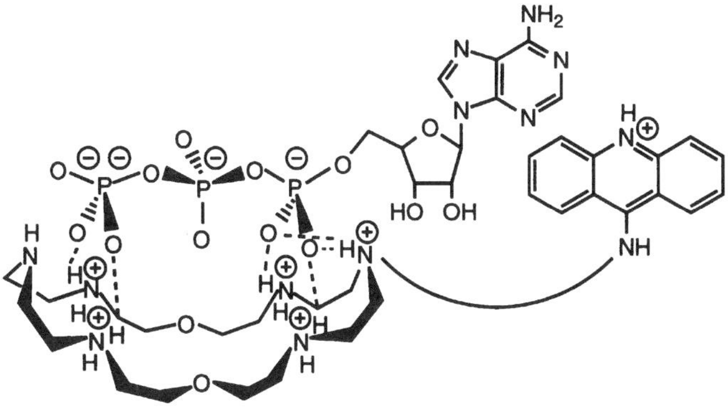

MOLECULAR RECOGNITION OF NUCLEOTIDES

Fig. 11.5. Schematic representation of ATP binding by **11.2** through electrostatic interactions between charged centers and stacking between the acridine moiety of the receptor and adenine.

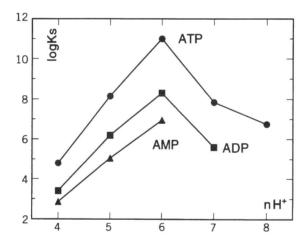

Fig. 11.6. Stability constants of ADP and ATP complexes with protonated **11.1** in aqueous solution; n is the number of H^+ involved in the complexation pattern.

11.2.3 Structure of the Complex

Although solution thermodynamics and stoichiometry determination show that **11.1**-nH^+ forms stable and selective 1 : 1 complexes with ATP and ADP, they do not allow one to define the geometry of these complexes. However, molecular models indicate that the complexes detected are at least compatible with the size and shapes of the macrocycle and of the nucleotide. A solid-state analysis of (**11.1**-$6H^+$, $6Cl^-$) salt showed that among the six Cl^- anions present, the fully

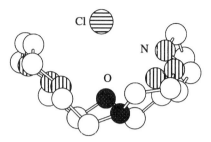

Fig. 11.7. Solid-state structure of Cl^- binding by **11.1**-$6H^+$.

protonated **11.1**-$6H^+$ forms an inclusion complex with one of them, the Cl^- being located almost in the center of the macrocycle (Fig. 11.7).[50] It is worth noticing that the overall "cup"-type shape of the protonated macrocycle resembles the hypothetical structure drawn in Fig. 11.7. A computer modeling study using molecular dynamics revealed that the "cup"-type conformation of the macrocycle may also be the preferred one in aqueous solution.[50]

11.2.4 Selectivity Enhancement

Compound **11.1** contains only positively charged binding sites capable of interaction with the polyphosphate chains of the substrates. In order to reach better recognition of nucleotides, one may, in addition to the anion binding sites, introduce other sites capable of interactions with the sugar moiety and/or the nucleic base of nucleotides. Interactions with the nucleic base may be achieved either by stacking or by sites capable of forming complementary hydrogen-bonding patterns.[55–59] In order to obtain a receptor molecule that would interact simultaneously with both the polyphosphate chain of the nucleotide and their nucleic base moiety, an acridine derivative, known to associate with the nucleic bases by stacking, was covalently attached via a side arm to the macrocycle **11.1**, yielding compound **11.2** (Fig. 11.5).[53] The multi-functional receptor molecule **11.2** has been designed and synthesized in order to achieve higher molecular recognition and reaction selectivity via multiple interaction with bound substrates. It combines three functional subunits: two recognition sites—a macrocyclic polyammonium moiety as anion binding site and an acridine side chain for stacking interactions—as well as a catalytic amino group in the macrocycle for facilitating hydrolytic reactions. Compound **11.2** binds mono- and dinucleotide polyphosphates by simultaneous interactions between its macrocyclic polycationic moiety and the polyphosphate chain, as demonstrated by ^{31}P-NMR spectroscopy, and by stacking between its acridine derivative and the nucleic base of nucleotides, as observed by both ^{1}H-NMR spectroscopy and by fluorescence spectrophotometry. Compound **11.2** also binds strongly to DNA plasmid pBR 322 at 10^{-6} M, probably via a double type of interaction, involving both intercalation and electrostatic interactions with the phosphate groups.

11.2.5 Polymer-Supported Nucleotides Receptor Molecules

As discussed above, amongst various macro(poly)cyclic polyamines examined for their complexation of nucleotides in aqueous solution, the protonated hexaaza macrocycle **11.1** has been shown to bind strongly AMP, ADP, ATP, and polyphosphates such as pyro and triphosphate with well-defined 1:1 stoichiometry. The stability constants determined for the binding of ADP and ATP by protonated **11.1** were the highest reported. Moreover, the binding ability of compound **11.1** could be controlled by varying the pH of the solution, that is by controlling the number and the location of protonated amino groups on the receptor and the ionic phosphate moiety of the substrate. Therefore, this rather versatile macrocycle was an attractive candidate for covalent attachment to a polymeric support yielding a nucleotide receptor polymer.[54] This polymer supported receptor (Fig. 11.8) should, after the uptake of the substrate at appropriate pH, release it back into solution under alkaline conditions and thus be regenerated.

Spectrophotometric methods are usually well suited for measuring low concentrations in solution. Since ADP and ATP are both weakly fluorescent, their 1, N^6-etheno derivatives (εADP, εATP)[61] which show strong emission at 410 nm when excited at 300 nm were used to model ADP and ATP in uptake experiments.

In solution at pH 4, the polymer-supported macrocycle was shown to uptake nucleotide polyphosphates such as adenosine di- and triphosphate. The uptake is due to the binding of the negatively charged polyphosphate chain of the nucleotide by the polyprotonated macrocyclic part of the polymer. Kinetic studies indicated that maximum uptake is reached within an immersion period of 1–3 min of the functionalized polystyrene beads in the solution containing the nucleotide. At pH 11, the unprotonated polymer releases the bound nucleotide within 30–60 sec. Competition experiments between ADP and ATP revealed that the modified polystyrene balls bind ATP stronger than ADP. Both the parent macrocycle **11.1** and the polymer-supported macrocycle shows similar affinity for ATP.

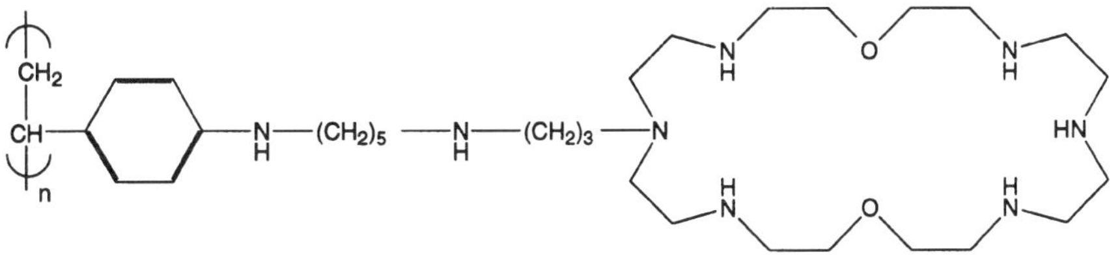

Fig. 11.8. Structure of the polystyrene supported macrocycle **11.1**

11.2.6 Receptor Molecules for Extraction of Nucleotides

Compound **11.1** and its structural analogues were also functionalized with one, two, or six hexadecyl aliphatic chains (Fig. 11.9).[60]

Extraction studies of nucleotides are under investigation.

11.3 FROM RECOGNITION TO TRANSFORMATION

Complexation and transformation of biologically relevant substrates are particularly important. Although bond-cleavage reactions have been extensively studied,[9–11,13–17] the bond-formation processes, leading to synthetic reactions, still remain a challenge.[12] Indeed, in order to realize "bond-making" rather than

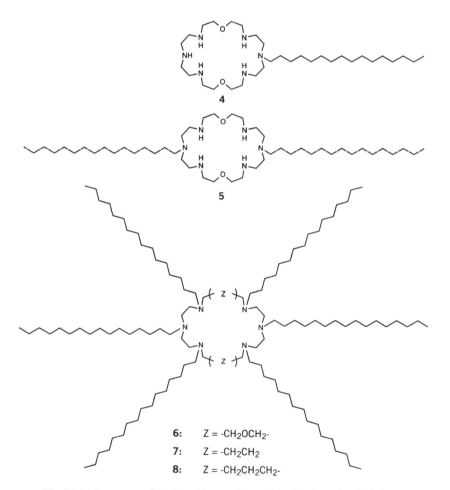

Fig. 11.9. Structure of **11.1** and its analogues bearing hexadecyl chains.

"bond-breaking" processes, the presence of several binding and reactive groups is essential. This may be the case for co-receptor molecules in which subunits may cooperate for binding and transformation of the substrate yielding a co-catalyst.

Phosphoryl transfer processes, which play a fundamental role in the bioenergetics of all living organisms, usually involve anionic species, and can take place in bond formation as well as in bond cleavage reactions.[62,63] In this regard, nucleotide polyphosphates are of special interest, since they undergo both dephosphorylation (bond-breaking) and phosphorylation (bond-making) processes.[62]

As discussed above, protonated macrocyclic polyamines indeed strongly bind nucleotide polyphosphates, in particular, ADP and ATP. Furthermore, they catalyze both their hydrolysis[22,64–73] and synthesis.[73–75,78]

11.3.1 Dephosphorylation Processes (P–O–P Bond Breaking)

Hydrolysis of Adenosine Triphosphate. The hydrolysis of ATP occurs in biological systems via highly efficient and specific enzymes called ATPases. These natural catalysts play important roles in numerous biological processes such as photosynthetic phosphorylation, oxidative phosphorylation, muscle contraction, active transport, etc. There is, therefore, considerable interest in analyzing the controlling factors and the mechanism of these reactions. Furthermore, construction of abiotic catalysts mimicking the natural enzymes may also lead to new catalysts useful in terms of their synthetic applications.

The uncatalyzed hydrolysis of ATP takes place almost exclusively at the terminal phosphate, by the attack of a water molecule, leading to ADP and phosphate (P). The pH profile of the reaction indicates that the P–O–P cleavage is rapid in acidic solution and becomes significantly slower at neutral and alkaline media.[63]

Although biogenic polyamines such as putrescine, spermidine, and spermine bind nucleotides,[51] they have virtually no effect on the rate of ATP hydrolysis.[79,80] Linear polyamines such as pentaethylene hexaamine, on the other hand, produce significant rate enhancements.[49d,79,80] The hydrolytic reaction of nucleotides is found to be catalyzed by a series of macrocyclic polyamines.[22,49,64–72] Among various polyamines investigated, compound **11.1** behaves in a rather interesting manner, and therefore, it was extensively studied. Thus, in the following paragraphs we will mainly focus on compound **11.1**.

The hydrolysis of ATP occurs in the presence of **11.1** with a rate enhancement of ca. 100 under neutral conditions.[64–68] The course of the reaction was monitored by ^{31}P-NMR spectroscopy (Fig. 11.10), which revealed the formation of a transient species identified to be the phosphorylated macrocycle: the phosphoramidate PN (Fig. 11.11).

The observed rate constant k_{obs} for the catalyzed reaction in the presence of **11.1** is almost invariable over a wide range of pH (from 2.5–8.5). In fact, the

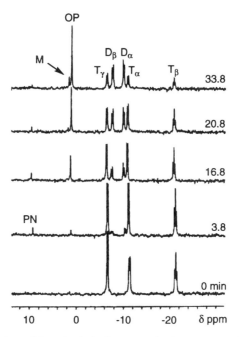

Fig. 11.10. Observation of ATP hydrolysis at pH 7 in the presence of protonated **11.1** by ^{31}P-NMR spectroscopy (see Fig. 11.11 for PN).

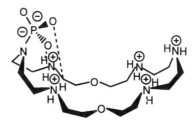

Fig. 11.11. The structure of the phosphoramidate intermediate.

catalyzed reaction at pH 8.5 occurs at almost the same rate as the uncatalyzed reaction at pH 0.5, a difference of 10^8 in H$^+$ concentration.

The hydrolysis of ATP in the presence of **11.1**-nH$^+$ is a true catalytic process. Indeed, it was observed that for the ratio of [ATP]/[**11.1**-nH$^+$] = 10, the change in [ATP] was linear with time in the early period. This point is important since a major limitation, in particular, for the supramolecular catalysis involving cationic substrates, is the product inhibition of the process. In the present case, ATP is more strongly complexed than both ADP and P; hence, no such inhibition takes place over most of the course of the reaction. Based on detailed kinetics and mechanistic studies, at neutral pH, the catalytic cycle shown in Fig. 11.12 is proposed for ATP hydrolysis catalyzed by **11.1**-nH$^+$.[66]

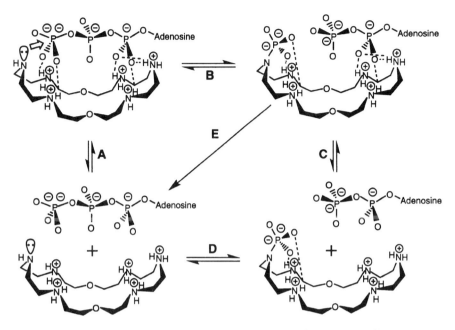

Fig. 11.12. Catalytic cycle for ATP hydrolysis in the presence of **11.1**.

The elementary steps of this cycle are as follows. Step A: formation of (**11.1**-nH^+, ATP^{4-}) complexes. Due to the structure of the macrocycle **11.1**, in particular the distances separating the six amine functionalities, at pH 7 only 4 or 5 of these groups are protonated, yielding the ammonium binding sites; hence, at least one of the central amines remains unprotonated. Step B: due to the presence of this reactive amine (nucleophile) and the proper disposition of the polyphosphate chain within the "pocketlike" cavity of the macrocycle, the terminal phosphate of ATP is positioned close to the nucleophile. The result is an intracomplex phosphoryl transfer from the substrate to the catalyst, yielding the covalent phosphoramidate intermediate (PN). The phosphoramidate PN was isolated and its structure elucidated by NMR experiments (Fig. 11.11).[75]

There are two possible pathways for regenerating the catalyst. Step C: decomplexation of ADP followed by hydrolysis of the reactive phosphoramidate by a water molecule to give P and **11.1**-nH^+ (Step D), or the breakdown of the PN bond by the attack of H_2O followed by the decomplexation of ADP (Step E).

Enzyme Kinetic Analysis. Compound **11.1** behaves truly as an enzymelike catalyst in the dephosphorylation reaction of ATP and ADP.[72] Let us consider the dephosphorylation of ATP to ADP and P catalyzed by **11.1**. The initial velocities (V_{el}) of this reaction were measured and corrected for water hydrolysis, with $k_0 = 1.33 \times 10^{-4}$ min^{-1} according to $V_{el} = V_{obs} - k_0$ [ATP].

The V_{max} and K_m values determined are 0.064 mmole min^{-1} and 11×10^4 M, respectively. The initial velocity of the reaction of excess ATP with **11.1** shows steady-state kinetics confirming the preassociation of the substrate and the catalyst in a reactive Michaelis-type complex.

Although the turnover number of 0.064 min^{-1} is substantially lower than the typical value of 1000 min^{-1} for the natural ATPase, the K_m value is similar to those reported for the enzyme.[77]

Competitive inhibition by triphosphate anion, an analogue of ATP that competes for the same binding site, takes place with a K_i value of 3.3×10^{-4} M and a V_{max} of 0.058 mmole min^{-1}, a value reasonably close to that found for the uninhibited reaction.

A turnover number or K_{cat} ($V_{max}/[$**11.1**$]$) of 0.064 min^{-1} is calculated for the process. Taking into account the difference in the temperature between the **11.1**-catalyzed reaction (70°C) and those catalyzed by ATPase (ca. 37°C), the turnover number obtained for **11.1** is 10^5 times smaller than the average turnover of 3.2×10^4 min^{-1} for ATPase.[77] Alternatively, the apparent second-order rate constant k_{cat}/k_m for the catalyzed reaction of ATP is 6.4×10^2 L mol^{-1} min^{-1}, whereas the corresponding value for a typical ATPase with a k_m of 1×10^{-4} M and a k_{cat} of 3.2×10^4 min^{-1} is 3.2×10^8 L mol^{-1} min^{-1}. While this difference of 10^6 is quite substantial, compound **11.1**, although much less effective than ATPase but also structurally more simpler than the natural enzyme, shows remarkable analogies with ATPase.

Another important factor is the selectivity of catalysts between ADP and ATP. Whereas ATPases usually catalyze only the breakdown of ATP, compound **11.1**-nH$^+$ catalyzes both ATP and ADP hydrolysis, with a selectivity of ca. 3 in favor of ATP. In order to attain higher values, compound **11.2** was prepared and tested.[53,54] This compound behaves in a similar manner as compound **11.1**; in particular, it shows nucleophilic catalysis of the phosphoryl transfer leading to the same type of phosphoramidate intermediate. Moreover, due to the simultaneous binding of the polyphosphate chain and the nucleic base moiety of the substrates, probably leading to a better positioning of ATP, this compound gives a selectivity factor of catalysis of 9 in favor of ATP.

Activation Parameters. The determined energies of activation (E_a) for the macrocycle **11.1**-catalyzed hydrolysis of ATP at pH 3.6 and 7.6 are 22.3 and 23.4 kcal mol^{-1} respectively.[72] The calculated entropies of activation ($\Delta S^{\ddagger}$) are -11 e.u. at lower pH and -8.7 e.u. at pH 7.6. The same study employing acetyl phosphate (AcP) and **11.1** (see below) at pH 7 gave an E_a of 15.7 kcal mol^{-1} for AcP loss, and a calculated entropy of activation of -24.9 e.u. The E_a value for the disappearance of PN intermediate is 16.0 kcal mol^{-1} with a $\Delta S^{\ddagger}$ of -26.4 e.u.[72] The free energies of activation ($\Delta G^{\ddagger}$) for all reactions are in the range of 23–26 kcal mol^{-1}, whereas the enthalpies of activation ($\Delta H^{\ddagger}$) are ca. 2.2 kcal mol^{-1} for ATP hydrolysis at both pH 3.6 and 7.6 and 15 kcal mol^{-1} for AcP and the intermediate PN disappearence.[72] Comparable analysis of the activation parameters has been helpful in distinguishing the major mechanisms of phosphate ester,[81] acyl phosphate,[82] and pyrophosphate[83] hydrolysis. In the present case, the rather negative $\Delta S^{\ddagger}$ values do not support an elimination reaction but seems rather to indicate either nucleophilic attack to give a pentacovalent oxyphosphorane, or a concerted mechanism. Nevertheless, since more than one mechanism may operate (i.e., nucleophilic attack by the centrally located amine and direct attack of water), no definitive conclusions may be drawn.

Associative or Dissociative Mechanism. ATP analogues APPNHP, APCH$_2$PP, and the dinucleotides APPPPA and APPCCl$_2$PPA are interesting mechanistic probes.[84–86] Substrates with labile β,γ-phosphate bridges are hydrolyzed to give APXPY and P ($X = $ O, CH$_2$; $Y = $ O, NH$_2$), where both other compounds are slowly cleaved at the α,β bridge, yielding AMP. In the presence of **11.1** at pH ca. 6 and 70°C, the labile imido-diphosphate APPNHP is cleaved at comparable rate, only three times slower than ATP.[67] The methylene bisphosphonate (APCH$_2$PP), which has a poorer leaving group ability because of its phosphonate oxyanion, is hydrolyzed almost an order of magnitude slower than is ATP.

However, the most striking result is the relative stability to hydrolysis of the (β,γ-difluoromethylene derivative, the dinucleotide AP$_4$A and its β,β'-dichloromethylene analogue.[67] These results suggest that the macrocycle **11.1** preferentially catalyzes the cleavage of terminal phosphate groups. This statement is further supported by the fact that in the course of ATP hydrolysis catalyzed by **11.1**, no pyrophosphate is detected,[64,65] and that the hydrolysis of AP$_4$ in the presence of **11.1** proceeds through ATP, which in its turn undergoes a β,γ cleavage, yielding ADP.[87] The two possible mechanisms that may operate in this exolytic-type activity are associative S_N2P or dissociative S_N1P processes.

Assuming an associative path and a symmetrical binding mode of the nucleotide to the macrocycle **11.1**, which itself implies that both α- and γ-phosphoryl centers may be equally subject to nucleophilic attack by the central amino group of the receptor, two possible origins for the observed selectivity of nucleophilic attack, which occurs only at the γ-terminal position, may be envisaged. (1) Requirement for proton transfer from the attacking amino center to a phosphate oxygen either prior to attack or in the phosphorus trigonal bipyramidal transition state (TBP).[88,89] (2) Existence of some steric hindrance due to the formation of a TBP with an adenosyl group in the equatorial position giving rise to an unfavorable equatorial/apical interaction.[90] Although the first requirement is difficult to gauge, the second one may be ruled out, since the adenosyl group appears remote in this case.

Alternatively, the observed selectivity may be interpreted in terms of a dissociative mechanism. The reaction proceeds through the formation of metaphosphate, which cannot be generated from nonterminal phosphates. An S_N1P mechanism has been proposed, arguing for hydrolysis of phosphate monoester monoanions[91] and of ATP.[92] Although in general there is no convincing evidence for the existence of a metaphosphate intermediate in dilute aqueous solution, a metaphosphate with a lifetime too short to allow diffusion cannot definitely be excluded.[93] Furthermore, in the present case, the amine group is located in close proximity to the incipient metaphosphate within the complex. Thus no important distinction can be made between a true dissociative process and a preassociative mechanism, which requires some involvement for the nitrogen nucleophile in the transition state.

It has also been suggested that binding of the nucleotide to the macrocycle may not be totally symmetrical.[73] Rather the nucleotide may approach from a vertical direction with the γ-phosphate entering the macrocyclic cavity. MM2 calculations in conjunction with docking studies indicated that a complex resulting from such binding could be stabilized by hydrogen bonding interactions between the γ-phosphate and both triamine units of the macrocycle.[73a] Such an interaction could explain the site selectivity of hydrolysis. Cited as further support for this proposal are studies of the influence of macrocyclic size on rates of ATP hydrolysis.[49c,73b] Results indicate a distinct correlation between size and rate.

The Effect of Metal Ions. Biologically relevant metal ions, Ca^{2+}, Mg^{2+}, and Zn^{2+}, strikingly affect the rates of **11.1**-catalyzed dephosphorylation reactions of ADP and ATP.[69,70] Only Ca^{2+} and La^{3+} show distinct rate acceleration; Mg^{2+} does not affect the rate of the reaction, whereas Zn^{2+} and Cd^{2+} cations slow significantly the dephosphorylation process. On the other hand, ADP-catalyzed hydrolysis is slightly accelerated by Ca^{2+} and La^{3+} and retarded by Mg^{2+}. An explanation for this metal-dependent effect may be the formation of ternary macrocycle–metal–nucleotide complexes. The formation of such complexes was suggested by ^{31}P- and ^{13}C-NMR studies,[70] which showed that the stoichiometry of **11.1**-nH^+ : Ca^{2+} : ATP is 1:1:1. In the case of ternary complexes with Ca^{2+}, Mg^{2+}, and La^{3+} cations, the metal ions are probably chelated by ATP complexed within the polyammonium receptor molecule. Chelation of metal cations by ATP is well documented.[94–97] However, Zn^{2+} and Cd^{2+} probably interact strongly with the macrocycle in the ternary complexes and, thus, by competitive inhibition retard the dephosphorylation process. The formation of stable binary complexes between Zn^{2+}, Ni^{2+}, and polyprotonated **11.1** has been reported.[98]

An interesting finding when Mg^{2+}, Ca^{2+}, or La^{3+} was present during the **11.1**-catalyzed dephosphorylation of ATP at pH 7 is that the maximum percentage of the intermediate phosphoramidate PN increases from 17% in the absence of the metal cations to 30, 37, and 47%, respectively, in their presence. Moreover, whereas in the absence of Ca^{2+} the dephosphorylation reaction of ATP in the presence of **11.1** does not generate any pyrophosphate,[64] in its presence a small amount of pyrophosphate (4–6%) is formed. A tentative explanation for this reaction may be the following: Mg^{2+}, Ca^{2+}, and La^{3+} in the ternary complexes with **11.1** and ATP, by chelating to the leaving ADP, renders

ATP in the complex a more powerful phosphorylating reagent. Consequently, the concentration of the intermediate PN increases and the latter phosphorylates the inorganic phosphate present in solution. This follows the mechanism proposed for the **11.1** catalyzed formation of pyrophosphate, when acetylphosphate is used as the phosphoryl donor (see below).[74–76] In addition, since the formation of pyrophosphate has been shown to be inhibited by the addition of anionic species, in particular ADP,[74–76] the removal by chelation with the metal cations of the latter, which competes with incoming phosphate, may be a factor enhancing the rate of pyrophosphate formation. The rate of the disappearance of the intermediate when generated by the reaction of **11.1** with acetylphosphate is retarded from $0.097\,\text{min}^{-1}$, in the absence of Ca^{2+}, to $0.054\,\text{min}^{-1}$ in its presence indicating that the metal ion may stabilize PN and, thus, may enhance the amount of PP formed. Furthermore, a positional isotope exchange study revealed that the formation of ATP resulting from the phosphorylation of ADP by PN takes place in the presence of Ca^{2+} but not in its absence.[68]

Positional Isotope Effect. The results of a positional isotope exchange study[68] show that at pH 7.6 and 70°C in the absence of Ca^{2+}, the ADP formed during the catalyzed hydrolysis of $[\beta\text{-}^{18}O_2]$-ATP[99] by the macrocycle **11.1** does not attack the intermediate phosphoramidate PN supporting the water attack of the latter leading to the starting compound **11.1**. In contrast, in the presence of Ca^{2+} under the same conditions, substantial reaction of ADP with PN takes place.

An explanation as to the effect of Ca^{2+} on the phosphorylation process may be the following: Ca^{2+} has a predilection for oxygen atoms relative to nitrogen ligands. By binding to PN, it stabilizes the intermediate, increases electrostatic catalysis, and, by charge neutralization it lowers the barrier of the attack of PN by ADP. This is in agreement with the fact that during the **11.1**-catalyzed ATP dephosphorylation, the presence of Ca^{2+} causes the accumulation of the intermediate, which leads to the formation of small but significant amount of PP.[69,70] Formation of PP was also reported during AcP hydrolysis in the presence of a high concentration of NaCl (6 M).[93]

Hydrolysis of ATPγS. The observed first-order rate constants for the uncatalyzed hydrolysis of ATPγS at pH 8 and 70°C is $0.014\,\text{min}^{-1}$. In the presence of **11.1** in 100 mM Tris–HCl solution, the reaction is accelerated only by a factor of ca. 3 ($k_{obs} = 0.039\,\text{min}^{-1}$).[68]

A number of factors may be responsible for lowering of the catalytic activity of **11.1** with ATPγS relative to that shown with ATP.[65,66] (1) The presence of 100 mM Tris–HCl buffer may be a cause of the diminution of catalysis, since the **11.1**-catalyzed hydrolysis of ATP at neutral pH is specifically retarded in solution of high ionic strength by chloride ions.[66] (2) The uncatalyzed hydrolysis of ATPγS is an order of magnitude faster than the uncatalyzed hydrolysis of ATP. Compound **11.1** is, thus, less able to accelerate such a fast reaction. The same argument holds for **11.1**-catalyzed AcP hydrolysis, for which an acceleration factor of only 5 is obtained.[74,75] (3) Although the rate of the uncatalyzed hydrolysis of ATPγS is higher than the rate of ATP hydrolysis, the enzyme-catalyzed reactions of phosphorthioate monoesters proceed with a slower rate than phosphate monoester enzyme-catalyzed reactions, indicating that the enzyme-catalyzed reactions do not follow a free energy pathway, favored by mechanisms in which the uncatalyzed reactions proceed with a strong dissociative character.[100,101] If the macrocycle favors associative rather than dissociative mechanisms, then a similar effect may be operating.

Stereochemistry. The stereochemical investigation of the catalyzed hydrolysis of adenosine S'-$[\gamma(R)$-$^{17}O, ^{18}O$, thio]triphosphate[102,103] by **11.1** at pH 8 revealed that the inorganic [$^{16}O, ^{17}O, ^{18}O$] thiophosphate had the (S) configuration,[68] and, hence, like the uncatalyzed hydrolysis,[104] the reaction has proceeded with inversion of configuration at phosphorus.

Since no phosphothioamidate was observed in the ^{31}P-NMR spectrum of the reaction mixture, an adjacent attack on this hypothetical intermediate may be reasonably ruled out, and the simplest interpretation may be that the thiophosphoryl group is directly transferred to water. This indicates clearly that the hydrolysis of nucleotide triphosphate catalyzed by **11.1** proceeds at least by two mechanisms, the first involving nucleophilic catalysis via the intermediate PN, and the second involving direct attack of the terminal phosphate group by a water molecule. An explanation of why ATP and ATPγS behave so differently in their hydrolysis as catalyzed by **11.1** may be due to a switch in the dominant mechanism for the following reasons: (1) Thiophosphoryl groups react more readily by dissociative mechanisms than phosphoryl groups,[101] but less readily by an associative path.[105] (2) The reaction involving nucleophilic catalysis is associative in character and is slowed down on going from phosphoryl to thiophosphoryl groups; thus the competing mechanism of direct thiophosphoryl transfer to water, which proceeds via a dissociativelike mechanism, becomes the dominant route for catalyzed hydrolysis of ATPγS.

Dephosphorylation of Acetyl Phosphate. The hydrolytic reaction of acetyl phosphate (AcP) leading to acetate and phosphate has been extensively studied (Fig. 11.13).[106–114]

This reaction is catalyzed by a series of macrocyclic polyamines.[75,76] In particular, the reaction of AcP in the presence of compound **11.1** is accelerated and produces not only acetate and phosphate but also two other species that were identified to be the intermediate PN, observed during ATP-catalyzed hydrolysis, and pyrophosphate (PP).

Kinetics measurements revealed that the first-order rate constant of AcP disappearance was enhanced by a factor of ca. 5 in the presence of **11.1**-nH$^+$ at pH 7. For the uncatalyzed reaction, the observed rate of hydrolysis was rapid in both strongly acidic and alkaline media, and quite slow and invariable in the intermediate pH range. The **11.1**-nH$^+$-catalyzed reaction proceeds with a maximum rate at pH 7, indicating substantial nucleophilic catalysis.

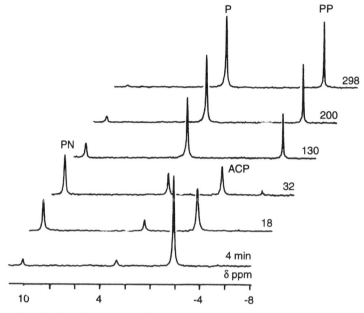

Fig. 11.13. Observation of AcP hydrolysis in the presence of **11.1**.

As for the uncatalyzed hydrolysis, the reaction in the presence of **11.1** takes place with P–O bond cleavage. No amide formation resulting from acylation of the macrocycle by AcP via a C–O bond rupture was observed.

Since in the present case up to 50% (Fig. 11.14) of the starting catalyst was transformed into the intermediate, the latter was isolated and purified, and its

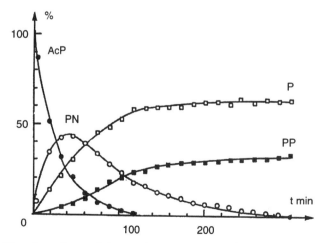

Fig. 11.14. Time dependence of the reaction components of AcP reaction in the presence of protonated **11.1**) in aqueous solution and at pH 7.

structure was elucidated by NMR spectroscopy. Furthermore, it was demonstrated that it was identical to the one generated during the reaction of ATP.

Dephosphorylation of Formyl Phosphate. The hydrolysis of formyl phosphate[115,116] has been studied in detail. The reaction proceeds 45% by C–O bond cleavage and 55% by P–O bond cleavage at 25°C and pH 7 with a rate constant of 8.8×10^{-3} min^{-1}. The entropy of activation $\Delta S^{\ddagger}$ was calculated to be -11.8 e.u., an intermediate value between that expected for a monomolecular and a bimolecular reaction. The rate of the hydrolysis varies little from pH 5 to 8, whereas it increases dramatically at low and high pH values. At extreme pH values 1 and 11, the mechanism involves 100% C–O bond cleavage by nucleophilic substitution at the carbonyl, with entropies of activation $\Delta S^{\ddagger}$ of -21.3 and -25.5 e.u., respectively, as expected for bimolecular reactions. In the presence of nucleophiles such as glycine, formyl phosphate formylates the primary amine with a 100% yield. Imidazole and pyridine also catalyze formyl phosphate breakdown. Although in the case of imidazole, no formylated product was observed, the reaction proceeds with C–O bond cleavage, whereas the reaction with pyridine takes place predominantly by P–O bond cleavage.

The aminolysis of formyl phosphate by linear and macrocyclic polyamines at pH 7 and 5°C proceeds by C–O bond cleavage, yielding the N-formylated amines.[117] In the presence of excess bromide ion, the reaction is reported to be second order with rate constants of 9.8 and 7.1 M^{-1} min^{-1} for **11.1** and the 18-membered hexaaza macrocyclic analog [18]N$_6$, respectively. However, in the absence of excess bromide, the reaction is first order with rate constants of 0.30 min^{-1} for **11.1** and 0.26 min^{-1} for the latter. Surprisingly, the linear tetraamine, 1,4,7,10-tetraazadecane, and the pentaamine, 1,4,7,10,13-pentaazatridecane, are more efficient than the two macrocycles studied.

The activation of carboxylate anions, leading to the formation of amide bonds, is believed to proceed via the formation of enzyme-bound carboxyl phosphate anhydride by enzymes such as glutamine synthetase,[118] glutathione synthetase,[119] and N^{10}-formyltetrahydrofolate synthetase.[120] These enzymes, in the presence of metal ions, usually Mg^{2+}, catalyze the transfer of the terminal phosphoryl group of ATP to the carboxylate. An exciting finding was the activation of formate in neutral aqueous solution by macrocyclic polyammonium cations.[121] Compound **11.1** activates formate in the presence of ATP and calcium(II) or magnesium(II) ions, the reaction product being the formylated macrocycle (Fig. 11.15) and phosphate.

The reaction sequence proposed proceeds via the phosphoramidate PN, which results from the reaction of **11.1** with ATP.[64-66] This intermediate PN, accumulating due to the presence of metal cations,[69,70] is the key species in the reaction and acts as a phosphorylating agent. The reaction of the phosphorylated macrocycle with formate anion leads to the formation of formyl phosphate, as a transient species, which formylates the macrocycle. This complex sequence of steps involving several substrates and bond cleavage as well as formation nicely

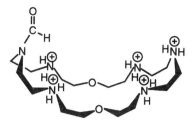

Fig. 11.15. Structure of the formylated macrocycle **11.1**.

patterns the chemistry of N10-formyltetrahydrofolate synthetase.[121] It also attests to the versatility of simple polyammonium macrocycles as models for complex enzymes.

11.3.2 Phosphorylation Processes (P–O–P Bond Formation)

Cocatalysis of Pyrophosphate Synthesis. As discussed above, the hydrolysis of ATP in the presence of protonated polyamines such as macrocycle **11.1** at pH 7 yields a covalent intermediate PN that subsequently undergoes a hydrolytic reaction. The phosphoramidate intermediate may also be considered a phosphorylating reagent, since it transfers its phosphoryl group to a water molecule. This intermediate may also bind another anionic substrate such as a phosphate molecule leading to the (PN,P) complex (Fig. 11.16), using its unmodified diethylenetriammonium subunit.

Due to the location of PN and P, an intracomplex phosphoryl transfer process, from the phosphorylated macrocycle to the precomplexed phosphate (P), may take place within the complex, yielding pyrophosphate (PP). Thus the catalyst **11.1** may act as an organic template, in the sense that it organizes a spatial arrangement of the two negatively charged species (entropic effect). Furthermore, the organic polycation should favor the reaction between them (enthalpic effect) by neutralizing the negative charges.

During ATP hydrolysis only 6–10% of PN was formed,[22,65,66] and so to increase the amount of the intermediate PN, other phosphorylating agents

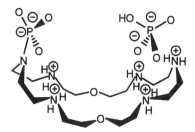

Fig. 11.16. Schematic representation of the phosphoramidate (PN)–phosphate complex leading to the synthesis of pyrophosphate.

440 SUPRAMOLECULAR CATALYSIS OF PHOSPHORYL ANION TRANSFER PROCESSES

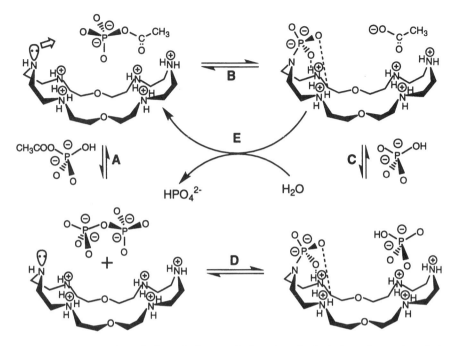

Fig. 11.17. Proposed catalytic cycle for acetyl phosphate hydrolysis and pyrophosphate synthesis catalyzed by **11.1**.

such as acetyl phosphate ($CH_3COOPO_3^{2-}$ = AcP) were tested.[74–76] Whereas the uncatalyzed hydrolysis of AcP yields acetate and phosphate in the presence of **11.1** at pH 7, the same reaction, after generating up to 50% of PN (Fig. 11.14), leads indeed to the formation of PP. The latter was not observed under the same conditions in the absence of **11.1**.

The rate of disappearance of PN was also determined, and the pH profile indicated that, as for simple phosphoramidate, the PN bond was rapidly cleaved at pH values lower than 5, whereas it was rather stable at higher pH.

The formation of PP is dramatically pH dependent. The percentage of PP formed is maximum at pH ca. 7 and decreased substantially at both higher and lower values. At pH 3, although no PN formation was observed, a few percent of PP were detected. At pH 9, where the highest amount of PN was formed, no synthesis of PP occurs. At pH 11 no PN or PP is detected.

Both the formation of PN and PP are under strong structural control. A comparative study of a series of linear macrocyclic and macrobicyclic polyamines reveals that the phosphoryl transfer process is geometrically controlled.

Based on a detailed kinetic and mechanistic investigation, the following cycle for PP formation at pH 7 is proposed (Fig. 11.17). The elementary steps involved in this cycle are: Step A: formation of the anion complex (**11.1**-5H$^+$, AcP^{2-}), followed by Step B: the attack of AcP by the unprotonated amino group of the catalyst leading to the intermediate PN. At this stage, there are two possibilities: Step C, a transcomplexation between AcO^- and HPO_4^{2-}, leading to the complex (PN, HPO_4^{2-}), followed by Step D: an intracomplex phosphoryl transfer from the intermediate to the precomplexed P, yielding PP; or Step E, the hydrolysis of the PN bond regenerating the starting catalyst.

The elementary steps discussed above are similar to those involved in enzymatic phosphorylation processes, and, thus, compound **11.1** may be considered a protokinase.

Cocatalysis of Triphosphate Synthesis. The second phosphoryl transfer leading to the synthesis of PP is generalized to a series of phosphate esters by two different approaches. The first alternative consists of using a mixture of aqueous and organic solvents. Indeed, since the formation of PP is in competition with the hydrolysis of PN, by decreasing the amount of water, the hydrolysis of PN leading to a nonproductive reaction (phosphoryl transfer to water) should substantially be reduced. Furthermore, since going to a less polar medium increases the interactions between charged species, the amount of (PN,P) complex should consequently increase. As predicted, in a mixture of $H_2O:DMSO = 3:7$, not only is the PP formation significantly enhanced, but also the formation of triphosphate (PPP) resulting from the phosphorylation of PP by PN is observed.[76]

Cocatalysis of Nucleotide Synthesis. A further step consists in the synthesis of biologically important pyro- or triphosphate-containing

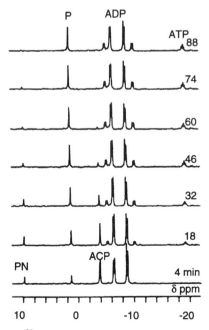

Fig. 11.18. Observation by ^{31}P-NMR of ATP synthesis from ADP mediated by **11.1**.

substrates. The intermediate PN may be used as a specific phosphorylating reagent, leading to the synthesis of ATP from ADP and ADP from AMP. In particular, the reaction of AcP with ADP in the presence of **11.1** at pH 7 and 40°C in a mixture of $H_2O:DMSO = 3:7$ lead to the synthesis of ATP (Fig. 11.18).

In this process, AcP plays the role of cofactor or phosphoryl source, ADP is the acceptor, and finally the catalyst may be considered as a molecular relay between the source (AcP) and the acceptor (ADP) of the phosphoryl group.[12,74–76] It may be noticed that these phosphoryl transfer processes are regio- and chemoselective. Indeed only the terminal phosphate of ADP located in the proximity of the phosphoramidate is phosphorylated. On the other hand, the hydroxy groups on the sugar moiety of the nucleotide, although chemically reactive, do not undergo any reaction. This specificity was used for preparing pyrophosphate derivatives of biological interest (Fig. 11.19).

A second alternative may be the use of a divalent metal (M) cation while performing the phosphoryl transfer processes in water. In this case the reaction takes place within a ternary (PN, acceptor, M) complex (Fig. 11.20).[78] Again, the phosphorylated macrocycle PN acts as an organic template in the sense that it brings in to close proximity its phosphoramidate moiety and the phosphoryl acceptor group.

The metal, on the other hand, bridges two negatively charged species, hence increasing the percentage of the reactive complex (PN, acceptor) and also by neutralizing the negative charges on both the phosphoramidate and the phosphate-containing substrate, thus promoting the phosphoryl transfer from PN to the acceptor. Thus the metal, which may be called promoter or cocatalyst, functions as a metallic template. Using the following system composed of AcP (cofactor or phosphoryl source), ADP (acceptor), M (promoter or cocatalyst)

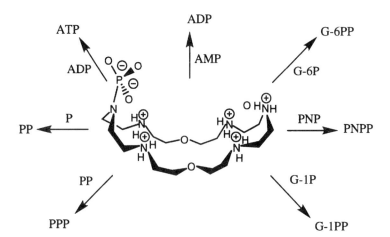

Fig. 11.19. Synthesis of biologically important polyphosphates using the intermediate phosphoramidate as a phosphorylating reagent.

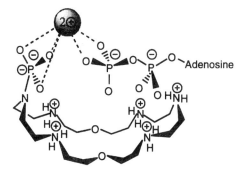

Fig. 11.20. Schematic representation of the ternary complex with divalent metal cations leading to the synthesis of ATP.

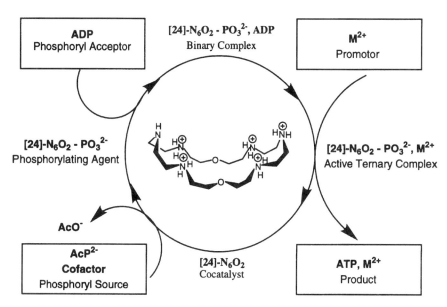

Fig. 11.21. Catalytic cycle for metal ions and **11.1** promoted synthesis of ATP from AcP and ADP.

and **11.1**-nH$^+$ (effector or transferor), ATP is synthesized in aqueous solution at pH 7 and at 40°C (Fig. 11.21.).

Indeed, by using a five-fold excess of AcP, up to 35% of ADP is transformed into ATP in the presence of one equivalent, with respect to **11.1**-nH^{2+} of Mg^{2+} or Ca^{2+}. The system was studied in great detail by varying the concentrations of AcP, ADP, **11.1**, and M^{2+}. The stoichiometry of M^{2+} is found to be dependent on the intermediate but not on AcP or ADP. These results also indicate that the reactive species is indeed the ternary complex.[77]

The above-mentioned ATP generating system has already been coupled to ATP-consuming natural enzymes, leading to the production of NADH.[122]

11.4 CONCLUSION

Protonated macrocyclic polyamines, after binding negatively charged nucleotides, promote by nucleophilic and electrostatic catalysis their hydrolytic reactions. The reaction at physiological pH proceeds through the formation of a covalent phosphoramidate intermediate. The latter may be used as a molecular phosphorylation reagent, leading to a series of biologically important substrates. Furthermore, in the presence of divalent cations such as Mg^{2+} or Ca^{2+}, acting as promoters, the reaction of a phosphorylating reagent (AcP) with nucleotides (AMP, ADP) catalyzed by **11.1**-nH^+ leads to a biomimetic synthesis of ATP.

Since the different steps involved in the catalytic cycle of ATP hydrolysis promoted by **11.1**-nH^+ and its analogues, as well as its synthesis, are rather similar to those involved for natural ATPases and kinases, these compounds may be considered as protoenzymes (proto-ATPase and proto-kinase). These compounds, although rather primitive, do in fact mimic the catalytic activity and features of the naturally occurring enzymes.

Development of other catalysts based on molecular organization of both binding and transformation sites may further lead to other chemical transformations such as amide hydrolysis, acyl transfer processes, sulforyl transfer processes, etc.

11.5 REFERENCES

1. (a) J.-M. Lehn, *Angew. Chem., Int. Ed. Engl.* **27**, 89 (1988); (b) J.-M. Lehn, *Science* **227**, 849 (1985).
2. M. W. Hosseini, in *Bioorganic Chemistry Frontiers*, H. Dugas, Ed., Springer-Verlag, Berlin Heidelberg, 1993, Vol. 3, p. 69.
3. J.-M. Lehn, *Pure Appl. Chem.* **51**, 979 (1979).
4. J.-M. Lehn, in *Design and Synthesis of Organic Molecules Based on Molecular Recognition*, G. Van Binst, Ed., Springer, Berlin, 1986.
5. R. C. Hayward, *Chem. Soc. Rev.* **12**, 285 (1983).
6. J. Rebek, Jr., *Science* **235**, 1478 (1987).
7. J. Rebek, Jr., *Top. Curr. Chem.* **149**, 189 (1988).
8. I. O. Sutherland, *Chem. Soc. Rev.* **15**, 63 (1986).
9. J.-M. Lehn, in *Biomimetic Chemistry*, Z. I. Yoshida, N. Ise, Eds., Kodansha, Tokyo/Elsevier, Amsterdam, 1983, p. 163.
10. J.-M. Lehn, *Ann. N. Y. Acad. Sci.* **471**, 41 (1983).
11. C. Sirlin, *Bull. Soc. Chim. Fr.*, 15 (1984).

REFERENCES   445

12. M. W. Hosseini, *La Recherche* **206**, 24 (1989).
13. R. Breslow, *Science* **218**, 532 (1982).
14. R. M. Kellog, *Top. Curr. Chem.* **101**, 111 (1982).
15. I. Tabushi and K. Yamamura, *Top. Curr. Chem.* **113**, 145 (1983).
16. Y. Murakami, *Top. Curr. Chem.* **115**, 107 (1983).
17. F. P. Schmidtchen, *Top. Curr. Chem.* **132**,101 (1986).
18. V. Balzani, *Supramolecular Photochemistry*, D. Reidel Publishing Co., 1987.
19. J. J. R. Frausto da Silva and R. J. P. Williams, *Struct. Bond.* **29**, 67 (1976).
20. J. L. Pierre and D. Baret, *Bull. Soc. Chim. Fr.*, 367 (1983).
21. F. Vögtle, H. Sieger, and W. H. Müller, *Top. Curr. Chem.* **98**, 143 (1981).
22. M. W. Hosseini, Ph.D. Dissertation, Université Louis Pasteur, Strasbourg, France (1983).
23. H. E. Simmons and C. H. Park, *J. Am. Chem. Soc.* **90**, 2428 (1968).
24. E. Graf and J.-M. Lehn, *J. Am. Chem. Soc.* **98**, 6403 (1976).
25. J.-M. Lehn, E. Sonveaux, and A. K. Willard, *J. Am. Chem. Soc.* **100**, 4914 (1978).
26. B. Dietrich, J. Guilhem, J.-M. Lehn, C. Pascard, and E. Sonveaux, *Helv. Chim. Acta* **67**, 91 (1984).
27. B. Dietrich, M. W. Hosseini, J.-M. Lehn, and R. B. Sessions, *J. Am. Chem. Soc.* **103**, 1282 (1981).
28. M. H. Hosseini and J.-M. Lehn, *J. Am. Chem. Soc.* **104**, 3525 (1982).
29. M. W. Hosseini and J.-M. Lehn, *Helv. Chim. Acta* **69**, 587 (1986).
30. E. Kimura, A. Sakonaka, T. Yatsunami, and M. Kodama, *J. Am. Chem. Soc.* **103**, 3041 (1981).
31. E. Kimura, M. Kodama, and T. Yatsunami, *J. Am. Chem. Soc.* **104**, 3182 (1982).
32. E. Kimura and A. Sakonaka, *J. Am. Chem. Soc.* **104**, 4984 (1982).
33. E. Kimura, *Top. Curr. Chem.* **128**, 113 (1985).
34. J. Cullinane, R. I. Gelb, T. N. Margulis, and L. J. Zompa, *J. Am. Chem. Soc.* **104**, 3048 (1982).
35. R. I Gelb, B. T. Lee, and L. J. Zompa, *J. Am. Chem. Soc.* **107**, 909 (1985).
36. R. I. Gelb, L. M. Schwartz, and L. J. Zompa, *Inorg. Chem.* **25**, 1527 (1986).
37. F. Peter, M. Gross, M. W. Hosseini, J.-M. Lehn, and R. B. Sessions, *J. Chem. Soc., Chem. Commun.*, 1067 (1981).
38. F. Peter, M. Gross, M. W. Hosseini, and J.-M. Lehn, *J. Electroanal. Chem. Interfacial Electrochem.* **144**, 279 (1983).
39. E. García-España, M. Micheloni, P. Paoletti, and A. Bianchi, *Inorg. Chim. Acta* **102**, L9 (1985).
40. M. E. Manfrin, N. Sabbatini, L. Moggi, V. Balzani, M. W. Hosseini, and J.-M. Lehn, *J. Chem. Soc., Chem. Commun.*, 555 (1984).
41. M. F. Manfrin, L. Moggi, V. Castelvetro, V. Balzani, M. W. Hosseini, and J.-M. Lehn, *J. Am. Chem. Soc.* **107**, 6888 (1985).
42. F. Pina, L. Moggi, M. F. Manfrin, V. Balzani, M. W. Hosseini, and J.-M. Lehn, *Gazz. Chim. Ital.* **119**, 65 (1989).
43. D. Heyer and J.-M. Lehn, *Tetrahedron Lett.* **27**, 5869 (1986).

44. T. Fujita and J.-M. Lehn, *Tetrahedron Lett.* **29**, 1709 (1988).
45. M. W. Hosseini and J.-M. Lehn, *Helv. Chim. Acta* **70**, 1312 (1987).
46. M. W. Hosseini and J.-M. Lehn, *Helv. Chim. Acta* **71**, 749 (1988).
47. M. W. Hosseini, J.-P. Kintzinger, J.-M. Lehn, and A. Zahidi, *Helv. Chim. Acta* **72**, 1078 (1989).
48. (a) J. F. Marecek and C. J. Burrows, *Tetrahedron Lett.* **27**, 5943 (1986); (b) J. F. Marecek, P. A. Fisher, and C. J. Burrows, *Tetrahedron Lett.* **29**, 6231 (1988).
49. (a) A. Bianchi, M. Micheloni, and P. Paoletti, *Inorg. Chim. Acta* **151**, 269 (1988): (b) A. Bencini, A. Bianchi, M. I. Burguete, A. Domenech, E. García-España, S. V. Luis, M. A. Niño, and J. A. Ramirez, *J. Chem. Soc., Perkin Trans 2*, 1145 (1991); (c) A. Bencini, A. Bianchi, E. Garcia-Espana E. C. Scott, L. Morales, B. Wang, T. Deffo, F. Takusagawa, M.P. Mertes, K. B. Mertes, and P. Paoletti *Bioorganic Chem.* **20**, 8 (1992); (d) A. Andrès, J. Arago, A. Bencini, A. Bianchi, A. Domenech, V. Fusi, E. García-España, P. Paoletti, and J. A. Ramirez, *Inorg. Chem.* **32**, 3418 (1993); (e) A. Andrès, C. Bazzicalupi, A. Bencini, A. Bianchi, V. Fusi, E. Garcia-Espana, C. Giorgi, N. Nardi, P. Paoletti, J. A. Ramirez, and B. Valtancoli, *J. Chem. Soc., Perkin Trans. 2*, 2367 (1994).
50. S. Boudon, A. Decian, J. Fisher, M. W. Hosseini, J.-M. Lehn, and G. Wipff, *J. Coord. Chem.* **23**, 113 (1991).
51. C. Nakai and W. Glinsmann, *Biochemistry* **16**, 5636 (1977).
52. W. H. Voige and R. I. Elliott, *J. Chem. Ed.* **59**, 257 (1982).
53. (a) M. W. Hosseini, A. J. Blacker, and J.-M. Lehn, *J. Chem. Soc., Chem. Commun.*, 596 (1988); (b) M. W. Hosseini, A. J. Blacker, and J.-M. Lehn, *J. Am. Chem. Soc.* **112**, 3896 (1990).
54. D. Cordier and M. W. Hosseini, *New J. Chem.* **14**, 611 (1990).
55. W. Saenger, in *Principles of Nucleic Acid Structure*, Springer-Verlag, New York, 1988.
56. D. Voet and A. Rich, *Prog. Nuc. Acid Res. and Mol. Biol.* **10**, 183 (1970).
57. A. D. Hamilton and D. VanEngen, *J. Am. Chem. Soc.* **109**, 5035 (1987).
58. A. D. Hamilton and N. Pant, *J. Chem. Soc., Chem. Commun.*, 765 (1988).
59. J. Rebek, Jr., B. Askew, P. Ballester, C. Buhr, S. Jones, D. Nemeth, and K. Williams, *J. Am. Chem. Soc.* **109**, 5033 (1987).
60. G. Brand, M. W. Hosseini, and R. Ruppert, *Helv. Chim Acta* **75**, 721 (1992).
61. N. J. Leonard, *Science* **175**, 646 (1972).
62. J. R. Knowles, *Ann. Rev. Biochem.* **49**, 877 (1980).
63. F. Ramirez, J. F. Marecek, *Pure Appl. Chem.* **52**, 1021 (1980).
64. M. W. Hosseini, J.-M. Lehn, and M. P. Mertes, *Helv. Chim. Acta* **66**, 2454 (1983).
65. M. W. Hosseini, J.-M. Lehn, and M.P. Mertes, *Helv. Chim. Acta* **68**, 818 (1985).
66. M. W. Hosseini, J.-M. Lehn, L. Maggiora, K. B. Mertes, and M. P. Mertes, *J. Am. Chem. Soc.* **109**, 537 (1987).
67. G. M. Blackburn, G. R. J. Thatcher, M. W. Hosseini, and J.-M. Lehn, *Tetrahedron Lett.* **28**, 2779 (1987).
68. R. C. Bethell, G. Lowe, M. W. Hosseini, and J.-M. Lehn, *Bioorg. Chem.* **16**, 418 (1988).

69. P. G. Yohannes, M. P. Mertes, and K. B. Mertes, *J. Am. Chem. Soc.* **107**, 8288 (1985).
70. P. G. Yohannes, K. E. Plute, M. P. Mertes, and K. B. Mertes, *Inorg. Chem.* **26**, 1751 (1987).
71. H. Jahansouz, Z. Jiang, R. H. Himes, M. P. Mertes, and K. B. Mertes, *J. Am. Chem. Soc.* **111**, 1409 (1989).
72. M. W. Hosseini, J.-M. Lehn, K. C. Jones, K. E. Plute, K. B. Mertes, and M. P. Mertes, *J. Am. Chem. Soc.* **111**, 6330 (1989).
73. (a) M. P. Mertes and K. B. Mertes, *Acc. Chem. Res.* **23**, 413 (1990). (b) L. Qian, Z. Sun, J. Geo, B. Movassagh, L. Morales, and K. B. Mertes, *J. Coord. Chem.* **23**, 155 (1991).
74. M. W. Hosseini and J.-M. Lehn, *J. Chem. Soc., Chem. Commun.*, 1155 (1985).
75. M. W. Hosseini and J.-M. Lehn, *J. Am. Chem. Soc.* **109**, 7047 (1987).
76. M. W. Hosseini and J.-M. Lehn, *J. Chem. Soc., Chem. Commun.*, 397 (1988).
77. C. F. Walsh, in *Enzymatic Reaction Mechanism*, W. H. Freeman, San Francisco, 1979, p. 33.
78. M. W. Hosseini and J.-M. Lehn, *J. Chem. Soc., Chem. Commun.*, 451 (1991).
79. T. Suzuki, T. Higashiyama, and A. Nakahara, *Bioorg. Chem.* **2**, 145 (1973).
80. T. Suzuki, T. Higashiyama, and A. Nakahara, *Bioorg. Chem.* **4**, 250 (1975).
81. G. J. Hutchings, B. E. C. Banks, M. Mruzek, J. H. Ridd, and C. A. Vernon, *Biochem.* **20**, 5809 (1981).
82. A. J. Kirby and W. P. Jencks, *J. Am. Chem. Soc.* **87**, 3209 (1964).
83. G. Di Sabato and W. P. Jencks, *J. Am. Chem. Soc.* **86**, 4400 (1965).
84. G. M. Blackburn, T. D. Perrée, A. Rashid, C. Bisbal, and B. Lebleu, *Chemica Scripta* **26**, 21 (1986).
85. G. M. Blackburn, F. Eckstein, D. E. Kent, and T. D. Perrée, *Nucleosides Nucleotides* **4**, 165 (1986).
86. G. M. Blackburn, R. H. Tattershall, G. E. Taylor, G. R. J. Thatcher, and A. McLennan, in *Phosphorus Chemistry Directed towards Biology*, K. S. Bruzik, W. J. Stec, Eds., Elsevier, Amsterdam, 1987, p. 451.
87. M. W. Hosseini and J.-M. Lehn, unpublished results.
88. F. H. Westheimer, *Acc. Chem. Res.* **1**, 70 (1970).
89. G. M. Blackburn and M. J. Brown, *J. Am. Chem. Soc.* **91**, 525 (1969).
90. R. R. Holmes, *ACS Monograph* **176**, Ch. 1 (1980).
91. F. Ramirez, J. F. Marecek, and J. Szamosi, *J. Org. Chem.* **45**, 4787 (1980).
92. A. J. Kirby and S. G. Warren, in *Reaction Mechanisms in Organic Chemistry*, C. Eaborn, N. B. Chapman, Eds., Elsevier, Amsterdam, 1967, Vol. 5, Ch. 10.
93. D. Herschlag and W. P. Jencks, *J. Am. Chem. Soc.* **108**, 7938 (1986).
94. J. L. Bock, *Biochem.* **12**, 119 (1980).
95. Y. J. Shvy, T. C. Tsai, and M. D. Tsai, *J. Am. Chem. Soc.* **107**, 3478 (1985).
96. P. Tanswell, J. M. Thoronton, and A. V. Korda, *Eur. J. Biochem.* **57**, 135 (1975).
97. K. H. Scheller, F. Hofstetter, P. R. Mitchell, B. Prijis, and H. Siegel, *J. Am. Chem. Soc.* **103**, 248 (1981).

98. R. J. Motekaitis, A. E. Martell, J.-P. Lecompte, and J.-M. Lehn, *Inorg. Chem.* **22**, 609 (1983).
99. G. Lowe and B. S. Sproat, *J. Chem. Soc., Perkin Trans. 1*, 1874 (1981).
100. P. Domanico, K. Mizrahi, and S. J. Benkovic, in *Mechanisms of Enzymatic Reactions Stereochemistry*, P. A. Frey, Ed., Elsevier, Amsterdam, 1986, p. 127.
101. R. Breslow and I. Katz, *J. Am. Chem. Soc.* **90**, 7376 (1968).
102. R. C. Bethell and G. Lowe, *J. Chem. Soc. Chem. Commun.*, 1341 (1988).
103. R. C. Bethell and G. Lowe, *Biochemistry* **27**, 1125 (1988).
104. S. P. Harnett and G. Lowe, *J. Chem. Soc., Chem. Commun.*, 1416 (1987).
105. J. Ketelaar, H. Geismann, and K. Koopmans, *Rec. Trav. Chim.* **71**, 1253 (1952).
106. D. E. Koshland, *J. Am. Chem. Soc.* **74**, 2286 (1952).
107. J. H. Park and D. E. Koshland, *J. Biol. Chem.* **233**, 986 (1958).
108. G. Di Sabato and W. P. Jencks, *J. Am. Chem. Soc.* **83**, 4393 (1961).
109. G. Di Sabato and W. P. Jencks, *Ibid.* **83**, 4400 (1961).
110. D. Herschlag and W. P. Jencks, *J. Am. Chem. Soc.* **108**, 7938 (1986).
111. D. R. Philips and T. H. Fife, *J. Am. Chem. Soc.* **90**, 6803 (1968).
112. D. R. Philips and T. H. Fife, *J. Org. Chem.* **34**, 2710 (1969).
113. J. L. Kutz and C. D. Gutsche, *J. Am. Chem. Soc.* **82**, 2175 (1960).
114. H. P. Lau and C. D. Gutsche, *Ibid.* **100**, 1857 (1978).
115. G. W. Smithers, H. Jahanouz, J. L. Kofron, R. H. Himes, and F. H. Redd, *Biochemistry*, **26**, 3943 (1987).
116. H. Jahansouz, K. B. Mertes, M. P. Mertes and R. H. Himes, *Bioorg. Chem.* **17**, 207 (1989).
117. Z. Jiang, P. Chalabi, K. B. Mertes, H. Jahansouz, R. H. Himes, and M. P. Mertes, *Bioorg. Chem.* **17**, 329 (1989).
118. A. Meister, in *The Enzymes*, P. D. Boyer, Ed., Academic, New York, 1974, Vol. 10, p. 699.
119. A. Meister, in *The Enzymes*, P. D. Boyer, Ed., Academic, New York, 1974, Vol. 10, p. 671.
120. D. H. Buttlaire, R. H. Himes, and G. H. Reed, *J. Biol. Chem.* **251**, 4159 (1976).
121. H. Jahansouz, Z. Jiang, R. H. Himes, M. P. Mertes, and K. B. Mertes, *J. Am. Chem. Soc.* **111**, 1409 (1989).
122. H. Finniri and J.-M. Lehn, *J. Chem. Soc., Chem. Commun.*, 1819 (1993).

INDEX

Abiotic catalysts, 429
Acetate complex
   crystal structure, 174, 175
Acceptor number (AN), 16, 17, 222–224, 312
N-Acetyl-D,L-alanine, 175
Acetyl coenzyme A, 75, 76
Acetyl phosphate hydrolysis, 436
Acid extraction (*see* extraction processes)
Acridine, 55, 426, 427
Activated carbon, 400
Activity coefficients, 305
Acyclic receptors, 150–161
Adduct formation, 304
Adenine–adenine dinucleotide binding, 357
Adenosine diphosphate, 422, 424, 435
Adenosine monophosphate, 422, 423
c-Adenosine monophosphate, 357, 359
   crystal structure with sapphyrin, 163
Adenosine triphosphate, 55, 91, 244, 421–448
   with methylated polyammonium macrocycles, 257–257
   dephosphorylation, 379
   hydrolysis, 90, 91, 429–435
   β,γ-, α-, and α,β-substituted ATP, 433
Adenosine triphosphatases, 429, 432
ADP (*see* Adenosine monophosphate)
Agricultural effluent, 397–400
AIDS (*see* Aquired Immune Deficiency Syndrome)
Alkalide salts, 204–207

Aluminum, 251
AMBER, 338, 346, 348
AMBER-OPLS, 337
Amides, 153, 181, 195
   hydrogen bonding interaction, 370
bis-Amidinium ion, 58
Amino acids, 97, 114, 127
   binding and transport, 369–375
Ammonium ion, 150
AMP (*see* Adenosine monophosphate)
Anion binding proteins, 180
Anion bridges, 156
Anion exchange, 2, 23–32
Anion receptors
   in analytical chemistry, 380–397
   in biology, 63–78, 207, 208, 355–378
   in catalysis, 378–397
   coordination equilibria, 278, 295–304
   in waste management, 397–401
Anion selective electrodes (*see* ion selective electrodes)
Anion selectivity (*see* selectivity)
Anion template reactions, 401, 402
Anion transfer (*see* transport)
Anionic radii (*see also* size considerations), 4, 6, 148
   effect on anion exchange, 25
   thermochemical, 4, 6, 28
1,8-Anthracenediethynyl*bis*(catechol boronate), 186
Anthraphyrin, 365–367
   chloride crystal structure, 368

449

Anthryl polyamines, 331, 357
Antimony
    SbCl$_5$, 221, 222
Anticrown chemistry, 118
Anticryptate effect, 125
Antiferromagnetic coupling, 205
Antiport transportation, 368
Antiviral activity, 356
Applications, 355–419
Acquired Immune Deficiency Syndrome (AIDS), 356
L-Arginine, 373
Arsenic compounds, 399, 400
Asmus method, 297
ATP (see Adenosine triphosphate)
ATPases (see Adenosine triphosphatases)
Axial ligand anion exchange, 381, 398
Azacrown ethers (see polyammonium receptors)
Azide, 52
    crystal structures, 199, 200, 168, 169
    with carboxypeptidase A, 69
    with Cu,Zn-SOD, 70
3′-Azido-2′-deoxythymidine (AZT), 356, 357
Azonia compounds (see polyammonium receptors),
AZT (see 3′-Azido-2′-deoxythymidine)

Back-electron-transfer reaction, 325, 326
Barbiturate binding, 395
Basicity constants, 256, 257
Beer–Lambert–Bouguer Law, 264
Benzene quadrupole moment, 235, 236
1,3,5-Benzenetricarboxylic acid, 290, 298
2-Benzyl-3-formyl-propanoic acid (BFP), 65
(-)2-Benzyl-3-p-methoxybenzoylpropionic acid (BMBP), 65
N-[[[(Benzyloxycarbonyl)amino]-methyl]hydroxy-phosphinyl]-1-phenylalanine, 65
Bias, 3
    effect of charge, 28
    effect of dielectric constant, 26
    effect of ion size, 25–28
Binding constants (see stability constants)
Binding stoichiometry (see stoichiometry)
Bioenergetics, 429

tris-Bipyridine iron helicate, 201–203
Bjerrum theory, 24, 219
Bond energies, 308
Boric acid
    theoretical studies with [18]aneN$_6$, 336, 337
Born equation, 310
Born–Haber cycle, 278
Born model, 5–15, 32, 33, 311
Borane receptors, 120, 121, 185–188, 251
    1,8-naphthalenedyl bis(dimethylborane), 185, 186
    theoretical studies, 348–350
Borane silyl receptors, 185
Borate, 5, 336, 337
Butanol, 19, 21

Calcium ion
    ternary nucleotide complexes, 424
    effect on nucleotide hydrolysis, 434–435
Calculi, 360–362
Calculi dissolution, 55
Calixarenes, 117, 118, 132–134
    bis-cobalticinium, 370, 371
    conformations, 346–348
    copper complex, 182
    halide selectivity, 365, 366
    iodide complex, 196, 197
    silver complex, 182
    sulfonamide derivatives, 180
    thiourea derivativized, 365, 366
    tetrahedral anion binding, 195–197
    uranyl complexes, 181, 182
    urea derivatives, 243
Carbonate
    molecular mechanics with [18]aneN$_6$, 336, 337
Carbon-paste electrode, modified (MCPE), 294
Carborane-based mercury receptors, 189–193
Carboxylates (see also polycarboxylates)
    activation of, 438
    α,ω-carboxylates, 231, 232
    benzenetricarboxylate, 53, 290, 298
    binding, 369–375
    chiral aromatic, 109
    citrate, 53
    dicarboxylates, 53, 91, 92, 169, 170, 246

fumarate, 53
maleate, 53
malonate, 53
mercury competition for receptor, 304–306
oxalate, 53
succinate, 53
tartrate, 53
transport, 369–375
Carboxypeptidase A (CPA), 64–69, 72, 76
Cascade anion binding, 50, 51, 125, 126, 199, 200
Catalysis, 109, 110, 378–397, 439–443
  in enzymes, 63–78
  of decarboxylation reaction, 96, 97
  of ester hydrolysis, 98, 100
  of Michael reaction, 109, 110
  of phosphoryl anion transfer processes, 153, 421–448
Catalysts
  abiotic catalysts, 429
  hydrolysis, 153
  phosphoryl transfer, 153
  cocatalysts, 439–443
Cations
  binding, 46
  hosts, 85–118
  radii, 32
Cellular nucleoside kinases, 357
Ceside, 205–207
Charge separation, 323
Charge transfer, 248
CHARMM, 339, 346
Chemical sensors, 278
Chemotherapeutic agents, 362
Chiral recognition, 53, 109, 110, 175, 198, 199, 373, 402
Chirality, 102, 108–110
Chloride
  binding and transport, 363–369
  cascade binding, 125
  crystal structures, 153, 165, 182, 368, 369
  helicate binding, 201–203
  molecular dynamics, 337–339
  selective electrodes for, 386
  solvation of, 13, 15
Chloride anion channel proteins, 363
Chromatographic methods in separations, 380–384
  affinity chromatography, 380, 381
  expanded porphyrin solid supports, 383, 384
  ion exchange chromatography, 380
  phenylborate matrices, 381
  reversed phase chromatography, 381
  metalloporphyrin solid supports, 381–383
Chromium compounds as pollutants, 400
Chronoamperometry, 282, 283, 297–299
Citrate, 74–76, 150, 151
Citrate synthase, 74–76
^{35}Cl NMR, 154
Cobalt EDTA complex
  photochemical reduction, 332
Cobalticenes, 117, 195, 365, 370
Competitive binding, 231, 306–308
Computer modeling, 82, 94, 238, 336–350
Coordination equilibria, 278, 295–304
  analysis for nonelectroactive anions, 304–307
  analysis from electrode potential data, 296, 297
  competitive methods, 304–307
Copper complexes, 125, 199, 200, 126
Corrins, 102
Cottrell equation, 282, 283, 304
Coulomb law, 81
Coupled chemical reactions, 282, 288–295
CPA (see Carboxypeptidase A)
Crown ethers, 46, 149, 195, 203–207
  electrides, 350–352
Cryptands, 86, 92, 150, 166–171, 177, 206, 207, 364
  lanthanide photochemistry, 326
  MM2 calculation of halide binding, 340–342
Crystal structures
  anthraphyrin chloride complex, 368
  calixarene receptors, 182–184
  copper cascade complexes, 199, 200
  cyclophane receptors, 177
  guanidinium receptors, 171–178
  helicate, 201–203
  iodide complex, 167
  Lewis acid receptors, 183–195
  mercury-containing receptors, 160, 165, 188–193

Crystal structures *(continued)*
  neutral receptors, 177–183
  nitrate complexes, 152, 154, 164
  organometallic receptors, 195–197
  oxalate complex, 155, 157
  perchlorate, 170, 171
  polyammonium receptors, 150–161
  polycyclic receptors, 161–171
  polyoxometallates, 200–201
  pyrophosphate complex, 155
  rosarin chloride, 369
  sapphyrins, 158, 162, 163
  tetrachloropalladate, 158
  tin-containing receptors, 193–195
  transition metal complexes, 198–200
  zinc complex with methylthymine, 198, 199
  uranyl complexes, 181–183
Curie–Weiss law, 205
Cyano complexes (*see* hexacyanometallates)
Cyclam
  in–out conformation, 229
  nickel complex, 314
Cyclic voltammetry, 283–285
Cyclodextrins, 115–116, 126, 247, 374, 395, 396
  amino-derivatized, 116, 358
  bis-imidazolyl, 116
  with naphthalene sulfonates, 247–248
Cyclophanes, 98–101, 169, 170, 177, 178
  pyridinium-based, 365
Cyclotriveratrylene (CTV), 197
Cystic fibrosis, 363, 364
Cystic fibrosis transmembrane regulator (CFTR) protein, 363, 364

DABCO based carriers, 358
DDI (*see* 2′,3′-dideoxyinosine)
Deoxyribonucleic acid (*see* DNA)
Dephosphorylation, 429–438
  of acetyl phosphate, 436
  of ATP, 379
  catalysts, 153
  of formyl phosphate, 436–438
Dicarboxylates (*see* carboxylates)
Dichloroethane, 19, 21
Ditopic receptors, 53, 54, 98–100, 110–113, 246

2′,3′-dideoxyinosine (DDI), 356
Dielectric constant, 16, 17, 18, 221
  effect on extraction, 26
Dielectric permittivity, 129
1,1′-Dimethyl-4,4′-dipyridinium dication, 324
Dimroth-Reichardt parameter, 16, 17, 18
Dinucleotides
  APPPPA, 433
  APPCCl2PPA, 433
Dipalmitoylphosphatidylcholine (DPPC), 391
Dipole interactions, 221, 236, 237
Dispersion interactions, 218, 237
Divinyl benzene, 38, 39
DNA
  plasmid binding with polyammonium macrocycle, 426
  plasmid pBR, 55
  plasmid pBR 322, 426
  templated synthesis of, 401
Docking experiments, 82
Donor number (DN), 16, 17, 221, 312
Dowex 1-X8, 35, 36
Dropping mercury electrode, 285, 286
DPPC (*see* Dipalmitoylphosphatidylcholine)

Effluent, 397–400
Electrides, 203–207
  theoretical studies, 350, 351
Electric field gradient, 70
Electroactive films, 313
Electrocatalysis, 278
Electrochemical sensors, 278
Electrochemical techniques, 92, 282–286
  chronoamperometry, 282
  cyclic voltammetry, 283–285
Electrochemically active cations, 195
Electrochemistry, 277–319
  coupled chemical reactions, 282
  irreversibility, 281
  reversibility, 281
  solvent effects, 310–313
  supporting electrolyte effects, 310–313
  structural significance, 308–310
  thermodynamics, 279–281, 308–310
Electrode potential, E_A, 279, 280
  in analysis of coordination equilibria, 297

Electrodes
  chemically modified, 390
  dropping mercury, 285, 286
  glass, 262
  glassy carbon, 385
  graphite, 391
Electron acceptor ability, 16
Electroneutral anion receptors, 118–135
Electronic energy transfer, 322
Electronic excited states, 321
Electron transfer
  models, 313
  model systems, 375
  photochemical, 322–324
  processes, 278
Electropolymerized films, 385, 390
Electroselectivity, 40
Electrostatic interactions, 69, 82, 98, 128–135, 218–220, 224, 227–236
Electrostatic model, 20
Enantioselectivity
  of amino acids, 114, 115, 126, 127, 175, 198, 199, 370, 373, 402
  of bilirubin, 102
  of carboxylates, 53, 108, 109
Encapsulation, 92
Endo/exo conformation, 229, 343, 344
Energetics of noncovalent interactions, 227–251
Energy minima, 336
Energy transduction, 356
Enhancement of selectivity, 426
Enthalpy of activation
  in catalysis of ATP hydrolysis, 432, 433
Entropy of activation
  in catalysis of ATP hydrolysis, 432, 433
Environmental contaminants, 397–400
Enzyme inhibitors, 66
Enzymes (*see also* proteins), 63–76
  ATP synthase, 75
  citrate synthase, 74–76
  Cu,Zn superoxide dismutase, 70–72, 76
  cofactors, 75
  RNAase, 75
Erdmannate anion, 30–32
Ester hydrolysis, 67, 98
Ethylenediamine
  citrate complexes, 150, 151
Excited states, 321

Expanded porphyrins, 50, 158, 359–361, 363, 366–368
Extraction, 22, 112–114
  of amino acids, 114
  dielectric constant effects, 26
  liquid-liquid, 2, 23, 32–34, 112–114
  of hexacyanoferrate, 28
  of nucleotides, 113, 428
  quaternary salts, 28
  of tetracyanometallates, 28
Extraction processes,
  acid extraction, 27
  in analytical chemistry, 2
  anion exchange, 2, 23–32
  hydrometallurgy, 2
  ion pair, 2, 23, 27, 32–34
  in nuclear industry, 2

"F-center" electronic defects, 351
Ferrocene, 195, 393–396
Fluorescence
  probes, 101, 102
  receptors, 394
  spectroscopy, 331
Fluoride (*see also* halide), 103, 104, 166–168, 187, 188, 365, 375–378, 398, 399
Fluorocrown ethers, 129
Formation constants
  from electrode potential data, 296, 297
Formyl phosphate hydrolysis, 436–438
Föster-type mechanism, 375
F-test, 263
Fumarate, 153

Gene replication, 356
Germanium, 251
Gibb's enthalpy of hydration, 81
Gibb's free energy of
  binding, 68
  hydration, 224
  salt partitioning, 33
  solvation, 4–9, 12, 148, 225
  transfer, 9, 10, 12–22, 33, 37, 225
Gibbs–Helmholtz equation, 219
Glass electrode (*see* electrodes),
Glassy carbon electrodes (*see* electrodes)
Glutamine synthetase, 438
Glutathione synthetase, 438

Gold surfaces, 392–397
    alkanethiols on, 392
    calixarenes on, 392–394
    ferrocenes on, 394
    cyclodextrines on, 395, 396
Gran's method, 262
Granulated activated carbon (GAC), 400
GROMOS, 345
Ground water purification, 397–400
GTP binding protein, 375
Guanidinium receptors, 53, 54, 105–115
    bicyclic, 108–110, 114
    crystal structures, 171–177
    functionalized monolayers, 357
    in hydrolysis reactions, 379
bis-Guanidinium receptors, 173, 175
    macrocyclic salts of, 106
    and protein tertiary structure, 171
Guest exchange rates, 84

Halides, 6, 7, 17, 47–52, 153
    cascade binding, 50
    history, 47–52
    octahedral coordination, 168
    with calixarenes, 243
    with carborane-based mercury receptors, 189–193
    with catapinands, 47
    with expanded porphyrins, 50
    with Lewis acid receptors, 51, 52, 120–124
    with quaternary ammonium tricycles, 48
    selectivities, 365, 366
    with spherical receptors, 47, 48
    tetrahedral coordination, 168
    transport, 364
    with *bis*-tren, 48–50
Heavy metal toxicity, 366
Helicates, 201–203
Helmholtz free energy, 345
Herpes simplex, 356
1,4,7,12,15,18-Hexaazacyclodocosane, 338
1,4,7,10,13,16-Hexaazacyclooctadecane (*see* hexacyclen),
1,4,10,13,16,22-Hexaaza-7,19-dioxa-cyclotetracosane, 337
Hexacyanoferrate (*see* hexacyanometallate)
Hexacyanometallate, 244
    electrochemistry, 286–288

hexacyanocobaltate, 56, 57, 90, 92, 153
hexacyanocobaltate photochemistry, 327–332
hexacyanoferrate, 56, 57, 286–288, 297–304
hexacyanoferrate equilibrium constants, 304
hexacyanoferrate coordination equilibria, 297–304
hexacyanoruthenate, 56, 57
Hexacyclen, 52, 152, 153
    boric acid complex, 336, 337
    carbonate complex, 336, 337
    silicic acid complex, 336
^{199}Hg NMR, 123, 124, 188
High performance liquid chromatography (HPLC), 382–384
Hildebrand solubility parameter, 16–18
Histidinato bridge, 71
Hofmeister series, 313, 385, 386
Hoogsteen hydrogen bonding arrays, 113
HPLC (*see* high performance liquid chromatography)
Hydration, 4–9
    of organic phase species, 22
Hydride anion, 51
Hydride sponge, 185
Hydrogen bond donor ability, 18, 19, 33
Hydrogen bonding, 129–134, 218, 223, 233, 237–247
    acceptor/donor functionality, 180
    anion recognition, 242
    carboxylate group, 240, 241
    complex stability, 239
    directionality, 239–241
    donor species, 241
    gas phase, 238
    solvation, 82
Hydrogen fluoride, 238, 239
Hydrogen peroxide, 70
Hydrogen sulfide, 400
Hydrolysis
    of acetyl phosphate, 426
    of ATP, 429–435
    of ATPS, 435, 436
    of esters, 67, 98
Hydrometallurgy, 2
Hydrophobic effect, 38, 223, 224, 247, 248
Hydroxylic solvents, 82

Induced dipoles, 236, 237
Industrial effluent, 397–400
Inhibitors, 66, 67, 72
bis-Intercaland receptors, 249–251
Intramolecular energy transfer, 322
Intramolecular photoreaction, 322
Iodide, 375–378
  crystal structure, 167
Ion channel
  chloride ion channel proteins, 363
  sensors, 391
Ion-dipole, 233–235
  binding, 128–135
Ion pair extraction (*see* extraction)
Ion pairs, 81, 86, 324–327
  electrostatic model, 218–221
  effects, 23–26
  ion pair charge transfer (IPCT), 325
Ionic radii, 81, 148
Ion selective electrodes (ISEs), 313, 384–392
Ion selective membranes, 313
Ion solvation, 15
  Born model, 5
  single shell model, 8
Ion transport (*see* transport)
Isophthalate, 58

Kalide, 205–207
Kamlet-Taft parameter, 14, 17, 18
Katapinands, 47, 149, 161–166, 177, 183, 187
  in–out equilibria, 228, 229
Kemp acid receptor, 113, 114, 129,
Kinase, 379
Kinetic discrimination, 253

Lactone
  Michael addition to, 109, 110
Langmuir–Blodgett molecular assemblies, 385, 391
Lanthanide ion
  effect on ATP hydrolysis, 434
Lasalocid A, 326, 327
Lattice energy, 4
Lewis acid–Lewis base interaction, 118
Lewis acid receptors, 50, 118–128, 183–195, 251–253, 366
Light energy

photochemical conversion and storage, 324
Linear potential sweep chronoamperometry, 283–285
Linear sweep voltammetry, 283–285
Liquid-liquid extraction (*see* extraction)
Litholytic agents, 362
London dispersion interactions, 221, 236
Low permittivity solvents, 28
Luminescence, 321
  quenching in lasolacid A complexes, 326, 327
Luminofores, 323

Macrocycles (*see also* specific type of macrocycle), 149–172, 228–234, 244–246, 249–251, 256, 327–331
  electrostatic interactions, 228–236, 244–246
  as enzyme mimics, 379
  catalysis of phosphoryl transfer, 421–448
  historical perspectives, 149–172
  in–out configuration, 228–230
  polycarboxylate binding, 256–259
  stacking effects, 248–251
  transition metal complexes, 327–331
Macrocyclic polyammonium cations (*see* polyammonium receptors)
Magnesium ion
  effect on ATP hydrolysis, 434
Magnetic susceptibility, 206
Mean diffusion coefficient, 299
Mechanism
  of ATP hydrolysis, 433, 434
  of ion transport, 344
  probes, 433
Membranes, 111, 247
Mercury containing receptors, 122–124, 188–193, 251
  crystal structure, 160, 165
  o-diphenylenedimercurial receptor, 188, 189
  o-phenylenedimercurial receptor, 188, 189
Mercury polyamine complex, 305
Metalloporphyrins, 102
  amino acid receptors, 374
  capped, 370, 371
  cobalt porphyrin, 314

Metalloporphyrins *(continued)*
  electrodes, 386
  rhodium porphyrin, 103
  zinc porphyrin, 126, 127
Metaphosphate, 433, 434
Methylviologen, 324
Micelles, 95, 247
Michael reaction, 109, 110
Michaelis type complex, 432
Microcalorimetry, 265, 266
MM2, 336–340, 346, 350
MM2P, 346
Modified surfaces, 392–397
Molecular assemblies, 58
  on surfaces, 392–397
Molecular boxes, 177
Molecular dynamics, 4
  of calixarenes, 347, 348
  of macrotricycle, 342–346
  of polyammonium macrocrocycles, 338–341
Molecular mechanics, 336
Molecular modeling, 173, 179
Molecular modeling software,
  AMBER, 338, 346
  AMBER-OPLS, 337
  CHARMM, 339, 346, 348
  GROMOS, 345
  MM2, 152, 336–342, 346, 350
  MM2P, 346
Molecular orbital calculations, 238
  of Lewis acid receptors, 120
Molecular sensory devices, 313
Monte Carlo calculations, 4
Molybdate, 39
Multiple site recognition, 181, 426
Muscle contraction, 429

NAD (*see* nicotinamide adenine dinucleotide)
1,8-Naphthalenedyl *bis*(dimethylborane) (*see* borane receptors)
Naphthalene sulfonates
  thermodynamic parameters with cyclodextrines, 247, 248
Natural anion receptors (*see* enzymes and proteins), 63–76
Nernst equation, 279
Neutral receptors, 50, 177–183, 370

  inorganic hosts, 181–183
  organic hosts, 177–180
  zwitterionic, 179
Nicotinamide adenine dinucleotide (NAD), 443
  electrochemistry, 289–295
  redox scheme, 293
Nitrate, 52, 53
  crystal structures, 152, 154, 156, 164
  molecular dynamics, 339–341
  in waste, 397
Nitrobenzene, 19, 20, 21
Nitrogen oxides, 314
6-Nitrobenzisoxalole-3-carboxylate, 96
NMR (*see* nuclear magnetic resonance)
NMR titration, 173
Noncovalent interactions, 227–251
Nuclear magnetic resonance (NMR) (*see also* ^{31}P NMR), 263, 264
  EQNMR program for fast exhange, 264
  ^{1}H of macrocyclic receptors with halides, 234, 235
  HypNMR program for stability constants, 264
  use of in anion binding constants, 263, 264
  ^{119}Sn NMR, 50, 121, 252, 253, 388
Nucleotides, 55, 100–104, 113, 114, 116, 168, 248–251
  binding and transport, 356–362
  binding to cyclodextrine, 116, 117
  binding to macrocyclic polyamines, 244, 245, 422
  binding to polymer supported receptors, 427
  derivatives, 427, 428, 433, 435
  extraction, 113, 428
  ^{31}P NMR studies, 423–426, 429–431
  polyphosphates, 55
  recognition, 422–428
  stacking effects, 248–251
  transformation, 428–443
Nucleotide synthesis, 441–443

Oligonucleotides, 114, 362, 363
Optical methods related to thermodynamics, 264, 265
Organometallic receptors
  crystal structures, 195–197

INDEX 457

Organotin doped membranes, 387
Oxalate
  crystal structure, 155, 157
Oxidative phosphorylation, 429
Oxo anions (*see also* specific anions), 6, 92, 105–113
Oxy acids, 37
Oxyphosphorane, 433

Parallel reactions, 288
Partitioning, 14
PdCl$_4^{2-}$, (*see* tetrachloropalladate)
Pentanethiol, 395
Peptide hydrolysis, 66, 67
Perchlorate, 56
  crystal structure, 170, 171
Perrhenate, 26, 30, 32
Phenanthridinium, 101
tris-Phenanthroline osmium complex, 326
Phenolic compounds, 400
Phenylalanine, 65–68, 374
Phenylboronate matrices, 381
o-Phenylenedimercurial receptor (*see* mercury-containing receptors),
L-Phenyllactate, 65–68
pH-Metric techniques, 261–263
Phosphates, 52, 54, 72–75, 86, 111, 397–400
  binding and transport, 356–362
  binding by guanidinium ion, 172, 173
  binding by uranyl complex, 127, 128, 183
  dibenzylphosphate crystal structure, 175, 176
  diphenyl phosphate crystal structure, 362
  phenyl phosphate crystal structure, 175, 176
Phosphate binding protein (PBP), 72–76, 208
Phosphodiester hydrolysis, 172379
Phosphodiesterase model, 116
Phosphinyl ion, 73
Phosphonyl ion, 73
Phosphoryl transfer, 56, 421–448
  geometric control, 440
  oxidative phosphorylation, 429
  phosphorylation, 439–443
Photoaquation reactions, 327–330

Photochemical electron transfer, 322
Photochemistry, 321–333
  of anthrylpolyamines, 331
  of Co(CN)$_6^{n-}$, 327–330
  of electron transfer, 322
  of ion pairs, 324–327
  of lanthanide cryptands, 326
  of sapphyrin, 331
Photodissociation, 322
Photoisomerisation, 322
Photophysics, 321–333
Photoresponsive spiropyrane, 374
Photostimulated anion transport, 332
Photosynthesis
  model systems, 375
  phosphorylation, 429
Physiologically important anions, 356–378
  carboxylates and amino acids, 369–375
  chloride, 363–369
  fluoride, 375–378
  iodide, 375–378
  oligonucleotides, 362, 363
  phosphates and nucleotide analogues, 356–362
Picrate, 18, 19
^{31}P-NMR, 221, 222, 244, 359, 360, 426, 436
  binding stoichiometries by, 423, 424
  in ATP hydrolysis, 429–431
Polar solvents, properties, 11
Polarizability, 221, 225, 236
Polarographic half-wave potentials, 296
Polarography, 285, 286, 299
Polyammonium receptors, 85–102, 305, 327–346, 365, 376, 388
  acyclic, 150–161
  carboxylate binding, 370
  macrobicycles, 46, 50, 340–342
  macromonocycles, 336–340
  macrotricycles, 47, 94–100, 134, 135, 342–346
  monocyclic, 100, 150–161, 309
  pK_as, 87
  polycyclic, 161–171
  zwitterionic, 366
Polyaza compounds (*see* polyammonium receptors)
Polycarboxylates, 53

thermodynamics of binding, 232, 233,
Polycyclic hosts, 161–171
Polymer matrix liquid membranes, 385
  polyvinyl chloride (PVC), 385
Polymer supported receptors, 427,
Polyoxometallic receptors, 201–203, 401
Polypyrrole macrocycles (see expanded porphyrins)
Polystyrene resins, 36
Porphyrins (see metalloporphyrins),
Positional isotope effect, 435
Positively charged anion hosts, 85–118
Potentiometric determination of stability constants, 261–263
Powdered activated carbon (PAC), 400,
Proteins (see also enzymes), 63–76
  phosphate binding (PBP), 72–74, 76
  sulfate binding (SBP), 72–74, 76
  transport, 72
Proton sponges, 120
Proton transfer potentials, 238
Pseudomonas aeruginosa infections, 363
π-Stacking (see stacking),
Pulmonary epithelia cells, 363
Putrescine, 429
Putrescine diphosphate, 151
tris-Pyrazolylborate receptor, 158, 160
Pyridinium N-phenolate betaine dye, 16
Pyridinium type cations, 324
Pyrophosphate
  crystal structure, 155
  formation in acetyl phosphate hydrolysis, 439, 440
  formation in nucleotide hydrolysis, 433–435
  synthesis, 439, 440

Quantum yields, 322, 327–329
Quenching effect, 327

Radiationless deactivation, 321
Recognition, 3
  chiral, 53, 175, 198, 199, 373, 402
  enantiomeric, 102, 108, 109, 114, 126, 127
  of halides, 182
  multiple, 181
  of nucleotides, 422–428
  by proteins, 68, 72

shape, 385
Redox catalysts, 313
Regioselectivity, 115
Renal irrigation fluids, 362
Resins
  Dowex 1–X8, 35, 36
  polystyrene-based, 36
  Type I alkyltrimethylammonium-based, 34–36, 38
  Type II alkydimethyl ethanolammonium-based, 34–36
Resin anion exchange, 2, 34–40
  ion exchange affinities, 35, 36
  selectivity, 35
  strong base resins, 34, 35
  structure and affinity, 38
  weak base resins, 34
RNA hydrolysis, 74
Rosarin
  chloride crystal structure, 369
Rubidide, 205–207
Rubyrin, 359, 366, 367

Sacrificial electron donors, 326
Salmonella typhimurium, 73, 208
Sapphyrins, 359, 366, 377, 383, 390
  carboxylate binding, 371
  complex with fluoride, 103, 104
  crystal structures, 158–163, 360, 363, 367
  photochemistry, 331
Second coordination sphere, 57, 246
Second sphere complexes, 246
Selectivity (see also bias), 253–259
  effect of anion radii, 34, 35
  effect of charge, 34, 39
  effect of hydration, 34, 35
  electroselectivity, 40
  enantiomeric, 53, 102, 108, 109, 114, 126, 127, 373, 402
  enhancement, 426
  kinetic, 83
  nucleotide enhancement, 426, 427
  thermodynamic, 83
  size, 161
  structural, 169
Selective electrodes, 384–392
Selectivity enhancement, 426
Self assembly, 201

Sensors
  electrochemical, 278
Separations (*see also* chromatographic methods), 104, 112–114
  of biologically important anions, 380
  of nucleotides, 381
  of transfer RNAs, 381
Sepulchrate, 325, 326
Signal transduction,
  membrane potential change, 385
  permeability change, 385
Silicic acid
  MM2 caculations, 336, 337,
SiF_6^{2-}, 56, 171
Silicon containing receptors, 185–188, 364
Simultaneous anion/cation complexation, 181
Single ion partition coefficient, 386
Single shell model, 8, 12, 20, 33
Size considerations, 40, 49, 52, 89–93, 253, 340, 342
^{119}Sn NMR (*see* nuclear magnetic resonance), 50, 121, 252, 253, 388
Soccer ball ligand, 149
SOD (*see* superoxide dismutase)
Sodide, 205–207
Solar energy,
  conversion and storage, 326
Solvation, 5, 82, 98, 119, 193, 221–227
  effect on thermochemical cycles, 310–313
  free energies of, 224–227
Solvatochromic polarity parameter, 16, 18, 19
Solvents
  apolar, 222–224
  dry solvents, 22
  effect on electrode potentials, 310–313
  hydroxylic, 82
  permittivity effect on extraction, 26–28
  polar, 10, 11
  polar aprotic, 222–224
  protic, 222–224
  scales, 16
  water, 85
  water immiscible, 19
Solvent acceptor number, 16, 222–224
Solvent properties
  dielectric, 221–224
  dipole moment, 221–224
  donor number, 221–224
  hydrogen bonding ability, 223
  molecular polarizability, 221–224
Solvophobic interactions, 98, 218
Spermine, 86, 424
Spermidine, 86
Spiropyrane
Squarate, 153
Stability constants, 260–265
  anions with calixarenes, 242
  ATP, 244
  coordination equilibria, 278, 295–304
  determination by NMR, 261–263
  determination by potentiometric methods, 263–265
  dicarboxylates, 246
  halides with tin macrocycles, 251–253
  halides with *bis*-tren, 253–256
  nucleotides with polyammonium macrocycles, 87–89, 91–93, 258, 424, 425
Stacking, 248–251
  charge transfer contribution, 248
  hydrophobic effect, 248
  in nucleotide complexes, 248–251
  π-stacking, 198, 218, 371
  van der Waals contribution, 248
Steady-state fluorescence spectroscopy, 331
Steric effects, 98
Stoichiometry, 297
  Asmus method, 297
  molar ratio method, 297
  ^{31}P NMR of nucleotide binding to polyammonium macrocycles, 423–425, 429–431
Stokes radius, 28
Structure, 147–209
  binding effects, 85–93
  ring size, 89–91
  selectivity, 85–93
Sugar moiety, 426
Sulfate, 55, 56, 72–74, 113, 180
Sulfate binding protein (SBP), 72, 76, 130, 208
Sulfonamide receptors, 180, 242, 243
Sulfonates (trifluoromethane-), 177, 178
Superoxide, 70, 71

Superoxide dismutase (Cu,Zn SOD), 70–72, 76
Supported liquid membranes, 397, 398
Supporting electrolyte effects, 310–313
Surfaces (see modified surfaces), 392–397
Synport transportation, 368
Synthesis
    guanidinium receptors, 108, 109
    polyammonium macrocycles, 86
    tricyclic ammonium macrocycles, 96

Template reactions, 401, 402
Terephthalate, 58
Ternary complexes, 424, 434
1,4,10,13-Tetraaza-7,16-dioxacyclooctadecane, 339
Tetraphenylborate, 19
Tetraphenylporphyrin (see metalloporphyrins)
Texaphyrin, 390
Theoretical studies, 335–354
    of calixarenes, 346, 347
    of electrides, 351
    of organoborane receptors, 348–350
    of polyammonium macrobicycles, 340–342
    of polyammonium macromonocycles, 336–346
    to polyammonium macrotricycles, 342–346
    of tin-containing receptors, 350
Thermodynamics, 217–266
    cycles, 280
    determination of parameters, 259–266
    electrolyte effects, 310–313
    microcalorimetry, 265, 266
    nmr methods, 263, 264
    of electrode processes, 279–281
    optical methods, 264, 265
    potentiometric methods, 261–263
    solvent effects, 310–313
Tetrachloropalladate, 56, 57, 158
Thiourea receptors, 173, 374
Thiocyanate receptors, 387
Thiolate monolayers, 392
Thyroid gland, 375
Tin containing receptors, 193–195121, 122
    chloride crystal structure, 193–195
    equilibrium constants with halides, 251–253
    MM2 studies, 350
p-Toluenesulfonic acid, 377, 378
Transition metal complexes, 198–200
Translational entropy, 219
Transport, 9–22, 356–378
    active, 429
    Born treatment, 5, 15, 20, 32, 33
    of carboxylates and amino acids, 369–375
    of chloride, 363–369
    electrostatic model, 9–15
    hydrogen bond donor ability in, 18, 19, 22
    of hydrophilic phosphates, 111
    mechanism of, 344
    nucleotide, 104
    of oligonucleotides, 362, 363
    in organic solvents, 18
    of phosphates and nucleotide analogues, 356–362
    photostimulated, 332
    of physiologically important anions, 356–378
    processes, 56
    protein, 72
    single layer model, 12, 15, 20, 33
bis-Tren, 48, 52, 150, 166–171, 254, 255
2,2,2-Trifluoroethanol, 19, 21
Triphosphate, 432, 440–441
Tungstate, 39
Turcasin, 369
Tyrosine phosphatase, 207

Uranyl ion
    anchor groups, 370
    phosphate binding, 127, 128
    salene complex, 127, 128
bis-Urea receptors, 173
Urinary lithiasis, 360–362

Van der Waals interactions, 98, 218, 236, 248
Vanadium, 200
Van't Hoff equation, 260, 261, 265
Viossat method, 299

Waste management, 397–400

Watson–Crick hydrogen bonding array, 113
Watson–Crick molecular recognition co-carriers, 359

Yeast phenylalanine transfer RNA, 151

Yersina protein, 207

Zinc complexes, 198, 199
Zinc porphyrin (*see* metalloporphyrins)
Zwitterionic host molecules, 231
Zwitterions, 97, 135